FIBER-FORMING POLYMERS

FIBER-FORMING POLYMERS

Recent Advances

Edited by J.S. Robinson

NOYES DATA CORPORATION

Park Ridge, New Jersey, U.S.A.

1980

Published in the United States of America by
Noyes Data Corporation
Noyes Building, Park Ridge, New Jersey 07656

Library of Congress Cataloging in Publication Data

Main entry under title:

Fiber-forming polymers.

    (Chemical technology review ; no. 150)
    Includes index.
    1. Textile fibers, Synthetic--Patents. 2. Poly-
mers and polymerization--Patents. I. Robinson, J. S.,
1936-   II. Series.
TS1548.5.F48   677'.4'0272   80-213
ISBN 0-8155-0791-7

FOREWORD

The detailed, descriptive information in this book is found in U.S. patents issued since January 1977 that deal with fiber-forming polymers.

This book is a data-based publication, providing information retrieved and made available from the U.S. patent literature. It thus serves a double purpose in that it supplies detailed technical information and can be used as a guide to the patent literature in this field. By indicating all the information that is significant, and eliminating legal jargon and juristic phraseology, this book presents an advanced commercially oriented review of fiber-forming polymers.

The U.S. patent literature is the largest and most comprehensive collection of technical information in the world. There is more practical, commercial, timely process information assembled here than is available from any other source. The technical information obtained from a patent is extremely reliable and comprehensive; sufficient information must be included to avoid rejection for "insufficient disclosure." These patents include practically all of those issued on the subject in the United States during the period under review; there has been no bias in the selection of patents for inclusion.

The patent literature covers a substantial amount of information not available in the journal literature. The patent literature is a prime source of basic commercially useful information. This information is overlooked by those who rely primarily on the periodical journal literature. It is realized that there is a lag between a patent application on a new process development and the granting of a patent, but it is felt that this may roughly parallel or even anticipate the lag in putting that development into commercial practice.

Many of these patents are being utilized commercially. Whether used or not, they offer opportunities for technological transfer. Also, a major purpose of this book is to describe the number of technical possibilities available, which may open up profitable areas of research and development. The information contained in this book will allow you to establish a sound background before launching into research in this field.

Advanced composition and production methods developed by Noyes Data are employed to bring these durably bound books to you in a minimum of time. Special techniques are used to close the gap between "manuscript" and "completed book." Industrial technology is progressing so rapidly that time-honored, conventional typesetting, binding and shipping methods are no longer suitable. We have bypassed the delays in the conventional book publishing cycle and provide the user with an effective and convenient means of reviewing up-to-date information in depth.

The table of contents is organized in such a way as to serve as a subject index. Other indexes by company, inventor and patent number help in providing easy access to the information contained in this book.

v

15 Reasons Why the U.S. Patent Office Literature Is Important to You —

1. The U.S. patent literature is the largest and most comprehensive collection of technical information in the world. There is more practical commercial process information assembled here than is available from any other source.

2. The technical information obtained from the patent literature is extremely comprehensive; sufficient information must be included to avoid rejection for "insufficient disclosure."

3. The patent literature is a prime source of basic commercially utilizable information. This information is overlooked by those who rely primarily on the periodical journal literature.

4. An important feature of the patent literature is that it can serve to avoid duplication of research and development.

5. Patents, unlike periodical literature, are bound by definition to contain new information, data and ideas.

6. It can serve as a source of new ideas in a different but related field, and may be outside the patent protection offered the original invention.

7. Since claims are narrowly defined, much valuable information is included that may be outside the legal protection afforded by the claims.

8. Patents discuss the difficulties associated with previous research, development or production techniques, and offer a specific method of overcoming problems. This gives clues to current process information that has not been published in periodicals or books.

9. Can aid in process design by providing a selection of alternate techniques. A powerful research and engineering tool.

10. Obtain licenses — many U.S. chemical patents have not been developed commercially.

11. Patents provide an excellent starting point for the next investigator.

12. Frequently, innovations derived from research are first disclosed in the patent literature, prior to coverage in the periodical literature.

13. Patents offer a most valuable method of keeping abreast of latest technologies, serving an individual's own "current awareness" program.

14. Copies of U.S. patents are easily obtained from the U.S. Patent Office at 50¢ a copy.

15. It is a creative source of ideas for those with imagination.

CONTENTS AND SUBJECT INDEX

INTRODUCTION

Synthetic fibers can be produced from a wide variety of substances: polyesters, polyamides, vinyls, etc. The same material can often be used in molding as well as spinning processes. Some of the polymers, especially those which are suitable thermoplastics, are largely of petroleum origin; others, such as the polylactones, represent an attempt to develop new materials from renewable and biodegradable resources.

Industrial production aims for continuous, one-stage, energy-efficient processes which can be conducted at reasonable temperatures. Conventional catalysts which accomplish this goal at acceptable rates and good yields unfortunately have a deleterious effect on other properties such as color, thermal stability, and mechanical properties. It is generally true that the improvement in one property is accomplished at the expense of another. Flame retardants may lead to discoloration and low tensile elongation at break point; carriers which improve affinity for acid or basic dyes may lead to inferior mechanical properties. Much of the material in this book deals with compositions producing improvement in one aspect without sacrificing other superior properties.

Each type of polymeric material has its own outstanding qualities and its own defects. Polyethylene terephthalate, the preferred commercial polyester, has superior mechanical properties and chemical resistance; however, its dimensional stability and transparency are poor because of low heat distortion temperatures and high rates of crystallization. Impact strength is also poor, making this polymer unsuitable for molded articles. Polycarbonates, on the other hand, have a high heat distortion temperature and superior transparency, but poor resistance to chemicals. Polyesters with improved tensile strength, flame resistance, and thermal stability, which still retain their strong abrasion resistance and have low enough melt viscosities for ease in processing, are eagerly sought.

Polyamides are outstanding in physical strength and toughness, and have superior mechanical, electrical, and shaping properties and good chemical stability. Atmospheric degradation and poor thermal stability at melt temperature are the problems with these resins, but degradation still leaves those qualities superior compared to competitative polymers.

Nylon fibers are stronger than any natural fibers, have an abrasion resistance four times that of wool, and are unaffected by dry cleaning solvents. But nylon does not always dye uniformly, and, since the mechanism of its pyrolysis is not known, it is difficult to predict the success of flame retardants. Low moisture absorption is a further problem, leading to static cling and general lack of comfort.

Polyimides have exceptional thermal stability, but cannot be melt processed. Acrylics and modacrylics also must be processed by wet or dry spinning. This not only requires solubility in a convenient solvent, but leads to microscopic cavities in the resulting filament and an accompanying tendency to shrink in hot water.

Copolymers, especially block copolymers, can provide the low processing temperature or other desirable attribute of one of the components coupled with some superior properties of the other.

The demand for flame-retardant fabrics has increased. Nonfusability is also an important safety factor, since it is possible to incur severe burns by contact with molten polymers. It is also desirable for smoldering textile goods to generate as little smoke or poisonous gas as possible.

There is an ongoing effort to duplicate, in the synthetics, some of the properties of natural fibers such as absorbency and water retention. This topic will be discussed at greater length in a forthcoming Noyes publication.

POLYESTERS

POLYCONDENSATION AND TRANSESTERIFICATION CATALYSTS

Linear aromatic polyesters are normally produced by first forming a monomeric ester by esterification of an aromatic carboxylic acid with ethylene glycol (so-called direct esterification) or by an ester interchange reaction consisting of reacting a glycol with an ethylene dialkyl ester of the acid followed by a polycondensation reaction.

Germanium Dioxide and Antimony Trioxide

The method developed by *R.P.L.V. Taubinger and R.B. Rashbrook; U.S. Patent 4,133,800; January 9, 1979; assigned to Imperial Chemical Industries Limited, England* is concerned with the direct esterification process.

Polyesters, e.g., polyethylene terephthalate, are produced by a direct esterification process. The esterification is performed in the presence of a germanium dioxide polycondensation catalyst dissolved in an alkaline ethylene glycol solution. The alkali in the catalyst solution acts as the softening point stabilizer. An antimony polycondensation catalyst is also employed.

Germanium compounds are particularly useful as they generally give polyesters of improved color compared with the use of antimony compounds but it has been found that the use of a combination of germanium and antimony compounds gives particularly good polyesters. The method is particularly suited to the production of polyesters for fiber or film manufacture.

Example 1: *Comparative* — 60.5 parts by weight of terephthalic acid and 38.8 parts by weight ethylene glycol were esterified under a pressure of 40 psi at a temperature of 240°C in the presence of 0.0035 part sodium hydroxide. 0.07 part triphenyl phosphate stabilizer was then added after esterification; 0.0035 part of amorphous germanium dioxide and 0.021 part antimony trioxide were added as polycondensation catalyst. Polycondensation was completed by raising the temperature to 290°C and applying a vacuum of 0.1 mm of mercury.

The resultant polymer had an intrinsic viscosity, as measured on a 1% solution in o-chlorophenol at 25°C, of 0.658, a yellowness of 30 units and softening point of 255.4°C. The yellowness figure quoted is a measurement of the color of the sample obtained on a "colormaster" differential colorimeter. The higher the yellowness figure, the poorer the color of the polymer.

Example 2: 60.5 parts terephthalic acid and 38.8 parts ethylene glycol were esterified under the conditions described in Example 1 but in the presence of 0.0035 part hexagonal germanium dioxide dissolved in an ethylene glycol solution containing the 0.0035 part sodium hydroxide. 0.07 part triphenyl phosphate stabilizer and 0.021 part antimony trioxide catalyst were added after esterification and polycondensation were completed as above in Example 1. The polymer of intrinsic viscosity 0.682 gave a yellowness of 27 units and a softening point of 256.7°C.

Example 3: 86.5 parts terephthalic acid and 42 parts ethylene glycol were esterified at a temperature of 241°C and a pressure of 2.8 atmospheres in the presence of 0.005 part sodium hydroxide. 0.21 part triphenyl phosphate as stabilizer and 0.005 part hexagonal germanium dioxide, dissolved in an ethylene glycol solution containing 0.002 part sodium hydroxide, and 0.03 part antimony trioxide as polycondensation catalyst were then added. Polycondensation was completed by raising the temperature to 290°C and decreasing the pressure to 0.1 mm mercury pressure. The polymer had an intrinsic viscosity of 0.658, a yellowness of 32 units and softening point of 258.8°C.

Antimony, Lead, and Calcium Combinations

According to *G. Morawetz and L. Buxbaum; U.S. Patent 4,080,317; March 21, 1978; assigned to Ciba-Geigy AG, Switzerland*, polyester precondensates are catalytically polycondensed in solid phase at a constant temperature with a linear increase (during polycondensation) in intrinsic viscosity, $[\eta]$. The linear increase makes possible an accurate advance determination of reaction time required to produce thermoplastic polycondensate of any desired intrinsic viscosity.

A melt condensate containing catalytic amounts of a catalyst combination consisting of antimony, lead and calcium very quickly reacts (when carrying out solid phase condensation) to form a higher molecular weight polycondensate which is nearly colorless. Contrary to other solid phase polycondensates containing (in addition to antimony) components, such as manganese, the solid phase condensates containing antimony, lead and calcium display only very minor chromatic aberrations from white. Aside from the external optical impression, this is evident from colorimetric measurements taken with the trichromatic colorimeter on polyester samples.

Producing a melt precondensate, catalysts can be employed as mixtures of metals or as alloys, as solutions of metals in ethylene glycol or as oxides or as salts from low molecular weight carboxylic acids. The catalyst is ordinarily employed in amounts of from 0.001 to 2% by weight of polyester precondensate with which it is preferably in intimate admixture.

The process is in no way limited to catalyst combinations consisting of antimony, lead and calcium; alternative suitable catalyst systems consist of: antimony, lead and zinc; antimony, lead, calcium and manganese; antimony and manganese;

titanium and calcium; polymeric germanium hydride and zinc; and potassium germanide, etc.

A melt condensate for carrying out solid phase condensation which is particularly suitable contains up to 0.05% by weight of antimony, up to 0.02% by weight of lead and between 0.005 and 0.1% by weight of calcium. The three metals mentioned surprisingly have a synergistic effect on each other; lead alone yields strongly colored solid phase condensates, and the combination of antimony trioxide and calcium acetate yields molecular weights which are inadequate for many types of application (comparative example).

Example: 112 g of a pulverized lead-antimony alloy (Pb:Sb = 3:7) sifted to a granular size of 40 μ and suspended in 500 ml of ethylene glycol and 160 kg of dimethyl terephthalate are added to 86 liters of ethylene glycol. Then, 80 g of calcium dissolved in 15 liters of ethylene glycol are added. Methanol distillation sets in at 105°C and the methanol formed is distilled off by means of a packed column with adjustable runback. The temperature in the reaction mixture rises to 201°C.

The thus-obtained ester interchange mixture is pressed through a filter with a mesh size of 40 μ into a second reactor and there is heated to 275°C. After distilling off the excess ethylene glycol, a vacuum of 0.5 torr is applied within 2 hours. After a further 3½ hours, an intrinsic viscosity of 0.75 dl/g is reached. The reaction is then stopped and (after quenching in water) the obtained precondensate is cut into granules of approximately cylindrical shape of 3 mm length and about 2.5 mm diameter.

400 kg of the obtained precondensate are predried in a tumble dryer of 1 m³ capacity for 3 hours at 50°C under a vacuum of 0.3 torr. The granules are then heated at constant vacuum within 5 hours to 175°C and kept at this temperature for 2½ hours. The density of these granules is 1.383 g/cm³, the concentration of carboxylic end groups is 0.83 val/mol ($[\eta]$ = 0.75). The temperature of the granules is then increased to 236°C within 6 hours at still constant vacuum. A sample taken now shows an intrinsic viscosity of 0.89 dl/g.

The granules are now kept for 7 hours at 236°±0.5°C, with samples being taken at 2-hour intervals and the intrinsic viscosity of the samples being determined. The final product has an intrinsic viscosity of 1.12 dl/g and a density of 1.416 g/cm³. The course of the solid phase condensation is shown by graphing intrinsic viscosity against the polycondensation time in hours. A straight line results, showing a linear increase in viscosity as the polycondensation proceeds.

Comparative Example: 208 g of calcium acetate and 48 g of antimony trioxide are added to 101 liters of ethylene glycol and 160 kg of dimethyl terephthalate. Methanol distillation sets in at 125°C and the methanol formed is distilled off by means of a packed column with adjustable runback. 66 liters of methanol are distilled off within 3 hours, and the temperature of the reaction mixture rises to 202°C.

The ester interchange mixture thus obtained is pressed into a second reactor and heated to 275°C. After distilling off the excess ethylene glycol, a vacuum of 0.5 torr is applied within 2 hours. After a further 2-hour period, an intrinsic viscosity of 0.60 dl/g is reached. The reaction is then stopped and, after quenching in water, the precondensate obtained is cut into granules of approximately cylindrical shape of about 3 mm length and about 2.5 mm diameter.

50 kg of the precondensate thus obtained are predried in a tumble dryer of 1 cubic meter capacity for 3 hours at 50°C under a vacuum of 0.3 torr. The granules are then heated at still constant vacuum within 3 hours to 175°C and kept at this temperature for 3 hours. The density of these granules is 1.398 g/cm^3, the carboxylic acid value is 0.89 val/mol ($[\eta]$ = 0.60). The density (after cooling off) of a sample tempered for 6 hours at 236°C is 1.429 g/cm^3.

The temperature of the granules is then increased at still constant vacuum within 4 hours to 236°C. A sample now taken shows an intrinsic viscosity of 0.62 dl/g. The granules are now kept at 236°±0.5°C for 36½ hours with samples being taken at various intervals to determine the intrinsic viscosity. The final product has an intrinsic viscosity of 0.96 dl/g and a density of 1.428 g/cm^3. The nonlinear course of the solid phase condensation and its coming to a standstill are evident when the intrinsic viscosity is plotted against the polycondensation time in hours.

Tetravalent Tin with Organotin Linkage

G.R. Chipman, M.G. Henk, J.A. De Boer and E.W. Blaha; U.S. Patent 4,014,858; March 29, 1977; assigned to Standard Oil Company describe the reaction of terephthalic acid and butanediol using a tetravalent tin catalyst having the organo-to-tin linkage.

The catalysts are tetravalent tin catalysts having one organo to each tin linkage such as the hydrocarbyl stannoic anhydrides having the structure

$$\underset{R-Sn-O-Sn-R,}{\overset{O \quad\;\; O}{\overset{\|\quad\;\;\;\|}{}}}$$

wherein each R contains 1 to 24 carbon atoms and can be alkyl, aralkyl, alkaryl or aryl; hydrocarbyl stannoic acids having the structure

$$\underset{R-Sn-OH}{\overset{O}{\overset{\|}{}}}$$

wherein R contains from 1 to 24 carbon atoms and can be alkyl, aralkyl, alkaryl or aryl; hydrocarbyl tin halides having the structure

$$\underset{\overset{|}{X_2}}{\overset{\overset{X}{|}}{R-Sn-X_1}}$$

wherein R contains from 1 to 24 carbon atoms and can be alkyl, aralkyl, alkaryl or aryl and X is a halogen group such as chlorine or bromine and X_1 and X_2 are halogen such as chlorine or bromine or hydroxy groups, etc. Suitable catalysts include methyl stannoic anhydride, ethyl stannoic anhydride, isopropyl stannoic anhydride, butyl stannoic anhydride, octyl stannoic anhydride, benzyl stannoic anhydride, p-octylphenyl stannoic anhydride, methyl stannoic acid, methyltin trichloride, butyltin dihydroxide monochloride, stearyltin dihydroxide monochloride, etc. Of these, best results have been obtained with the commercially available butyl stannoic anhydride.

Example 1: This example illustrates the production of polybutylene terephthalate using a 1.25 to 1 mol ratio of 1,4-butanediol to terephthalic acid. 166 g terephthalic acid (1 mol), 112.5 g butanediol (1.25 mols) and 0.55 g butyl stannoic anhydride in a one-liter resin kettle equipped with a knock-back condenser was immersed in a hot oil bath heated to 260°C.

After about 40 to 50 minutes, the reactants reached 220°C, as measured by an internal thermocouple, and the hot oil bath temperature was adjusted to maintain the internal temperature at 220°C. Seventy minutes after immersion in the oil bath, the first-stage esterification was completed, the knock-back condenser was removed and the polycondensation was carried out at 0.50 mm at 240°C for 2 hours. The molten syrup was removed from the reactor, cooled and ground to yield a polymer having an inherent viscosity of 0.88. The condenser system contained 0.08 to 0.09 mol tetrahydrofuran.

Examples 2 through 5: These examples illustrate the production of polybutylene terephthalate using butyl stannoic anhydride and butyl stannoic acid as contrasted to dibenzyltin oxide and stannous oxide. Example 1 was repeated using a 1.5 to 1 mol ratio of 1,4-butanediol to terephthalic acid and 0.25% by weight catalyst based on the theoretical molecular weight of the polymer. In these runs it took about 40 to 50 minutes to get the reactor temperature to 220°C. The results are shown below:

Ex. No.	Catalyst	Esterification Time (min)	Tetrahydrofuran Formed (mol)
2	Butyl stannoic anhydride	55	0.12
3	Butyl stannoic acid	56	0.10
4	Dibenzyltin oxide	120*	0.38
5	Stannous oxide	120*	0.34

*Did not clear.

The above results illustrate that tin compounds having one organo to each tin linkage are advantageous in producing polybutylene terephthalate from 1,4-butanediol and terephthalic acid.

The organotin catalysts of this process produce substantially less tetrahydrofuran in a much shorter first-stage esterification period.

Antimony Diglycollate with Silicon Compounds

K. Weinberg and G.C. Johnson; U.S. Patent 4,116,942; September 26, 1978; assigned to Union Carbide Corporation provide a process for producing polyesters and copolyesters, useful for making films and fibers, by the polycondensation of dicarboxylic acids and aliphatic glycols using mixtures of antimony diglycollate and silicon compounds as catalysts.

In the production of polyesters and copolyesters the reaction is generally considered a dual or two-stage reaction. In the first stage esterification or transesterification occurs and in the second stage polycondensation occurs. In the second stage, or the polycondensation, the catalyst complexes of this process are useful.

The silicon compounds that are used in conjunction with antimony diglycollate as the catalyst mixture of this process are represented by the generic formulas on the following page.

$$
(1) \qquad W-(COOC_nH_{2n})_m-\underset{\underset{R}{|}}{\overset{\overset{R}{|}}{Si}}-R'
$$

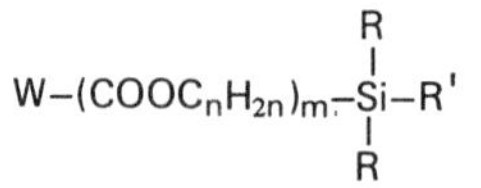

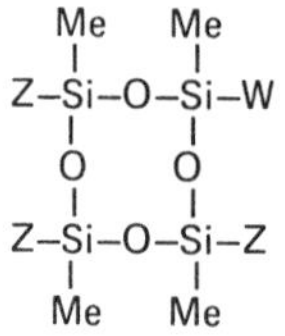

$$
(2) \qquad
\begin{array}{c}
\overset{Me}{\underset{|}{}} \quad \overset{Me}{\underset{|}{}} \\
Z-Si-O-Si-W \\
\overset{|}{O} \quad \overset{|}{O} \\
Z-Si-O-Si-Z \\
\underset{Me}{|} \quad \underset{Me}{|}
\end{array}
$$

$$
(3) \qquad R'''O-\left[\underset{\underset{Me}{|}}{\overset{\overset{Me}{|}}{SiO}}\right]_x-\left[\underset{\underset{Me}{|}}{\overset{\overset{W}{|}}{SiO}}\right]_y-R'''
$$

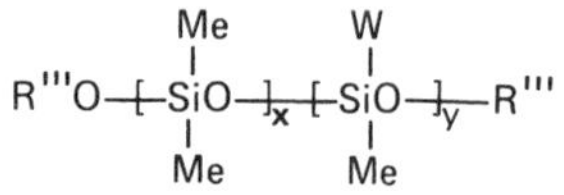

or

$$
(4) \qquad QCH_2CH_2SiR_3{}^{**}
$$

In these formulas:

W is $CH_2{=}CX-$ or $(R^*O)_2P(O)CH_2CHX-$ wherein

X is hydrogen or methyl and is methyl only when m is 1;

R^* is alkyl or haloalkyl having from 1 to 4 carbon atoms;

R^{**} is methyl, ethyl, butyl, acetoxy, methoxy, ethoxy or or butoxy;

R is methyl, ethyl, butyl, methoxy, ethoxy, butoxy, or trimethylsiloxy;

R' is methyl, methoxy, ethoxy, butoxy or trimethylsiloxy;

R'' is methoxy, ethoxy, butoxy, trimethylsiloxy or vinyldimethylsiloxy;

R''' is methyl, ethyl, butyl or trimethylsilyl;

Me is methyl;

Z is methyl or W;

Q is an NH_2CH_2-, $NH_2CH_2CH_2NHCH_2-$, $NC-$, $HS-$ or $HSCH_2CH_2S-$ group;

n is an integer having a value of from 2 to 5;

m is an integer having a value of 0 or 1;

x is an integer having a value of from 1 to 100; and

y is an integer having a value of from 1 to 100.

Subgeneric to Formula (1) are the compounds represented by the following subgeneric formulas:

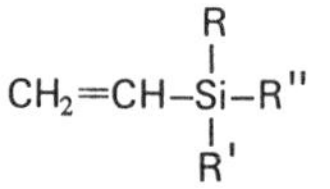

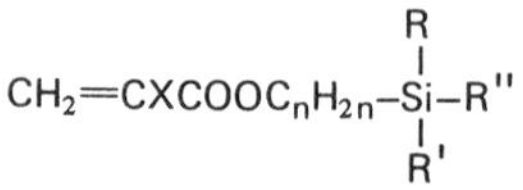

$$
CH_2{=}CH-\underset{\underset{R'}{|}}{\overset{\overset{R}{|}}{Si}}-R'' \qquad\qquad CH_2{=}CXCOOC_nH_{2n}-\underset{\underset{R'}{|}}{\overset{\overset{R}{|}}{Si}}-R''
$$

$$
\underset{\substack{\text{O} \\ \| \\ (R^*O)_2PCH_2CH_2-\underset{\substack{| \\ R'}}{\overset{\substack{R \\ |}}{Si}}-R''}}{} \qquad\qquad \underset{\substack{\text{O} \\ \| \\ (R^*O)_2PCH_2CHXCOOC_nH_{2n}-\underset{\substack{| \\ R'}}{\overset{\substack{R \\ |}}{Si}}-R''}}{}
$$

Examples: A mixture of 39.1 g of dimethyl terephthalate, 32.4 g of ethylene glycol, 0.0176 g of zinc acetate dihydrate as transesterification catalyst and a mixture of 0.020 g of antimony diglycollate and 0.053 g of beta-diethoxyphosphorylethyl methyl diethoxysilane as polycondensation catalyst were combined and heated at 178° to 190°C for 3 hours under argon to effect transesterification. The reaction mixture was then heated to 230°C and this temperature was maintained for 1 hour.

The temperature was thereafter raised to 280°C while the pressure was reduced to less than 1 mm of mercury and then maintained for 50 minutes at these temperatures and pressure conditions to carry out the polycondensation. The polycondensation was continued until the intrinsic viscosity of the polyester was 0.57.

Fibers produced from this polyester had a denier of 122, an elongation at break of 20% and a tenacity of 3.62 g/d.

For comparative purposes the same reaction was carried out under similar conditions using 736.4 g of dimethyl terephthalate, 533.2 g of ethylene glycol, 0.2711 g of zinc acetate dihydrate and 0.2726 g of antimony oxide as catalyst. The polycondensation time required was 60 minutes to achieve the same intrinsic viscosity, a period of time 20% longer.

Fibers produced from this polyester had a denier of 124, an elongation at break of 11.7% and a tenacity of 3.69 g/d.

Following the above procedure, polyester is produced by substitution of the following silicon compounds for the beta-diethoxyphosphorylethyl methyl diethoxysilane: 3-aminopropyl triethoxysilane, 2-cyanoethyl triethoxysilane, 2-mercaptoethyl triethoxysilane, vinyl methyl diethoxysilane, and tetramethyl divinyl diethoxysilane.

Germanium Dioxide with Tetraalkyl Ammonium Hydroxide

The process of *H.L. King and C.C. Wu; U.S. Patent 4,119,614; October 10, 1978; assigned to Monsanto Company* utilizes a prepolymer which is formed by reacting a glycol with a dibasic acid such as terephthalic acid in a known manner. Preferably, the prepolymer is a glycol phthalate. The prepolymer is subjected to a polycondensation reaction in the presence of a germanium dioxide/tetraalkyl ammonium hydroxide catalyst to form a high polymer.

Example 1: This example discloses a method for making the catalyst used in the process. 10½ g (0.1 mol) of germanium dioxide and 37 g (0.1 mol) of 40% aqueous tetraethyl ammonium hydroxide were stirred with 60 ml of ethylene glycol until solution was complete. The solution was heated to distill off water and additional ethylene glycol was added to bring the volume to 105 ml. The clear, colorless solution contained 0.10 germanium dioxide equivalent per ml of solution.

Example 2: This example discloses a method for making a glycol phthalate prepolymer which can be polycondensed to form a fiber-forming polyester. A slurry consisting of ethylene glycol and terephthalic acid in the molar ratio of 2:1 was reacted at 270°C and 20 pounds pressure in a continuous polyester esterifier. Water of reaction and some ethylene glycol were removed by distillation and the low molecular weight prepolymer product was collected continuously through a valve at the bottom of the reactor.

The rate of removal of prepolymer was such as to maintain a constant level in the reactor. The average residence time in the reactor was 135 minutes. The low molecular weight glycol phthalate prepolymer collected had an intrinsic viscosity of 0.08 and a carboxyl concentration of 659 μeq/g, both measured in a conventional manner.

Example 3: 210 g of the prepolymer of Example 2 were added to a 1-liter capacity, stirred batch autoclave. To the autoclave was added 0.43 ml (150 ppm germanium, based on polymer weight) of the catalyst of Example 1, the autoclave purged with nitrogen and heated to 280°C. Pressure in the system was reduced by applying a vacuum and polymerization to high molecular weight continued at pressures of less than 2 mm Hg until the desired molecular weight was obtained.

A polymer with an intrinsic viscosity of 0.61 was obtained in a period of 62 minutes from the beginning of pressure reduction. The polymer melted at 256°C and contained 1.32 mol percent diethylene glycol. Drawn fibers formed from the polymer had a purity value of 1.6 and a brightness value of 88 as measured by tristimulus values obtained with a GE recording spectrophotometer.

Example 4: *Comparative*– Example 3 was repeated except that the catalyst used was 0.05 g (150 ppm antimony, based on polymer weight) of antimony glycol oxide. Polymers with properties equivalent to Example 3 were obtained; however, the time required for polymerization to an intrinsic viscosity of 0.61 was 102 minutes.

Trivalent Antimony plus Selected Ethylenically Unsaturated Compounds

More particularly the catalyst system developed by *R.L. Muntz and F.A. Via; U.S. Patent 4,067,856; January 10, 1978; assigned to Stauffer Chemical Company* comprises an antimony-containing polycondensation catalyst and an ethylenically unsaturated compound having the formula:

$$(1) \quad \begin{array}{c} R^1 \; R^3 \; R^4 \\ | \quad | \quad | \\ C=C-C-OR^6 \\ | \quad \quad | \\ R^2 \quad \; R^5 \end{array}$$

wherein R^1, R^2, R^3, R^4 and R^5 are independently selected hydrogen, alkyl, aryl, substituted alkyl or substituted aryl and R^6 is hydrogen, alkyl, aryl, acyl, allyl or epoxypropyl. Typical substituents include hydroxyl, mercapto, amino, amido, imino, oxime, halo, aldehydo, alkoxy, alkylthio, keto, sulfinyl, sulfonyl, phosphato, phosphono, or phosphino; the substituted groups in R^4 being hydroxyl, mercapto, amino, amido, halo, carboxyl, carboxylate, haloformyl, carboxylic anhydride, aldehydo, alkoxy, alkylthio, keto, sulfinyl, sulfonyl, phosphato, phosphono or phosphino.

Preferably the alkyl and acyl groups in compound (1) have from 1 to 18 carbon atoms, and the aryl groups from 6 to 18 ring carbon atoms. All compounds having the formula (1) can be prepared according to the Williamson ether synthesis as described in *Org. Syn. Coll. Vol. I*, p. 75, 205, 258, 296 and 435, Wiley and Sons (1932).

Exemplary ethylenically unsaturated compounds include allyl alcohol, allyl acetate, allyl glycidyl ether, allyl phenyl ether, allyl hexyl ether, bis-1-octen-1-yl ether, (3-mercaptopropyl) allylether, 3-phenyl-1-octen-3-ol, (4-hydroxy-2-naphthyl) allyl ether, 1-methyl-1-propen-2-ol, 3-toluenesulfonyl-1-buten-4-ol, (4-diethylphosphono-2-buten-1-yl) glycidyl ether, methallyl (2-diethylphosphono-3-naphthyl) ether, ethyleneglycol diallyl ether, diallylterephthalate, etc. Preferred are those compounds (1) where R^1 through R^5 are hydrogen and R^6 is allyl or glycidyl.

Exemplary antimony compounds include the antimony acids and salts thereof, such as antimonous acid, magnesium antimonite, zinc antimonite, calcium antimonite, manganese antimonite, etc.; antimony alkoxides such as trimethyl antimonite, tributyl antimonite, trihexyl antimonite, tridodecyl antimonite, tricyclohexyl antimonite, diethylmethyl antimonite, diethylacetyl antimonite, diethylphenyl antimonite, etc.; antimony carboxylates such as antimony acetate, antimony butyrate, antimony benzoate, antimony tolylate, antimony formate, etc.; antimony halides such as antimony bromide, antimony fluoride, antimony chloride, etc.; antimony sulfide; antimony oxides such as antimony trioxide; and antimony glycoxides such as antimony ethylene glycoxide, antimony butylene glycoxide, etc. The preferred polycondensation catalysts are the antimony acids, antimony alkoxides, antimony acetate and the antimony glycoxides.

Example 1: Allyl glycidyl ether was screened for thermal stability towards graying in the following manner. A 0.03 mol portion of the ether was added to a 25 ml flask containing 0.01 mol tri-n-butyl antimonite dissolved in 15 to 20 g of triethylene glycol. Then the mixture was heated with a mantle while stirring magnetically. The decomposition temperature was that at which a gray color appears. A decomposition temperature of $>285°C$ was observed, as contrasted with a decomposition temperature of 240° to 250°C for a control containing only 0.01 mol tri-n-butyl antimonite.

Example 2: The screening test described in Example 1 was followed with the exception that tetraethylene glycol was employed as the solvent instead of triethylene glycol. A control employing only 0.01 mol of tri-n-butyl antimonite resulted in a decomposition temperature of 240° to 250°C. When allyl glycidyl ether was employed in three tests wherein the mol ratio of the ether to the antimonite was 3:1, 2:1 and 1:1, decomposition temperatures of $>300°C$, $>300°C$ and 260°C, respectively, were observed.

Example 3: Bis(hydroxyethyl) terephthalate (33.0 g), triphenyl phosphite (0.020 g), tri-n-butyl antimonite (0.30 g; 0.09 mmol) and allyl glycidyl ether (0.34 g; 0.30 mmol) were mixed in a 50 ml 2-neck flask equipped with a magnetic stirrer, a short-path distillation head and a thermometer. The mixture was heated to about 160°C and a vacuum applied. The polycondensation was carried out at a temperature up to 280°C and a pressure of 1 mm Hg over a period of 1 hour. Seven to eight grams of ethylene glycol were distilled out of the mixture.

The resultant polymer was poured onto a watch glass, cooled and ground to a fine powder in a high-speed blender. The powder was placed in a Petri dish to form a layer about ¼" thick and held in place with a large rubber stopper. The color of the sample was then measured on a Hunterlab Model D25 Color Difference Meter, employing Hunterlab Standard T400 for color difference meter 45° 0° geometry, CIE Illuminant C, ASTM D2244-647 standard; MgO (ASTM E-259). The polymer had an L value of 87.4.

Comparative Example: The polycondensation step of Example 3 was repeated with the exception that tri-n-butyl antimonite (0.30 g) was employed without allyl glycidyl ether. The color of five samples of the resultant polymer was measured, and an average L value of 81.7 was obtained. This indicates that the use of catalyst without stabilizer results in graying.

In a related patent, *R.L. Muntz; U.S. Patent 4,067,857; January 10, 1978; assigned to Stauffer Chemical Company* uses, as a stabilizer, an ethylenically unsaturated compound having the formula:

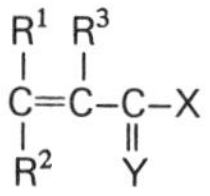

$$\begin{array}{cc} R^1 & R^3 \\ | & | \\ C=C-C-X \\ | & \| \\ R^2 & Y \end{array}$$

wherein X is hydrogen, $-OR^4$, $-SR^4$, amino, halogen, $-O_2CR^5$, $-NC(O)R^5$ or mercapto; Y is oxygen or sulfur; and R^1, R^2, R^3, R^4 and R^5 are independently selected hydrogen or substituted or unsubstituted alkyl, cycloalkyl, alkenyl, heterocyclic, aryl, aralkyl or alkaryl, the substituted groups in R^1, R^2, R^3 and R^5 being independently selected hydroxyl, mercapto, amino, amido, imino, oxime, halo, aldehydo, alkoxy, alkylthio, keto, sulfinyl, sulfonyl, phosphato, phosphono or phosphino; the substituted groups in R^4 being hydroxyl, mercapto, amino, amido, imino, oxime, halo, carboxyl, carboxylate, haloformyl, carboxylic anhydride, aldehydo, alkoxy, alkylthio, keto, sulfinyl, sulfonyl, phosphato, phosphono or phosphino.

Exemplary ethylenically unsaturated compounds suitable for use include: acrylic acid, methacrylic acid, methacrylamide, α-octadecylacryloyl chloride, acrylic propionic anhydride, 2-(diethylphosphato)ethyl acrylate, (ethyl phenylphosphino)-methyl acrylate, etc.

Antimony Alkoxide Reacted with Unsaturated Dicarboxylic Acid Anhydride

The catalyst developed by *R.L. Muntz; U.S. Patent 4,130,552; December 19, 1978; assigned to Stauffer Chemical Company* consists essentially of the reaction product of an unsaturated α, β-dicarboxylic acid anhydride having 4 to 12 carbon atoms and an antimony alkoxide.

Typical anhydrides which can be used in the process are those represented by the formulae:

$$(1) \quad \begin{array}{c} R-C-CO \\ \| \quad\quad\ \diagdown O \\ R^1-C-CO \diagup \end{array} \qquad (2) \quad \begin{array}{c} R^2CH=C-\!-CO \\ |\quad\quad\quad \diagdown O \\ CH_2-CO \diagup \end{array}$$

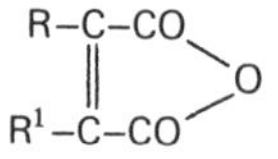

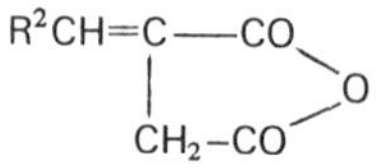

(3)

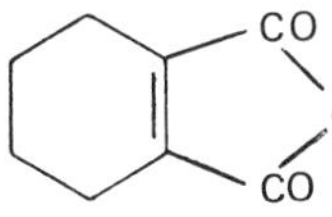

wherein R and R^1 are independently selected hydrogen, chlorine, or alkyl of 1 to 4 carbon atoms and R^2 is hydrogen, alkyl of 1 to 4 carbon atoms or phenyl. Exemplary anhydrides include maleic anhydride, tetrahydrophthalic anhydride, dimethylmaleic anhydride, dichloromaleic anhydride, methyl isobutyl maleic anhydride, di-n-butyl maleic anhydride, citraconic anhydride, itaconic anhydride, etc.

The antimony-containing polycondensation catalysts comprise anhydride derivatives of the known antimony alkoxides having the formula: $Sb(OR^4)(OR^5)(OR^6)$ where R^4, R^5 and R^6 are independently selected hydrogen, alkyl, acyl, aryl or substituted aryl, with the proviso that at least one of R^4, R^5 and R^6 must be other than acyl, the alkyl and acyl groups having 1 to 18 carbon atoms and the aryl groups having 6 and 18 carbon atoms, and the substituted aryl having substituents such as phenyl, naphthyl, methylphenyl, ethylphenyl, chlorophenyl, dimethylphenyl, methoxyphenyl, etc.

Exemplary alkoxides include trimethyl antimonite, tributyl antimonite, trihexyl antimonite, diethylphenyl antimonite, etc. The preferred polycondensation catalysts are the derivatives of the trialkyl antimonites, and particularly triethyl antimonite, tributyl antimonite and tripropyl antimonite.

The components of the catalyst are employed in a mol ratio of the acid anhydride to the antimony alkoxide of 0.5-3:1, and preferably in a ratio of 1-2:1. The reaction product is generally a mixture of compounds having the formula: $(ZO)_x Sb(OOCACOOZ)_y$ where Z is R^4, R^5, or R^6 as previously described; A is

$$-CR=CR^1-, \qquad \begin{matrix} -C-CH_2- \text{ or} \\ \| \\ CHR^2 \end{matrix}$$

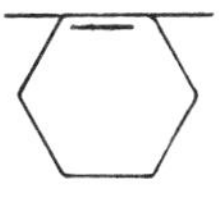

wherein R, R^1 and R^2 are as previously described; y is an integer from 1 to 3 and x is 0 to 2, with the proviso that x + y = 3.

Example 1: *(a) Reaction Product Preparation* — Maleic anhydride (196.0 g, 2.00 mols) was added in about 25 g portions to tri-n-butyl antimonite (341.0 g, 1.00 mol) in a 1-liter, 3-necked flask equipped with a thermometer, magnetic stirrer and nitrogen inlet. An exotherm was observed giving a maximum pot temperature of 80°C. The product was a clear light yellow liquid.

Analysis was consistent with a formula for the major component of the reaction product of: $CH_3CH_2CH_2CH_2OSb(O_2CCH=CHCO_2C_4H_9)_2$.

(b) Polycondensation — Bishydroxyethyl terephthalate (33.0 g), triphenyl phosphite (0.020 g) and 0.47 g of the maleic anhydride reaction product prepared in part (a) were mixed in a 50-ml 2-neck flask equipped with a magnetic stirrer, a short-path distillation head and a thermometer. The mixture was heated to about 160°C and a vacuum applied. The polycondensation was carried up to a temperature of 280°C and an absolute pressure of 1 mm Hg over a period of 1 hour. Seven to eight grams of ethylene glycol were distilled out of the mixture.

The resultant polymer was poured onto a watch glass, cooled and ground to a fine powder in a high-speed blender. The powder was placed in a Petri dish to form a layer about ¼" thick and held in place with a large rubber stopper. The color of the sample was then measured on a Hunterlab Model D25 Color Difference Meter, employing Hunterlab Standard T400 for color difference meter 45° 0° geometry, CIE Illuminant C, ASTM D2244-647 Standard; MgO (ASTM E-259). An average L value of 87.4 for two measurements was obtained.

Comparative Example: The polycondensation step of Example 1 was repeated with the exception that tri-n-butyl antimonite (0.30 g) was employed instead of the reaction product. The color of five samples of the resultant polymer was measured, and an average L value of 81.7 was obtained. This indicates that the use of catalyst without stabilizer results in graying.

Aromatic Titanates

S. Hashimoto, N. Okumura and K. Kazama; U.S. Patent 4,131,601; December 26, 1978; assigned to Teijin Limited, Japan describe a process for the preparation of substantially linear, highly polymerized polyesters by polycondensing a glycol ester of an aromatic dicarboxylic acid and/or a low molecular weight condensate thereof, the process being characterized by using as a polycondensation catalyst a preformed titanate compound obtained by reacting a titanic acid ester represented by the formula $Ti(OR)_4$, where R is an alkyl group having from 1 to 5 carbon atoms, with an aromatic acid selected from the group consisting of trimellitic acid, trimellitic anhydride, hemimellitic acid, hemimellitic anhydride or a mixture of one or more thereof at a molar ratio of from about 0.5 to about 2.5 mols of the aromatic acid per 1 mol of the titanic acid ester, to form polyesters which have a high softening point and good color tone.

Examples 1 through 3: *(A) Preparation of Polycondensation Catalysts* — A predetermined amount of trimellitic acid as shown in the table below was dissolved in 25 parts ethanol. A predetermined amount of tetra-n-butyl titanate as shown in the table below was added to the ethanol solution of trimellitic acid. The mixture was heated at a temperature of 80°C for 60 minutes under normal atmospheric pressure to remove the n-butyl alcohol formed.

After cooling, 15 parts acetone was added to the reaction mixture at room temperature to precipitate the obtained titanate compound. The precipitated titanate compound was filtered from the mother liquor and dried at a temperature of 100°C for 2 hours. According to infrared spectrum and elementary analysis of the preformed titanate compound of Example 1, it was found that about 2 of the original substituent groups of tetra-n-butyl titanate were replaced by trimellitic acid groups.

(B) Preparation of Polyesters — A reaction vessel fitted with a stirrer and a

rectifying column was charged with 970 parts dimethyl terephthalate, 640 parts ethylene glycol and 0.20 part (about 0.01 mol percent calculated in terms of titanium atoms based on dimethyl terephthalate) of the preformed titanate compound obtained in A, following which an ester-exchange reaction was carried out by heating the mixture for 3 hours at 140° to 230°C while removing 320 parts methanol formed.

This was followed by adding 0.18 part trimethyl phosphate (a stabilizer) and 4.85 parts titanium dioxide (a delustering agent) to the ester-exchange product. Thereafter, the ester-exchange product was transferred into a polycondensation vessel fitted with a stirrer and a condenser for removing ethylene glycol, following which the polycondensation reaction was carried out by heating the product for 3.5 hours raising the temperature from 230° to 285°C under a reduced pressure of 1 mm Hg.

The physical properties of the obtained polymers were as shown in the table below.

	. . . . . .Preformed Titanate Compounds.								
	Trimellitic		Tetra-n-Butyl			. . . . . .Properties			
Ex. No.	. . .Acid. . .		. .Titanate. .		Titanium		SP		
	Parts	Mols	Parts	Mols	(% by wt)	$[\eta]$	(°C)	L	b
1*	1.20	6	0.64	2	10.1	0.451	262.0	81.5	3.1
1	0.80	4	0.64	2	11.5	0.652	261.8	82.0	3.2
2	0.40	2	0.64	2	12.0	0.648	261.8	81.0	3.0
3	0.20	1	0.64	2	11.8	0.605	259.8	81.3	4.3
2*	0.08	0.4	0.64	2	13.2	0.600	259.5	80.0	5.3

*Comparative.

The polyesters of Examples 1 through 3 had high softening points (more than 258°C) and good color tone (L values were more than 80 and b values were less than 5). The polycondensation rates in Examples 1 through 3 were rapid so that the obtained polyesters had high intrinsic viscosities (more than 0.600). On the other hand, the polyester of Comparative Example 1 had a low intrinsic viscosity, which means a slow polycondensation rate was involved. The polyester of Comparative Example 2 possessed a yellow color (a 5.3 b value).

Ester-Interlinking Agent

A widely employed method of increasing the rate of polycondensation in polyester-forming processes is by the addition of a reactive material capable of interlinking polymer chains. Such ester-interlinking materials provide efficient and economical means for producing polymers of high molecular weight.

Y. Okuzumi; U.S. Patent 4,017,463; April 12, 1977; assigned to The Goodyear Tire & Rubber Company discloses a high melting N,N'-terephthaloyl bisphthalimide, a method for its preparation and the use of this high melting N,N'-terephthaloyl bisphthalimide as an ester-interlinking agent for polyesters.

Example 1: The apparatus employed to prepare the high melting N,N'-terephthaloyl bisphthalimide consists of a 2,000-ml glass reaction vessel equipped with a stirrer, thermometer, condenser, nitrogen inlet and a valved outlet located at the bottom of the reaction vessel. The glass reaction vessel was compartment-alized by a sintered glass partition which was located immediately above the valved outlet. The reaction vessel was attached to a receiving flask, equipped with a vacuum outlet, through the valved outlet.

To this reaction vessel were added 37.4 g (0.203 mol) of potassium phthalimide in a mixture of 250 ml of dioxane and 500 ml of diphenyl ether. Then 20.3 g (0.1 mol) of terephthaloyl chloride were dissolved in 250 ml of dioxane and this mixture slowly added to the reaction vessel with constant stirring. The contents of the reaction vessel were gradually heated to reflux temperature (be-tween 110° and 115°C) and maintained at reflux for 1 hour.

At the end of this time the dioxane solvent component was slowly distilled from the reaction mixture through a fractional distillation column. After the dioxane was completely removed, the reaction mixture was maintained at the boiling point of the diphenyl ether (about 254°C) for an additional hour. The hot solution, containing suspended potassium chloride by-product, was then quickly filtered, at the boil and under vacuum, into the receiving flask. White crystals of N,N'-terephthaloyl bisphthalimide (TBP) product immediately formed in the filtrate.

When the filtrate had cooled to about 100°C the white crystals were collected by filtration and dried under heat and vacuum. The dried crystals (Product 1) weighed 36.5 g, representing a yield of 87.3% and had a melting point of 340° to 345°C. As the filtrate cooled to about 60°C more white crystals formed which were collected and dried under heat and vacuum. These crystals (Product 2) had a melting point of 330° to 336°C. The precipitate collected below 60°C (Prod-uct 3) was also dried under heat and vacuum and its melting point found to be 315° to 325°C.

A portion of Product 1 was recrystallized from diphenyl ether. The recrystal-lized material had a melting point of 354° to 357°C. A portion of this recrys-tallized material was then again recrystallized from diphenyl ether and the melt-ing point of the material found to be 365° to 367°C. An analysis of this latter material gave 68.06% carbon, 2.85% hydrogen and 6.35% nitrogen. The calcu-lated values for the compound having the above formula are 67.93% carbon, 2.85% hydrogen and 6.60% nitrogen.

In view of the above analysis the product is believed to have the structure:

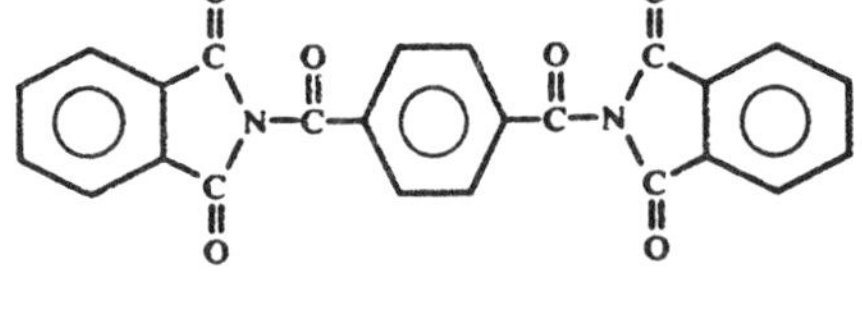

which has the chemical name N,N'-terephthaloyl bisphthalimide. For convenience, this is abbreviated and referred to hereinafter as TBP.

Example 2: A glass reaction tube approximately 35 cm long having an inside diameter of 38 mm, equipped with a side arm, a nitrogen gas inlet tube and stirrer was charged with 50 g of dimethyl terephthalate, 40 g of ethylene glycol and 0.01% (calculated as metal) of manganese octoate. Nitrogen gas was slowly passed into the reaction tube and over the mixture. The mixture was stirred and heated by means of a vapor bath, which surrounded the tube, having a temperature of 240°C.

After completion of the transesterification reaction, polycondensation was commenced by adding 0.0123 g (calculated as metal) of antimony trioxide, increasing the temperature of the mixture to 280°C and gradually reducing the pressure in the tube to 0.05 mm of mercury pressure. The polycondensation reaction was carried out for 50 minutes and the mixture sampled to determine its intrinsic viscosity at that point. 1.40 parts by weight (per 100 parts dimethyl terephthalate) of the twice recrystallized TBP of Example 1 was then added to this mixture and the polycondensation reaction continued for an additional 2 minutes.

Example 3: Employing the same apparatus, materials and amounts thereof as in Example 2, a comparative example was run in which no high melting TBP was added in order to illustrate the normal polycondensation rate. The results show that the addition of small amounts of pure TBP greatly accelerates the polycondensation rate over that of the normal polycondensation reaction rate.

Calcium Acetate with Cobalt Acetate

Y. Omoto, T. Konishi, S. Ichihara and H. Murai; U.S. Patent 4,058,507; Nov. 15, 1977; assigned to Teijin Limited, Japan disclose a process for preparing a polyester which comprises reacting a di-lower alkyl ester of a difunctional carboxylic acid at least 90 mol percent of which consists of terephthalic acid with a polymethylene glycol containing 2 to 10 carbon atoms at least 90 mol percent of which consists of ethylene glycol in the presence of an ester-interchange reaction product, and then polycondensing it in the presence of a polycondensation catalyst, wherein a substantially uniform solution in ethylene glycol of (A) 20 to 150 mmol percent, based on the di-lower alkyl ester of the difunctional carboxylic acid, of calcium acetate and (B) 2 to 25 mmol percent based on the di-lower alkyl ester, of cobalt acetate is used as the ester-interchange reaction catalyst.

Example 1: *(a) Preparation of a Solution of Calcium Acetate and Cobalt Acetate in Ethylene Glycol* — A tank located within a circulating circuit of a homogenizer was charged with 30 kg of the ethylene glycol, 1.22 kg (6.925 mols) of calcium acetate monohydrate and 0.18 kg (0.723 mol) of cobalt acetate tetrahydrate, and these materials were homogenized at room temperature for 45 minutes. The mixture was then transferred into a catalyst-preparing tank held at 80°C and stirred for 2 hours.

The solution was allowed to cool to room temperature to prepare a catalyst solution which was a reddish violet clear solution with a water content of 1.05% by weight.

(b) Ester-Interchange Reaction — An ester-interchange reaction tank equipped with a rectification column, a condenser, a stirrer and a heating device was charged with 2,000 kg of dimethyl terephthalate and 1,250 kg of ethylene glycol,

and then with stirring, all the catalyst solution prepared in (a) was added. Then the mixture was heated.

Methanol began to distill off when the temperature of the inside of the reaction tank reached about 150°C, and in 140 minutes when the inside temperature reached 220°C, 820 ml of the distillate was obtained. At this time, 0.69 g of phosphorous acid was added to the reaction system, and the mixture was stirred. Then, 0.82 kg of antimony trioxide was added and the mixture was stirred at 240° to 260°C for 30 minutes. Then, 10 kg of anatase-type titanium dioxide was added as an ethylene glycol slurry, followed by stirring for about 5 minutes.

In order to examine the conversion of ester-interchange at the time when phosphorous acid was added, a part of the reaction mixture was sampled, and its terminal methyl group concentration (ester methyl group) was measured and found to be 110 eq/10^6 g. This corresponded to an ester-interchange conversion of 98.6%.

(c) Filtration of the Ester-Interchange Product — The ester-interchange reaction product obtained in (b) above was passed in the molten state in an atmosphere of nitrogen at 1.5 kg/cm²g through a filter screen (400 Tyler's mesh) with a filtration area of 1,700 cm² which was located intermediate between the ester-interchange reactor and a polycondensation reactor to be described below. In about 5 minutes, all of the reaction product passed through the filter screen.

(d) Polycondensation Reaction — A stainless steel polycondensation reactor equipped with an ethylene glycol distilling condenser, a stirrer, a vacuum generating device and a heating device was charged with the ester-interchange reaction product obtained after filtration in (c) above. The ester-interchange reaction product was first heated at 260° to 270°C in an atmosphere of nitrogen at atmospheric pressure for 15 minutes. Then, the pressure was reduced gradually, and at the same time, the temperature was raised gradually. When the pressure reached 0.3 mmHg and the temperature of the inside of the reactor reached 289°C, at the end of 160 minutes after methanol began to distill off, the polycondensation reaction ended.

(e) Properties of the Resulting Polyethylene Terephthalate — Terminal methyl group concentration, 5 eq/10^6 g; [η] 0.64; SP 262.5°C; DEG content, 0.75% by weight; weight loss, 8.3% by weight; and color, L value 68.5, b value 3.5.

Example 2: The ester-interchange reaction and the polycondensation reaction were repeated using the same ester-interchange reactor and the same polycondensation reactor as used in Example 1 and repeating the procedures of (a) through (d) of Example 1.

In order to examine the spinnability and drawability of the resulting polyesters, 1.0 ton of the polyester (1.8 tons as obtained) obtained in the third batch ([η] = 0.61) was spun at a melting temperature of 285°C through a spinneret with 30 holes each having a diameter of 0.35 mm, a take-up rate of 1,100 m/min and a discharge rate of 64 g/min. The filaments were wound up as 10-kg packages. At the beginning of the spinning, the pack pressure was 200 kg/cm²g, and it was 220 kg/cm²g at the end of the spinning (the 7th day). 100 of these packages were drawn at a draw ratio of 3.67 at a drawing temperature of 82°C and a setting temperature of 230°C (noncontact type). The draw speed was 800 m/min. The drawn filaments were taken up on a bobbin in an amount of 2.5 kg.

No trouble was observed during the spinning, and the wrap percentage during drawing was 0.25%.

Manganese Antimony Glycoxide

High molecular weight, colorless polyesters suitable for shaping into filaments and films are obtained by *J.F. Kenney; U.S. Patents 4,104,263; August 1, 1978; and 4,122,107; October 24, 1978; assigned to M & T Chemicals, Inc.* employing catalysts selected from the group consisting of reaction products of specified antimony or zirconium(IV) compounds with a carboxylate of calcium, manganese or zinc and an acid anhydride, alcohol or glycol. The molar ratio of the antimony or zirconium compound to the aforementioned carboxylate is between 1:1 and 1:6. The catalysts are active in both transesterification and polycondensation. The rates of these reactions are more rapid than can be achieved using prior art catalysts.

Example 1: A reaction vessel equipped with a nitrogen inlet, agitator, thermometer and water-cooled condenser was charged with 107 parts manganese acetate tetrahydrate, 250 parts ethylene glycol and 29 parts antimony trioxide. This mixture was heated to the boiling point. A total of 128 parts of the refluxing liquid was collected, during which time the temperature of the reaction mixture increased from 145° to 189°C. The distillate was a mixture containing ethylene glycol, water and acetic acid.

When no further increase of reaction mixture temperature with time was noted, the reaction mixture was allowed to cool to ambient temperature. The reddish-brown solid present in the reaction vessel was isolated, washed using anhydrous methanol and then dried at 75°C under reduced pressure. The solid was found to contain 18.4% manganese and 23.6% antimony. The thermogram obtained using differential thermal analysis (DTA) exhibited no endothermic melting peak and no exothermic peak characteristic of decomposition within the range from 25° to 450°C. These data indicate that the compound is amorphous and polymeric in nature.

Example 2: The product, manganese antimony glycoxide, of Example 1 was used as the sole catalyst to prepare polyethylene terephthalate. A reactor was charged with 194 parts dimethyl terephthalate, 168 parts ethylene glycol and 0.16 part of manganese antimony glycoxide. The reactants were heated at 155° to 200°C with stirring under a nitrogen atmosphere until the calculated amount of methanol required for complete transesterification had been collected. The time required for this phase of the reaction was 30 minutes. The temperature of the reaction mixture was then raised to 250°C and the reactor was evaluated using a vacuum pump.

The pressure in the reactor gradually decreased to 0.3 mm of mercury as the unreacted ethylene glycol was distilled off. The temperature of the reaction mixture was then raised to 280°C and the pressure maintained below 0.3 mm of mercury for 3 hours. The resultant highly viscous molten polyethylene terephthalate was removed from the reactor. A sample of this polymer exhibited an inherent viscosity of 0.5 and appeared white. The yellowness index value was

12. A thermogram of the polymer obtained using differential scanning calorimetry (DSC), exhibited an endotherm at 255°C, indicative of a melting point. The carboxyl content of the polymer was 31 eq/10^6 g.

The yellowness index value of the polyester was determined using a Meeco Colormaster Differential Colorimeter. This instrument measures the percent of incident green, red and blue light reflected from a plaque measuring about 0.1 inch (0.3 cm) in thickness that was prepared by allowing molten polymer to cool between two stainless steel plates. The yellowness index (YI) is calculated using the equation YI = (% red reflectance − % blue reflectance)/% green reflectance.

For purposes of comparison with the present catalysts, polyethylene terephthalate was prepared using the foregoing procedure and a prior art catalyst system. The catalysts employed were 0.08 part manganese acetate tetrahydrate for the transesterification reaction and 0.25 part antimony triacetate for the polycondensation step. The transesterification reaction required 60 minutes to complete, at which time 0.14 part of tris-nonylphenyl phosphite was added as the sequestering agent, followed by 0.25 part of antimony triacetate as the polycondensation catalyst. The polycondensation reaction was carried out for 3 hours.

The resultant polymer exhibited an inherent viscosity of 0.73, a gray color and a yellowness index of minus 3. The DSC melting point was 250°C and the carboxyl content of the polymer was 26 eq/10^6 g.

Combinations of Manganese, Magnesium, Titanium and Antimony Compounds

A. Kohler, H. Pelousek, H. Ohse, H. Westermann and K.-H. Magosch; U.S. Patent 4,128,533; December 5, 1978; assigned to Bayer Aktiengesellschaft and Faserwerke Huls Gesellschaft mit beschrankter Haftung, Germany have developed a process for the production of a high molecular weight polyester, which comprises reacting a dicarboxylic acid or derivative thereof with a diol in the presence of a combination of compounds of manganese, magnesium, titanium and antimony as a catalyst.

Example 1: 97 g of dimethyl terephthalate and 93 g of ethylene glycol are introduced into a glass autoclave. 8 ppm of magnesium in the form of magnesium acetate tetrahydrate, 60 ppm of manganese in the form of manganese acetate tetrahydrate and 10 ppm of titanium in the form of isopropyl titanate are added as catalysts. Transesterification is carried out over a period of 3 hours at 200°C. The temperature is then increased to 220°C, followed by precondensation for 30 minutes. Polycondensation then takes place over a period of 2 hours at 275°C/<1 torr. The product is then spun off and granulated. The polyethylene terephthalate obtained has the following properties: melting point, 256°C; relative viscosity, 1.162; and number of carboxyl groups, 27 val/t.

Example 2: The procedure and conditions are the same as in Example 1, except that the catalysts used are as follows: 8 ppm of magnesium in the form of magnesium acetate tetrahydrate; 60 ppm of manganese in the form of manganese acetate tetrahydrate; 300 ppm of antimony in the form of Sb_2O_3.

The granulate obtained has the following properties: melting point, 256°C; relative viscosity, 1.163; and number of carboxyl groups, 24 val/t.

Example 3: The test is carried out under the same conditions as in Example 1, except that the catalysts used are as follows: 60 ppm of manganese in the form of manganese acetate tetrahydrate; 8 ppm of magnesium in the form of magnesium acetate tetrahydrate; 5 ppm of titanium in the form of isopropyl titanate; and 150 ppm of antimony in the form of Sb_2O_3.

The granulate obtained has the following properties: melting point, 256°C; relative viscosity, 1.169; number of carboxyl groups, 20 val/t.

Although only half the titanium as in Example 1 and half the antimony as in Example 2 were used, so that the same relative viscosity as in these two tests could be expected, a higher molecular weight was reached, and the number of carboxyl groups was improved relative to Examples 1 and 2. Both effects are attributable to the synergistic effect of the Mn/Mg/Ti/Sb catalyst system.

Combinations of Manganese, Magnesium and Titanium Compounds

The catalyst system developed by *A. Kohler, H. Pelousek, H. Ohse, H. Westermann, and K.-H. Magosch; U.S. Patent 4,128,534; December 5, 1978; assigned to Bayer Aktiengesellschaft and Faserwerke Huls Gesellschaft mit beschrankter Haftung, Germany* has in particular the following advantages.

This catalyst combination of compounds of manganese, magnesium and titanium is suitable both for the direct esterification process and for the transesterification process, there being no need to add a separate polycondensation catalyst.

The formation of polymanganese glycolate, which cannot be avoided where manganese is used without the addition of magnesium and which, in a continuously operated reactor, leads to deposits and coarse crosslinked particles, is largely avoided. Accordingly, the reactor need be cleaned less often and the filters are not blocked as quickly during the spinning process.

In batch-type operation, the polycondensation temperature is 275°C, relatively low, in cases where this catalyst combination according to the process is used. This means that, in continuous operation, the spinning temperature (in the case of direct spinning) can be freely selected from this temperature upwards.

The low polycondensation temperature of 275°C has the further advantage that, in cases where it may be desired to utilize the reactor capacity to a greater extent in continuous operation, the polycondensation temperature may be further increased by 20° to 295°C, giving an increase in throughput of from 30 to 35%.

Another advantage of this low polycondensation temperature of 275°C is that the polyester granulate obtained is very light in color (as measured by remission) and, in addition, still contains a small number of carboxyl groups.

Example: 80 kg of dimethyl terephthalate and 77 kg of ethylene glycol (molar ratio 1:3) were introduced into an autoclave. 8 ppm of magnesium (5.64 g magnesium acetate tetrahydrate), 60 ppm of manganese (22.3 g manganese acetate tetrahydrate) and 5 ppm of titanium (2.4 g of isopropyl titanate) were added as catalyst. Transesterification was carried out over a period of 3 hours at 200°C under normal pressure. Precondensation was carried out over a period of 0.5 hour at a temperature of 220°C. Polycondensation was subsequently carried out over a period of 2.5 hours at 275°C under a pressure of <1 torr, followed by spinning and granulation. 72 kg of polyethylene terephthalate with the following properties were obtained: melting point, 256°C; relative viscosity, 1.163; and number of terminal carboxyl groups, 23 val/t. The granulate can be processed without difficulty into filaments and fibers.

Manganese or Cobalt Salts plus Titanium Alkoxides, Antimony Compound, and Phosphate Ester

The catalyst-inhibitor system developed by *N.C. Russin, R.A. Tershansy and C.J. Kibler; U.S. Patent 4,010,145; March 1, 1977; assigned to Eastman Kodak Company* comprises a combination of organic or inorganic salts of manganese and cobalt; titanium alkoxides; an antimony compound; and a phosphate ester.

The catalyst-inhibitor system accelerates ester-interchange and polycondensation and produces poly(ethylene terephthalate) having excellent color. This process involves conducting the ester-interchange reaction in the presence of a catalyst system comprising a mixture of a titanium alkoxide such as acetyl triisopropyl titanate and organic or inorganic salts of manganese and cobalt and an antimony compound. The manganese salt should be present in the amount of 25 to 110 ppm manganese; cobalt salt should be present in the amount of 10 to 100 ppm cobalt; the titanium should be present in the amount of 20 to 60 ppm titanium; and the antimony compound should be present in the amount of 50 to 300 ppm of antimony. All parts by weight are based on the acid fraction of the polymer weight to be produced.

The preferred manganese salt is manganous benzoate tetrahydrate and the preferred cobalt salt is cobaltous acetate trihydrate. The preferred antimony compound is antimony triacetate.

After the ester-interchange reaction a phosphate ester is added to the reaction product and the reaction product is polycondensed. The preferred phosphate ester has the formula:

$$O=\overset{\displaystyle OR}{\underset{\displaystyle OR}{\overset{|}{\underset{|}{PO}}}}\!-\!(C_2H_4O)_{\overline{n}}\!-\!\overset{\displaystyle OR}{\underset{\displaystyle OR}{\overset{|}{\underset{|}{P}}}}\!=\!O$$

wherein n has an average value of 1.5 to about 3.0 with about 1.8 being most preferred and each R is hydrogen or an alkyl radical having from 6 to 10 carbon atoms with octyl being most preferred, the ratio of the number of R groups of hydrogen atoms to the number of phosphorus atoms being about 0.25 to 0.50 with about 0.35 being most preferred; and the ester having a free acidity equivalent of about 0.2 to 0.5; the ester being present in the amount to provide phosphorus in the amounts of 13 to 240 parts per million based on the acid fraction of the polyester to be produced.

Other useful phosphate esters include ethyl acid phosphate, diethyl acid phosphate, triethyl acid phosphate, aryl alkyl phosphates, tris-2-ethylhexyl phosphate and the like. The phosphate ester may be used in an amount to provide phosphorus at a concentration such that the atom ratio of the amount of phosphorus to the sum of the amounts of cobalt, manganese, and titanium is between 1.0 and 2.0.

The process and catalyst-inhibitor system provides for the manufacture at high production rates of high-quality poly(ethylene terephthalate) polyester having excellent properties for the fabrication of fibers and films.

The poly(ethylene terephthalate) produced has excellent color (whiteness), low concentration of diethylene glycol (ether linkages), excellent stability against thermooxidative, hydrolytic, and ultraviolet radiation degradation effects, and, when melt spun into fibers or filaments, results in essentially no deposits on spinneret faces and which will minimize bathochromic shift of dye color during dyeing. The polymer made using this catalyst-inhibitor system is also characterized by improved solid state polymerization activity.

Titanium Tetrabutylate plus Phosphate Stabilizer

A process for the production of high molecular weight low color polyethylene terephthalates using a particularly effective polycondensation catalyst is disclosed by *P. Bier and R. Binsack; U.S. Patent 4,115,371; September 19, 1978; assigned to Bayer Aktiengesellschaft, Germany.* Titanium tetrabutylate is added to the reaction mixture after the initial esterification or transesterification reaction is about 95% complete and the polycondensation is completed at temperatures between about 200° and 270°C under reduced pressure. Metaphosphoric acid or its alkali or alkaline earth metal salts are added with the titanium tetrabutylate to control the color.

Commonly accepted catalysts may be employed for the initial esterification or transesterification. Polymers with intrinsic viscosities in excess of 0.7 dl/g and yellowness numbers less than 6 can be produced.

Examples: 5,826 g (30 mols) of dimethyl terephthalate are transesterified with 4,104 g (66 mols) of ethylene glycol, with the elimination of methanol, in the presence of 3.3 g of zinc acetate, while stirring and passing nitrogen over the mixture at 200° to 220°C.

When the transesterification has ended, 0.0009 to 0.009 mol of a titanium compound and 0.006 to 0.6 mol of a phosphorus compound, relative to 1 mol of the dicarboxylic acid component in each case, are added. The temperature is then raised to 250°C in the course of a further hour and, at the same time, a vacuum (<1.0 mm Hg) is applied to the apparatus. Finally, the mixture is stirred at 250° to 280°C and under a pressure of less than 0.5 mm Hg until the desired viscosity is reached. The polyester melt is then spun off through a water bath and granulated.

Results are shown in the table below.

Ex. No.	Catalyst	Stabilizer	Yellow* Shade	Intrinsic Viscosity** (dl/g)
1	Titanium tetrabutylate	Na metaphosphate	5.4	0.98
2	Titanium tetrabutylate	Metaphosphoric acid	5.1	0.95
3	Titanium tetraisopropylate	Na metaphosphate	10.3	0.86
4	Titanium tetrabutylate	Triphenyl phosphate	8.6	0.82

*For the crystalline product by the Gardner method.
**Measured at 0.5 g/dl in phenol/tetrachloroethane (weight ratio 1:1) at 25°C.

Examples 1 and 2 describe polyethylene terephthalates which have an intrinsic viscosity of $\geqslant 0.7$ dl/g and a b value (= yellow shade) of <6 prepared according to this process. Examples 3 and 4 are comparison examples and show that the catalyst and the stabilizer must be balanced in order to obtain polyethylene terephthalates which are light colored and, at the same time, of high molecular weight.

Metal Acetates

A melt-processable resorcinol phthalate polyester having a terephthalic acid content not greater than about 30 mol percent is prepared by *C.E. Kramer; U.S. Patent 4,127,560; November 28, 1978; assigned to Celanese Corporation* by a two-stage process utilizing an alkali or alkaline earth metal acetate catalyst. The first stage yields a nonvolatile prepolymer which is then melt polymerized to the resorcinol phthalate polyester. The resulting polyester exhibits high char and low flammability in addition to being melt processable.

Example: 29.9 g (0.18 mol) isophthalic acid, 3.3 g (0.02 mol) terephthalic acid, 39.4 g (0.20 mol) 1.5% excess resorcinol diacetate, 10 ml diphenyl ether, 1 ml acetic anhydride and 0.27 g (0.002 mol) sodium acetate trihydrate catalyst are weighed into a dry three-neck 300 ml flask. The acetic anhydride is used as a water scavenger and acetylating agent for any unreacted or partially acetylated resorcinol. The diphenyl ether solvent is not required for the reaction but is used for convenience to wash the sides of the vessel during the early stages of the condensation.

The flask is fitted with a nitrogen inlet, Servodyne mechanical stirrer with rpm and torque readouts, condenser, 50 ml graduated receiver and vacuum adaptor. The flask is purged with dry, oxygen-free nitrogen and brought to 240°C over 0.5 hour with stirring. The reaction is held at 240°C for 3 hours while a theoretical amount of acetic acid is collected.

At this point, vacuum is applied to remove the diphenyl ether, and the reaction temperature is brought to 280°C. The mixture is held at 280°C and $\leqslant 1.0$ mm Hg pressure for 3 hours and the increase in viscosity is monitored by observing stirrer torque. The reaction is then cooled under nitrogen, the flask broken and polymer ground to approximately 5 mesh.

Molded shaped articles formed from such a resorcinol phthalate polyester generally exhibit a superior tensile strength, flex strength and impact strength. Also, the appearance of the resulting molded articles generally is superior with the resulting molded articles being commonly clear and exhibiting a light yellow to amber color and an attractive smooth surface.

Silicon Compounds

K. Weinberg and G.C. Johnson; U.S. Patent 4,077,944; March 7, 1978; assigned to Union Carbide Corporation describe a process for producing polyesters and copolyesters, useful for making films and fibers, by the polycondensation of dicarboxylic acids and aliphatic glycols using certain silicon compounds as catalysts.

The dicarboxylic acid compounds used in the production of polyesters and copolyesters are well known to those skilled in the art and illustratively include

terephthalic acid, isoterephthalic acid, p,p'-diphenyldicarboxylic acid, p,p'-dicar-
boxydiphenyl ethane, p,p'-dicarboxydiphenyl hexane, p,p'-dicarboxydiphenyl
ether, p,p'-dicarboxyphenoxy ethane, and the like, and the dialkyl esters thereof
that contain from 1 to about 5 carbon atoms in the alkyl group thereof.

Suitable aliphatic glycols for the production of polyesters and copolyesters are
the acrylic and alicyclic aliphatic glycols from 2 to 10 carbon atoms, especially
those represented by the general formula $HO(CH_2)_pOH$, wherein p is an integer
having a value of from 2 to about 10, such as ethylene glycol, trimethylene
glycol, tetramethylene glycol, pentamethylene glycol, decamethylene glycol, and
the like.

Other known suitable aliphatic glycols include 1,4-cyclohexanedimethanol, 3-ethyl-
1,5-pentanediol, 1,4-xylylene glycol, 2,2,4,4-tetramethyl-1,3-cyclobutanediol, and
the like. One can also have present a hydroxylcarboxyl compound such as 4-hy-
droxybenzoic acid, 4-hydroxyethoxybenzoic acid, or any of the other hydroxylcarboxyl compounds known as useful to those skilled in the art.

The first-stage esterification or transesterification reaction is carried out tradi-
tionally by heating the mixture of reactants at about 150° to 250°C, preferably
175° to 250°C. During this stage any of the well-known esterification or transesterification catalysts can be used. In the second stage, or the polycondensation,
the silicon catalysts are used.

Illustrative of suitable silicon compounds are the following: beta-cyanoethyl
triethoxysilane, gamma-mercaptopropyl triethoxysilane, gamma-aminopropyl triethoxysilane, diethoxyphosphorylethyl methyl diethoxysilane, vinyl triethoxysilane, gamma-methacryloxypropyl trimethoxysilane, beta-cyanoethyl trimethyl-
silane, vinyl methyl diethoxysilane, tetramethyl divinyl disiloxane, and diethoxyphosphorylethyl methyl diethoxysilane.

Example: A mixture of 39.1 g of dimethyl terephthalate, 32.3 g of ethylene
glycol, 0.0179 g of zinc acetate dihydrate as transesterification catalyst and
0.0172 g of 2-cyanoethyl triethoxysilane as polycondensation catalyst was heated
under argon for 3 hours at 168° to 200°C.

During this first-stage transesterification reaction, methanol was distilled from
the reactor. The temperature was then raised to 218° to 230°C and maintained
for about one hour to complete the transesterification. Thereafter, the temperature was increased while the pressure was gradually reduced to below 1 mm of
mercury and a temperature of 273° to 284°C was maintained for 2 hours to
carry out the second-stage polycondensation reaction. A clear, white polyester
having good draw and fiber properties was produced.

Phosphites and/or Phosphates

According to *L. Buxbaum; U.S. Patent 4,101,526; July 18, 1978; assigned to
Ciba-Geigy Corporation,* the reaction for the manufacture of linear polyesters
by polycondensing dicarboxylic acids, hydroxycarboxylic acids or the esters and
diols thereof is speeded up by the addition of 1 to 25% by weight of esters of
phosphoric and/or phosphorous acid.

Suitable aromatic dicarboxylic acids are: terephthalic acid, isophthalic acid, o-phthalic acid, 1,3-, 1,4-, 2,6- or 2,7-naphthalenedicarboxylic acids, 4,4'-diphenyl-dicarboxylic acid, 4,4'-diphenylsulfonedicarboxylic acid, 1,1,3-trimethyl-5-car-boxyl-3-(p-carboxyphenyl)-indane, 4,4'-diphenyl ether-dicarboxylic acid, bis-p-(car-boxylphenyl)-methane.

It is preferred to use at least 80 mol percent of terephthalic acid and at least 80 mol percent of aliphatic diols containing 2 to 10 carbon atoms, in particular 2 to 4 carbon atoms, or 1,4-dihydroxymethylcyclohexane, referred to the pure polyester. In particular, the polyesters contain linear aliphatic diols containing 2 to 4 carbon atoms and terephthalic acid.

A further advantageous group of diols, dicarboxylic acids or hydroxycarboxylic acids comprises those which contain halogen atoms, in particular bromine or chlorine atoms. These halogenated monomers are used for providing the poly-esters obtained therefrom with a flame-resistant finish, while the known syner-gism in combining halogenated organic compounds and compounds of the ele-ments of the 5th main group of the Periodic Table can be very advantageously exploited in the polyesters obtained from halogenated monomers. Examples of such halogenated monomers are: tetrabromo-bisphenol-A-diglycol ether, 4,4'-dihydroxydecabromobiphenyl, tetrabromo-p-xylylene glycol, 2,5-chloro-terephthalic acid or 1,3-bis(hydroxyethyl)-4,5,6,7-tetrachloro- or tetrabromo-benzimidazolone.

Example: A 10-liter reactor, equipped with stirrer, nitrogen inlet, cooler and thermometer, is charged with 3,880 g of dimethyl terephthalate (DMT), 2,480 g of ethylene glycol, 1.31 g of zinc acetate and 1.4 g of antimony trioxide and the mixture is heated to 145°C. While stirring and introducing nitrogen, 99% of the theoretical amount of methanol is distilled off over 4 hours, and the tem-perature of the reaction mixture rises to 226°C. The reaction mixture is then conveyed to a second reactor and heated to 241°C.

A vacuum of 50 torrs is then applied in the course of half an hour by means of a water jet pump and the reaction temperature is simultaneously raised to 275°C. The vacuum is increased to 0.7 torr with a vacuum pump at the same reaction temperature in the course of half an hour and kept thereat for a further 1:35 hours. A sample taken at this time after eliminating the vacuum by means of nitrogen exhibits a degree of polycondensation ($\overline{P}n$) of 98. Then 5% by weight of triphenyl phosphate is added and a vacuum of 1 torr is applied immediately again. The $\overline{P}n$ rose to 130 over 35 minutes.

The melt is pressed through spinnerets to filaments, which are chilled in water and comminuted to cylindrical granules measuring 2 x 3 mm. 300 g of these granules are dried first for 2 hours under a vacuum of 0.3 torr at 140°C in a 1-liter rotary evaporating flask, which is immersed in an oil bath, and crystal-lized. The reaction is thereafter continued for 6 hours in the solid phase under the same vacuum at 235°C. The end product has a $\overline{P}n$ of 234.

MANUFACTURING PROCESSES

Prevention of Stickiness

In order to avoid thermal degradation, it is known to produce high-viscosity polyethylene terephthalate in the solid phase at temperatures below the melting point of the condensate and under an inert gas blanket or vacuum. The drawbacks of this improvement are the large efforts and expenditure required to prevent the granulate from sticking in the course of the solid state polycondensation.

H.J. Rothe, H. Heinze, B.D. Whitehead and G. Priepke; U.S. Patent 4,064,112; December 20, 1977; assigned to Zimmer Aktiengesellschaft, Germany provide a process wherein the sticking of the granulate during the solid state polycondensation is reliably prevented by only a small processing effort and without the addition of anticaking agents, while simultaneously avoiding the disadvantages of the known prior art processes.

Prior to the start of the solid state polycondensation reaction, the granulate is crystallized to a density of at least 1.390 g/cm^3 under forced motion at temperatures of 220° to 260°C in an inert gas atmosphere. The crystallized granulate is then transferred, at the same temperature or at a lower temperature, while avoiding the addition of air, to a continuously operating, fixed bed reactor where it is polycondensed at a temperature equivalent to, or lower than, the crystallization temperature.

According to a preferred embodiment of the process, the crystallization is effected at a temperature of 230° to 245°C and a density of 1.403 to 1.415 g/cm^3. These preferred conditions produce a satisfactory reaction rate during the crystallization and prevent thermal degradation. The solid state polycondensation is preferably effected at a temperature of 230° to 245°C, which results in an optimum reaction rate at the lowest possible degree of thermal degradation.

The apparatus for determining the relative sticking characteristics is illustrated in the schematic diagrams of the unit. Referring to Figure 1.1, the measuring flask comprises a stainless steel vessel **10** surrounded by a heating jacket **11** to which a heating medium such as recycle oil enters through the pipe connections **11a** and **11b**. The temperature of the heating medium may be set by a thermostat-controlled heating device **12**. The heating jacket is coated with an insulating envelope **13** indicated by a broken line in the figure.

An agitating device **14** equipped with several extensions **15** protruding from the rotation axis is positioned coaxially within the vessel. The agitating device is attached to the lower end of a driving shaft **16** which leads upwards to a driving motor **17**. The housing of the motor is pivotally suspended and connected with a recording or plotting device **18** comprising a lever arm **18a** with a recording pen **18b** and a recording roll **18c**. The driving shaft is connected with the driving motor by a disengaging clutch **19** in order to be able to reset the agitating device into a no-load condition in an axial direction after each measurement.

A stationary, circular, disk-shaped wall element **20** having openings therein is positioned in the lower part of the vessel. Positioned above, and parallel to the wall element is a second, vertical mobile wall element **21** having substantially the same dimensions as wall element **20**. The driving shaft passes through wall element **21** so that the agitating device is placed between the two wall elements. The upper mobile wall element **21** rests against a hollow strut **22** which coaxially surrounds the driving shaft. The upper end of the hollow strut is connected by joint **23** with lever arm **24** which is variably loadable through a horizontally displaceable weight **25**. The lever arm has a fulcrum **26** which is vertically adjust-

able by means of an operating screw **27**. An indicating device **28** comprising a
pointer **28a** and a scale **28b** serves to determine the horizontal alignment of the
lever arm.

Figure 1.1: Apparatus for Determining Relative Sticking Characteristics

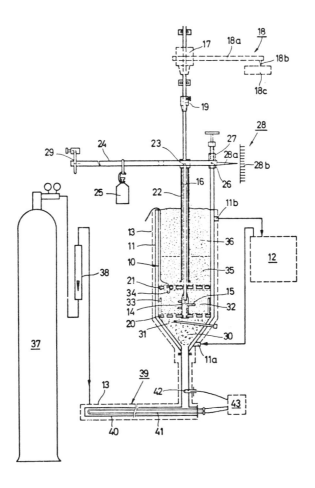

Source: U.S. Patent 4,064,112

By means of the operating screw, it is possible to align the lever arm **24**, depending upon the bulk material volume, horizontally between the wall elements **20**
and **21**. The distance between the wall elements coincides with the layer height
and may be read on the indicating device. The transfer of force from the lever
arm to the mobile wall element **21** is effected via the hollow strut **22** in con-
formity with the principle of leverage. An adjustable guide **29** is attached to the

free end of the lever arm **24**. Below the lower wall element **20**, there is a packing **30** of, for instance, glass rings to ensure a uniform temperature distribution across the whole cross section of the vessel **10**. A temperature feeler **31** is introduced in this packing to measure the process temperature. Between the wall elements **20** and **21**, there is a bulk material layer **32** of the granulate to be analyzed. In the marginal zone of this packing, a number of baffle plates **33** are positioned on the periphery in order to prevent a rotation of the packing.

A temperature feeler **34** extends through the wall element **21** into this packing. Above the upper wall element **21**, there is another packing **35** of the same granulate to be analyzed. Finally, on top of this granulate there is a packing **36** of glass wool for reducing heat losses and preventing the ingress of atmospheric oxygen.

A gas source **37** is positioned upstream of the vessel for feeding, for instance, nitrogen to the vessel. The gas volume passed to the vessel is measured by a flow meter **38**, by way of example, a rotameter. The gas flows then into a gas heater **39** comprising a horizontal cylinder **40** and a heating element **41**. The gas temperature is controlled by means of a temperature sensor **42** and an output regulator **43** for the heating element.

As a result of positioning the extensions **15** on an agitating device **14** at an angle of 120 degrees to one another as well as being positioned on different heights, the granulate particles are forced to move at the beginning of each measurement and after approximately 0.3 revolution of the agitating device. After one full revolution, almost all of the granulate particles have moved relative to one another. Any previously existing agglomeration will be largely removed after one revolution so that the remaining torque is due only to the friction of the granulate particles against each other and against the surface of the agitator. The stickiness between the granulate particles was measured by using the apparatus described above as follows.

850 g of polyethylene terephthalate granulate used for a measurement were placed in the apparatus and wall element **21** was placed on top of the granulate. After the glass wool was put in place and the hollow strut **22** slipped on, the lever arm was positioned. The lever arm was loaded with a weight **25** to 80 kp to equal a static pressure (at the bottom) of polyethylene terephthalate having a granulate height of 7.1 m. The cross-sectional area of the granulate in layer **32** was 122 cm^2. The loading of 80 kp remained the same for all tests.

The height of the layer was determined by means of indicating device **28**. The recording device **18** was set to zero and the lever arm **18a** was loaded with a tension spring acting in its longitudinal direction (not shown). The size of the tension spring was selected in accordance with the expected torque. Springs were used with an elongation of 28 mm, 50 mm and 118 mm for each kp tensile strength. The lever arm had a length of 28 cm. The disengaging clutch **19**, was tightened and the first measurement was taken at room temperature (25°C).

When measurements were taken, the motor **17** and the recording device were actuated at the same time. The agitator speed was set at 2 rpm. The paper advance was adjusted to 26 cm/min. The clutch was then disengaged and a gas stream of 6 Nm^3 N$_2$/h was set. The heating device **12** was connected and the relevant thermostat adjusted to an oil temperature at which the measurements of friction resistance and the stickiness were taken. The N_2-heater was connected and a predetermined temperature was set on the temperature sensor **42** and which could be reached on the temperature sensors **31** and **34** within approximately 40 minutes. This predetermined temperature was not to be exceeded on temperature sensor **31** by more than 1°C. The temperature of the temperature sensor **42** as a result was higher by about 7° to 15°C than that of the temperature sensor **31**.

After taking the last measurement, the temperature was lowered to less than 200°C, and all the heating elements and N_2 supply disconnected. Upon removal of the glass wool and the other component parts of the measuring apparatus, and after the cooling of the granulate to room temperature, the granulate was withdrawn and its intrinsic viscosity and density were determined on a representative sample.

The torque plottings recorded by the recording or plotting device **18** were then used to determine the stickiness of the granulate. The plottings set forth torque curves at a distance from the abscissa proportional to the effective torque at the agitator. The number of revolutions of the agitator were selected as the abscissa since this best reflects how the stickiness of the granulates to be measured depends upon their movement during the measurement.

The relevant torques were recorded in N-m on the ordinate. The curves have a characteristic course, namely a high torque at the beginning of the measurement, i.e., during the "breaking" of possible existing agglomerations, and followed by subsequent lower torque.

For purposes of evaluation, the peaks of the plotted curves were connected with one another. A straight line was then drawn parallel to the abscissa through the obtained curve so that the sectional areas above and below the straight line were, as much as possible, of equal size. In this way, an average of the peaks is obtained which is mainly the result of the friction and is called for short, the R-characteristic.

In addition, a second straight line is drawn parallel to the abscissa analogous to the initially occurring higher peaks. In this way, an average is obtained, but a curve section of a different size must be taken into account depending on the formation of the peaks. This curve section is in practice generally within the limit of 0.3 to 1.1 revolutions of the agitating device. The R-characteristic is exceeded at a correspondingly high temperature. This is attributed to the stickiness of the bulk material. The difference between the two straight lines drawn through the different peaks is denoted as the V-characteristic. The V-characteristic primarily represents the intensity or the power of stickiness of the granulate.

Homogeneous Distribution of Reactants in Batch Process

H. Heinze and E. Hackel; U.S. Patent 4,077,945; March 7, 1978; assigned to Zimmer Aktiengesellschaft, Germany describe a batch process for esterifying ethylene glycol and terephthalic acid at elevated temperature and pressure in which a part of the ethylene glycol required for the complete esterification reaction is placed in an autoclave and heated under pressure, and the remainder of the glycol is added to the heated glycol in the form of a paste with terephthalic acid. Preferably the glycol initially present in the autoclave amounts to 0.3 to 0.8 mol per mol of terephthalic acid while the glycol added as a paste amounts to 1.4 to 0.9 mol per mol of terephthalic acid. The glycol initially in the autoclave may be glycol reclaimed from the mixture volatilized during the reaction and may contain up to 20% water, preferably 2 to 6% water.

The process prevents agglomeration of terephthalic acid and provides homogeneous distribution of reactants and minimal formation of undesired byproducts.

A process diagram, according to which the following example was carried out, is explained in detail below with the aid of Figure 1.2.

Figure 1.2: Batch Process for Making Linear Polyesters

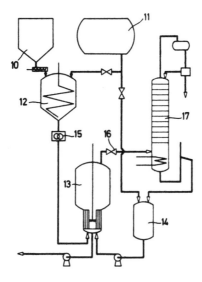

Source: U.S. Patent 4,077,945

Supply hopper **10** contains terephthalic acid. Storage tank **11** contains ethylene glycol. The reactants are fed into an agitating container **12** and homogenized to a readily flowable paste. Glycol flows from the tank to a heated container **14**, and thence to autoclave **13**. The paste is fed by means of a pump **15** into the preheated glycol in the autoclave under agitation. The vapor mixture rising during esterification passes through a relief valve **16** into a distillation column **17**, where the water is separated off. The glycol deposited in the sump of the column passes first into the container and from there back into the autoclave.

The esterification product is then conveyed from the bottom of the autoclave into a polycondensation reactor, not shown. It can, if desired, be first prepolymerized to a degree of from 3 to 20 in the autoclave—as is known—through further reduction of the pressure, before it is transferred to the polycondensation reactor. Then it is polycondensed until the desired end viscosity is reached.

Example: In an autoclave a mixture of 238 g of glycol and 13 g of water under a nitrogen atmosphere was heated until a pressure of 5.0 atm had built up. Then a paste consisting of 1,142 g of terephthalic acid, 427 g of glycol, 3.4 diisopropylamine and 0.75 g of antimony triacetate was fed in continuously over a period of 90 minutes under steady agitation. During this time the temperature was maintained at 260°C and the pressure was maintained constant at 5.0 atmospheres by operation of the pressure relief valve.

A mixture of 36 g of glycol and 2 g of water was then fed in at the bottom of the autoclave over a period of 30 minutes and the temperature was raised to 270°C. Fifteen minutes after completion of the feeding in of the paste, the pressure was lowered continuously over 25 minutes from 5.0 atmospheres gauge to atmospheric pressure. 10 minutes later the esterification reaction was arrested. The esterification product had an acid number of 22.2 mg KOH/g and a diethylene glycol content of 1.07% by weight.

After a 3-hour polycondensation reaction at 275°C and a pressure of 0.1 torr, the esterification product was converted to a polyethylene terephthalate having an intrinsic viscosity of 0.89, a content in −COOH end groups of 24 eq/10^6 g and a diethylene glycol content of 1.11% by weight.

Semicontinuous Process

E.P. Brignac and B.T. Hanvey; U.S. Patent 4,079,046; March 14, 1978; assigned to Monsanto Company describe a semicontinuous polyester esterification process employing continuous feeding and discontinuous withdrawal combined with a plurality of batch polycondensation processes to produce outstanding polyester polymer with improved efficiency and flexibility of process.

The economy and ease of operation inherent in continuous acid-glycol polyester esterification is combined with the known relative efficiency in attaining uniform molecular weight distribution inherent in polyester batch polycondensation processes.

Example: A continuous esterifier reactor is operated while being fed continuously from a feed make-up tank with 1,620 g terephthalic acid, 1,520 g of ethylene glycol, 1.0 g of antimony glycoloxide, 19.7 g of an alkoxy polyoxyalkylene glycol chain terminator having a general formula: $R-O[G-O]_x-H$ where R = 14 to 15 and x = 14, 0.02 g pentaerythritol, and 0.04 g lithium acetate per ten 316-g polycondensation batches.

The mixture is esterified at a pressure of 15.1 pounds per square inch (psi) (½ psi gauge) (785.5 mm mercury) by heating for an average holding time of 90 minutes at a temperature of 240°C during which time water and ethylene glycol are continually removed. The resulting common prepolymer has an intrinsic viscosity 0.081 and carboxyl level of about 370 μeq/g. A one-tenth portion of the abovedescribed common prepolymer is rapidly removed, filtered and supplied to a batch finisher having a variable speed agitator and variable temperature jacket in which is contained 37.6 g of a 24% solution in ethylene glycol of the bisglycol ester of 5-sulfoisophthalic acid, sodium salt.

Temperature is increased in the finisher to remove excess ethylene glycol and polymerization is completed under nitrogen at a temperature of 280°C at a pressure of less than 1 mm. The polymer obtained after 180 minutes of polycondensation time has a specific viscosity of 0.275, an intrinsic viscosity of 0.55 and a DTA (differential thermal analysis) melting point of 250°C; and contains 0.65 weight percent diethylene glycol (DEG). The polymer is spun and drawn 5.1 times to a yarn of excellent whiteness. Microscopic examination reveals no aggregates of the dye additive present in the yarn. The fiber dyes to a dark shade with Sevron Blue 2G (CI Name: Basic Blue 22; no CI number) cationic dye.

A second one-tenth portion of the common prepolymer is batch polycondensed to a specific viscosity of 0.275 (intrinsic viscosity of 0.55) under the same conditions as above but without combining with it a minor element of starting material for the batch polycondensation, (such as the 5-sulfoisophthalic acid sodium salt with which the first portion of the common prepolymer was combined). The product had a DTA melting point of 247°C, a DEG content of 1.95 weight percent and good brightness and whiteness.

Continuous Distillation of Excess Glycol

J. Rebhan and H.G. Matthies; U.S. Patent 4,039,515; August 2, 1977; assigned to BASF Aktiengesellschaft, Germany describe an improved process for the continuous manufacture of linear high molecular weight polyesters by direct esterification of terephthalic acid with excess glycols in a medium comprising terephthalic acid glycol esters or their oligomers, under superatmospheric pressure at temperatures above 200°C, followed by polycondensation in the presence of catalysts at from 260° to 290°C.

The improvement is that the mixture fed to the esterification contains terephthalic acid and glycols in the molar ratio of 1:1.15-1.5 and that during the esterification from 20 to 30% by weight of the excess glycols is continuously distilled from the reaction mixture.

Example: Per minute, a mixture of 0.34 kg of terephthalic acid and 0.13 kg of ethylene glycol (molar ratio 1:1.4) is fed to a stirred autoclave of 70 l capacity, fitted with a reflux condenser (for partial condensation) and a descending condenser (for condensing the overhead product). The above mixture is mixed, in the weight ratio of 1:1, with oligomeric terephthalic acid glycol ester of degree of polymerization 2 to 3. A temperature of 265°C and a pressure of 8 bars are maintained in the stirred autoclave. 28% of the excess glycol, together with the water produced in the reaction, are distilled from the reaction mixture, a residence time of 45 minutes being maintained. The esterification product which is taken off continuously has a degree of conversion of 84%, based on terephthalic acid introduced, and contains 1.8 mol percent of diglycol.

The mixture thus obtained is subjected to further condensation, after addition of 0.07% by weight of antimony triacetate, for 90 minutes at 285°C, while lowering the pressure from 760 to 0.1 mm Hg. The polyester thus obtained contains 2.2 mol percent of diglycol and melts at 252°C.

Comparative Example: The above procedure is followed except that only 15% excess glycol is distilled off. The polyester obtainable after further condensation contains 4.2 mol percent of diglycol and melts at 245°C.

Multistage Continuous Production of Polybutylene Terephthalates

H. Strehler, L. Beer, E. Heil, F. Urbanek and H. Fischer; U.S. Patent 4,056,514; November 1, 1977; assigned to BASF Aktiengesellschaft, Germany provide a continuous process which, as a result of employing relatively short residence times and mild reaction conditions, gives polybutylene terephthalate having a high degree of polycondensation, showing little thermal degradation, and having a good color. Loss of terephthalic acid and production of undesirable by-products, especially tetrahydrofuran, are kept to a minimum.

This is achieved in a multistage continuous process by transesterification of dimethyl terephthalate with 1,4-butanediol, if desired in the presence of up to 40 mol percent of other starting materials which form linear polyesters, and in the presence of transesterification catalysts, at elevated temperatures, followed by polycondensation at elevated temperatures, subatmospheric pressure and in the presence of polycondensation catalysts, by (a) transesterifying dimethyl terephthalate with 1,4-butanediol in the molar ratio of 1:1.2-1.5 in a plurality of successive stages at temperatures rising from 160° to 230°C in the presence of transesterification catalysts, and distilling methanol off continuously.

(b) The mixture of dihydroxybutyl terephthalate and its oligomers thus obtained is subsequently passed, at from 230° to 270°C and from 20 to 2 mm Hg, upward through a bundle of stationary heated tubes, and the resulting mixture of vapor and liquid passed through a hold tank located immediately above the tube ends, the volume of the hold tank being at least ¼ and at most 2.5 times the volume of the tube bundle and the mean residence time in the tube bundle and hold tank together being from 10 to 60 minutes.

The predominant part of the amount of heat required to heat the reaction mixture and vaporize the volatile constituents of the reaction mixture is supplied to the reaction mixture in the tubes, and the mixing of the reaction mixture is effected predominantly by the vapor bubbles rising from the tubes. (c) The precondensate thus obtained is polycondensed at from 240° to 260°C and from 0.1 to 2 mm Hg, with continuous formation of thin films.

Example: (a) 100 kg of dimethyl terephthalate, (DMT), 70 kg of 1,4-butanediol (molar ratio of 1:1.5) and 0.05 kg of tetra-n-butyl o-titanate (0.05% by weight based on DMT) per hour were metered into a cascade of 4 kettles. On passing through the cascade, the reaction mixture is heated stepwise from 160° to 220°C. The vapors from the reaction mixture are passed through a joint pipeline to a column where methanol and small amounts of tetrahydrofuran are removed at the top while entrained 1,4-butanediol and terephthalate ester are returned to the reaction mixture. The residence time in the cascade is 120 minutes. A transesterification mixture of bishydroxybutyl terephthalate and its oligomers, having a mean degree of condensation of from 2 to 3, is obtained.

(b) The transesterification mixture from (a) is passed upward through a bundle of vertical stationary heated tubes, at the top end of which is a hold tank. The bundle comprises 7 tubes having an L:D ratio of 125:1 (length 500 cm, internal diameter 4 cm). The ratio of the total volume of the tube contents to the volume of the hold tank is 1:2. The reaction mixture is heated to 250°C in the tubes. The bubbles of 1,4-butanediol vapor which rise up ensure adequate mixing. The total residence time is 20 minutes. In the hold tank, which is under a pressure of 3 mm Hg, the vapor/liquid mixture separates. The vapor constituents are drawn off and condensed, while the liquid precondensate is discharged. It has a relative viscosity of 1.18. This precondensate is again mixed with 0.15 kg of tetrabutyl titanate, based on DMT.

(c) The precondensate thus obtained is passed through an apparatus of the type described in German Laid-Open Application No. 2,244,664. The apparatus is divided into 9 chambers and equipped with 2 complete discs, 3 perforated discs and 4 spoked wheels. The discs rotate at 3 rpm. A temperature of 250°C and a pressure of 0.8 mm Hg is maintained during condensation and the residence time is 90 minutes. The polyester melt is subsequently filtered, and passed in the form of strands through a water bath. The strands are suitably granulated and the granules obtained (about 3 x 3 mm) are dried at elevated temperature. A polycondensate having a relative viscosity of 1.71, a reference color number (measured on the granules) of Fs = 80% and a carboxyl end group content of 30 meq/kg is obtained.

The loss of terephthalic acid is 0.5%. A total of 3 kg of tetrahydrofuran is produced per 100 kg of polybutylene terephthalate.

High-Pressure Esterification in the Presence of Amines

W.C.L. Wu and R. Eichenbaum; U.S. Patent 4,096,124; June 20, 1978; assigned to Mobil Oil Corporation have discovered that terephthalic acid can be more rapidly esterified with a glycol by operating at temperatures above the normal boiling point of the glycol reactant, i.e., at superatmospheric pressures. It also has been discovered that the presence of an alkaline organic substance during esterification inhibits ether by-product formation. Usually a polycondensation catalyst, such as antimony trioxide, is added, and the intermediate glycol terephthalate formed by direct esterification is polycondensed to the final polyalkylene terephthalate product.

The method comprises heating in a first stage 1 mol of terephthalic acid, with at least about 1.2 mols of the alkylene glycol and 0.01 to 1.0% based on the weight of the acid of a volatile organic base at a temperature between 240° and 320°C and at a pressure above the vapor pressure of the reactant alkylene glycol at the reaction temperature until about 75 to 85% of the acid groups initially present in the terephthalic acid have been esterified, and then maintaining the reaction mixture from the first stage in a second stage at a temperature between 240° and 320°C at a pressure less than the pressure of the first stage until at least about 95% of the acid groups initially present in the terephthalic acid have been esterified. More specifically, the pressure in the second stage is reduced to substantially atmospheric.

In a preferred embodiment, particularly when use is made of a relatively low mol ratio of the alkylene glycol to the terephthalic acid, the second stage is carried out by reducing the pressure, incrementally or continuously, until a substantially atmospheric pressure is attained while the temperature is maintained in the 240° to 320°C range.

About 0.01 to 1.0% by weight of the organic base (molar ratio about 0.000116 to 0.0164:1 for triethylamine and tripropylamine) with respect to the terephthalic acid is added to the glycol-terephthalic acid reaction mixture prior to esterification. Superior results are obtained when the base is homogenized with the reactant glycol and acid. When less than 0.01% by weight of the base per weight of terephthalic acid initially present is employed, its effect and the pH of the reaction mixture are generally too low to minimize effectively the formation of ether by-products. On the other hand, while more than 1.0% of the base is also effective, there is a tendency for an undesirable amount of residual base to be present in the ultimate polymer and subsequent discoloration of the ultimate polymer.

Example: In a continuous process for producing hydroxyethyl terephthalate prepolymer, terephthalic acid (1 mol) and ethylene glycol (1.5 mols) were separately fed to and mixed in a mixer and from which there was continually removed a slurry of acid and ethylene glycol in the mol ratio. The slurry was continuously fed to the first of a series of four reactors of which the first and second reactor comprised the first-stage reaction as embodied herein.

For the first-stage reaction, the first of the first two reactors served to preheat the slurry to reaction temperature and to initiate the first-stage esterification reaction. The pressure in the first reactor was controlled by the pressure in the second reactor. Operating pressures in the last three (termed for convenience as the second, third and fourth reactors) were 100, 30, and 2 psig respectively. Temperature was 270°C, and residence time was 15 minutes in each reactor.

During operation of the continuous process, diisopropylamine (DIPA) in fairly small amount was introduced into the slurry prior to its entrance into the first reactor while, at times, the operation was carried out without addition of that amine.

When the process was carried out without addition of the amine, the percent esterification of the first reactor product was markedly lower than for the corresponding operation carried out in the presence of a small amount of the amine (0.28% based on weight of the acid). Expressed otherwise, at a pressure of 100 psig in the first reactor, 67.7% esterification was obtained in the absence of the amine whereas 77.3%, or about a 14% increase in esterification was obtained under the same conditions but in the presence of the amine.

Thus, for the same period of time, the amine functioned to markedly accelerate the esterification reaction. Taking into consideration that the uncatalyzed reaction occurred at a substantially fast rate under the defined operating conditions, it should be apparent that the aforesaid increase in rate of esterification is a very significant and surprising result as to the catalytic effect of the amine with respect to the esterification reaction.

The presence of the amine not only markedly increased the rate of desired esterification whereby a higher percentage of esterification was obtained in the same or similar period of time but, for both runs in which the amine was present, the esterification product was of markedly lower content of undesired ether than were the corresponding runs in the absence of the amine. Additional runs using a higher concentration of the amine indicate that the catalytic effect of the amine on the rate of esterification generally increases as the amine concentration is increased.

Multistage Direct Esterification

A continuous process for producing high molecular weight polyester of fiber-forming quality is described by *P. Schaefer, P.A. Mason and W.H. Yates; U.S. Patent 4,100,142; July 11, 1978; assigned to Fiber Industries, Inc.* The process includes the direct esterification of dicarboxylic acid with a glycol to yield a prepolymer mixture of ester and low molecular weight polyester condensable under conditions of elevated temperature and vacuum into polymer of high intrinsic viscosity.

Figure 1.3 shows a schematic representation of a continuous process for producing polyester by the direct esterification method. The drawing is described with reference to the production of polyethylene terephthalate.

Dicarboxylic acid and glycol reactants are continuously metered into a mixer **1** through its inlets **2** and **3**, respectively. The glycol is added downstream from the particulate acid, as in the case of terephthalic acid and ethylene glycol, to avoid formation of large agglomerates which may clog the discharge line and slow the esterification reaction, thus interrupting the continuous process. The mixer may be of any suitable type capable of forming a pumpable paste of diacid-glycol. For purposes of illustration, a longitudinally oriented vessel which may, for example, be of a vaned agitator or flighted screw type is shown.

Figure 1.3: Continuous Process for Polyester Production by Direct Esterification

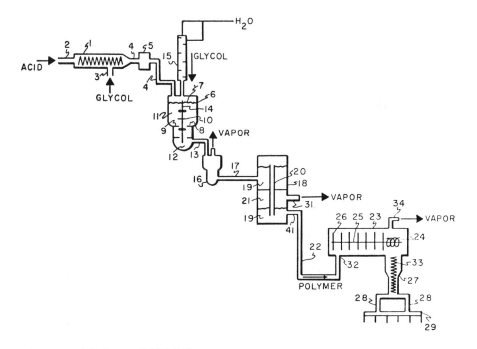

Source: U.S. Patent 4,100,142

The paste, due to its high viscosity is positively conveyed by a pump **5** through a conduit **4** into a primary esterifier **6** above a liquid level **7**. Considering the primary esterifier in greater detail, there is seen a semicompartmentalized reactor with baffle means **8, 9** and **10** forming upper and lower reaction zones **11** and **12**. Monomer is continuously withdrawn through a line **13**. An agitator **14** functions to rapidly dissolve incoming paste in liquid monomer as well as aiding the esterification reaction by physically causing molecular interaction and adequate heat transfer. Although the primary esterifier need not be compartmentalized, it preferably contains 2 or more semicompartments, i.e. 2 to 6 compartments, separated by suitable baffles to prevent excessive passage of undissolved, particulate dicarboxylic acid either into the polymerization stage of the integrated process or where employed, into the secondary esterifier.

The primary esterifier is operated above the partial vapor pressure (the pressure generated by the quantity of glycol in the primary esterification system) of the glycol at the temperature employed to insure maximum retention of glycol in the system for esterification while allowing maximum removal of water, i.e., the pressure should be regulated so that it does not exceed about the sum of the partial vapor pressures of water and glycol within the system, i.e. less than about autogenic pressure. Autogenously generated steam is sufficient to maintain the desired pressure range within the reactor.

Water is removed through a distillation column **15** wherein glycol is condensed and washed as it flows back into the esterifier. Thus, the primary esterification process step, which can be a multistep design, is essentially self-compensating with respect to glycol. If this were not so, a large molar excess of glycol would have to be supplied to the reactor to reach an acceptable degree of esterification. For practical purposes it is nearly impossible to achieve over 95% esterification, that is, to esterify over 95% of carboxylic groups initially present, under pressures as described, because of retention of water within the system.

It has been found highly desirable to insure that an essentially 100% dicarboxylic-reacted effluent, corresponding to at least 50% esterification, is fed into the polycondensation stage. Product and process advantages, particularly with respect to low ether formation, rapid molecular weight development and polymer product uniformity are evident where an effluent free of unreacted diacid is produced at this stage of the overall process.

To this end, intermediate from the primary esterifier is passed to a secondary esterifier **16** operating under a lower pressure than the first esterifier. As a result, process monitoring benefits due to a gradual decrease in pressure are realized. Temperature in vessel **16** may be maintained at about that employed in the primary esterifier since the paramount purpose of the second stage is the maximum removal of water and excess glycol leading to a further increase in percent esterification.

With the direct esterification process, about a 50% esterification is sufficient for production of fiber-forming material of high intrinsic viscosity since chain stoppers are not inherently present in the first product of esterification. In such instance the effluent may be transferred directly to a polymerizer.

The second esterifier where in use is indicated generally at **16** since the design thereof can correspond, for example, to that of the primary esterifier. The liquid inlet is positioned above, although it may be below, the liquid level in the reactor. Agitation is not required since the lower relative pressure therein causing vapor emanation, especially of glycol, creates sufficient agitation within the liquid. Since the objective is to complete the esterification step therein, excess glycol as well as water is desirably removed from the reactor; hence, the distillation column is eliminated, and vaporous effluent can be separated by suitable distillation into a water fraction which is passed to sewage and a glycol fraction to be recycled by means of a glycol recovery system not shown to the mixer 1.

The product of esterification is then fed to the polycondensation phase of the continuous process wherein a single polymerizer may be employed or, as presently recommended and preferred, a multistage "low" or first polymerizer and a single-stage "high" or final polymerizer are sequentially employed. A "low" polymerizer 18 can be constructed of a plurality of chambers 19 connected by means of a plurality of conduits 20 positioned with inlets slightly above the liquid level in the upper or first compartment (i.e. a series of separate vessels with appropriate tubing could be employed) and outlets dipping below the liquid level of the next compartment, i.e. if a third compartment is desired a second series of transfer tubes would arise from below the liquid level of the third compartment up into a vapor space 21, being positioned comparably to the upper series of transfer tubes 20.

In the preferred embodiments of the process, two to three reaction zones or stages are employed in the low polymerization system with temperature being increased step-wise in each reaction zone from inlet to outlet while pressure is being reduced, i.e. vacuum is progressively increased in stages from feed inlet to low polymerizer product outlet. This pressure drop, assisted by the turbulence of the system where present during the initial stages of polymerization under subatmospheric pressure, causes liquid as well as evolved vapors consisting primarily of water-glycol mixture, to flow from an upper or previous zone to the next succeeding zone of lower pressure.

Since the outlets of the transfer tubes dip well below the surface level of liquid, material from the previous zone passes substantially through the entire liquid level before flowing to the final outlet or a succeeding reaction zone. The vaporous effluent is removed from the final reaction zone. If this procedure was not maintained, low molecular weight oligomer could be lost as part of discarded vaporous effluent due to the entrainment thereof in the "flashing" glycol-water mixture at subatmospheric pressure.

Stabilization of oligomer occurs through redissolution of vapor in a subsequent reaction zone wherein the oligomer attains a molecular weight sufficiently high to virtually eliminate sublimation and/or entrainment of product under existing temperatures and vacuum conditions. The product of "low" polymerization exits as overflow by means of gravity through an outlet 41. Liquid level may be regulated by location of the overflow. Rate and degree of polymerization in the low polymerizer may be maintained constant as desired by proper regulation of inventory and feed rate, and temperature and pressure differentials throughout the system, to allow sufficient residence time for production of low molecular weight polymer of desired characteristics within the liquid fraction of the system.

The product of the low polymerizer is transferred to a second polymerizer **23** through a conduit **22**. In the illustrative embodiment, a single "high" polymerizer is depicted, one of the process advantages being the use of a single vessel in which the intermediate as more fully defined hereinafter from the "low" polymerizer, i.e. one having an intrinsic viscosity of about 0.2, is converted into high molecular weight, fiber-forming product.

The "high" polymerizer is preferably a cylinder-shaped vessel, longitudinally aligned with inlet **32** in one end and a rotatable shaft **25**, coinciding with the longitudinal axis, carrying a plurality of formaminous members **26** and, adjacent gravity-feed product discharge **33**, a forwarding spiral **24**. Evolved glycol from polycondensation exits through an outlet **34** in the top of the reactor. A funnel-shaped screw extractor **27** forces the viscous product through dual, symmetrically arranged outlets **28** into a spinning manifold **29**.

Temperature is increased in the "high" polymerizer, accompanied by a further reduction in pressure by means not shown. As the temperature of the molten mass approaches the polymer degradation temperature, accurate temperature control both within the polymerizer and extractor is required. Further, residence time is controlled to be as short as practical, since polymer degradation is a function of time versus temperature, but commensurate with production of high molecular weight product.

The rotating formaminous members, with increasing mesh sizes, i.e. percent void space, and progressively longer distances between consecutive formaminous members from inlet to outlet form thin polymeric films creating a large surface area for glycol evaporation in an upper vapor space, the reactor being only partially, i.e., $1/10$ to $1/5$, full.

The formaminous members which may be in the form of screens, perforated discs, spoked wheels and the like, are of such diameter, typically slightly less than the internal diameter of the cylinder, to expose maximum polymer surface area to the upper vapor space, thus facilitating removal of volatiles and further condensation into high molecular weight product. Members **26** are spaced to eliminate significant bypassing passing and bridging thereinbetween with percent void space, and the positioning thereof, increasing sufficiently to allow an uninterrupted flow of polymer toward discharge.

Computer-Controlled Apparatus

A.C. Moody, Jr., R.G. Kenward and J.W. Shiver; U.S. Patent 4,106,098; Aug. 8, 1978; assigned to Allied Chemical Corporation describe a computer-controlled apparatus for the preparation of high-molecular-weight, high-quality polyethylene terephthalate polyester suitable for processing into fibers, films, and other shaped articles. The computer is interrelated with a hardware system in such a manner that it is capable of initiating process changes capable of producing a predetermined intrinsic viscosity polymer during the polycondensation reaction stages.

Referring more particularly to Figure 1.4, element **1** is a first polycondensation reactor; element **2** is a first viscometer; element **3** is a second polycondensation reactor, and element **4** is a second viscometer. Each viscometer is provided with means for measuring polymer temperature therein.

Figure 1.4: Computer-Controlled Apparatus for Preparation of Polyethylene Terephthalate

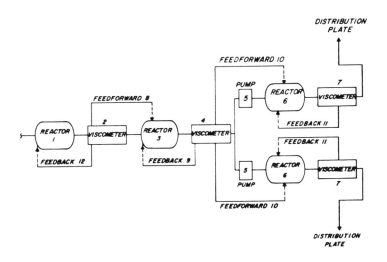

Source: U.S. Patent 4,106,098

Elements **5,5** are pumps for feeding third polycondensation reactors **6,6**. The polyester from reactors **6,6** passes to viscometers **7,7** and then to the distribution plate and spinneret. Line **12** indicates information flow from viscometer **2** to reactor **1**. Lines **8, 9, 10,10** and **11,11** indicate information flow from viscometer **2** to reactor **3**, from viscometer **4** to reactor **3**, from viscometer **4** to reactors **6,6** and from viscometers **7,7** to reactors **6,6**.

Analog inputs via lines **8, 9, 10, 11,** and **12** to the computer represent vapor pressure, melt viscosity, and temperature, for controlling reactors **1,3** and reactors **6,6**. Inputs pertaining to reactors **6,6** include additionally the pump speeds. The computer control system controls the vapor pressure in each of the reactors.

Example: 41½ parts by weight per hour of purified terephthalic acid, 23 parts per hour of ethylene glycol and 0.88 part per hour of diisopropylamine are continuously fed to a paddle wheel mixer where they are converted to a paste. The paste mixture is then pumped from the mixer by a feed pump to the inlet of a circulating pump. The paste mixture is pumped with 40 parts of recirculating mixture by the circulating pump through a multiple tube and shell heat exchanger where it is heated to 260° to 270°C.

After leaving the heat exchanger, the mixture enters an esterifying reactor maintained at 260° to 270°C by conventional heating means, and 90 to 110 psig pressure by means of an automatic vent valve. The reaction mixture leaving this reactor is split, with part returned to the inlet of the circulating pump where it is combined with fresh paste, and part is reduced in pressure to near atmospheric

in a series of three reactors maintained at 260° to 270°C, then flowed for polycondensation to the first in a series of reactors arranged and controlled as dia-
grammatically illustrated in Figure 1.4. Each of the final (ultimate) reactors **6,6**
in the series of reactors is an essentially horizontal totally enclosed cylindrical
reactor having an essentially horizontal polyester flow, a pool of polyester in its
lower portion, and driven wheels to create high surface area in the polyester to
facilitate evaporation of volatiles from the polymer. The pool of polyester in
final reactors **6,6** is maintained by pumps **5,5**. A change in inventory of polyester in the reactors may be estimated by observing the change in level on a
conventional level recorder (not shown on drawing).

The polyethylene terephthalate polyester issuing from the last reactor stage has
average intrinsic viscosity of 0.96 unit. It is passed at a temperature of 295°C
through a filter distribution plate to a 192-hole spinneret, and processed into
1,300 denier yarn. Quality of the yarn produced is excellent, i.e., tenacity is
9.2 gpd and elongation at break is 14%.

Continuous Process with Recycling of Ethylene Glycol

A continuous process is disclosed by *R.R. Edging and L.-M. Lee; U.S. Patent
4,110,316; August 29, 1978; assigned to E.I. Du Pont de Nemours and Company* for direct esterification of terephthalic acid with ethylene glycol, followed
by low-pressure polymerization in a continuous polymerizer and in a finisher
polymerizer to produce polyester suitable for melt-spinning into yarn for textile
uses.

Vapors from the polymerizers are condensed in spray condensers to recover glycol liquid, which is recycled to the direct esterification reaction. Efficient operation of the spray condensers at absolute pressures of less than 60 mm of mercury
is achieved by addition of substantially dry ethylene glycol to lower the water
content of condensate used in the sprays.

As shown in Figure 1.5, terephthalic acid is supplied from bin **10** by screw
feeder **12** to mixing tank **14**. Ethylene glycol is supplied to the mixing tank
from glycol tank **16** at a rate within the range of 1.5 to 4.0 (preferably 1.8 to
3.0) mols of glycol per 1.0 mol of terephthalic acid. The reactants are mixed
to form a slurry by mixer **18** which is driven by motor **20**. Slurry pump **22**
feeds the slurry to the direct esterification reactor **24**. The reactants are heated
in reaction kettle at about 280° to 315°C and about atmospheric pressure to
esterify the terephthalic acid and form an ester reaction product having an average degree of polymerization of 2 to 10.

Reaction vapors are conducted to a glycol recovery system **26**, which may be
any suitable equipment for removing water and organic by-products of the reaction from the glycol, and recovered glycol is recycled to the glycol tank **16**.

Liquid ester reaction product is withdrawn continuously from the reactor and
pumped through transfer line **28** to continuous polymerizer **30** by stuffer pump
32. Additives may be mixed with the ester reaction product passing through
the transfer line. For this purpose, a slurry or solution of one or more additives
in ethylene glycol is introduced into the transfer line at a rate determined by
passage of the ester reaction product through meter pump **34**. A mixer **36** and
a bulge **38** having 1 to 10 minutes' holdup time are used to completely incorpo-

rate the additives. An amount of ethylene glycol is preferably used which provides a rate of glycol addition of about 0.25 to 0.60 mol of glycol/mol of ethylene terephthalate in the polymer produced.

The feed passes upward into the polymerizer column through back-pressure valve **40** and heater **42**, which maintains a temperature of from 250° to 300°C. Prepolymer having a relative viscosity of about 5 to 8 is withdrawn near the top of the column and passes through transfer line **44** to finisher polymerizer **46**. The finisher vessel is heated to maintain a temperature of from 270° to 305°C. Polymer with a relative viscosity of 12 to 29 is withdrawn through screw pump **48** for spinning into textile filaments. The process can be operated to produce higher relative viscosity polymer for spinning into industrial filaments.

Vapors from the continuous polymerizer **30** pass from the top of the column through vapor line **50** to condenser **52**. The vapors contain ethylene glycol, about 2 to 35 wt % water, about 1 to 4 wt % of organic materials which form solids at lower temperatures, and small amounts of other organic impurities. Condensate from the vapors flows into hotwell **54** where it is combined with ethylene glycol containing less than 3 wt % water to form a mixture containing less than 10 (preferably 2 to 7) wt % water for use as spray in the condenser. Pump **57** supplies the mixture to the top of the condenser through line **58** and cooler **60**. In order to operate at the desired pressures, the cooled spray mixture should be at a temperature of less than 60°C, and preferably less than 50°C.

The cooled mixture is sprayed on the vapors in the condenser by spray head **61**. Vacuum line **62** leads to vacuum pumping means (not shown) for maintaining the condenser and the continuous polymerizer at an absolute pressure of 5 to 60 mm mercury (0.67 to 8.0 kPa). A glycol spray containing 2 to 7 wt % water is highly efficient in condensing the vapors to minimize the load on the vacuum pumping means. The spray also prevents solids from depositing on surfaces to cause fouling.

Vapors from the finisher polymerizer pass from the finisher vessel through vapor line **64** to condenser **66**. Condensate from the vapors flows into hotwell **68** where it is combined with ethylene glycol containing less than 1 wt % water, from glycol line **70**, to form a mixture containing less than 3 wt % water for use as spray in the condenser. The low water content is needed because the pressure in the finisher vessel and its condenser is lower than in the continuous polymerizer vessel and its condenser **52**. Pump **72** supplies the mixture to the top of condenser **66** through line **74** and cooler **76**. In order to operate at the desired pressures, the cooled mixture should be at a temperature of less than 60°C, and preferably less than 50°C.

The cooled mixture is sprayed on the vapors in the condenser by spray head **78**. Vacuum line **80** leads to vacuum pumping means (not shown) for maintaining the condenser and the finisher polymerizer at an absolute pressure of 0.5 to 10 mm of mercury (0.07 to 1.3 kPa) which is also lower than the pressure in the continuous polymerizer.

Ethylene glycol liquid containing less than 3 wt % water overflows from finisher hotwell **68** to hotwell **54** through line **82**. Ethylene glycol liquid containing less than 10 (preferably 2 to 7) wt % water is transferred from hotwell **54** to glycol tank **16** through line **58**, recycle pump **84** and line **86**. The rate of transfer is determined by the rate at which ethylene glycol from glycol tank **16** is used in the esterification reaction. The rate of transfer is controlled by valve **88**, which can be a conventional automatic valve actuated by level sensing device **90**. The rate of addition of glycol to finisher hotwell **68** is similarly determined by the level of liquid in hotwell **54**.

The ethylene glycol liquid containing less than 1 weight percent water used in the finisher condenser system should be added at a rate of about 0.4 to 1.15 mol of glycol/mol of ethylene terephthalate in the polymer produced.

The process is preferably operated so that ethylene glycol is recycled to the esterification reaction from the polymerizer condenser systems at a rate of about 0.75 to 1.50 mols of glycol/mol of ethylene glycol in the polymer produced.

Figure 1.5: Polyethylene Terephthalate Process with Recovery of Ethylene Glycol

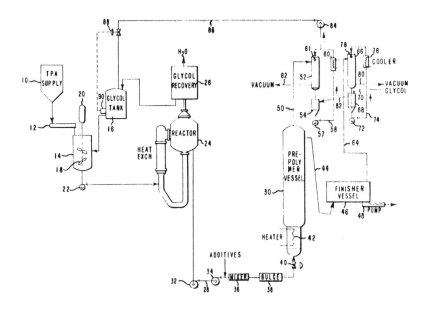

Source: U.S. Patent 4,110,316

POLYMERS HAVING SUPERIOR THERMAL OR CHEMICAL STABILITY

Polyethylene terephthalate fibers and cords are known to exhibit excellent dimensional stability, i.e., low extension or growth during service, as well as to have a high resistance to thermal degradation; however, in pneumatic tires and industrial belts under high speed conditions under heavy load, loss of tensile strength is experienced due to high temperature conditions emanating under such conditions. Efforts to remedy this problem have all too often been ineffective. Most research in this field has been directed to producing a high molecular weight linear polyester having a low content of free carboxyl groups. The following patents are pertinent:

U.S. Patent 3,051,212 to W.W. Daniels relates to reinforced rubber articles and to textile cords and fibers for reinforcing such articles. This patent discloses that a linear terephthalate polyester having a concentration of free carboxyl groups of less than 15 equivalents per million grams may be prepared in a number of different ways. One effective procedure is to treat the filaments, after they have been formed, with a chemical reagent which reacts with and caps the free carboxyl group. One such agent is diazomethane.

U.S. Patent 3,627,867 to E.C.A. Schwarz discloses a process and apparatus for melt spinning high molecular weight polyethylene terephthalate into high-performance fibers under conditions which reduce the normally high viscosity of such polyester. Ethylene oxide or other low-boiling oxirane compound is injected under pressure into molten polyester before it is fed to the metering pump of the melt-spinning machine. The fibers are characterized by low free-carboxyl content and freedom from voids which might be expected from injection of the volatile material.

U.S. Patent 3,657,191 to R. Titzmann et al is directed to a process for the manufacture of linear polyesters having an improved stability with respect to compounds with active hydrogen. Polyesters of this type are obtained by reacting polyesters with ethylene carbonates or monofunctional glycidyl ethers. The reaction is first carried out within a temperature range lying 10° to 60°C below the softening point of the polyester and is then terminated during the melting and melt-spinning process.

U.S. Patent 3,869,427 to R.W. Meschke et al discloses a process of preparing polyester filaments having low free-carboxyl-group contents which give superior performance in pneumatic tires and other reinforced rubber articles where heat-degradation is a problem. Reduction of free carboxyl groups is achieved by mixing with the molten polyester, prior to melt-spinning, 1,2-epoxy-3-phenoxypropane or 1,2-epoxy-3-n-hexyloxypropane.

U.S. Patent 4,016,142 to W. Alexander et al discloses preparation of a fiber-forming polyester wherein the number of free carboxyl end groups present in the polymer may be reduced by adding to the polymerized polyester a glycidyl ether which reacts with the carboxyl end groups present to form free hydroxyl end groups.

Although the aboveidentified patents directed to stabilized polyesters are of major interest, certain of the proposed polyester modifiers are known to be highly toxic and/or hazardous to use on commercial scale; others are relatively less effective in terms of reducing the carboxyl end group concentration of the polyester. Accordingly, considerable research continues in an attempt to solve or mitigate the long standing problem of producing high molecular weight polyester stabilized against deterioration under high temperature operating conditions.

Reaction with Epoxy Compound

According to *S.D. Lazarus and K. Chakravarti; U.S. Patents 4,115,350; September 19, 1978; and 4,130,541; December 19, 1978; both assigned to Allied Chemical Corporation*, high molecular weight linear condensation polyesters are stabilized against deterioration by heat by reacting the polyester in molten form with an epoxy compound having 5 to 25 carbon atoms in the molecule and selected from the group consisting of

$$R_1-\overset{\displaystyle O}{\underset{\displaystyle R_2}{C}}\!\!-\!\!\overset{}{\underset{\displaystyle R_3}{C}}(CH_2)_n-\overset{}{\underset{\displaystyle R_4}{N}}-\overset{\displaystyle O}{\overset{\|}{C}}-R, \qquad R_1-\overset{\displaystyle O}{\underset{\displaystyle R_2}{C}}\!\!-\!\!\overset{}{\underset{\displaystyle R_3}{C}}-(CH_2)_n-O-\overset{\displaystyle O}{\overset{\|}{C}}-R, \text{ and}$$

$$R_1-\overset{\displaystyle O}{\underset{\displaystyle R_2}{C}}\!\!-\!\!\overset{}{\underset{\displaystyle R_3}{C}}-(CH_2)_n-S-\overset{\displaystyle O}{\overset{\|}{C}}-R$$

where R represents the radical remaining after removal of the carboxyl group from a monocarboxylic acid, R_1, R_2, R_3 and R_4 represent hydrogen or hydrocarbon radicals, and n is an integer that can be 0 to 3.

Example 1: About 41.5 lb/hr of terephthalic acid, 27.9 lb/hr of ethylene glycol, 65 g/hr of diisopropylamine and 16 g/hr of antimony acetate are continuously fed to a paddle mixer where they are converted to a paste. The paste mixture is then pumped from the mixer by a feed pump to the inlet of a circulating pump. The paste mixture is pumped with 40 parts by weight per part of paste mixture of recirculating mixture by the circulating pump through a multiple tube and shell heat exchanger where it is heated to 260° to 270°C.

After leaving the heat exchanger, the mixture enters an esterification reactor which is maintained at 260° to 270°C by conventional heating means, and 90 psig pressure by means of an automatic vent valve. The recirculating mixture leaving this reactor is split, with part being returned to the inlet of the circulating pump where it is combined with fresh paste and part flowed to a series of three reactors where further esterification takes place at 270° to 275°C. Total esterification time is about 3 hours.

Following esterification, the reaction mixture is fed into a polycondensation reactor operating at 275°C and 30 torr pressure, with a residence time of 60 minutes. The resulting polyester polymer is fed to a polycondensation reactor operating at 275°C and 2 torr pressure, with a residence time of 120 minutes. Then, the polyester polymer is processed in a final polycondensation at 278°C and 0.5 torr pressure for 130 minutes. The polyester polymer melt at about 278°C is pumped from the final polycondensation reactor by means of a screw pump and conducted to gear pumps for transfer to a spinning machine where polymer temperature is increased to about 300°C. Between the screw pump and the gear pump, 0.255 lb/hr of N-(2,3-epoxypropyl)benzamide is added to the polyester polymer as stabilizer and intimately mixed with the polymer by means of a conventional stationary mixer. The polyester polymer is reacted with the N-(2,3-epoxypropyl)benzamide for 3 to 20 minutes at about 278° to 300°C until the polymer is spun at the rate of 48 lb/hr through a 192 hole spinnerette. Yarn is continuously spun and drawn to form 1,300 denier, 192 filament yarn. The undrawn yarn from the spinnerette has an intrinsic viscosity of 0.80 to 0.90 dl/g and about 12 equivalents of carboxyl end groups per 10^6 g of polyester. The drawn yarn has 15.9% ultimate elongation and 8.5 g per denier tensile strength. The drawn yarn retains 87% of its strength after exposure to pure ammonia gas for 3 hours at 150°C.

This test shows that the yarn is very stable to both heat and ammonia, which is indicative of a good tire yarn. The drawn yarn is overfinished with a lubricating composition, twisted into 3 ply, 9 tpi tire cord, woven into a fabric, dipped in a blocked diisocyanate-epoxide emulsion, stretched at 420°F, dipped in a resorcinol-formaldehyde-vinyl pyridine polymer emulsion, stretched at 440°F, and calendered with rubber to make rubberized fabric for tire building. Tires made with this fabric are characterized by excellent durability when run on the wheel test stand.

Similar results are obtained when equivalent amounts of N-(2,3-epoxypropyl)-stearamide, N-(epoxyethyl)benzamide, glycidyl benzoate or S-(glycidyl)thiobenzoate are used in place of the N-(2,3-epoxypropyl)benzamide.

Example 2: This example demonstrates the use of 4-dimethylaminopyridine as a catalyst to accelerate the reaction of polyethylene terephthalate with an epoxy compound of this process.

About 48 lb of polyethylene terephthalate chips having an intrinsic viscosity of 0.95 are mixed with 0.255 lb of N-(2,3-epoxypropyl)benzamide and 0.01 lb of 4-dimethylaminopyridine by tumbling in a can. The mixture is then melted and spun at about 300°C through a 1" extruder into 48 filament yarn which is plied and drawn at a draw ratio of 6.05 to 1 into 1,300 denier, 192 filament yarn. The undrawn yarn from the spinnerette has an intrinsic viscosity of 0.84 and 9 equivalents of carboxyl end groups per 10^6 g. The drawn yarn has 14.5% ultimate elongation and tensile strength of 8.4 g per denier. The drawn yarn retains 90% of its strength after exposure to pure ammonia gas for 3 hours at 150°C. This yarn is converted into tire cord as in the first example. The cord is characterized as having excellent fatigue and durability properties.

Example 3: *Comparative* — Example 1 is repeated, except that no N-(2,3-epoxypropyl)benzamide is added to the polyester polymer. The undrawn yarn from the spinnerette has an intrinsic viscosity of 0.80 to 0.90 and 30 equivalents of carboxyl end groups per 10^6 g of polyester. The drawn yarn has 16.7% ultimate elongation and tensile strength of 8.2 g per denier. The drawn yarn retains only 59% of its strength after exposure to pure ammonia gas for 3 hours at 150°C. These data in comparison with the data of Examples 1 and 2 demonstrate the beneficial effect on number of carboxyl end groups and strength retention of the polyester yarn of adding the stabilizer compound.

1,2-Epoxy-3-Phenoxypropane with Triphenylphosphite as a Stabilizer

In the preparation of a fiber-forming polyester by *W. Alexander, A.G. Causa, and J.G. Fraser; U.S. Patents 4,016,142; April 5, 1977; and 4,122,063; October 24, 1978; both assigned to Millhaven Fibers, Ltd., Canada*, the number of free carboxyl end groups present in the polymer are reduced by adding to the polymerized polyester a glycidyl ether which reacts with the carboxyl end groups present to form free hydroxyl end groups.

The resultant reaction product comprises a modified polyethylene terephthalate molecule wherein a carboxyl end group has been converted into an ester containing a single free hydroxyl end group. Surprisingly, the esterified molecule then reacts further with other molecules to produce a polymer of higher molecular weight.

The glycidyl ether is represented by the formula:

$$R-O-CH_2-CH \underset{\displaystyle O}{\underline{\qquad\qquad}} CH-R'$$

where R is selected from the group consisting of aliphatic, cycloaliphatic, and aromatic groups and R' is selected from the group consisting of hydrogen, aliphatic, cycloaliphatic and aromatic groups. 1,2-epoxy-3-phenoxypropane is exemplary.

According to a preferred embodiment, polyethylene terephthalate polymer is prepared by a process including addition to the melt of 0.45 up to 0.9% by weight of 1,2-epoxy-3-phenoxypropane, and triphenyl phosphite as a stabilizer.

Example 1: A quantity of a standard polyethylene terephthalate polymer was prepared according to the following procedure. 1,550 g of dimethyl terephthalate and 1,100 ml of ethylene glycol were placed in a distillation flask fitted with a reflux column. The mixture was heated to 160°C and 0.015% by weight of zinc acetate dihydrate was added as an ester interchange catalyst.

The ester interchange reaction was continued until 640 ml of methanol had been evolved and collected. 0.04% by weight of antimony trioxide as polycondensation catalyst was added together with 0.5% by weight of titanium dioxide as a delustrant and the batch was then transferred to an autoclave. The temperature of the batch was raised to 285°C and the autoclave was simultaneously evacuated to a vacuum corresponding to 0.1 mm of mercury. The polycondensation was allowed to continue until the electric power requirements of the stirrer motor indicated that a suitable molecular weight had been reached. 0.64 ml of triphenyl phosphite was added to the batch as a stabilizer and the batch stirred under nitrogen gas at atmospheric pressure for 60 minutes.

The batch was then extruded in ribbon form from the autoclave and the solidified polymer cut into cubes. Upon examination, the polymer was shown to have an intrinsic viscosity of 0.63, a softening point of 262.6°C and a free carboxyl end group concentration of 32.6 equivalents per million grams of polymer.

Example 2: A polymer batch was prepared under the conditions outlined in Example 1. At the end of the polymerization period, 7.0 g (0.45% w/w) of 1,2-

epoxy-3-phenoxypropane was added to the melt and the melt stirred for 10 minutes under dry nitrogen at atmospheric pressure. The autoclave was then evacuated to remove excess epoxide and the batch extruded in ribbon form. Examination of the polymer showed the intrinsic viscosity to be 0.668 and the free carboxyl end group concentration was 18.6 equivalents per 10^6 g of polymer.

Example 3: A polymer batch was prepared under the conditions listed in Example 1. At the end of the polymerization period, the autoclave was pressurized to 30 psig with dry nitrogen and 14.0 g (0.90% w/w) 1,2-epoxy-3-phenoxypropane was added to the melt. After 10 minutes stirring under pressure, the autoclave was evacuated to remove excess epoxide and the batch extruded in the form of a ribbon. Upon examination, the polymer was found to have an intrinsic viscosity of 0.671 and a free carboxyl end group concentration of 6.0 equivalents per 10^6 g of polymer.

The polymers from the above three examples were converted into 75 denier, 33 filament yarns using techniques familiar to those skilled in the art. Tire cord samples were prepared by plying together 28 ends of each 75 denier, 33 filament yarn. Six samples of each of these plied yarns were exposed to hydrolytic conditions in steam at 15 psig and 120°C for 72 hours, and six samples were retained unexposed to the steam to determine the initial strength. After exposure, the strength loss of each sample was measured by means of an Instron tensile tester using a crosshead speed of 20 cm/min, and a sample length of 20 cm. The following table summarizes the results obtained and it can be seen from the results that the percentage strength retention, after exposure to hydrolytic conditions, bears a direct relationship to free carboxyl end group concentration of the polymer.

Polymer Example	Polymer Carboxyl End Group equiv/10^6 g	Yarn Carboxyl End Group equiv/10^6 g	Percent Strength Retention
1	32.6	36.7	62
2	18.6	25.3	72
3	6.0	13.0	83

Example 4: 700 lb molten dimethyl terephthalate and 46 gal of ethylene glycol were mixed in a suitable ester-interchange vessel. 0.015% zinc acetate dehydrate was added to the mixture as a catalyst and the temperature raised from 150° to 225°C at which point the evolution of methanol ceased and the theoretical quantity of methanol had been removed from the mixture via a distillation column. 0.08% antimony trioxide catalyst was added to the melt together with 0.05% titanium dioxide delustrant and the melt was transferred to an autoclave.

Polymerization was carried out by increasing the autoclave temperature to 296°C and lowering the pressure below 1 mm mercury. The increase in molecular weight was observed by watching the increase in power required to agitate the polymer melt. When the power consumption reached 4 kW, sufficient triphenyl phosphite stabilizer was added to give a phosphorus content of 100 ppm. Maximum molecular weight was achieved in 3 hours, 35 minutes at which point a plot of the power consumption of the agitator with time had leveled off at 11.7 kW. At this point, the polymer was extruded and diced. The intrinsic viscosity was found to be 0.864 with a carboxyl end group content of approximately 40 equivalents per 10^6 g polymer. A repeat polymerization gave polymer with an intrinsic viscosity of 0.865 and a carboxyl end group content of 41.0 equivalents per 10^6 g. In this case, the power consumption was 12.0 kW and the polymerization time 3 hours, 35 minutes.

Polyesters from Bisphenols

Polyester fibers, such as those spun from polyethylene terephthalate, have found wide commercial acceptance. These fibers, however, generally have limited thermal stability. A major pathway for thermal decomposition of known aliphatic-aromatic polyesters is thermal elimination at the site of aliphatic beta-hydrogens.

A.H. Frazer; U.S. Patent 4,065,432; December 27, 1977; assigned to E.I. DuPont de Nemours and Company has prepared thermally stable, high tenacity, high modulus polyesters prepared from aromatic dibasic acids and thermally stable, rigid bisphenols containing no aliphatic beta-hydrogen atoms.

Thermally stable, rigid bisphenols of the formula:

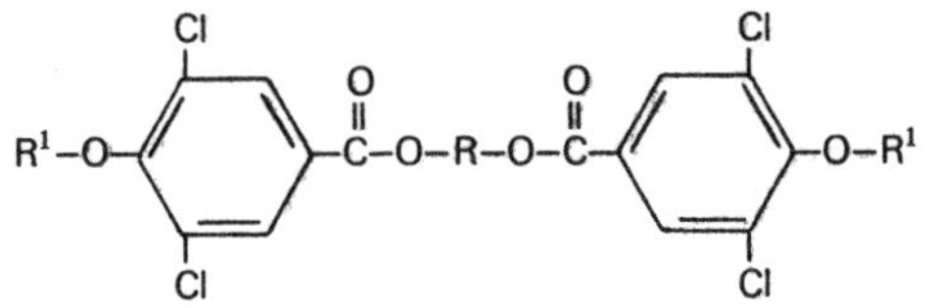

where R is

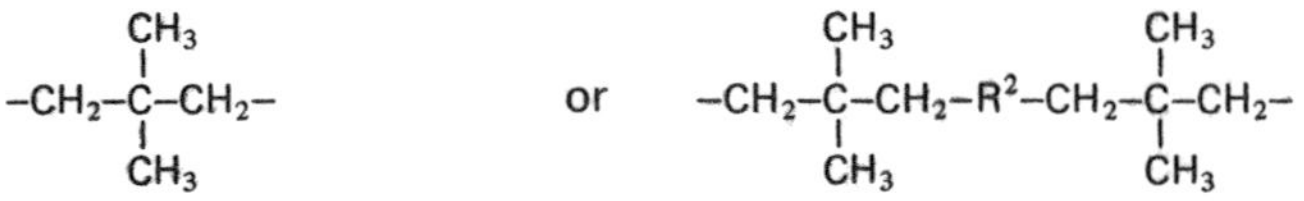

where R^2 is arylene or substituted arylene, and R^1 is hydrogen or acyl, are used to prepare thermally stable, rigid, ordered polyesters of the formula:

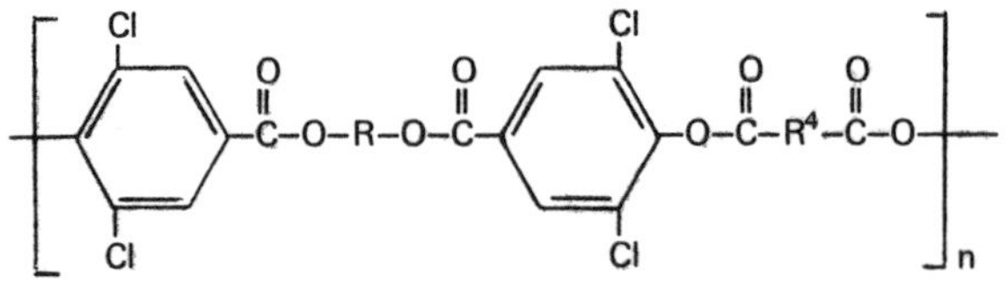

where R^4 is arylene or substituted arylene, and n is at least about 10.

Example: *Part (A)* — A mechanically stirred mixture of 0.6 g of calcium acetate, 0.6 g of antimony oxide, 15.7 g (0.15 mol) of 2,2-dimethyl-1,3-propanediol, and 132 g (0.6 mol) of methyl 3,5-dichloro-4-hydroxybenzoate was heated under a nitrogen atmosphere at 200°C for 18 hours. The reaction mixture was washed three times with benzene in a Waring blender and extracted for 5 days with 2 liters of hot benzene. The extract, on cooling, yielded 50.5 g of 4,4'-[2,2-dimethyl-1,3-propanediylbis(oxycarbonyl)] bis-(2,6-dichlorophenol) melting at 180° to 181°C. The infrared spectrum was consistent with the indicated structure. The compound contained 4.115 meq of OH/g, indicating a molecular weight of 486 for the bisphenol.

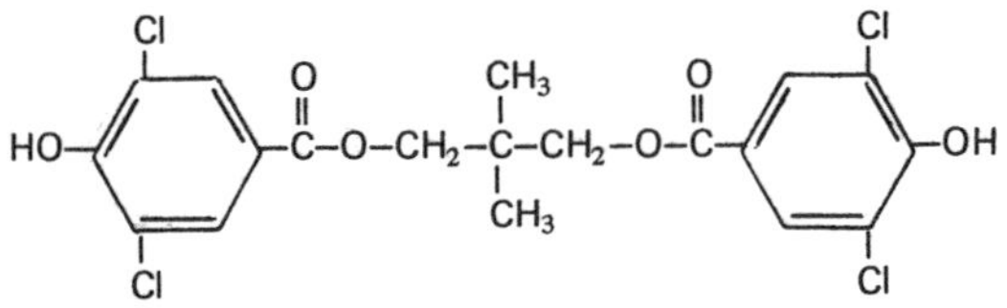

Part (B) — A mixture of 50 g of the bisphenol from Part (A) and 400 ml of acetic anhydride was heated to the boil for 1 hour, concentrated to 200 ml, and filtered hot. On cooling, the reaction mixture yielded the bisphenol diacetate which after three washes with distilled H_2O and drying at 120°C in a vacuum oven overnight weighed 45.6 g and had a melting point of 121° to 122°C. The infrared spectrum was consistent with the proposed structure and the compound was found to have a molecular weight by mass spectroscopy of 564.

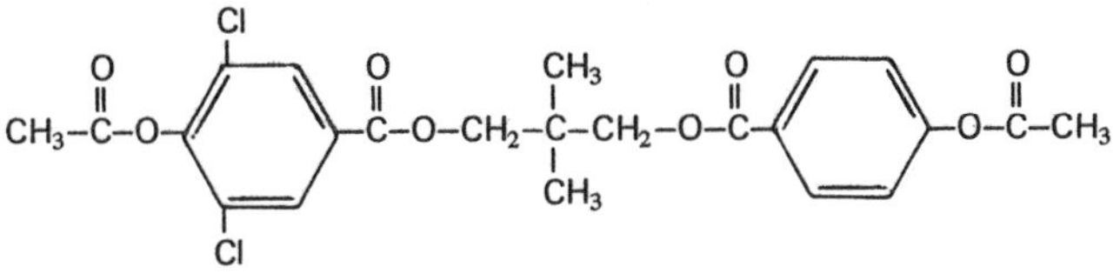

Part (C) — To a glass reactor with a nitrogen bleed and sidearm was added 1.044 g (0.00185 mol) of the bisphenol diacetate of Part (B) and 0.34 g (0.0021 mol) of terephthalic acid. The resulting mixture under a nitrogen atmosphere was heated for 18 hr at 275°C and for 5.5 hr at 305°C, followed by 2 hr at 305°C and less than 0.05 mm Hg pressure. The resulting poly[oxycarbonyl(1,4-phenylene)carbonyloxy(2,6-dichloro-1,4-phenylene)carbonyloxy(2,2-dimethyl-1,3-propanediyl)-oxycarbonyl(3,5-dichloro-1,4-phenylene)], which had a polymer melt temperature (temperature at which a polymer sample leaves a wet, molten trail when stroked with moderate pressure across a clean, heated metal surface) of 320°C, could be manually spun into fiber at approximately 270°C, had an inherent viscosity of 0.43 at 0.5% in a 40/60 weight % mixture of 1,1,2,2-tetrachloroethane and phenol at 30°C, and was amorphous by x-ray diffraction.

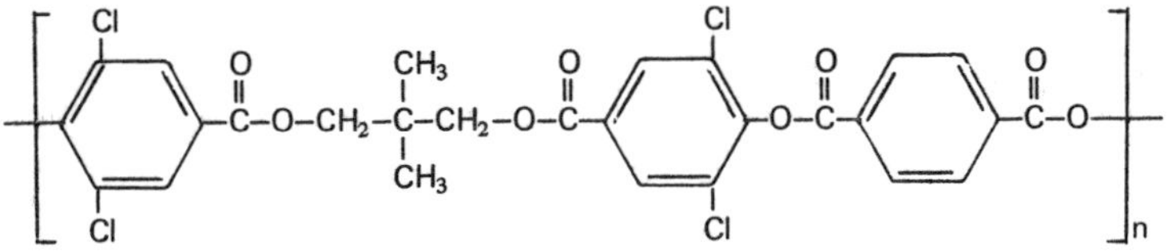

Rigid Dibasic Acids plus Aromatic Dihydroxy Compounds

According to *A.H. Frazer; U.S. Patent 4,085,091; April 18, 1978; assigned to E.I. DuPont de Nemours and Company*, thermally stable, rigid dibasic acids of the formula:

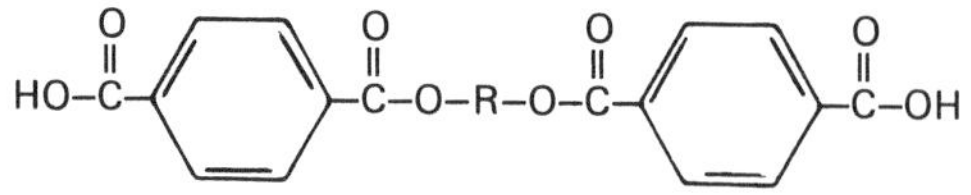

$$HO-\underset{\overset{\|}{O}}{C}-\text{⟨benzene⟩}-\underset{\overset{\|}{O}}{C}-O-R-O-\underset{\overset{\|}{O}}{C}-\text{⟨benzene⟩}-\underset{\overset{\|}{O}}{C}-OH$$

where R is

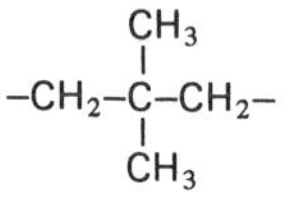

$$-CH_2-\underset{\overset{|}{CH_3}}{\overset{\overset{CH_3}{|}}{C}}-CH_2-$$ or $$-CH_2-\underset{\overset{|}{CH_3}}{\overset{\overset{CH_3}{|}}{C}}-CH_2-R^1-CH_2-\underset{\overset{|}{CH_3}}{\overset{\overset{CH_3}{|}}{C}}-CH_2-$$

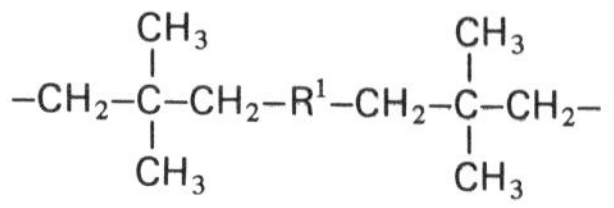

where R^1 is arylene or substituted arylene, are used to prepare thermally stable, rigid polyesters of the formula:

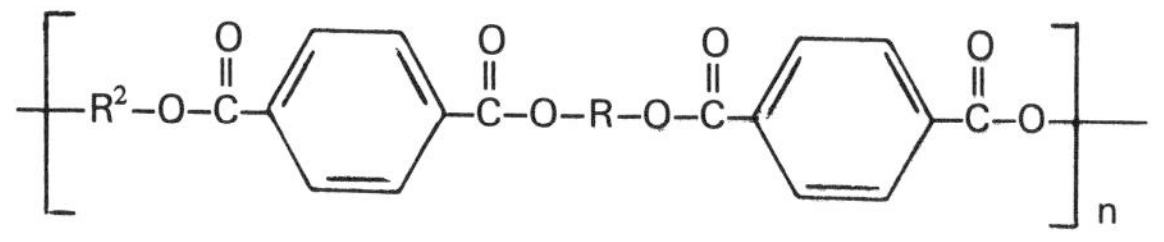

$$\left[-R^2-O-\underset{\overset{\|}{O}}{C}-\text{⟨benzene⟩}-\underset{\overset{\|}{O}}{C}-O-R-O-\underset{\overset{\|}{O}}{C}-\text{⟨benzene⟩}-\underset{\overset{\|}{O}}{C}-O-\right]_n$$

where R^2 is arylene or substituted arylene, and n is at least 10.

In the following examples, polymer melt temperature (PMT) is as defined in the preceding patent. The standard fiber test designation T/E/Mi refers to tensile strength in grams per denier, elongation in percent, and initial modulus in grams per denier.

Example 1: A mechanically stirred mixture of 15.6 g (0.15 mol) of 2,2-dimethyl-1,3-propanediol, 92.25 g (0.5 mol) of 4-carboxybenzoyl chloride and 500 ml of pyridine was refluxed under a nitrogen atmosphere. The resulting mixture was concentrated to dryness under reduced pressure, dispersed in 500 ml of distilled water, and acidified to pH 1 with concentrated hydrochloric acid. The precipitate was filtered off, washed three times with distilled water, and dried overnight at 100°C in a vacuum oven. The product was extracted for 5 days with 2 liters of hot benzene.

The extract, on cooling, yielded 48.2 g of 4,4'-[2,2-dimethyl-1,3-propanediyl)bis-(oxycarbonyl)] bis(benzoic acid) melting at 287° to 288°C. The infrared spectrum was consistent with the proposed structure and the compound was found to contain 5.13 meq of CO_2H/g which corresponded to a molecular weight of 390 for the diacid.

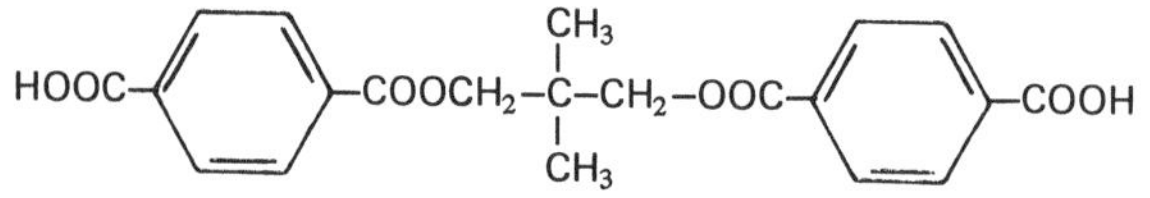

$$HOOC-\text{⟨benzene⟩}-COOCH_2-\underset{\overset{|}{CH_3}}{\overset{\overset{CH_3}{|}}{C}}-CH_2-OOC-\text{⟨benzene⟩}-COOH$$

Example 2: To a glass reactor with a nitrogen bleed and sidearm was added 21.84 g (0.04 mol) of the diacid from Example 1 and 8.53 g (0.041 mol) of methyl-hydroquinone diacetate. The resulting mixture, under a nitrogen atmosphere, was heated for 36 hours at 242°C, and for 6 hours at 275°C, followed by 3 hours at 275°C at less than 0.05 mm Hg pressure. The resulting poly[oxycarbonyl(1,4-phenylene)carbonyloxy(2,2-dimethyl-1,3-propanediyl)oxycarbonyl(1,4-phenylene)carbonyloxy(2-methyl-1,4-phenylene)] had a PMT of approximately 400°C, an inherent viscosity of 0.75, and showed low crystallinity by x-ray diffraction.

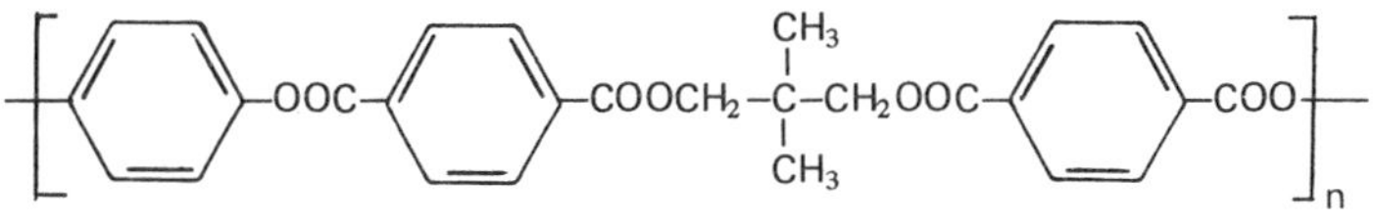

Example 3: Polymer from Example 2 was spun at a spinneret temperature of 370°C and the fiber was wound up at 250 yd/min. The fiber could not be drawn and, after being heated at 250°C for 18 hours under a nitrogen atmosphere under restrained conditions, had an orientation angle of 21°, was of medium crystallinity, had a T/E/Mi at room temperature of 7.2/2.1/310 and a T/E/Mi at 150°C of 5.8/2/248.

Polyesters from Mixtures of Bisphenols and Hydroquinone

A linear copolyester having a reduced specific viscosity of at least 0.5 is prepared by *H. Inata and S. Kawase; U.S. Patent 4,064,108; December 20, 1977; assigned to Teijin Limited, Japan* from terephthalic or isophthalic acid, a bisphenol and hydroquinone. The ester unit derived from terephthalic or isophthalic acid and hydroquinone is contained in the copolyester in a proportion of 5 to 35 mol %. The copolyester has superior crack resistance, thermal stability, transparency and chemical resistance.

The following example and comparative example illustrate the process in greater detail. In these examples, all parts are by weight. The various properties given in these examples were measured by the following methods. Reduced specific viscosity was measured at 35°C using a mixed solvent consisting of phenol and tetrachloroethane in a weight ratio of 60:40 while maintaining the concentration at 1.2 g/dl.

To measure crack resistance, an injection-molded plate-like article of the polymer, about 3.2 mm thick, about 10 mm wide and about 100 mm long, is placed on two edges apart from each other by 80 mm. Then, a load of 2 kg is exerted on the central part of the molded article, and the entire molded article is immersed in carbon tetrachloride while it is under flexural stress. The time that elapses until cracks occur in the molded article is measured, and expressed in seconds.

To measure light transmittance retention, an injection-molded plate-like article, about 3.2 mm thick, is used as a sample. The light transmittance of the sample is measured by a Poic integral spherical ultrafine turbidimeter (SEP-TU-type, a product of Nippon Seimitsu Kogaku Kabushiki Kaisha). Then, the sample is immersed in acetone at room temperature for 1 day, and then its light transmittance is measured. The light transmittance retention (%) is calculated from the light transmittance of the sample before immersion in acetone and that after immersion in acetone.

The amount (mol %) of hydroquinone copolymerized in the polymer is determined by high resolving power nuclear magnetic resonance spectroscopy.

Heat distortion temperature is measured in accordance with ASTM D-648.

Example: A reactor equipped with a stirrer was charged with 318.0 parts of diphenyl terephthalate, 22.0 parts of hydroquinone, 193.8 parts of bisphenol A and 0.071 part of stannous acetate, and they were reacted at about 280°C for 60 minutes. The phenol generated was distilled off from the reaction system. The pressure of the inside of the reaction system was reduced gradually, and in 15 minutes, the pressure was adjusted to about 0.5 mm Hg (absolute). At this reduced pressure, the reaction was carried out for another 30 minutes. During this time, the reaction product solidified. It was taken out, and pulverized to a size of about 20 mesh on a Tyler mesh. The pulverized solid product was subjected to solid-phase polymerization at about 0.5 mm Hg (absolute) for 30 minutes at 260°C, and then for 4 hours at 280°C. The resulting polymer had a reduced specific viscosity of 0.93. 19 mol % hydroquinone was polymerized.

The polymer was melted at about 370°C, and injection-molded through a die at about 100°C to form a sample plate having a size of about 3.2 x 10 x 50 mm. The resulting molded article was transparent, and when it was allowed to stand in acetone at room temperature for 1 day, it remained stable without any change. The heat distortion temperature was 172°C, light transmittance was 81%, light transmittance retention was 88%, and crack resistance was >300 seconds.

Comparative Example: Using 254.4 parts of diphenyl terephthalate, 63.6 parts of diphenyl isophthalate, 239.4 parts of bisphenol A, and 0.071 part of stannous acetate, a polymer was prepared in the same way as in the above example. The resulting polymer had a reduced specific viscosity of 0.95. An injection-molded article of this polymer obtained in the same way as in the above example was transparent. However, when it was immersed in acetone, its surface whitened in 5 minutes, and the molded article lost transparency. The heat distortion temperature was 171°C, light transmittance was 80%, light transmittance retention was 14%, and crack resistance was 40 seconds.

Esters from Hydroquinones and Bis(4-Hydroxyphenyl)Ether

J.J. Kleinschuster and T.C. Pletcher; U.S. Patent 4,066,620; January 3, 1978; assigned to E.I. DuPont de Nemours and Company disclose a class of copolyesters derived from terephthalic acid and derivatives of methyl- or chlorohydroquinone and bis(4-hydroxyphenyl)ether, the fibers and other shaped articles prepared therefrom, and the optically anisotropic copolyester melts from which these fibers can be prepared.

Measurements and Tests: *X-Ray Orientation Angle* — The orientation angle (OA) values reported herein are obtained by the procedures described in Kwolek, U.S. Patent 3,671,542, using Method Two of that patent. For fibers of this process, the arc used for orientation angle determination occurs at about 20° for 2θ value. The 2θ value is shown parenthetically after the OA value.

Inherent Viscosity — Inherent viscosity (η_{inh}) is defined by the following equation:

$$\eta_{inh} = \frac{\ln(\eta_{rel})}{C}$$

wherein (η_{rel}) represents the relative viscosity and C represents a concentration of 0.5 g of the polymer in 100 ml of solvent. The relative viscosity is determined by dividing the flow time in a capillary viscometer of a dilute solution of the polymer by the flow time for the pure solvent. The dilute solutions used herein for determining relative viscosity are of the concentration expressed by C above; flow times are determined at 30°C; the solvent is p-chlorophenol.

Fiber Tensile Properties — Filament and yarn properties are measured by the procedures shown in Morgan, U.S. Patent 3,827,998, using fibers that have been conditioned for at least 1 hour. Tenacity (T) and Modulus (Mi) are given in grams per denier. Elongation (E) is given in percent. At least three breaks are averaged.

It should be noted that different values are obtained from single filaments (filament properties) and from multifilament strands (yarn properties) of the same sample. Unless specified otherwise, all properties given herein are filament properties.

Optical Anisotropy — The melt-forming copolymers useful for fibers are considered to form anisotropic melts according to the thermooptical test (TOT) if, as a sample is heated between crossed (90°) polarizers to temperatures above its flow temperature, the intensity of the light transmitted through the resulting anisotropic melt gives a trace whose height (1) is at least twice the height of the background transmission trace on the recorder chart and is at least 0.5 cm greater than the background transmission trace, or (2) increases to such values.

Example: This example illustrates preparation of copoly[methyl-1,4-phenylene/oxybis(1,4-phenylene)terephthalate], (7/3), and an optically anisotropic melt thereof. Strong fibers with desirable properties are demonstrated.

Part (A) — In a 250 ml round-bottom, 3-necked flask equipped with a glass stirrer, bleed tube for nitrogen, and a distillation head are placed methylhydroquinone diacetate (29.1 g, 0.14 mol), bis(4-acetoxyphenyl)ether (17.2 g, 0.06 mol), and terephthalic acid (36.5 g, 0.22 mol). These stirred ingredients, under nitrogen, are heated between 265° to 280°C for about 1 hour in a Wood's metal bath; the acetic acid by-product distills out and is collected. During the next 1¼ hours the temperature is gradually increased to 320°C. Then, the nitrogen flow is halted and the reaction mixture placed under a reduced pressure for the next 35 minutes; the reaction temperature is maintained at 320°C. The heating bath is removed and the reaction system, still under vacuum, is allowed to cool. The copolymeric product is collected. It is optically anisotropic in the melt above 307°C (TOT); η_{inh} = 0.85.

Part (B) — A plug of the aboveprepared copolymer is placed in a melt spinning cell and extruded into air through a 5-hole spinneret (diameter of each hole = 0.023 cm; spinneret temperature range = 328° to 340°C; melt zone temperature 324° to 330°C). A bobbin of yarn is collected at a windup speed of 571 m/min. Filaments from this bobbin exhibit these properties: T/E/Mi/Den = 3.2/2.7/187/4.4; OA = 35° (18.8°).

A yarn sample collected from the above fiber is wound on a Fiberfrax-wrapped bobbin and is heated relaxed under nitrogen in an oven under these successive temperatures (1 hour at each temperature, oven initially 25°C) 280°C, 290°C, 300°C. The treated yarns exhibit these properties: T/E/Mi = 9.9/3.6/248; OA = 22° (19.3°).

Phenylene Terephthalate Copolymers

In a related patent, *J.R. Schaefgen; U.S. Patent 4,075,262; February 21, 1978; assigned to E.I. DuPont de Nemours and Company* describes melt spinnable co-polyesters capable of forming an anisotropic melt, the copolyester consisting essentially of units of the formula:

(1)

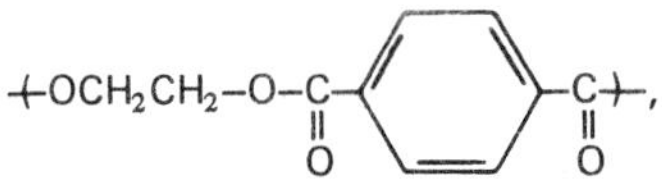

$$+OCH_2CH_2-O-\underset{\underset{O}{\|}}{C}-\text{\textless phenylene\textgreater}-\underset{\underset{O}{\|}}{C}+,$$

(2)

$$+O-Ar-O+, \text{ and}$$

(3)

$$+\underset{\underset{O}{\|}}{C}-\text{\textless phenylene\textgreater}-\underset{\underset{O}{\|}}{C}+$$

wherein units (2) and (3) are present in substantially equimolar amounts; Ar is selected from the group of chloro-, methyl-, 2,6-dichloro- or 2,6-dimethyl-1,4-phenylene or chloro-4,4'-biphenylene radicals; the copolyester containing from 15 to 70% by weight of unit (1).

Example: This example illustrates the preparation of copoly(ethylene terephthalate/chloro-1,4-phenylene terephthalate) (41/59% by weight) which exhibits melt anisotropy and forms strong fibers. Tests and abbreviations are identical to those in the preceding patent.

In a polymer tube are combined poly(ethylene terephthalate) (28.8 g, 0.15 mol, η_{inh} = 0.95), chlorohydroquinone diacetate (34.4 g, 0.15 mol), and terephthalic acid (24.9 g, 0.15 mol). The stirred ingredients, under nitrogen, are heated in a 100°C vapor bath for 15 minutes, then allowed to cool. They are then heated (under nitrogen) in a 283°C vapor bath for 1 hour 50 minutes; the by-product acetic acid is collected. The nitrogen bleed is removed and the reactants are heated at 283°C/3⅓ hours under reduced pressure of 2.0 to 0.20 mm Hg. The yield of copolymer is 55.9 g, η_{inh} = 0.78 (Method 2). The copolymer flows at 230°C and exhibits anisotropy above that temperature (TOT).

The copolymer is spun through a 5-hole spinneret at 252°C. The hole diameter is 0.018 cm, and the windup speed is 139 m/min.

A melt-spun fiber exhibits T/E/Mi = 4.1/2.4/283 and a denier/filament of 17; OA = 19° (19.1°).

Hindered Phenol Stabilizer

F.E. Carevic and A. Labriola; U.S. Patent 4,011,196; March 8, 1977; assigned to FMC Corporation describe a process for making improved stabilized fiber and film-forming polyethylene terephthalate polyester resins having greater resistance to deterioration by heat and moisture. A zinc esterification catalyst is used, and the heat stabilizer or antioxidant is a hindered phenol that is added to the monomers before they are reacted or to the reaction product or partial polymerization product thereof any time before the completion of the desired polyester polymer. The hindered phenols that may be used are represented by the following formulas:

(1)

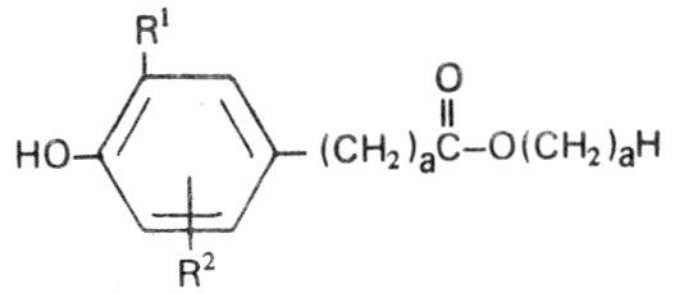

in which a has a value of zero to 18, inclusive;

(2)

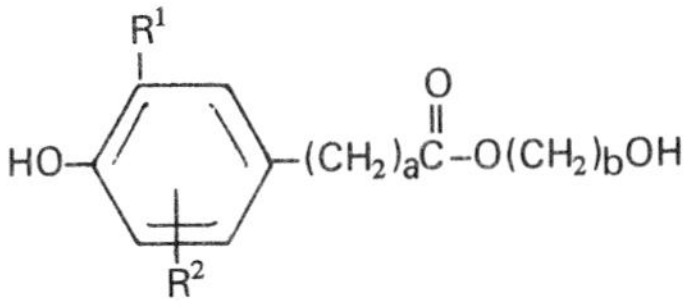

where a has a value of 0 to 18, inclusive; b has a value of 2 to 8, inclusive;

(3)

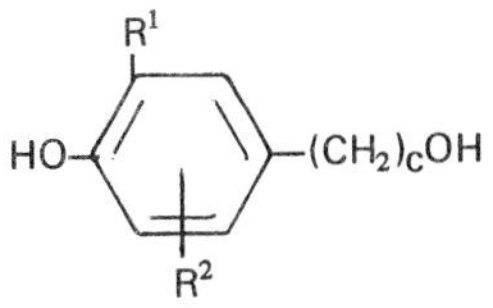

where c has a value of 1 to 18, inclusive;

(4)

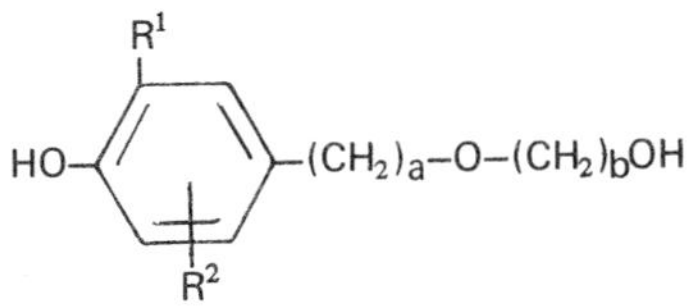

where b has a value of 2 to 8, inclusive; a has a value of 0 to 18, inclusive;

(5)

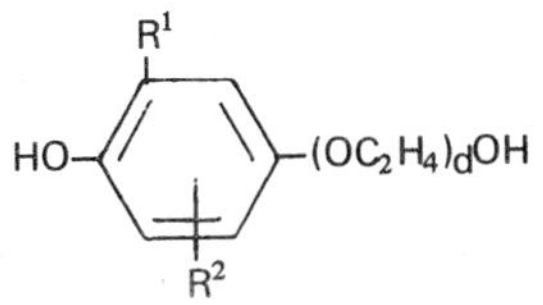

where d has a value of 1 to 10, inclusive;

(6)

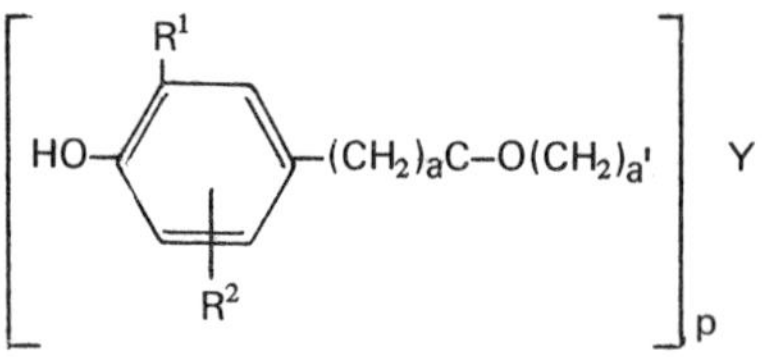

$$\left[HO-\underset{R^2}{\overset{R^1}{\bigcirc}}-(CH_2)_aC-O(CH_2)_{a'} \right]_p Y$$

where a has a value of 0 to 18, inclusive; a' has a value of 0 to 18, inclusive; p is either 2 or 4; Y may be $(CH_2)_k$; k has a value of 1 to 8; inclusive; Y may be C; Y may be $O(CH_2CH_2O)_e$, and e has a value of 1 to 10, inclusive.

R^1 is in the 3 position or with respect to the hydroxyl is in the ortho position. R^2 may be in either the 5 or 6 position or with respect to the hydroxyl may be in the other ortho position or meta position. R^1 and R^2 are each lower alkyl groups and may be isopropyl, tertiary butyl, or neopentyl.

Example: The following compounds in the amounts by weight indicated are mixed together in a reaction vessel: 100 lb dimethyl terephthalate, 67 lb ethylene glycol, and 13.6 g of zinc acetate dihydrate.

The reaction mixture is heated from room temperature up to 400°±5°F and methanol is distilled off. Then, 13.6 g of antimony trioxide is added as a condensation or polymerization catalyst and 19.6 g of triphenyl phosphite as a melt stabilizer is also added. At this point, 90.8 g of octadecyl 3-(3',5'-ditert-butyl-4'-hydroxyphenyl) propionate is added.

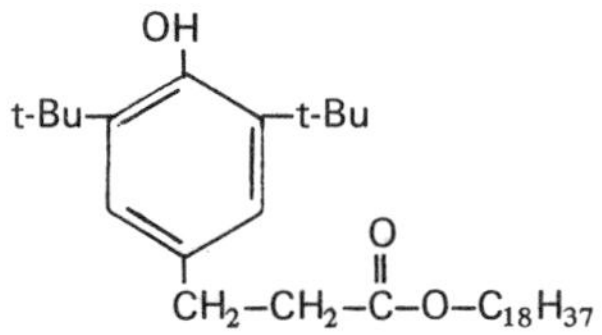

The reaction mixture is then heated up to 500°F at atmospheric pressure. Excess ethylene glycol is removed and, when the mixture reaches 510°F, a vacuum is applied and the pressure is gradually reduced over a 40-minute period to a pressure of 1.5 mm of mercury. The remaining ethylene glycol is removed under maximum vacuum, that is, a pressure of 0.5 to 0.7 mm of mercury and the temperature is increased up to 540° to 550°F. The composition is held at this temperature and this pressure until the condensation or polymerization is carried out to such an extent that the intrinsic viscosity reaches the desired amount. The intrinsic viscosity of polyester resin for fiber may be 0.6 and for film may be 0.7.

The films made of polyester resin containing the hindered phenol stabilizer added to the polyester resin before the completion of polymerization have improved resistance to deterioration by heat and moisture over films not containing the hindered phenol and films made of polyester resin to which a hindered phenol is added after the resin has reached its desired state or degree of polymerization and is present as such in the resin.

Substituted Ammonium Salts of Terephthalic Acid

G.R. Ure; U.S. Patent 4,028,307; June 7, 1977; assigned to Fiber Industries, Inc.
has discovered salts which when present in the esterification of ethylene glycol
and a dicarboxylic acid aid in substantially eliminating the production of diethylene glycol and on subsequent polymerization provide a polymer with outstanding
thermal stability properties as well as high softening point properties of the polymer product. These salts are the salts of carboxylic acids or derivatives thereof
and substituted quaternary ammonium bases wherein the substituents are selected
from the class consisting of alkyl, cycloalkyl, aralkyl, alkaryl and aryl radicals.

The preferred salts for the esterification reaction of terephthalic acid and ethylene
glycol are the salts of terephthalic acid and the substituted quaternary ammonium
bases wherein the substituents are alkyl radicals and contain from 1 to 4 carbon
atoms, at least one of the radicals and no more than two of the alkyl radicals
being substituted.

Example 1: Terephthalic acid (16 parts) is slurried in methanol (80 parts) at
0°C. A 45% solution of 2-hydroxyethyltrimethylammonium hydroxide (choline
base) in methanol (10 parts) is added to the slurried terephthalic acid. After
stirring for 30 minutes, the excess terephthalic acid is removed by filtration. The
filtrate is concentrated under reduced pressure and cooled, giving the mono-2-hy-
droxyethyltrimethylammonium salt of terephthalic acid as a precipitate which is
isolated by filtration.

Example 2: In a continuous process, 753 lb/hr of terephthalic acid and 450
lb/hr of ethylene glycol in the presence of 0.02 wt % sodium hydroxide based
on the weight of polymer were continuously added to an esterification unit maintained at a temperature of 243°C and a pressure of 40 lb/in^2 gauge. The residence time of the esterification product is sufficient to remove bis(2-hydroxy-
ethyl) terephthalate and derivative products such as low molecular weight products, hereinafter referred to as equilibrium monomer, in the same proportion as
the starting materials are fed to the esterification reactor.

The equilibrium monomer is pumped to a low polymerizer maintained at temperatures in the range from 255° to 275°C and at pressures from 20 to 100 mm
mercury. At the same time, 0.027 wt % trimethyl phosphite and 0.06 wt % antimony trioxide are added continuously to the equilibrium monomer.

The materials remain in the low polymerizer for a period of time sufficient to
provide a polymer of intrinsic viscosity in the range of about 0.2 to 0.3 dl/g.
This material is pumped to a high polymerizer at a temperature at 280°C and at
pressures of about 0.5 to 2.0 mm of mercury to provide further polymerization
to an intrinsic viscosity in excess of 0.4 dl/g and in the range from 0.4 to 1.0
dl/g. In this case, an intrinsic viscosity in the range of 0.9 dl/g is obtained. The
polymer is extruded into filaments, drawn, collected and used to produce textile
articles or industrial products such as tire cord. During beaming of the yarn produced from the polymer of this example, the number of major defects obtained
per 10^6 yd is 19.1.

Example 3: Example 2 is repeated, except that the sodium hydroxide is replaced
with 0.02 wt % of 2-hydroxyethyltrimethylammonium hydroxide (choline base)
in ethylene glycol. Both the sodium hydroxide and the 2-hydroxyethyltrimethyl-

ammonium hydroxide (choline) are introduced into the esterifier feed line as soon as possible after the addition of the ethylene glycol to the terephthalic acid. Thus, ample time is allowed for the reaction of 2-hydroxyethyltrimethylammonium hydroxide with terephthalic acid to form a salt.

During the spinning (extruding) of the polymer of this example, lower pack pressure and fewer failing filaments are obtained, compared with Example 2. Upon beaming yarn produced from the polymer of this example, 11.7 major defects per 10^6 yd are obtained. This represents a reduction in major defects of 39%, a significant improvement. This improvement was obtained without altering yarn or cord properties.

Polyester Yarn with Stable Internal Structure

An improved high performance polyester (at least 85 mol % polyethylene terephthalate) multifilament yarn is provided by *H.L. Davis, M.L. Jaffe, H.L. LaNieve, III, and E.J. Powers; U.S. Patent 4,101,525; July 18, 1978; assigned to Celanese Corporation*. The multifilament yarn possesses a high strength (at least 7.5 g per denier) and an unusually stable internal structure which renders it particularly suited for use in industrial applications at elevated temperatures.

The subject multifilamentary material exhibits unusually low shrinkage and hysteresis characteristics (i.e., work loss characteristics) coupled with the high strength characteristics normally associated with polyester industrial yarns. Accordingly, when utilized in the formation of a tire cord and embedded in a rubber matrix, a highly stable tire may be formed which generates significantly less heat upon flexing.

The stable internal structure is evidenced by the following combination of characteristics:

(a) A birefringence value of +0.160 to +0.189;

(b) A stability index value of 6 to 45 obtained by taking the reciprocal of the product resulting from multiplying the shrinkage at 175°C in air measured in percent times the work loss at 150°C when cycled between a stress of 0.6 g per denier and 0.05 g per denier measured at a constant strain rate of 0.5 in/min in inch-pounds on a 10 inch length of yarn normalized to that of a multifilament yarn of 1,000 total denier; and

(c) A tensile index value greater than 825 measured at 25°C and obtained by multiplying the tenacity expressed in grams per denier times the initial modulus expressed in grams per denier;

(d) A crystallinity of 45 to 55%;

(e) A crystalline orientation function of at least 0.97;

(f) An amorphous orientation function of 0.37 to 0.60;

(g) A shrinkage of less than 8.5% in air at 175°C;

(h) An initial modulus of at least 110 g per denier at 25°C;

(i) A tenacity of at least 7.5 g per denier at 25°C; and

(j) A work loss of 0.004 to 0.02 in-lb when cycled between a stress of 0.6 and 0.05 g per denier at 150°C measured at a constant strain rate of 0.5 in/min on a 10 inch length of yarn normalized to that of a multifilament yarn of 1,000 total denier.

The process is described with reference to the figure. As illustrated in Figure 1.6a, the polyethylene terephthalate polymer while in particulate form was placed in hopper 1 and was advanced toward spinneret 2 by the aid of screw conveyor 4. Heater 6 caused the polyethylene terephthalate particles to melt to form a homogeneous phase which was further advanced toward spinneret 2 by the aid of pump 8. The spinneret 2 had a standard conical entrance and a ring of extrusion holes, each having a diameter of 10 mils.

Figure 1.6: Formation of Polyester Filament Yarn

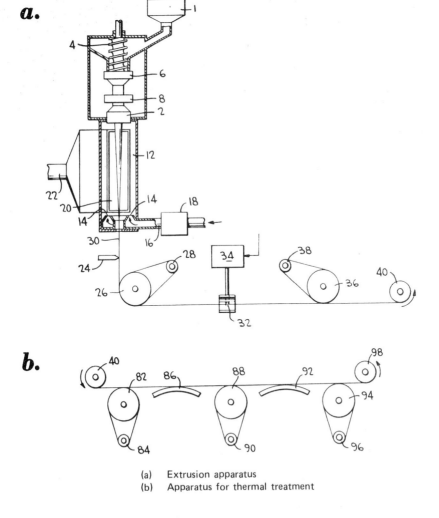

(a) Extrusion apparatus
(b) Apparatus for thermal treatment

Source: U.S. Patent 4,101,525

The resulting extruded polyethylene terephthalate passed directly from the spinneret 2 through solidification zone 12. The solidification zone 12 had a length of 6' and was vertically disposed. Air at 10°C was continuously introduced into solidification zone 12 at 14. The air was supplied via conduit 16 and fan 18. The air was continuously withdrawn from solidification zone 12 through elongated conduit 20 vertically disposed in communication with the wall of solidification zone 12, and from there was continuously withdrawn through conduit 22. While passing through the solidification zone, the extruded polyethylene terephthalate was uniformly quenched and was transformed into a continuous length of as-spun polyethylene terephthalate yarn. The polymeric material was first transformed from a molten to a semisolid consistency, and then from a semisolid consistency to a solid consistency while passing through solidification zone 12.

After leaving the exit end of solidification zone 12, the filamentary material lightly contacted lubricant applicator 24 and was continuously conveyed to a first stress isolation device consisting of a pair of skewed rolls 26 and 28, and was wrapped about these in four turns. The filamentary material was passed from skewed rolls 26 and 28 to a first draw zone consisting of a steam jet 32 through which steam tangentially was sprayed upon the moving filamentary material from a single orifice. High pressure steam at 25 psig initially was supplied to superheater 34 where it was heated to 250°C, and then was conveyed to steam jet 32. The filamentary material was raised to a temperature of about 85°C when contacted by the steam and drawn in the first draw zone. The longitudinal tension sufficient to accomplish drawing in the first draw zone was created by regulating the speed of a second pair of skewed rolls 36 and 38 about which the filamentary material was wrapped in four turns. The filamentary material was next packaged at 40.

Figure 1.6b illustrates the equipment arrangement wherein the subsequent thermal treatment was carried out. The resulting package 40 subsequently was unwound and passed in four turns about skewed rolls 82 and 84 which served as a stress isolation device. From skewed rolls 82 and 84, the filamentary material was passed in sliding contact with hot shoe 86 having a length of 24" which served as a second draw zone and was maintained under longitudinal tension exerted by skewed rolls 88 and 90 about which the filamentary material was wrapped in four turns. Hot shoe 86 was maintained at a temperature above that experienced by the filamentary material in the first draw zone.

The filamentary material, after being conveyed from skewed rolls 88 and 90, was passed in sliding contact with hot shoe 92 having a length of 24" which served as the zone wherein the final portion of the thermal treatment was carried out. Skewed rolls 94 and 96 maintained a longitudinal tension upon the filamentary material as it passed over hot shoe 92. The filamentary material assumed substantially the same temperature as hot shoes 86 and 92 while in sliding contact with the same. The differential scanning calorimeter peak melting temperature of the filamentary material was 260°C in each example, and no filament coalescence occurred during the thermal treatment illustrated in Figure 1.6.

End Capped Polyalkylene Carbonates

According to *D.D. Dixon, M.E. Ford, and G.J. Mantell; U.S. Patent 4,066,630; January 3, 1978; assigned to Air Products and Chemicals, Inc.*, polycarbonates, of the type formed by reacting an aliphatic or cycloaliphatic 1,2-monoepoxide with carbon dioxide and having substantially alternating units of epoxide and carbon dioxide, are improved in thermal stability by reacting the free hydroxyl groups with a hydroxyl-reactive organic compound.

Classes of hydroxyl reactive organic compounds which can be used for end capping the hydroxyl groups include: alkylating agents, e.g., organohalides such as methyl chloride, bromide, and iodide, ethyl bromide, allyl chloride, and chlorocyclohexane; carboxylic acid halides such as lower alkyl C_{1-12} acid halides, e.g., acetyl chloride, phosgene, propionyl chloride, benzoyl chloride, cyclohexanoyl chloride; acids such as fumaric, maleic, acetic, malonic, succinic and their anhydrides and the like; epoxides such as ethylene oxide, propylene oxide, styrene oxide, cyclohexane epoxide, etc; unsaturated hydrocarbyls, e.g., divinyl sulfone, acrylonitrile, lower alkyl (C_{1-6}) esters of acrylic acid, e.g., methyl methacrylate, ethyl acrylate, 2-ethylhexyl acrylate, and the corresponding acids, e.g., acrylic and methacrylic acid, methylvinyl ketone, ethylvinyl ketone, phenylvinyl ketone, acrylyl chloride, methacrylyl chloride, and the like; reactive methylol compounds, e.g., those generally having conjugated unsaturation with respect to the methylol group, e.g., N-methylolacrylamide and trimethylol phenol, and methylol melamines; and isocyanates such as toluene diisocyanate, phenyl isocyanate, phenyl diisocyanate, hexamethylene isocyanate, hexamethylene diisocyanate, isocyanato diphenylmethane, methyl isocyanate, and so forth, urea and urethanes.

Example: A 7.5 g sample of a polyethylene carbonate (PEC) formed by the procedure of Example 1 in U.S. Patent 3,900,424 having alternating units of ethylene oxide and carbon dioxide, a molecular weight of 100,000 and a melting point of about 190° to 200°C was dissolved in 100 ml chloroform. The polyethylene carbonate had free hydroxyl groups as evidenced by the presence of frequencies in the infrared spectrum at 3,480 cm^{-1} and 3,640 cm^{-1}. The free hydroxyl groups, assuming 1.5 x 10^{-4} mol hydroxyl group in the polyethylene carbonate, were end capped by adding 2.85 g (0.05 mol) methyl isocyanate to the polyethylene carbonate solution and then stirring at ambient temperature (20° to 25°C) for about 40 hours. The reaction medium then was added slowly to methanol and a precipitate formed which was removed by filtration. Residual solvent in the precipitate was removed by placing the polymer in a vacuum. Infrared analysis of the resulting polymer showed that hydroxyl groups were reacted and capped with methyl urethane units.

Untreated polyethylene and polypropylene carbonate, i.e., those having terminal hydroxyl groups and end capped polyethylene and polypropylene carbonate resins, were evaluated for thermal stability by heating a 0.005 g sample of the resin at a rate of 2½°/min from a temperature of 120°C to the decomposition temperature as indicated by a weight loss of 5% based on the weight of the sample. The atmospheric environments used in the evaluation of the samples were air and nitrogen.

The results show that, in each case where an oxygen-carbon bond was formed and the active hydrogen atom replaced, the decomposition temperature of the polycarbonate increased substantially both in air and in nitrogen. The results also show that the different hydroxyl end capping agents in themselves had little influence on the decomposition temperature.

Thermally Stable Naphthalate Polyesters

According to *I. Hamana, Y. Fujiwara, and S. Kumakawa; U.S. Patent 4,001,479; January 4, 1977; assigned to Teijin Limited, Japan*, electrically-insulating material can be produced by heat-treating a fabric consisting mainly of naphthalate polyester fibers. It was found that by heat-treating the fibrous cloth under conditions which meet the following two equations (where T is the heat treating temperature in °C and t is the heat treating time in seconds) there can be obtained a cloth of naphthalate polyester fibers which have superior heat resistance and mechanical strength, and also flatness, dimensional stability against heat and low shrinkage, and which have uniform texture and are especially suitable as electric insulating materials.

$$(1) \qquad T - 200 \geqslant 70e^{-2\log t}$$

$$(2) \qquad T - 200 \leqslant [1 - e^{-2(4-\log t)}]$$

Example 1: Polyethylene-2,6-naphthalate (usually prepared by reacting a naphthalene-2,6-dicarboxylic acid with ethylene glycol) having an intrinsic viscosity of 0.65 was melt-spun at 320°C through a spinneret having circular orifices with a diameter of 0.46 mm, and wound up at a rate of 800 m/min. A heating cell, 20 cm long, beneath the spinneret, kept the atmosphere under the spinneret at 315°C. The resulting 200 denier per 24 filament undrawn yarn having a birefringence of 0.0120 was wrapped through eight turns around a heated feed roller with a diameter of 90 mm, preheated on the roll, and then drawn to 4.0 times, followed by drawing and heat-treatment by a slit heater heated at 250°C. The resulting yarn was wound up at a rate of 530 m/min.

Example 2: The procedure of Example 1 was repeated except that the temperature of the feed roller was changed to 130°C to form pirns with 50 denier.

Thirty-nine pirns were doubled, and subjected to a weaving process. In a warp step, bobbin delivery, twisting, roller sizing, and drawing in were performed, and in a weft step, bobbin delivery, twisting and pirn winding were performed, thereby to form a woven fabric 101 cm wide. The warp and weft densities at this time were 72 x 31/inch. The woven fabric obtained was boiled in loop in warm water held at 90° to 100°C, and the amount of the residual size was reduced to less than 0.2%. Then, the fabric was roller dried at 120°C.

The woven fabric was heat-set in a pin tenter 15 m long at various temperatures and speeds, and there was obtained a naphthalate polyester woven cloth 100 cm wide suitable for an electrical insulating material. The warp and weft densities at this time were 74 x 32.5/inch.

The fabric obtained in a run in which the heat-treatment time was shorter than that defined in equation (1) had low tenacity, large shrinkage, poor flatness and an R value of 0.02, and proved unsuitable for an electrical insulating material. In contrast, in runs which satisfy the equations (1) and (2), the fabric obtained had good heat resistance, tenacity retention, low shrinkage, and good flatness with an R value of 0.50, and proved superior as an electrically insulating material. R value indicates the diffraction intensity ratio.

In a run in which the heat-treating time was longer than that specified in equation (2), the properties of the fabric obtained were not inferior, but it had poor processability. Furthermore, such a long period of heat-treatment is not economically feasible.

Naphthalates with Thermal and Chemical Stability

T. Kuratsuji, S. Kawase, and T. Shima; U.S. Patent 4,060,516; November 29, 1977; assigned to Teijin Limited, Japan provide naphthalate polyester filaments, yarns or fibers having superior chemical properties along with superior mechanical properties and thermal stability.

Naphthalate polyester filaments, yarns or fibers consist essentially of a naphthalate polyester which has an intrinsic viscosity of 0.3 to 3.5 and a softening point of at least 200°C and in which at least 89 mol % of the total recurring units consist of units of the following formula:

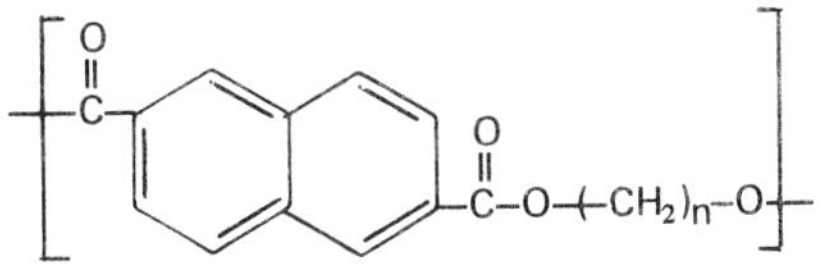

wherein n is 4 or 6, and having at least one diffraction peak at a Bragg Scattering angle $2\theta = 16.3°$ to $16.7°$, and/or $2\theta = 25.3°$ to $25.8°$ in their x-ray diffraction.

Example: Poly(tetramethylene-2,6-naphthalate) having an intrinsic viscosity of 0.86 was melted at 280°C and then extruded into the air through a spinneret having one spinning orifice of circular shape. The filament was immediately led into water kept at 0°C. The solidified filament was wound up, then drawn at a ratio of 4.6 in a bath of ethylene glycol kept at 70°C and heat-treated at 180°C at constant length to form a bristle-like filament having an intrinsic viscosity of 0.81 and a denier size of 610.

The properties of this bristle-like filament and those of a commercially available polyethylene terephthalate bristle (T-PRN of Hoechst AG, 600 denier) were measured, and the results are shown in the table below.

	Naphthalate Bristle	Commercially Available Bristle
Normal strength, g/den	5.9	4.3
Normal elongation, %	21	49
Knot strength, g/den	3.5	3.7
Knot elongation, %	12	30
Wet heat resistance, days	30	8
Resistance to oxidation, %	100	88
Alkali resistance, %	92	40
Shrinkage in boiling water, %	1.0	2.4

A paper-making canvas was produced using each of the bristles shown above. No abnormal phenomenon was seen in the canvas produced from the naphthalate bristle after continuous use for 2 months, whereas in the canvas produced from the commercially available bristles, several bristles were seen to break on the 20th day.

Hydrolytically Stable Polyester Resins

A.B. Conciatori and R.W. Stackman; U.S. Patent 4,031,063; June 21, 1977; assigned to Celanese Corporation describe high molecular weight polyester resin compositions which are uniquely stable under hydrolytic conditions. The polyester resin compositions are produced by the condensation of a dihydric polyol with a sterically-hindered neo-acid derivative such as $\alpha,\alpha,\alpha',\alpha'$-tetramethyl-p-phenylenediacetic acid.

Example 1: *Preparation of $\alpha,\alpha,\alpha',\alpha'$-Tetramethyl-p-Phenylenediacetyl Chloride* — To a 250 ml, three-necked flask equipped with stirrer, reflux condenser with drying tube, and nitrogen inlet is added 10 g $\alpha,\alpha,\alpha',\alpha'$-tetramethyl-p-phenylenediacetic acid and 50 g thionyl chloride. The mixture is heated at reflux for a period of 8 hours. The excess thionyl chloride is removed by distillation and the residue is distilled under reduced pressure to yield $\alpha,\alpha,\alpha',\alpha'$-tetramethyl-p-phenylenediacetyl chloride, BP 135°C at 5 mm Hg pressure.

Example 2: *Preparation of Hydroxyl-Terminated Polyester from $\alpha,\alpha,\alpha',\alpha$-Tetramethyl-p-Phenylenediacetyl Chloride and Ethylene Glycol* — To a 100 ml, three-necked flask equipped with a nitrogen inlet tube, stirrer and exit tube is added 108.5 g (0.5 mol) $\alpha,\alpha,\alpha',\alpha'$-tetramethyl-p-phenylenediacetyl chloride and 37.2 g (0.6 mol) ethylene glycol. The mixture is stirred rapidly as the exothermic reaction evolves hydrogen chloride. As the mixture becomes solid, heat is applied in order to keep the reaction mixture liquid. The temperature is held at about 150°C until the hydrogen chloride evolution is complete (2 hours). At the end of this period, the reaction mixture is cooled and a hard white solid is recovered which is hydroxyl-terminated poly(ethylene-$\alpha,\alpha,\alpha',\alpha'$-tetramethyl-p-phenylenediacetate).

Example 3: *Preparation of $\alpha,\alpha,\alpha',\alpha'$-Tetramethylpimeloyl Chloride* — To a 500 ml, one-necked flask is added 100 g $\alpha,\alpha,\alpha',\alpha'$-tetramethylpimelic acid and 250 g thionyl chloride. The mixture is refluxed for 4 hours during which time hydrogen chloride and sulfur dioxide are evolved. At the end of this period, the excess thionyl chloride is removed by distillation and the residue recrystallized twice from cyclohexane to yield pure $\alpha,\alpha,\alpha',\alpha'$-tetramethylpimeloyl chloride.

Example 4: *Preparation of Acid Chloride-Terminated Polyester from $\alpha,\alpha,\alpha',\alpha'$-Tetramethylpimeloyl Chloride and 1,12-Dodecanediol* — To a 250 ml, three-necked flask equipped with a nitrogen inlet tube, stirrer and exit tube is added 151.8 g (0.6 mol) $\alpha,\alpha,\alpha',\alpha'$-tetramethylpimeloyl chloride and 100 g (0.5 mol) 1,12-dodecanediol. The mixture is stirred rapidly, and the temperature is slowly increased to about 120°C in order to maintain hydrogen chloride evolution.

After the evolution of hydrogen chloride has ceased, the reaction mixture is cooled to room temperature. A pale yellow viscous material is recovered which is acid chloride-terminated poly(1,12-dodecane-$\alpha,\alpha,\alpha',\alpha'$-tetramethylpimelate).

Example 5: *Preparation of Polyester Block Copolymer* — The products from Example 2 [hydroxy-terminated poly(ethylene-$\alpha,\alpha,\alpha',\alpha'$-tetramethyl-p-phenylenediacetate)] and from Example 4 [acid chloride-terminated poly(1,12-dodecane-$\alpha,\alpha,\alpha',\alpha'$-tetramethylpimelate)] are mixed together and heated to a melted mixture. The temperature is slowly increased to about 250°C while a nitrogen purge is maintained and hydrogen chloride is evolved. At the end of about 3 hours at 250°C, the hydrogen chloride evolution has ceased. The polymer melt is then cooled.

The recovered polymer is a pale yellow solid which yields flexible films when compression molded. The polyester block copolymer has an inherent viscosity of 0.64 dl/g (measured as a 0.1% solution in a mixture of 10 parts phenol and 7 parts trichlorophenol).

FLAME-RESISTANT POLYMERS

With the growth in the use of synthetic materials in the textile, construction, automobile, household appliance and aircraft industries, there has been increasing concern for the potential and real danger that results from the incidence of damage to property and personal injury due to the high flammability of these synthetic materials. Concern for public safety has prompted several Government agencies to make inquiries and to propose and enact legislation to impose stricter flammability standards for these synthetic materials. For example, children's sleepwear must meet rigid flammability standards. Garments which can pass this test usually have an LOI (Limiting Oxygen Index) of 26 or more.

In response to Government pressure, the makers of synthetic materials are increasing their efforts to impart flame-retardant properties to their products. The manufacturers have in the past used additives containing phosphorus, nitrogen or halogen compounds, the compounds being physically admixed, baked on or affixed to the synthetic materials to impart fire retardancy. However, such additives tend to impart additional and often undesirable properties such as a decrease in strength, and increase in stiffness or an increase in weight of the synthetic materials. Further, it has been observed that these additives may wash off during home laundering in the case of textile synthetics and may be incompatible in the case of synthetic materials used in molding applications.

Another method for imparting flame retardancy is to produce a synthetic material incorporating a flame-retarding agent whereby the flame-retarding agent is made an integral part of the chemical structure of the synthetic material. One example of this method is the use of dibromopentaerythritol incorporated into the chemical structure of a polyester to impart flame retardancy. However, such synthetic materials tend to be thermally unstable leading to undesirable color formation during preparation. Accordingly, there is a need for a method of producing flame-retardant polyesters.

Poly(Cyclohexanedimethylene Dibromoterephthalate)

J.P. Nelson; U.S. Patent 4,028,308; June 7, 1977; assigned to Standard Oil Company provides a fire-retardant polyester having good mechanical fiber properties, consisting essentially of a dicarboxylic acid component and 1,4-cyclohexanedimethanol component wherein the dicarboxylic acid component comprises from 15 to 100 mol % 2,5-dibromoterephthalic acid moieties and correspondingly 85 to 0 mol % terephthalic acid moieties. The polyesters have an LOI of at least 26 and have excellent mechanical fiber properties. When the 2,5-dibromoterephthalic acid moiety is between about 55 to 100 mol % and the terephthalic acid moiety is correspondingly between about 45 to 0 mol %, the polyester fiber can be used in fire-retardant hydrophilic yarn blends. The homopolymer has maximum fire-retardancy and highest melting point. Other diols should be avoided to obtain maximum fiber properties.

The polyesters can be produced by reacting 1,4-cyclohexanedimethanol with the free acids or dimethyl esters of the appropriate acids at up to about 290°C.

Example: A mixture of 17.6 g of dimethyl 2,5-dibromoterephthalate and 15.86 g of trans-1,4-cyclohexanedimethanol were combined in a small glass reactor. The reactor head was constructed in such a way as to allow a nitrogen purge to be introduced at the bottom of the reactor and a vacuum maintained. The reactor assembly was suspended in an electrically heated oil bath. The bath temperature was increased to 180° to 200°C and held for 2 hours. When the charge was molten, 50 μl of tetraisopropyl titanate catalyst, dissolved in 1 cc of n-butanol, was injected through the nitrogen inlet system. The temperature was then raised to 250°C for 20 minutes, and then to 280° to 290°C for 1.6 hours. After this time, the reactor was cooled. A light yellow opaque polymer resulted. The polymer was somewhat brittle, but was self-extinguishing.

Analysis revealed that the polymer contained 29.8% bromine, compared with a calculated value of 36.1%. This material was then dissolved in hot phenol/tetra-chloroethane, filtered to remove gel, and precipitated with ether. The resulting white, crystalline polyester had an intrinsic viscosity of 0.30 and a bromine content of 35.57% compared to a calculated value of 36.1% Br. Differential thermal analysis showed a polymer Tm of 175 and 240°C, and a Tg of 100°C.

The polymeric melting point, Tm, is determined by a thermoanalysis method and is defined as that point at which the polyester changes from a plastic consistency to molten liquid. The glass transition point, Tg, is defined as that temperature where the polyester changes from glassy to brittle, to leathery, flexible, plastic, but not flowing.

The polymer melted in the hot stage at 258° to 261°C. Percent crystallinity as determined by x-ray diffraction was found to be 36%.

The polyester can be spun, drawn and textured into fibers in the same manner as polyethylene terephthalate. The polyester can also be blended with hydrophilic fibers in the same manner as polyethylene terephthalate.

Addition of a Halogenated Diphenylether plus a Diepoxide

According to *T. Aoyama, H. Okasaka, and H. Kodama; U.S. Patent 4,010,219; March 1, 1977; assigned to Toray Industries, Inc., Japan*, a flame-retardant polybutyleneterephthalate composition, having excellent flame resistance with reduced tendency to drip during combustion, is provided by adding a halogenated diphenylether and a diepoxide compound to a polyester which contains at least 80 mol % of terephthalic acid based on the total mols of dicarboxylic acid, and at least 80 mol % 1,4-butanediol based on the total mols of diol. Decabromodiphenyl is an optional additive.

Example 1: Polybutylene terephthalate having an intrinsic viscosity of 1.20 was mixed with decabromodiphenyl, hexabromodiphenylether, antimony trioxide and

Epikote 815 (a bisphenol A diglycidylether, Shell Chem. Co., Ltd.) using a Henschel Mixer, and extruded at 240°C into pellets using a 65 mm diameter extruder. These pellets were dried at 130°C for 5 hours in a vacuum and subsequently molded into about a 1 mm thick press sheet on a hot plate controlled at 250°C. Dumbbell-type tensile test specimens were punched out from this press sheet, and tested for tensile elongation at break. Components in the blend were: 100 phr polybutyleneterephthalate, 10 phr decabromodiphenyl + hexabromodiphenyl-ether, 5 phr Sb_2O_3, and 0 to 3.0 phr Epikote 815. The chemical structure of Epikote 815 is given below, where n is normally zero.

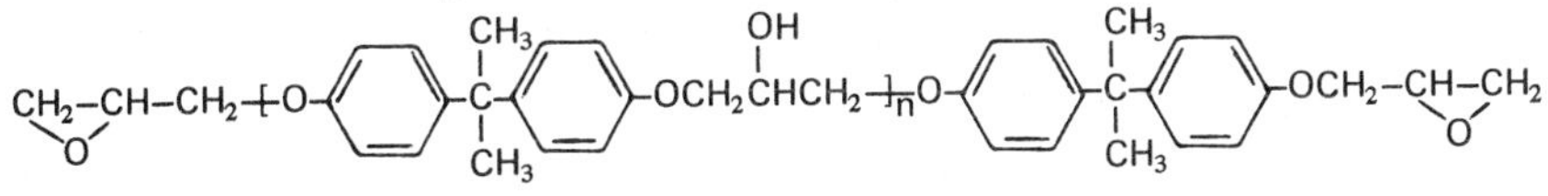

Epikote 815 was shown to improve tensile elongation at break of a flame-resistant polybutyleneterephthalate excellently in a large extent of decabromodiphenyl (DBB)/hexabromodiphenylether (HBE) compositions, particularly with more than 30 wt % HBE content. While using more than 70 weight % DBB based on the total weight of the flame retardants, poor tensile elongation is obtained.

Example 2: The pellets obtained in Example 1, where 0.5 wt % of diepoxy compounds were added, were injection molded into 5" x ½" x ⅛" flammability test bars using an in-line screw-type injection molding machine at 250°C on the cylinder and nozzle and 40°C on the mold.

Underwriters Laboratory's No. 94 vertical flame resistance test method showed that flame resistance was excellent, with extinguishing times of no greater than 2 seconds. When more than 70 wt % DBB based on the total weight of the flame retardants was used, many spots in brown or black colors were formed in the shaped articles and the surface appearance was rough and poor.

Incorporation of 2-Methyl-2,5-Dioxo-1,2-Oxaphospholane

U. Bollert, E. Lohmar, and A. Ohorodnik; U.S. Patent 4,033,936; July 5, 1977; assigned to Hoechst AG, Germany disclose an improvement in the process of U.S. Patent 3,941,752, according to which flame-retarding linear polyesters are manufactured by reaction of dicarboxylic acids capable of forming high molecular weight linear polyesters or the esters thereof with lower aliphatic alcohols, with diols, and phosphorus compounds of the formula:

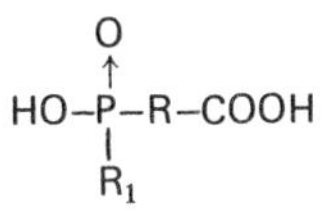

$$HO-\underset{\underset{R_1}{|}}{\overset{\overset{O}{\uparrow}}{P}}-R-COOH$$

or the esters thereof with lower aliphatic alcohols or also the cyclic anhydrides thereof. In the improvement, the cyclic anhydrides are used after having been dissolved previously at elevated temperature in a diol suitable for the polyester manufacture, preferably in the diol used as main component. The symbols of the formula have the following meanings: R = saturated open-chain or cyclic alkylene, arylene or aralkylene, R_1 = alkyl having up to 6 carbon atoms, aryl or aralkyl; R as well as R_1 optionally contains 1 or more hetero atoms, preferably F, Cl, Br, O or S. By dissolving the cyclic carboxyphosphinic acid anhydrides in a diol before adding them to the polyester manufacturing batch, the formerly necessary comminution of the hygroscopic cyclic anhydrides with absolute exclusion of moisture may be omitted, and, furthermore, it is possible to expose the anhydrides to the elevated polycondensation temperatures for a short time only.

For example, when the cyclic anhydride of 2-carboxyethyl-methylphosphinic acid, i.e., 2-methyl-2,5-dioxo-1,2-oxaphospholane, is dissolved in ethyleneglycol, the following reaction occurs:

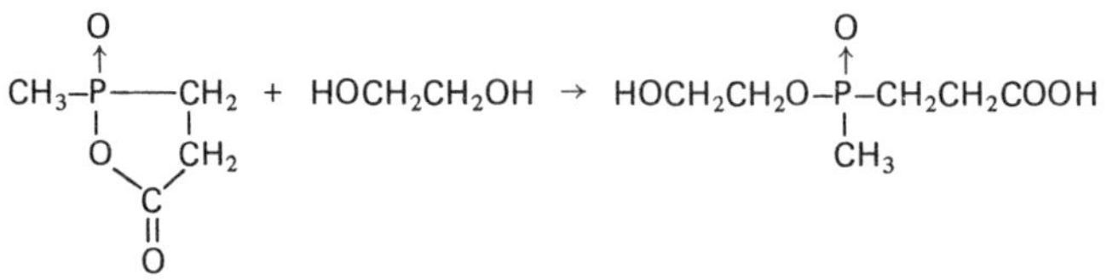

$$CH_3-\overset{\overset{O}{\uparrow}}{\underset{\underset{\overset{\displaystyle C}{\underset{\displaystyle O}{\|}}}{O}}{P}}\!\!\!\!\underset{\underset{CH_2}{}}{\overset{}{-\!\!-\!\!-CH_2}} + HOCH_2CH_2OH \;\rightarrow\; HOCH_2CH_2O-\overset{\overset{O}{\uparrow}}{\underset{\underset{CH_3}{}}{P}}-CH_2CH_2COOH$$

which is a preferred embodiment of the improved process.

Example: *Preparation of a Solution of 2-Methyl-2,5-Dioxo-1,2-Oxaphospholane in Ethyleneglycol* — 10 kg of ethyleneglycol are introduced into a vessel provided with agitator and a cooling jacket, the vessel being series-connected to the purification-distillation step of the oxaphospholane obtained according to V.K. Chajrullin et al, *Zh. Obshch. Khim.* 37 (1967), No. 3, pp 710–714. From the receiver, 10 portions of 870 ml each (= 1 kg) of the oxaphospholane cooled to 120°C are fed to the above vessel after having passed a calibrated intermediate vessel. During the feeding operation, a temperature of 120°C is maintained in the vessel provided with agitator. After having added the 10th and last portion, the mixture is either processed directly or cooled to room temperature and stored until further use.

Manufacture of the Polyester — 1,000 g of dimethyl terephthalate are transesterified with 720 ml of ethyleneglycol in the presence of 230 mg $Mn(OCOCH_3)_2 \cdot 4H_2O$ as catalyst, the transesterification taking place under nitrogen at temperatures of from 170° to 220°C. After completion of the separation of methanol, 200 g of the above oxaphospholane solution are added at 220°C.

After having added 350 mg of Sb_2O_3, the reaction vessel is further heated and, simultaneously, evacuated slowly to a pressure of 1 mm Hg at 250°C (internal temperature). Polycondensation is carried out at 0.2 mm Hg and at 275°C until a relative viscosity (measured as a 1% solution in dichloroacetic acid at 25°C) of 1.85 is attained. The product has a melting point of 240° to 244°C, and a phosphorus content of 1.98%.

The polycondensate is spun from the melt to filaments and the filaments are drawn. The filaments are self-extinguishing after removal of a Bunsen burner flame, and they are readily dyeable by means of disperse dyestuffs.

Polyesters Containing Bromine and Phosphorus

Polyesters containing bromine and phosphorus are provided by *K. Moedritzer; U.S. Patent 4,087,408; May 2, 1978; assigned to Monsanto Company* by the copolymerization of a ternary system based upon an alcoholic component and an acidic component. The third component is a bromo bisphenol alkyl or aryl phosphine oxide

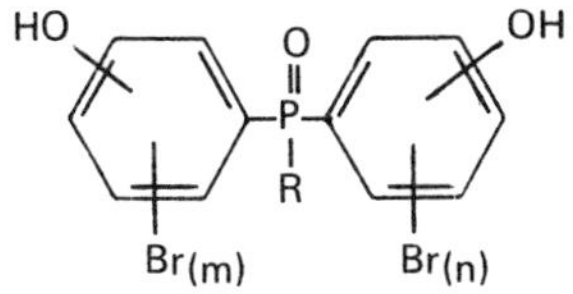

The hydroxyl groups are present in either the para or meta positions, m plus n is from 1 to 8, preferably 1 to 4 and R is an alkyl radical of 1 to 10 carbon atoms, or an aryl radical of 6 to 10 carbon atoms. The polyesters which are obtained have flame-retardant properties.

Example 1: A Grignard compound is prepared from 935 g (5 mol) of p-bromo-anisol and 116.5 g (5 g atoms) of magnesium in tetrahydrofuran. To this solution is added 234 g (2 mols) of methylphosphonous dichloride in 500 ml of tetra-hydrofuran. After completion of the addition, the mixture is refluxed for 2 hours and then hydrolyzed with dilute HCl. The mixture is extracted with $CHCl_3$ to yield 682 g of $(CH_3OC_6H_4)_2PCH_3$. The latter is oxidized to the corresponding phosphine oxide by dissolution in water and addition of 285 ml of 30% H_2O_2.

The water-insoluble product is separated and distilled in vacuo to give 361 g (1.3 mol) of $(CH_3OC_6H_4)_2P(O)CH_3$ in 65% yield. The phosphine oxide (252 g, 0.41 mol) is demethylated to the corresponding bisphenol by reacting it with 330 g (1.3 mol) of borontribromide at –70°C in methylene chloride. Work up of the reaction solution gives 140 g (0.56 mol) of bisphenol methyl phosphine oxide, MP 254°C. A quantity of 74.5 g (0.3 mol) of bisphenol methyl phosphine oxide in 230 ml butanol and 90 ml water is brominated by the addition of 192 g (1.2 mol) of bromine to give 150 g (0.267 mol) of the tetrabromobisphenol methylphosphine oxide, MP 310°C.

Example 2: In order to show the improvement brought about by the use of bromine-substituted compounds in comparison to the same diol without the bro-mine-substitution, the following two diols are used in the preparation of poly-esters

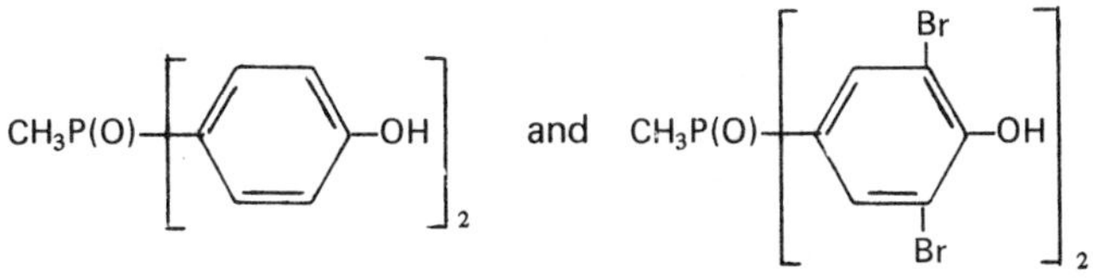

In separate reaction vessels, the aforesaid diols are ethoxylated by chloroethanol to give the dihydroxyethoxylated product. Each of the aforesaid reactions is carried out at about 25° to 150°C.

The formation of the ester employs the dihydroxyethoxylated compounds, re-spectively, (although similar results are obtained with the acetylated derivatives to permit ester interchange).

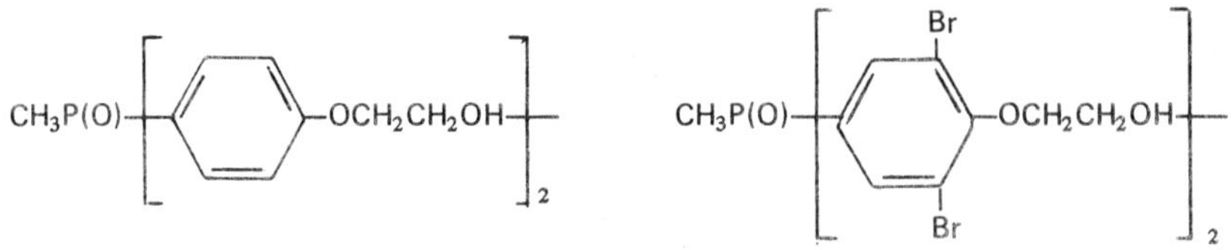

Each of the above is employed in the proportion of 8 g together with 100 g of a partially polymerized polyethylene terephthalate having an intrinsic viscosity of 0.09. The mixture is heated at 240°C with stirring for 30 minutes after which the polymerization is concluded at 280°C under high vacuum (about 1 mm Hg). The final bromine- and phosphorus-containing polymer has an intrinsic viscosity of 0.57, and contains 0.33 wt % phosphorus and 3.20 wt % bromine.

The two respective polymers are spun and drawn into fibers for testing.

The bromine- and phosphorus-containing polymer has an oxygen index of 23.1 while the unmodified polyethylene terephthalate based upon the diol (without bromine or phosphorus) has an oxygen index of 20.1. The oxygen index improvement is therefore 3.1 units.

When phosphorus alone is present, the oxygen index improvement is only 2.7 units.

When bromine alone is present [e.g., tetrabromo-(2,2-isopropylidene diphenol)-bishydroxyethoxylate], the oxygen index improvement is only 2.0 units.

Thus, the combination of phosphorus with bromine shows an unexpected superiority.

Polyarylene Esters Using Potassium Carbonate Catalyst

M.R. Ort; U.S. Patent 4,101,517; July 18, 1978; assigned to Monsanto Company describes a process for the preparation of polyarylene esters from aromatic dicarboxylic acids and diesters of a diphenol admixture containing 1,2-bis(4-hydroxyphenyl)ethane. The polymerization process is catalyzed by a potassium compound and yields polymer of improved color and fire safety performance.

Example 1: *Preparation of Poly-1,2-Bis(4-Hydroxyphenyl)Ethane Isophthalate* — A charge consisting of 167 parts of isophthalic acid and 298 parts of 1,2-bis-(acetoxyphenyl)ethane and 0.005 mol potassium carbonate per mol of 1,2-bis-(acetoxyphenyl)ethane is placed in a reaction vessel equipped with a stirrer, condenser and receiver. The vessel is evacuated and purged with nitrogen three times. During the reaction, a nitrogen blanket is maintained in the reactor. The temperature is raised to about 245°C and acetic acid distills. The rate of distillation during the initial 67% of reaction is 1.82 ml mol^{-1} min^{-1}. The distillation rate tapers after 87 parts of acetic acid have distilled in 55 minutes. The temperature rises slowly to 270°C during the initial stage. Thereupon the vessel is evacuated to a pressure of about 200 torrs for about 5 minutes and the pressure is then slowly reduced to about 1 torr while the temperature is increased to about 290°C.

When the amperage on the stirrer motor increases by 0.02 to 0.04 A, the stirrer motor is switched off and the vacuum is released with nitrogen. A total of

112 parts of acetic acid (98% of theoretical) is collected. The rate of evolution of acetic acid during the vacuum stage is 1.67 ml mol^{-1} min^{-1}. The prepolymer is extruded from the vessel under slight nitrogen pressure and is reduced in a mill to a powder of particle size in the range of from 0.1 to 0.25 mm. The inherent viscosity of the prepolymer is 0.28. The prepolymer is off-white in color.

The powder is charged to a reaction vessel which is then purged with nitrogen. The pressure is reduced to 0.1 to 0.2 torr and the temperature is raised to about 10° to 15°C below the melting point of the polymer. Heating is continued for 6 hours to complete the solid state polymerization. The vessel is cooled and the polymer is discharged. The inherent viscosity of the polymer is 0.90. It melts at about 305°C.

Example 2: Example 2 is carried out for comparative purposes. It is a repeat of Example 1 with sodium carbonate in place of the potassium carbonate. The prepolymer is off-white in color and has an inherent viscosity of 0.32. 113 parts of acetic acid (99% theoretical) distills. The rate of acetic acid evolution in the two stages of prepolymer formation is 1.98 and 1.25 ml mol^{-1} min^{-1}, respectively. After solid state polymerization, a crystalline polymer of inherent viscosity 0.90 is obtained.

Example 3: Example 3 is carried out for comparative purposes. It is a repeat of Example 1 with lithium carbonate in place of potassium carbonate. 96 parts of acetic acid (84% theoretical) distills. The rate of acetic acid evolution is 2.0 ml mol^{-1} min^{-1}. The prepolymer is amber in color and has an inherent viscosity of 0.22. The polymer obtained by solid state polymerization has an inherent viscosity of 0.88.

Example 4: Example 4 is carried out for comparative purposes, repeating the process of Example 1 without potassium carbonate catalyst. Approximately 84 parts of acetic acid distills in the first stage of the reaction in 117 minutes. The rate of distillation is 0.77 ml mol^{-1} min^{-1}. The rate of distillation in the second or vacuum stage is 0.69 ml mol^{-1} min^{-1}, and a total of 111 parts of acetic acid (97% theoretical) is collected. The prepolymer is light amber in color and has an inherent viscosity of 0.29. The crystalline polymer obtained therefrom by solid state polymerization has an inherent viscosity of 0.87.

Samples of the polymers of Examples 1, 2, 3 and 4 are molded in a thickness of 0.80 mm. The samples are then subjected to the UL-94 test and the average time of flame out and average time of afterglow is determined. The data are presented below and demonstrate the much shorter period of afterglow of the polyesters of Example 1 containing potassium and Example 4 without catalyst compared with the polyesters of Examples 2 and 3 containing sodium and lithium, respectively.

Ex. No.	Metal Ion	Average Flame-Out Time, Second Burn . . . . (sec)	Average Afterglow
1	Potassium	6	1
2	Sodium	6	23
3	Lithium	4	8
4	None	6	0

Glycol Phthalate plus Phosphine Oxide

H.L. King and C.C. Wu; U.S. Patents 4,127,565; November 28, 1978; and 4,127,566; November 28, 1978; both assigned to Monsanto Company prepare a flame-retardant, easily dyeable polyester by polycondensing a glycol phthalate prepolymer with a phosphine oxide and dye improver selected from the group consisting of $R-O(G-O)_x-H$ where R is an alkyl group containing an average of from about 8 to 20 carbon atoms; G is a hydrocarbon radical selected from the group consisting of ethylene, propylene, butylene and isomers thereof, and mixtures of the above, and x has a value of about 8 to 20, and

where R is O, $(CH_3)_2$ or H_2 and the sum of x and y is 4 to 30.

The polymer will be made up of 82 to 98 wt % of the prepolymer, 8 to 1 wt % of the dyeability improver and 10 to 1 wt % of the phosphine oxide and is preferably formed in the presence of a catalyst which is the reaction product of germanium dioxide and a tetraalkyl ammonium hydroxide.

Example 1: This example illustrates a method for making the catalyst used in this process. In making this catalyst, 1.05 g (0.01 mol) of germanium dioxide in crystalline form and 1.82 g (0.01 mol) of solid, hydrated tetramethylammonium hydroxide (50% active) were stirred with 3.3 g of ethylene glycol. The solid GeO_2 dissolved almost immediately. The solution was then heated to remove water and the remaining liquid weighed 5.25 g and contained 0.01 mol of catalyst.

1.05 g germanium dioxide (0.01 mol) in crystalline form and 3.7 g (0.01 mol) of 40% aqueous tetraethylammonium hydroxide (TEAH) were stirred with 8 g of ethylene glycol. The solid GeO_2 dissolved almost immediately. The solution was heated to remove 2.25 g of water. The remaining liquid was a clear, colorless solution weighing 10.5 g and containing 0.01 mol of catalyst.

Example 2: *Comparative* — 200 g of ethylene terephthalate prepolymer prepared as described above and 0.75 ml of GeO_2/TEAH catalyst solution of Example 1 were added to a 1 liter capacity, stirred, and batch autoclaved, the autoclave then being purged with nitrogen and heated to 275° to 280°C. Pressure on the system was reduced by vacuum and polymerization to form high molecular weight poly-(ethylene terephthalate) was carried out at a temperature of 275° to 280°C and a pressure of less than 2 mm Hg until the desired molecular weight was obtained.

A polymer with a melting point of 255°C and an intrinsic viscosity of 0.64 was obtained in a period of 69 minutes from the start of pressure reduction. The polymer was then spun into filaments and the dyeability of the filaments was determined. Fiber color was good but dye-bath exhaustion was only 13% after 2 hours of dyeing at 100°C.

Example 3: 200 g of ethylene terephthalate prepolymer, 2.0 g (1.0 wt %) of a monofunctional alkoxy poly(oxyalkylene) glycol having the structural formula $C_{14-15}H_{29-31}-O(CH_2CH_2O)_{14}H$, and 0.75 ml of the GeO_2/TEAH catalyst solution of Example 1 were added to a 1-liter capacity, stirred batch autoclave, which was then purged with nitrogen and heated to 230° to 250°C. Molten bis(carboxyethyl) methyl phosphine oxide, 10.5 g, was added and polymerization was completed as in Example 2. A polymer with an intrinsic viscosity of 0.65 was obtained in a period of 72 minutes after starting pressure reduction. The purity of drawn filaments formed from the polymer was 2.9 and brightness was 90.2. The filaments contained 0.77 wt % phosphorus and melted at 243°C. The filaments had an oxygen index of 22.3. When dyed for 2 hours with C.I. Disperse Blue 61, the dyebath exhaustion was 57%.

Examples 4 through 8: Polymers were prepared as in Example 3, and fibers from these polymers were tested for purity, brightness and oxygen index. The table below shows the result.

Ex. No.	Weight Phosphine Oxide	Intrinsic Viscosity	Melting Point (°C)	Purity	Brightness	OI
4	10.5	0.64	243	3.2	90.6	22.1
5	10.5	0.67	242	3.1	90.4	22.1
6	0	0.66	251	2.2	89.1	19.7
7	10.5	0.66	243	2.9	88.2	22.3
8	0	0.64	251	1.8	91.2	19.8

Examples 6 and 8 were included as comparative examples and no phosphine oxide was used. It will be noted that the OI of the fiber from these examples was less than 20.

Halogenated Oligomeric Styrenes

According to *R. Neuberg, K. Penzien, and H.G. Matthies; U.S. Patent 4,151,223; April 24, 1979; assigned to BASF AG, Germany*, a chlorinated and/or brominated oligomeric styrene as the flameproofing agent is added to a linear polyester, along with a conventional synergistic agent such as antimony trioxide on kaolin.

The chlorinated and/or brominated oligostyrenes can be introduced into the linear, saturated polyester fibers or filaments at any stage of their process of manufacture. In general, they are added before production of the fiber, toward the end of the manufacture of the polyesters, or during the spinning process. In a preferred process of manufacture, the brominated oligostyrenes are added to the molten polyester during melt spinning and the molten mixture is spun in the conventional manner to give filaments or fibers.

The degree of polymerization of the oligomers is preferably from 3 to 100, especially from 3 to 20, and cyclic oligomers may also be used. According to a preferred embodiment, the chlorinated and/or brominated oligomeric styrenes to be employed have the formula shown below, where X = Cl or Br (individual X's being identical or different), R = H or an aliphatic radical, especially an alkyl radical, e.g., CH_3 or C_2H_5, m is an integer from 1 to 5, and n is the degree of polymerization.

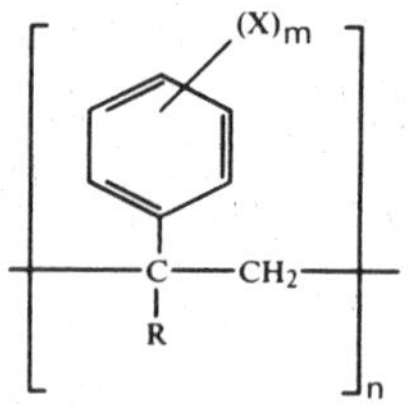

Limiting oxygen index values ranged from 24.5 for a polymer containing 2.5% bromine to 30.0 for a polymer containing 15.7%.

ADDITIVES WHICH IMPROVE DYEABILITY

Since it is preferred for reasons of cost not to process spun-dyed filaments for the production of textiles, the fiber-making process has to be followed by a dyeing process. This dyeing process frequently involves difficulties because textiles based on polyester fibers, especially those produced from polyesters based on terephthalic acid and ethylene glycol and/or 1,4-bis(hydroxymethyl)cyclohexane, are not easy to dye in a normal dyeing process.

Accordingly, the following methods of dyeing have been developed for polyester fiber materials and for blends of polyester fibers with other fibers: (1) the carrier dyeing method carried out at boiling temperature; and (2) the high temperature (HT) dyeing method carried out at 120° to 135°C for polyester fibers and their blends with cellulose fibers, and at 104° to 106°C for polyester fiber/wool blends.

Carriers have to be added in cases where dyeing is carried out at boiling temperatures and at temperatures in the range of from 104° to 106°C. In cases where dyeing is carried out at temperatures in the range of from 120° to 135°C, the diffusion rate of the dyes is normally increased to such an extent that, under these conditions, relatively high percentages of dye are attached to the polyester fibers fairly quickly even in the absence of carriers. It has been found, however, that even in the high temperature process, it is advantageous to use relatively small quantities of carrier for certain disperse dyes and also for polyester fibers having a relatively low affinity for dyes.

The carriers used are inter alia combinations based on o-phenyl phenol, chlorinated benzenes or phenols. In most cases, these compounds are physiologically not without risk. In other words, they should not be breathed in, should not come into contact with the skin and, in addition, should be removed as far as possible from the effluent, because they can never be completely biologically degraded, if at all.

Accordingly, it is desirable, both for eliminating health risks and for reasons of cost not only in regard to dyeing, but also in regard to effluent treatment, to provide polyesters, especially based on terephthalic acid and ethylene glycol, 1,4-butane diol and/or 1,4-bis(hydroxymethyl)cyclohexane, which can be dyed in the absence of carriers.

Bis-Ethoxylated Tetramethyl Bisphenol A

A. Kohler, H. Pelousek, and E. Frohberg; U.S. Patent 4,067,850; January 10, 1978; assigned to Bayer AG, Germany have found that filaments and fibers produced from polyesters which contain from 4 to 15% by weight of bis-ethoxylated tetramethyl bisphenol A (1) in chemically co-condensed form, show excellent textile properties in addition to an outstanding affinity for dyes in the absence of carriers.

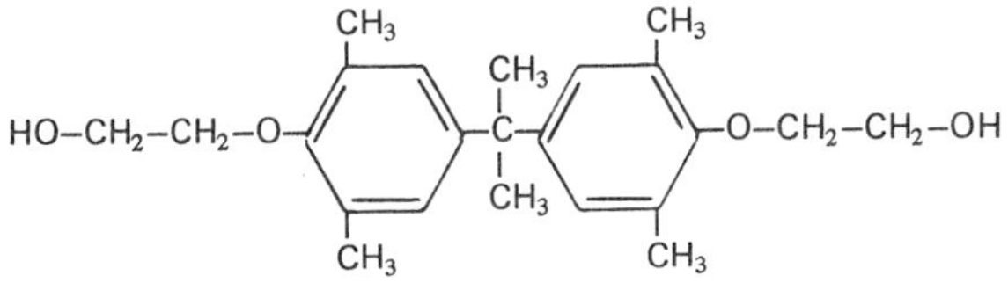

For the comparison measurements, dyeing was carried out by the following methods (cf H. Ludewig, *Polyesterfasern*, Akademie Verlag, Berlin, 1965, page 346).

Method 1: The fibers are thoroughly washed before dyeing. The goods to liquor ratio amounts to 1:20. For dyeing in the presence of a carrier, 4 g/l of a standard commercial-grade carrier (Levegal PT) are added to the solution. A pH value of from 4.5 to 5.5 is then adjusted with monosodium phosphate and acetic acid. This is followed by the addition to the solution of 2% by weight of the disperse dye:

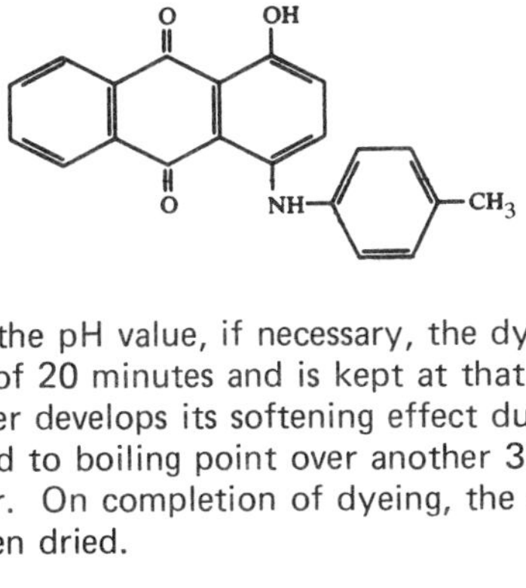

After readjustment of the pH value, if necessary, the dye bath is heated to $80°$ to $85°C$ over a period of 20 minutes and is kept at that temperature for 15 to 20 minutes. The carrier develops its softening effect during this residence time. The bath is then heated to boiling point over another 30 minutes and left at that temperature for 1 hour. On completion of dyeing, the dyed material is warm-rinsed in water and then dried.

Method 2: For dyeing in the absence of a carrier, the dyeing is carried out in exactly the same way as in Method 1, except that no carrier is added to the solution. The tests reproduced in the examples show that the staple fibers according to the process disclosed can be dyed as deeply with basic dyes or with disperse dyes in the absence of carriers as fibers of the pure polyester can in the presence of carriers.

The results of these tests were checked more accurately by determining the color valence which consists of three color values and which unequivocally defines one color. The reference system is the internationally agreed CIE system which is equivalent to the Standard Valence System defined in DIN 5033. Under the CIE system, the color values are denoted X, Y and Z.

For measurement, the fibers are introduced into a round cuvette. The three-range color measuring process is then carried out with an Elrepho filter photometer. In this process, the degree of remission of the test specimen is measured with three special color measuring filters and the color values X, Y and Z are calculated in a simple manner from the remission values R_x, R_y, R_z in accordance with the formulas: for standard light type C, $X = 0.782 R_x + 0.198 R_z$, $Y = R_y$, and $Z = 1.181 R_z$.

Example: Before the transesterification of dimethyl terephthalate (DMT) with ethylene glycol, 10% by weight of bis-ethoxylated tetramethyl bisphenol A, based on the DMT, are added to the reaction mixture. Transesterification takes place over a period of 2.75 hours, during which the temperature rises from 180° to 220°C. This is followed by precondensation over a period of 0.75 hour at a temperature of 250°C. The temperature is then increased to 270°C and the pressure adjusted to 1 torr for polycondensation.

Polycondensation is terminated after 2.5 hours. The copolyester is then spun off from the autoclave into water and granulated. The granulate thus obtained has a relative viscosity of 1.70 and a melting point of 243°C. It contains 10% by weight of the glycol of formula (1) in co-condensed form.

The granulate is melt-spun through a multibore spinnerette into filaments having an individual denier of 40 dtex which are wound up at a rate of 1,100 m/min. The spun filaments are doubled into a tow with a total denier of 100,000 dtex, stretched in a ratio of 1:3.20 in hot water at 90°C and then in a steaming duct, crimped in a stuffer box, dried in the absence of tension in hot air and subsequently cut into staple fibers 60 mm long. The fibers show a strength of 3.5 p/dtex, an elongation of 30% and a boiling-water induced shrinkage of 1%.

The fibers are then dyed with the disperse dye by the dyeing process described above (Method 2, without carrier). On completion of dyeing, the fibers are deep blue in color. The staple fibers are introduced into the cuvette and the three-range color measuring process described above is carried out.

The color values observed are as follows: $X = 14.4$, $Y = 13.0$, and $Z = 30.0$.

If the entire manufacturing process is carried out with a granulate of pure polyethylene terephthalate (relative viscosity 1.67) and if these fibers are dyed in the same way as described above, but in the presence of the carrier Levegal PT (Method 1), deep blue fibers are again obtained, their color values being as follows: $X = 14.1$, $Y = 12.8$, and $Z = 29.3$.

The fibers containing 10% by weight of bis-ethoxylated tetramethyl bisphenol A in co-condensed form can be dyed just as well and as quickly in the absence of carriers as fibers of pure polyethylene terephthalate can in the presence of carriers.

Polytetrahydrofuran Diol

C.-R. Bernert, E. Radlmann, and G. Nischk; U.S. Patent 4,042,570; August 16, 1977; assigned to Bayer AG, Germany describe high molecular weight polyesters from terephthalic acid and 1,4-cyclohexanedimethanol containing in polycondensed form a polytetrahydrofuran diol. Fibers and filaments produced from these polyesters can be dyed carrier-free.

Suitable polyether diols are polytetrahydrofurans with molecular weight of from 400 to 5,000. If the molecular weight is too low, the resulting polyesters show inferior mechanical properties, in addition to which the melting point of the polymer is drastically lowered. If the molecular weight is too high, the polyether diol may only be co-condensed at one end, if at all, on account of its considerable chain length.

Example: 158 parts by weight of 1,4-bis(hydroxymethyl)cyclohexane, 194 parts by weight of terephthalic acid dimethyl ester, 30.4 parts by weight of a polytetrahydrofuran diol with a molecular weight of 1,000 and 1.0 part by weight of tetraisopropyl titanate, accommodated in a reaction vessel equipped with a flat-blade anchor stirrer, a gas inlet pipe, a dephlegmator, condenser, vacuum adapter and receiver, are heated under nitrogen as quickly as possible by means of an oil bath or metal bath preheated to 300° to 315°C, and transesterified for about 1 hour. The introduction of nitrogen is then interrupted and the pressure slowly reduced over a period of 1 hour to 0.1 torr.

On account of the increase in the viscosity of the melt, the speed of the stirrer, approximately 150 rpm, is reduced shortly after melting to approximately 20 rpm. The colorless, homogeneous melt may be processed into shaped articles such as, e.g., filaments. The polyester contains 10% by weight of polytetrahydrofuran diol units, based on the polyester (3 mol %, based on total diol units). The filaments, which have a softening range of 290° to 295°C and a relative solution viscosity of 1.72, are stretched in a ratio of 1:2.7 by drawing over a plate heated to 130°C, and are then dyed by the open method on a water bath. Dyeing is carried out with a dispersion, acidified with acetic acid to pH 4.5, of 1.1% by weight of a dye corresponding to the formula:

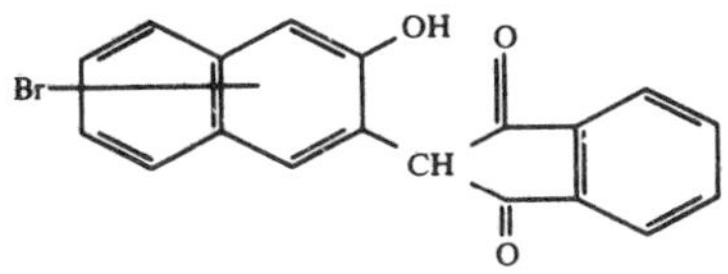

0.76% by weight of a dye corresponding to the formula:

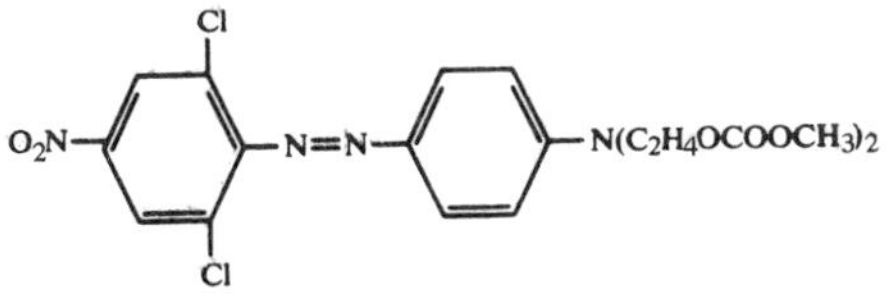

0.15% by weight of a dye corresponding to the formula:

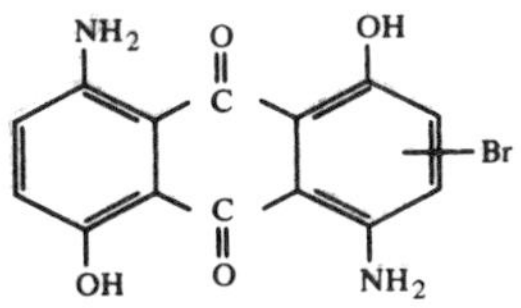

and 2% by weight of an anionic dispersant in water, in which the fibers are held for 1.5 hours at boiling temperature. The filaments take on a very good, uniform and deep dye finish which is highly washproof.

When a comparison polyester sample, to which no polytetrahydrofuran diol was added during condensation, is treated in a dye bath of the same composition, all that is obtained after 1.5 hours at 98°C is a superficial dye effect. Although this sample can also be effectively dyed by the addition of 2% of a carrier, e.g., o-cresotinic acid methyl ester, to the dye bath, it surprisingly does not have the depth of color obtained with the filaments modified with polytetrahydrofuran diol.

1,8-Bis(Dialkylamino)Naphthalene

D.J. Sikkema; U.S. Patent 4,029,637; June 14, 1977; assigned to Akzona Incorporated describes a filament-forming polyester or copolyester and fibers thereof composed of at least 80% of ethylene terephthalate units containing a tertiary amine as an additive to improve the receptivity for acid dyes. The additives comprise 1,8-diaminonaphthalenes in which the two amino groups have four alkyl substituents with 1 to 4 carbon atoms and the nucleus may have H, alkyl or aralkyl substituents. The preferred additive is N,N,N',N'-tetramethyl-1,8-diaminonaphthalene. A preferred copolyester contains 1 to 10% ethylene azelate units. The additives improve the exhaustion of the dye bath and the wash-fastness of the dyed filaments.

The wash-fastnesses of the dyed samples were determined by washing with 2 g/l Ultravon JU or 5 g/l Marseilles soap, each being at a liquor ratio of 1:40 and a temperature of 40°C. The change in color of the washed material with respect to unwashed material was evaluated according to NEN 5202 and 5203. The evaluation was expressed in indices from 1 to 5, with 5 representing no change and 1 considerable change in color. For practical purposes, a wash-fastness using this criterion is acceptable when assigned a 4 or 5 at evaluation. Lower figures indicate unsatisfactory wash-fastness.

Example: Polyethylene terephthalate was prepared from ethylene glycol and dimethyl terephthalate with addition of 340 ppm manganese acetate tetrahydrate, 300 ppm GeO_2 and 135 ppm trimethyl phosphate. The polycondensate had a relative viscosity of 1.64; upon drying, it was cut into chips and mixed in a rotating drum with 2 mol % N,N,N',N'-tetramethyl-1,8-diaminonaphthalene in powder form. The resulting mix was spun to filaments using an extrusion melt spinning machine and a temperature of about 270°C. The filament was subsequently drawn on two hot pins at a temperature of 85° and 155°C, respectively, at a drawing ratio of 6.0 to 35/8 (dtex) yarn.

The yarn thus obtained was designated Yarn A. Three other yarns were spun in a similar manner, but instead of N,N,N',N'-tetramethyl-1,8-diaminonaphthalene, N,N,N',N'-tetramethyl-1,5-diaminonaphthalene (Yarn B), pentaphenyl guanidine (Yarn C) and no amine (Yarn D), respectively, were added to the polymer. Skeins of Yarns A, B, C, and D were made up and dyed as indicated with the acid dye Nylosan blue FGLB. The wash-fastness was determined by washing the dyed skeins of Yarns A and B with Ultravon JU and Marseilles soap. Dyeing and washing results are listed in the following table.

Yarn	Bath Exhaust (%)	. . . Wash-Fastness Index. . . .	
		Ultravon JU	Marseilles Soap
A	>90	4-5	4
B	50	2	2
C	10	–	–
D	0	–	–

Yarn A, according to the process, is far superior to each of the Yarns B, C and D lying outside the scope of the process. In view of the low exhaustion level of the dye bath, pentaphenyl guanidine (Yarn C), known to be a strong base, hardly contributes to the affinity for acid dyes.

The wash-fastness of Yarn A is also superior to that of the known Yarn B. Because of its wash-fastness index of 4 or higher, Yarn A meets the demands posed by household and professional cleaning treatments. Yarn B (wash fastness index 2) does not comply with these requirements.

Sulfonamide Sulfonates

According to *H.R. Penton, Jr., U.S. Patent 4,049,633; September 20, 1977; assigned to Akzona Incorporated,* the disperse dyeability of a fiber- or film-forming polyester resin is improved by dispersing substantially uniformly in a polyester resin in which it is at least partially soluble a water-soluble organic sulfonamide sulfonate having the formula:

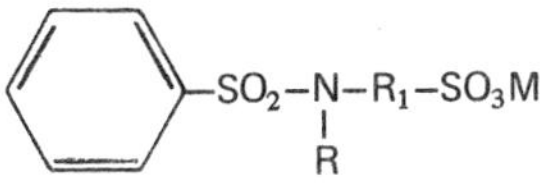

or

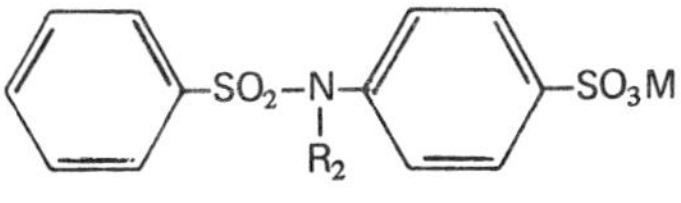

wherein R is benzyl or phenethyl, R_1 is ethylene or propylene, R_2 is hydrogen or lower alkyl and M is an alkali metal.

Example 1: About 300 g of dimethylterephthalate (DMT), about 241 g of ethylene glycol and about 0.149 g of manganous benzoate were heated for 70 minutes in a nitrogen atmosphere, the temperature rising to 220°C, with evolution of methanol. The sodium salt of N-benzyl-N-propylsulfonate benzenesulfonamide (about 24 g) 8 wt % based on starting amount of DMT, about 0.238 g of antimony tributylate and about 0.268 g of trimethyl phosphate were then added and the pressure was reduced to 0.1 mm of mercury and the temperature increased to 280°C. After 75 minutes, a polyester was obtained having an intrinsic viscosity of 0.63 and a melting point of 249°C.

The polyester was spun and drawn as 30/6 yarn. The yarn exhibited good thermal and light stability having 83% retained tenacity after being heated at 180°C for 120 minutes and 93% retained tenacity after being exposed in a Fade-Ometer for 80 hours.

Example 2: Example 1 was repeated, except that no organic sulfonamide sulfonate was added to the reaction vessel. The intrinsic viscosity of the polyester was 0.65 and the melting point was 251°C.

Knit fabrics, prepared from the yarns of Examples 1 and 2 having the same weight and construction, were dyed in the same dye bath at 100°C with 1.0% by weight based on the weight of the fabric of Duranol Blue G Grains-Disperse Blue 26 and 5% by weight, based on the weight of the fabric, Carolid 3F carrier, a commercially available carrier which is primarily biphenyl with at least 10% by weight polyethylene glycol dispersing agent.

The depth of shade of the dyed fabrics was then measured photometrically yielding a color differential value of 35.9 between the fabrics of Examples 1 and 2. The color differential is the difference in photocell values between the polyester dyed samples. The greater this value the greater is the depth of shade of the experimental sample compared to that of the normal polyester sample by weight, based on the weight of the fabric. AATCC IIIA wash-fastness test at 75°C and light-fastness tests in the Carbon-Arc Fade-Ometer at 10, 20, 40 and 80 hours were performed on the experimental dyed sample and good to excellent ratings were obtained. Hence, the fabric made with yarns of Example 1 had a much deeper shade than those made from yarns of Example 2.

N-(β-Hydroxyethyl)-N-(3-Carbomethoxybenzenesulfonyl)Taurine

G.W. McNeely; U.S. Patent 4,029,638; June 14, 1977; assigned to Akzona discloses cationic dyeable copolyester polymer and yarns made therefrom having as an integral part of the polymer chain, sulfonamide groups each containing a dye sensitizing unit, derived from alkaline earth or alkali metal salts of sulfonated sulfonamides such as the sodium salt of N-(β-hydroxyethyl)-N-(3-carbomethoxybenzenesulfonyl)taurine.

Example 1: N-(β-hydroxyethyl)-2-aminoethanesulfonic acid,

$$HOCH_2CH_2NCH_2CH_2SO_3H$$
$$\underset{H}{\overset{|}{}}$$

was prepared as follows: 2,250 parts water was saturated with sulfur dioxide at 10°C in a three-necked, round-bottom flask which was fitted with an electrical stirrer, thermometer, dropping funnel, and a gas inlet tube, 261 parts 1-aziridineethanol dissolved in 750 parts water was added via the dropping funnel while sulfur dioxide was bubbled into the flask and the temperature maintained at 10°C. After the addition was completed, the contents of the flask were concentrated to 600 parts and added to 3,000 parts ethanol (F-30).

The resulting solid was washed with 2,000 parts ethanol, filtered, and dried to give 340 parts of the desired product.

Example 2: The sodium salt of N-(β-hydroxyethyl)-N-(3-carbomethoxybenzenesulfonyl)taurine was prepared as follows: 253.5 parts of the N-(β-hydroxyethyl)-2-aminoethanesulfonic acid, prepared as in Example 1, 60.0 parts of 5.92 N sodium hydroxide, 79.5 parts sodium carbonate, 875 parts water, and 200 parts acetone were placed in a three-necked, round-bottom flask which was fitted with a stirring rod, thermometer and dropping funnel. 352 parts methyl benzoate-3-sulfonyl chloride dissolved in 1,300 parts acetone were added via the dropping funnel at such a rate that the flask temperature did not increase above 40°C. After the addition was completed, 2 hours, the resulting slurry was stirred for 2 hours, concentrated to 1,600 parts, then added to 5,000 parts isopropyl alcohol. This slurry was cooled to 10°C and filtered to give 506 parts of the desired product after drying under reduced pressure at 70°C. This product had the formula:

$$CH_3-O-\underset{\underset{O}{\|}}{C}-\text{(benzene ring)}-\underset{\underset{O}{\overset{O}{\|}}}{S}-N-CH_2CH_2OH$$

with the side chain: $N-CH_2-CH_2-SO_3Na$

Example 3: 4.32 parts of the sodium salt of N-(β-hydroxyethyl)-N-(3-carbomethoxybenzenesulfonyl)taurine prepared according to Example 2 were added to 108 parts of dimethyl terephthalate, 81 parts of ethylene glycol, 0.364 part of manganous benzoate, and 0.110 part of calcium acetate, the ratio of the added comonomer to dimethyl terephthalate being about 2.0 mol %.

The mixture was heated for 2 hours in a nitrogen atmosphere, the temperature rising to 200°C with evolution of methanol, 0.0864 part of antimony tributylate and 0.351 part of the glycol ester of phosphoric acid were added and the pressure then reduced to 0.1 mm of mercury and the temperature was increased to 270°C. After 1.6 hours, a polymer was obtained having an intrinsic viscosity of 0.49.

The Gardner b value, a measure of polymer yellowness, was 4.8 determined on polymer at the beginning of the extrusion and 6.0 at the end of the extrusion.

This compares with a Gardner b value of 7.0 determined on a polymer at the beginning of the extrusion and 10.7 at the end of the extrusion which was prepared according to the preferred embodiment of U.S. Patent 3,856,753 by incorporating the sodium salt of N-methyl-N-(3,5-dicarbomethoxy)benzenesulfonyl taurine into a polyester polymer. The improved color of the copolyester polymer of this process demonstrates the effect of incorporating the sulfonamide group directly in the polymer chain rather than having the sulfonamide group as a side chain.

The polymer was spun and drawn as yarn of 70 denier, 32 filaments (70/32 yarn) which had the following properties: intrinsic viscosity, 0.49; tenacity (g/den), 3.41; breaking elongation (%), 25.4; and meq COOM/kg, 60.

The yarn exhibited good thermal and light stability having 81.3% retained tenacity after being heated at 180°C for 120 minutes and 91.3% retained tenacity after being exposed in a Fade-Ometer for 160 hours. Knit tubes, prepared from this yarn, were dyed in separate-but-equal dye baths with 1.0% (owf) of each of the following dyes: (1) Astrazon Red BBL, Basic Red 23; (2) Sevron Orange G, Basic Orange 21; and (3) Resolin Blue FBLD, Disperse Blue 56.

The final dye bath exhaustions were measured spectrophotometrically and the percent exhaustion was better than 90% for each basic dyestuff. IIIA AATCC wash tests at 160°F and light-fastness tests in the carbon arc Fade-Ometer at 10, 20 and 40 hours exposure were performed on each dyed sample and fair to excellent ratings were obtained.

Highly Oriented Filaments Containing Sulfonate Groups

H. Yasuda and S. Omori; U.S. Patent 4,076,783; February 28, 1978; assigned to Toyobo Co., Ltd., Japan provide a method for producing polyester fibers having excellent dyeability with basic dyes or disperse dyes, which comprises melt-spinning a polyester comprising predominantly (at least 80% by mol) repeating units of ethylene terephthalate and containing 0.5% by mol or more, preferably 1.5 to 5% by mol of an ester unit derived from a dicarboxylic acid or diol containing at least one metal sulfonate group under the condition of being highly oriented so as to give filaments having a birefringence of 0.015 to 0.100 and drawing the resulting filaments at a fixed temperature and in a fixed draw ratio. The polyester fibers can be dyed with basic dyes or disperse dyes at a temperature of less than 100°C at a sufficiently high dye adsorption rate to give dyed fibers.

The intrinsic viscosity of the polymer and the filaments is measured at 30°C in a mixed solvent of phenol:tetrachloroethane of 3:2. The dye exhaustion (i.e., rate of dye adsorption) is calculated by measuring the color difference between the original dye solution before the dyeing and the remaining solution after the dyeing by colorimetry, wherein the original dye solution is prepared so that the content of the dye is 5% by weight on the basis of the weight of the fibers to be dyed, and when the remaining solution becomes completely colorless, the dye exhaustion is 100%, when the color concentration of the remaining solution is half of that of the original dye solution, the dye exhaustion is 50% and when the color concentration of the remaining solution is the same as that of the original dye solution, the dye exhaustion is 0%.

The birefringence is measured by a conventional method (J.F. Rund & R.B. Andrews, *Journal of Applied Physics*, Vol. 27, page 996, 1956).

Example: A mixture of dimethyl terephthalate and ethylene glycol is subjected to polycondensation by an ester exchange reaction in the presence of sodium 3,5-di(carbomethoxy)benzenesulfonate (2% by mol on the basis of dimethyl terephthalate) to give a polyester having an intrinsic viscosity of 0.38 and a melting point of 258°C.

The polyester thus obtained is melted and extruded through a spinneret having 300 orifices of 0.3 mm in diameter at a spinning temperature of 270°C, and the emerged filaments are rapidly cooled by blowing cooled air at a speed of 0.9 m/sec just under the spinneret and are then wound up on bobbins at a rate of 2,500 m/min. The spun filaments have a birefringence of 0.021 and an intrinsic viscosity of 0.37.

The spun filaments are bundled in the form of tows and are drawn 1.4 times at 140°C to give drawn tows comprising filaments of a denier of 2.59 d, a tenacity of 2.56 g/d and an elongation of 16.1%. The tows thus obtained are dyed by boiling with a basic dye C.I. Basic Orange 22 at 100°C for 90 minutes to give filaments dyed with a deep orange color. The dye exhaustion is 94%.

In the above procedure, when the tows are drawn at 80°C instead of 140°C, the characteristics of the drawn tows are a tenacity of 2.07 g/d and an elongation of 12.5%, and the orange color after the boil dyeing is fairly reduced (the dye exhaustion is 62%). When the tows are drawn at 160°C, the filaments are cut very often because of the fusing thereof and the desirable drawing can not be carried out.

Besides, when the tows are drawn at 140°C and in a rate of 1.9 times instead of 1.4 times, the drawn tows are not dyed to a deep color by the boil dyeing (the dye exhaustion is 42%).

Moreover, when the drawing step of the tows is omitted in the above procedure, the fibers are fairly well dyed by the boil dyeing with a basic dye (the dye exhaustion is 71%), but the characteristics of the filaments are impractically inferior (a tenacity of 1.15 g/d and an elongation of 182.0%).

When the same polyester used in the above example is melt-spun by a conventional method and the resulting filaments are wound up on bobbins at a rate of 1,300 m/min, the spun filaments have a birefringence of 0.009 and an intrinsic viscosity of 0.37. The filaments are drawn 3.0 times at 80°C and then subjected to a boil dyeing in the same manner as described in the above example. As a result, merely faintly orange colored filaments are obtained. The dye exhaustion is 21%.

Tetramethylpiperidine Groups Attached Through Oxyalkylene Linkages

According to *M.S. Tanikella and O. von Susani; U.S. Patents 4,001,189; January 4, 1977; and 4,001,190; January 4, 1977; both assigned to E.I. DuPont de Nemours and Company*, textile fibers or filaments prepared from polyesters modified with a minor proportion of 2,2,6,6-tetramethylpiperidine groups attached through oxyalkylene linkages to ends of linear polyester polymer chains have improved dyeability with acid dyes. A variety of alkylene oxide condensates of tetramethylpiperidine compounds are shown to have adequate heat-stability and compatibility for incorporating in polyesters under conditions of condensation polymerization to form desirable products.

Example 1: The preparation of 1,2,2,6,6-pentamethyl-4-hydroxypiperidine condensed with ethylene oxide in a molar ratio of 1:20 proceeds as follows: A 5-liter flask fitted with a thermometer and a capillary nitrogen inlet is weighed, and 766 g of a composition consisting mostly of 1,2,2,6,6-pentamethyl-4-hydroxypiperidine is introduced. The setup is warmed to 90°C and held at 1 to 2 torrs for 0.5 hour to remove water and volatile acidic components, such as carbon dioxide. A mechanical stirrer and bearing is inserted in the center neck, and a gas feed tube is inserted in place of the nitrogen bleed. The setup is then retared. A vertical condenser having a bubbler attached to its exit tube is then attached to the flask. The apparatus is purged with nitrogen for 10 minutes, and then heated to a temperature of 80°C to remelt the residue. With the temperature below 90°C, 1.52 g of sodium hydride dispersed in mineral oil is added. 2,719 g ethylene oxide gas is added to the stirred liquid at such a rate that no gas passes through the bubbler.

After the addition is complete, the flask is swept with nitrogen and cooled below 50°C. The product has a calculated average of 20.3 oxyethylene units. Acetic acid (1.9 g) is then added and the mixture is stirred for 10 minutes to quench the catalyst. The product is referred to as PHP-20EO.

Example 2: This example illustrates a preferred method for preparing the acid-dyeable polyester fibers and dyeing them in a multidye bath.

A molten stream of PHP-20EO having a temperature of 120° to 130°C is injected continuously into a molten stream of polyethylene terephthalate having a relative viscosity of 3 to 5. Injection is at such a rate as to provide 0.075% nitrogen in the final polymer. The combined ingredients are then passed into a finisher vessel held at 275°C and 1.3 torrs. Into the finisher vessel near its midpoint is injected 2.2%, based on final polymer weight, of diphenyl terephthalate to accelerate the polymerization. The flow of the polymer out of the finisher is controlled to give a holdup time of about 2 hours. The finisher polymer has a relative viscosity of 20 and is spun at 274°C through a 360-hole spinneret (orifice diameter 0.381 mm) to produce filaments which are collected as a yarn at 900 yd (825 m) per minute.

A creel stock is prepared from 25 of these yarns and the yarns are combined to form a tow which is drawn at a draw ratio of 4. The tow is drawn at 100 yd (92 m) per minute in an aqueous draw bath at 80°C. The tow is then crimped, relaxed and cut to staple fibers. The copolyester fibers have a denier of about 3.4, a tenacity of 4.0 g/den, and an elongation of 23%.

These fibers are then spun into a yarn, as are commercial fibers prepared from a basic-dyeable copolyester and commercial fibers prepared from poly(ethylene terephthalate). The basic-dyeable copolyester is prepared from ethylene glycol, terephthalic acid (98 mol % of acid units) and sodium 3,5-di(carbomethoxy) benzene sulfonate (2 mol % of acid units) according to U.S. Patent 3,018,272. The fibers are spun into three separate singles yarns on the midfiber spinning system to give yarns having a cotton count of 21 and 18 turns of Z twist. The yarns are knit into a fabric on a Lawson knitting machine. The fabric has three bands of approximately equal width, with each of the above yarns being knit into a separate band. A 5 g swatch of this fabric containing approximately the same amount of each of the above yarns is scoured and dyed in a size 14 Gaston County Pressure Dyeing Machine using the following procedure, where the percentages are based on weight of fabric.

The fabric is scoured 20 minutes at 82.2°C in a bath containing 1% of a nonionic surfactant based on the ethylene oxide condensate of a long chain linear alcohol and 1% of tetrasodium phosphate. The fabric is then rinsed well. The rinsed fabric is then placed in a set bath having a temperature of 37.8°C. Then, 1% of the nonionic surfactant described above, 1% of an amphoteric dyeing assistant based on the ethylene oxide adduct of a fatty amine, sodium sulfate and sodium formate are added. The sulfate is added to give 6 g/l and the formate 4 g/l.

The bath is then agitated for 5 minutes and 20% of dye carrier added, the bath agitated another 5 minutes and 0.15% C.I. Basic Blue 77, 0.35% C.I. Basic Blue 87 and 0.05% C.I. Basic Red 15 added. The bath is then agitated 5 minutes and 0.2% C.I. Disperse Yellow 54 (C.I. 47020) is added and the bath agitated another 5 minutes and the pH readjusted to 3.5 with formic acid. The bath is then heated to 76.7°C at 1.1°C/min. Then, 0.17% C.I. Acid Red 151((C.I. 26900) and

0.25% C.I. Acid Orange 128 is added and the bath agitated 5 minutes. The pH of the bath is then readjusted to 3.5 and the bath temperature increased to 87.8°C at 1.1°C/min and the dyer sealed. The temperature is then raised to 121.1°C at 1.7°C/min and the bath agitated for 2 hours at 121.1°C. The dye bath is allowed to cool to 71.1°C and the fabric rinsed well in hot tap water. The dyed fabric is then scoured 20 minutes at 71.1°C in a bath containing 1% of the nonionic surfactant described above. The fabric is rinsed in hot tap water and then dried. The dyed fabric is then heat-set for 1 minute at 176.7°C in a laboratory high temperature oven (Electric Hotpack Corp.) and then removed and the heat-setting step repeated. The poly(ethylene terephthalate) fibers are dyed yellow, the basic-dyeable polyester fibers are dyed blue and the fibers of this process are dyed red.

Quaternary Ammonium Compounds

Compositions of polyesters having improved affinity for acid dyes are produced by *Y. Vaginay; U.S. Patent 4,035,341; July 12, 1977; assigned to Rhone-Poulenc-Textile, France* by incorporating in the compositions of polyester-forming dicarboxylic acid and dihydric alcohols quaternary ammonium compounds, such as the p-toluene sulfonates of methyltrioctylammonium, tetrabenzylammonium, N,N'-dimethyl-N,N'-dibenzylpiperazinium; and the bis-p-toluene sulfonate of N,N,N',N'-tetramethyl-N,N'-distearyl hexamethylenediammonium or N,N,N,N',N',N'-hexamethyldodecamethylenediammonium; dimethylstearylbenzylammonium chloride, dimethylpalmitylbenzylammonium chloride, or trioctylbenzylammonium chloride.

Example: Tetrabenzylammonium p-toluene sulfonate is prepared from benzyl p-toluene sulfonate and tribenzylamine. All parts are by weight.

3,320 parts terephthalic acid and 2,838 parts hexanediol-1,6 are charged simultaneously, with agitation, into a stainless steel reactor. The mixture is gradually heated for 2½ hours to a temperature of 260°C. Thereupon, 5 parts of butyl orthotitanate are added and the pressure is gradually lowered to 1 mm mercury within 50 minutes; the polycondensation is carried out 1¼ hours at 265°C under a pressure of 1 mm Hg.

The vacuum is then broken with nitrogen and 220 parts of tetrabenzylammonium p-toluene sulfonate are added and agitated with the reaction mass for 15 minutes under vacuum.

There is obtained a homogeneous, slightly opalescent product which is cast and granulated. Its 1% specific viscosity in o-chlorophenol is 1.10 and its softening point 156°C.

After drying for 2 hours at 140°C, the molten granulates are extruded through a spinneret of 23 orifices of a diameter of 0.23 mm and then drawn on pin and plate under the following conditions: temperature of the pin, 70°C; temperature of the plate, 100°C; and draw ratio 3 X.

The thread obtained has the following properties: count, 32 dtex/23 ends; tenacity, 3.12 g/tex; and elongation, 12%. A hank of 20 g of this thread is placed in a bath of the following composition: C.I. Acid Blue 120, 0.4 g; diammonium phosphate, 1 g; condensate of stearyl- and oleylamine on ethylene oxide, 0.5 g; water, qs 1,000 ml; and acetic acid, qs pH 3–4.

After boiling for 30 minutes at 98°C, the bath is exhausted. The blue thread obtained is washed with water at 60°C containing 1.5% of an alkyl sulfate and then rinsed and dried. The thread is of a blue color and its color does not disappear upon extraction for 20 minutes in water and methylene chloride.

As control, a hexamethylene glycol polyterephthalate, prepared without addition of tetrabenzylammonium p-toluene sulfonate, obtained and treated in an identical manner, assumes only a very light blue color which disappears completely after washing and extraction under the same conditions.

Pyridine Dicarboxylic Acid Compounds

The esterification of an aromatic dicarboxylic acid with an alkylene glycol containing 2 to 10 carbon atoms is described by *S.D. Lazarus and J.D. DeCaprio; U.S. Patent 4,079,045; March 14, 1978; assigned to Allied Chemical Corporation* under conditions of direct esterification wherein the resultant polyester has low ether content. The dicarboxylic acid is esterified with the alkylene glycol in the presence of a compound of the formula:

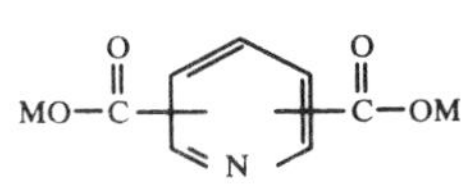

wherein M is H, an alkali metal selected from the group consisting of lithium, sodium and potassium or an alkaline earth metal selected from either calcium or magnesium.

Example 1: The reactor used is a 1 gal stainless steel autoclave equipped with a double spiral agitator turning at 30 rpm. The reactor is preheated to 95°C, and 1,800 g of terephthalic acid, 1,345 g of ethylene glycol, 9 g of pyridine dicarboxylic acid (2,5), and 1.5 g of antimony triacetate are charged. The reactor is purged with nitrogen, heated to 200°C, and pressurized with 75 psig of nitrogen. The reactor is then heated until the reaction mixture reaches 270°C. Generally, vapor is vented to the atmosphere to maintain a pressure of 75 psig in the reactor.

When reactor pressure no longer shows a tendency to rise above 75 psig, the residual pressure is slowly vented and the reaction mixture is swept with nitrogen for 15 minutes. At this point, the esterification reaction step is considered complete, and the polycondensation step is begun. Vacuum is applied to the reactor until the reaction pressure reaches 0.1 mm Hg, and the reaction temperature is increased to 290°C. This reaction temperature and pressure are maintained for 2 hours, after which nitrogen is introduced to adjust the pressure to atmospheric. The resulting polymer is extruded through a valve at the bottom of the reactor into a water bath and the polymer is pelletized in a Wiley Mill. For convenience, this polymer is called Polymer A.

A control polymer is made in the same manner as Polymer A except without the addition of pyridine dicarboxylic acid (2,5). For convenience, this control polymer is called Polymer B.

Polymer A had a melting point of 254°C, an intrinsic viscosity of 0.95 dl/g, and contained 1.47% diethylene glycol. Corresponding values for Polymer B are: 241°C, 0.96 dl/g, and 4.82%.

Solvent extraction of Polymer A with boiling ethyl alcohol to remove unreacted pyridine dicarboxylic acid, followed by infrared analysis, shows that the pyridine compound becomes part of the polymer chain.

Example 2: Polymer A and Polymer B of Example 1 are dried in a vacuum oven at 160°C for 16 hours and spun on a 1" extruder. The extrusion and spin block temperatures are about 295°C. Polymers A and B are drawn at a draw ratio of 4.0 to 1 into 70 denier, 16 filament yarns, which are knitted into Sleeve A and B, respectively. A swatch from each is dyed in conventional manner with Acid Blue 17. Visually, Sleeve A is observed to dye to a much greater extent than Sleeve B.

In additional tests, the pyridine dicarboxylic acid (2,5) of Example 1 was incorporated into a polyester formed by conventional ester interchange reaction of dimethyl terephthalate and ethylene glycol. Yarn made from this polymer was found to have enhanced dyeability comparable to yarn made from Polymer A.

Polyoxyethylene Glycol

R. Gilkey, S.D. Hilbert, B.J. Sublett, and T.H. Wicker, Jr.; U.S. Patent 4,056,356; November 1, 1977; assigned to Eastman Kodak Co. have disclosed a textile fiber comprised of a polyoxyethylene-glycol-modified poly(1,4-cyclohexylenedimethylene terephthalate)-type polyester containing phosphorus, a stabilizer, an organic titanium compound, and optionally a manganous ion. The fiber exhibits an unobvious combination of commercially acceptable mechanical properties, lightfastness, gasfastness, and dye rate.

The fiber can be described as a deep dyeable or light dyeable fiber comprised of:

    (A) A polyetherester of:

        (1) Terephthalic acid; and

        (2) A diol component comprised of:

            (a) 1,4-cyclohexanedimethanol; and

            (b) From 3 to 12 wt %, based on the weight of the polyetherester, of polyoxyethylene glycol having a molecular weight in the range of 300 to 800;

    (B) Based on the weight of the polyetherester, from 0 to 150 weight ppm of manganous ion;

    (C) Based on the weight of the polyetherester, from 20 to 500 weight ppm phosphorus derived from certain phosphorus compounds described below;

    (D) Based on the weight of the polyetherester, from 1,000 to 5,000 weight ppm of a stabilizer effective to reduce oxidative degradation of the polyetherester; and

(E) Based on the weight of the polyetherester, from 50 to 200 weight ppm titanium derived from a soluble tetravalent organic titanium compound.

As will be recognized, the amount of polyoxyethylene glycol is related to the depth of shade that will be achieved when the fiber is dyed without a carrier disperse dye. Generally speaking, the more polyoxyethylene glycol that is used, the darker will be the shade of the dyed fiber and the less polyoxyethylene glycol that is used, the lighter will be the shade of the dyed fiber. Accordingly, in the broadest embodiment, there is provided a deep dyeable or light dyeable fiber wherein the amount of polyoxyethylene glycol is from 3 to 12 wt %. In one specific embodiment, there is provided a light dyeable fiber wherein the amount of polyoxyethylene glycol is from 3 to 6 wt %. In another specific embodiment, there is provided a deep dyeable fiber wherein the amount of polyoxyethylene glycol is from 8 to 12 wt %.

In one aspect of this process, the phosphorus compound can be phosphorous acid, phosphoric acid, pyrophosphoric acid or polyphosphoric acid. In another aspect, the phosphorus compound corresponds to the formula:

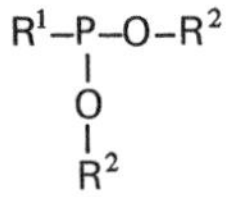

$$R^1-P-O-R^2$$
$$|$$
$$O$$
$$|$$
$$R^2$$

where R^1 is $-H$, a monovalent alkyl radical having 1 to 18 carbon atoms, a monovalent aryl or substituted aryl radical having 6 to 15 carbon atoms, or $-O-R$ where R is the same as R^1. R^2 is $-H$, monovalent alkyl radical having 1 to 18 carbon atoms, or a monovalent aryl or substituted aryl radical having 6 to 15 carbon atoms, provided that when R^2 is alkyl, at least one of R^2 has a chain of at least two unsubstituted methylene groups attached to the oxygen atom attached to phosphorus.

A preferred antioxidant used to prevent oxidative degradation is pentaerythritol tetrakis[3-(3,5-tert-butyl)-4-hydroxyphenyl] propionate (Irganox 1010) and corresponds to the structure:

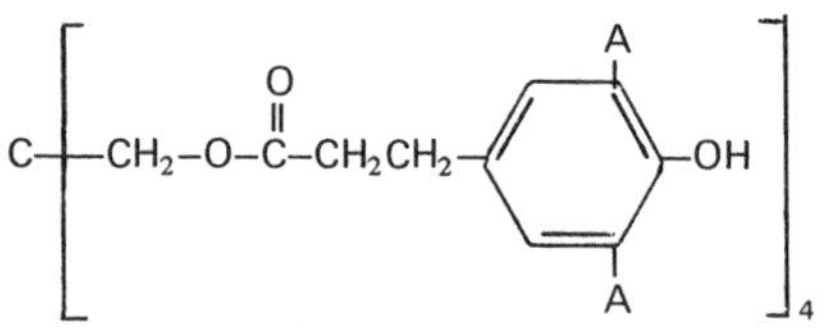

where A is $-C(CH_3)_3$.

Among the other hindered phenols which are useful are 4,4'-butylidenebis(6-tert-butyl-m-cresol), 1,3,5-trimethyl-2,4,6-tris(3,5-di-tert-butyl-4-hydroxybenzyl)benzene, tris(3,5-di-tert-butyl)-4-hydroxyphenyl phosphate, and dioctadecyl 3,5-di-tert-butyl-4-hydroxybenzyl phosphonate.

Hydantoin Derivatives with Polyalkoxylated Groups Linked to Ring Nitrogens

Fiber-forming polyesters are modified by E.L. Lawton, II; U.S. Patent 4,122,072;

October 24, 1978; assigned to Monsanto Company by the incorporation in the chain of hydantoin derivatives having polyalkoxylated groups linked to the ring nitrogens. These polyesters exhibit desirable properties such as improved dyeability, thermal oxidative stability, light stability and dyed lightfastness.

A preferred embodiment of the hydantoin derivative has the following formula:

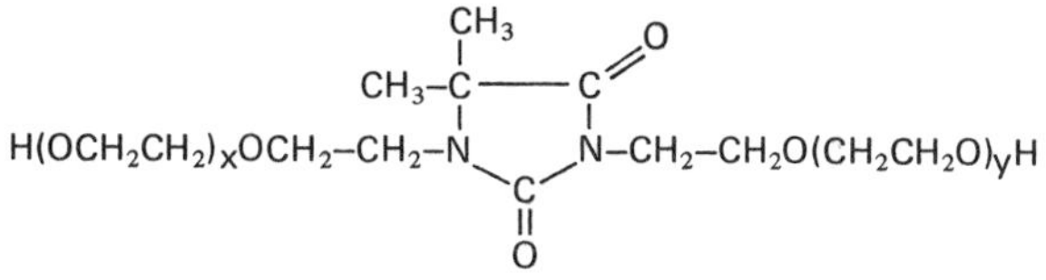

$$H(OCH_2CH_2)_xOCH_2-CH_2-N \underset{\underset{O}{\overset{\parallel}{C}}}{\overset{\overset{CH_3}{\underset{|}{C}}-C\overset{\nearrow O}{}}{}} N-CH_2-CH_2O(CH_2CH_2O)_yH$$

wherein the sum of x and y is an integer from 1 to 20 and more preferably an integer from 4 to 15.

Example: *Evaluation of Dye Depth* — Knit tubing of the fibers was scoured using a 40:1 liquor to fabric ratio in an aqueous scour bath containing 2 g/l of Varsol, 1 g/l of Igepal CO-630 and 0.5 g/l of sodium hydroxide. Tubing was scoured for 20 minutes at 93°C and rinsed in hot tap water. The tubing was then rinsed in an aqueous bath containing 0.5 g/l of acetic acid with 40:1 liquor to fabric ratio for 10 minutes at 43°C. The fabric was rinsed in hot tap water and dried at 60°C. The fabrics were dyed in capped glass tubes tumbled in an oil bath maintained at 100°C for 80 minutes.

The aqueous dyebath contained 4.0% owf brilliant blue standard commercial dye and 8.0% owf Igepon T-33, and dyeing were conducted at 40:1 liquor to fabric ratio. Dyed fabrics were scoured for 15 minutes at 60°C at 40:1 liquor to fabric ratio in an aqueous solution containing 0.5% owf of Igepal CP-710 and 0.5% owf trisodium phosphate. The dyed fabrics were then rinsed in hot tap water and dried at 60°C. The dye depth of tubings were measured by (K/S) reflectance values at 680 nm from the Kubelka and Munk equation as described in Deane B. Judd and Gunter Wysjecki, *Color in Business, Science, and Industry,* Third Edition, John Wiley, New York, 1975, pages 420-438.

This key property, dye depth is illustrated in Figure 1.7. Figure 1.7 shows the relationship between dye depth and amount of structural modifying unit. In this figure, dye depth (K/S) is the ordinate and the weight percent of structural modifier unit is the abscissa. Four curves are shown on this figure.

Curve 1 is representative of those compositions consisting of ethylene terephthalate copolymers prepared from polyethoxy diols. These compositions exhibit the best dye uptake with the lowest percentage of modifier. Although these compositions exhibit the best dyeability they are deficient to the compositions of this process in other areas, namely, thermal oxidative stability, light stability and dyed lightfastness.

Curve 2 is representative of compositions of this process having a structure shown by the above formula. This curve illustrates the beneficial effect of the polyalkoxylated hydantoin modifier when the x + y component falls within the prescribed range.

Curve 3 represents compositions modified with a hydantoin derivative, however, the x + y component is not within the prescribed range of this process.

Curve 4 is representative of polymers having compositions similar to that disclosed in U.S. Patent 3,856,754, also containing a hydantoin derivative. These compositions are shown to exhibit the poorest dyeability.

Figure 1.7: Dye Depth as a Function of Percent Modifier

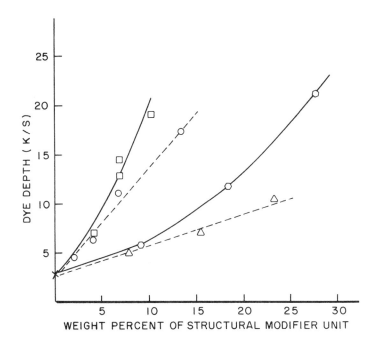

CURVE

1. □

2. ○

3. ○

4. △

Source: U.S. Patent 4,122,072

IMPROVEMENT IN OTHER PROPERTIES

Enhanced Solvent Resistance

J.R. Costanza and G.L. Collins; U.S. Patent 4,101,399; July 18, 1978; assigned to Celanese Corporation describe the irradiation of compositions comprising nonethylenically unsaturated brominated and chlorinated aromatic polyester fibrous materials with low intensity ultraviolet light radiation for relatively brief periods of time under appropriate conditions to yield a product of enhanced solvent resistance. The irradiation of the halogenated polyesters (as defined) generates free radicals within the fibrous material which react to produce sufficient crosslinking to enhance solvent resistance without a significant detrimental change of other product properties.

The period of exposure of the fibrous substrate possessing fibers of one-eighth inch thickness to ultraviolet light emitted from a lamp with a source intensity of about 5.0 watts per linear inch is about 20 minutes. A lamp with a source intensity of from 100 to 200 watts per linear inch may require an exposure of time of from 1 to 10 seconds or less (e.g., 0.1 to 1.0 second for a lamp with an intensity of 200 to 300 watts per linear inch). The fibers of a substrate which are one-quarter inch in thickness may require a period of exposure of 2 seconds or less with a lamp with a source intensity of 200 to 400 watts per linear inch.

Overexposure to a lamp with an intensity much greater than 200 watts per linear inch may result in degradation which takes the form of discoloration, brittleness due to excessive crosslinking, lessened strength, dullness, and even charring.

It is preferred to use a lamp with a source intensity of 200 watts per linear inch, a fiber thickness of one-eighth inch, and an exposure time of one second.

Example: A chlorinated aromatic polyester containing chlorine chemically bound to an aromatic ring is formed by reacting 190.9 parts by weight of tetra-chlorobisphenol A and a mixture of 75.6 parts by weight isophthaloyl chloride and 32.4 parts by weight terephthaloyl chloride. The resulting chlorinated poly-ester possesses the structure shown below where X and Y are chlorine, groups R and R' are methyl groups, and n = about 80.

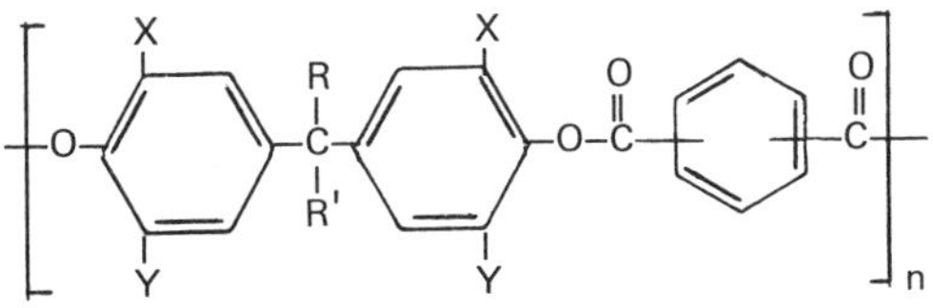

The chlorinated aromatic polyester has a chlorine content of about 27 wt %, and an inherent viscosity of about 0.8 dl/g determined at a concentration of 0.1 wt % in a solvent which is a mixture of 10 parts by weight phenol and 7 parts by weight trichlorophenol.

The chlorinated aromatic polyester is solution spun from a methylene chloride spinning solvent into an evaporative air atmosphere to form a filamentary material one-eighth inch in thickness which is hot drawn, and cut into 1½ inch lengths, and provided in a staple configuration. The chlorinated polyester so prepared is tested for its solvent resistance before and after exposure to the ultraviolet radiation. Thus several nonirradiated samples of the abovedescribed chlorinated polyester each weighing 5 g are placed in 95 cc of chloroform (solvent) and maintained at a temperature of 25°C and a pressure of 760 mm. At the same time several other samples of the same product each weighing 5 g are irradiated under conditions of a standard pass as described in the patent and are then placed in 95 cc of chloroform (solvent) under the same conditions of temperature and pressure.

The nonirradiated samples dissolve completely in 60 seconds, while the irradiated samples fail to dissolve to any substantial degree over a period of six months.

Improved Water Absorbency

According to *A.S. Forschirm; U.S. Patent 4,063,887; December 20, 1977; assigned to Celanese Corporation* the water absorbency of normally hydrophobic polyester fibers is improved by contacting (e.g., by immersing) these fibers in an aqueous solution of a hydroxyamine selected from the group consisting of monohydroxyamines, dihydroxyamines, and trihydroxyamines, and mixtures thereof.

Example: A 100% polyethylene terephthalate polyester fabric weighing 7.7 oz/yd^2 is first washed three times in clear hot water in a washing machine to remove oils. Then four swatches of about 2 g each are cut out and placed in separate pressure containers with 80 g of water. 0.2 g each of either ethanolamine, diethanolamine, or 2-diethyl-aminoethanol are added to the first three containers respectively yielding solutions having a pH of about 7 to about 9. The fourth container contains water and no additives and serves as a control. The containers are sealed and heated in a water bath at 120°C for 2 hours. The samples are removed, washed one time in clear hot water in a washing machine, and tumble dried.

The water absorptivity of the samples is tested by allowing a drop of water to fall from a dropper onto the fabric which is on a flat surface. Such test is similar to the procedure outlined in the AATCC wettability test 39-1971. The time is noted at the initial contact of the drop with the fabric and again upon disappearance of the drop as determined by an observer. The swatches are then washed successively in clear water and tested after 4, 6, 8, 10 and 12 washings. The results of the tests are given in the table below.

| | Number of Clear Water Washings | | | | | |
| | 1 | 4 | 6 | 8 | 10 | 12 |
Treatment	Water Absorption (sec)					
None	210	210	270	300	360	900
Ethanolamine	*	*	30	10	30	20
Diethanolamine	20	20	60	40	80	90
2-Diethylaminoethanol	*	*	*	*	5	8

*Instantaneous

Polyalkylene Terephthalates Which Crystallize Rapidly

There has been no lack of attempts to provide polycondensates in which the good properties of both polyethylene terephthalate and polypropylene terephthalate or polybutylene terephthalate are combined. Thus it is known, for example, that the tendency of polyethylene terephthalate to crystallize can be improved by nucleation with suitable nucleating agents and/or by increasing the rate of diffusion within the melt by adding lubricants. However, these measures are not suitable for increasing the rate of crystallization of polyethylene terephthalate to such an extent that it can be processed at low mold temperatures and short molding times similar to those used in the case of polybutylene terephthalate.

P. Bier, R. Binsack and H. Vernaleken; U.S. Patents 4,107,149; August 15, 1978 and 4,086,212; April 25, 1978; both assigned to Bayer AG, Germany have found that terephthalic copolyesters, the diol component of which mainly consists of ethylene glycol units, crystallize more rapidly than pure polyethylene terephthalate and can be molded in a way comparable to polybutylene terephthalate.

Examples 1 through 7: 97.1 g (0.5 mol) of dimethyl terephthalate are transesterified with 1.045 mols of ethylene glycol in the presence of 58 mg of zinc acetate for 2 hours at 200°C and for 1 hour at 220°C. When transesterification is complete, 0.6 ml of GeO_2 solution (5% strength in ethylene glycol), 103 mg of triphenyl phosphate and the corresponding codiol are added. The temperature is raised to 250°C in the course of one hour and, at the same time, the apparatus is evacuated (< 1.0 mm Hg). Polycondensation is complete after a further 45 to 60 minutes. A clear viscous melt of the copolyester is obtained and on cooling this solidifies to a white crystalline mass.

Examples 1 through 4 describe the copolyesters wherein 1.0 to 10 mol percent of the diol component is 3-methyl-pentane-2,4 diol. A slight deterioration in the properties can be observed when the proportion of 3-methylpentane-2,4-diol is increased to above 7 mol percent. Comparison examples 6 and 7 relate to polyethylene terephthalate and polybutylene terephthalate respectively, whereas Example 5 describes a copolyester wherein the proportion of the codiol, 3-methylpentane-2,4-diol, is well above the maximum specified for this process.

Ex. No.	Ethylene Glycol (mol %)*	3-Methyl-pentane-2,4-Diol (mol %)*	η (dl/g)	ΔH_m (cal/g)	T_m (°C)	ΔH_c (cal/g)	T_c (°C)	T_m-T_c (°C)
1	99	1	0.55	12.5	254.5	12.0	202.5	52
2	95	5	0.57	13.2	255	13.2	208.5	46.5
3	93	7	0.53	13.0	256	13.0	204	52
4	90	10	0.50	10.6	256	9.9	197	59
5	80	20	0.63	8.9	254	8.9	183	71
6	100	—	0.55	11.2	256.5	4.7	164	92.5
7**	—	—	0.55	14.7	229	11.8	181	48

*Relative to the diol component.
**Polybutylene terephthalate.

In the table, η denotes intrinsic viscosity in phenol/tetrachloroethane, 1:1, measured in a Ubbelohde capillary viscometer; polymer concentration: 0.5 g/dl, temperature: 25°C; ΔH_m denotes enthalpy of melting; T_m denotes melting temperature; ΔH_c denotes enthalpy of crystallization; and T_c denotes crystallization temperature, measured with a DSC 2 (Perkin Elmer) using a sample weight of about 10 mg at a heating and cooling rate of 20°C/min.

The samples were characterized by their intrinsic viscosity and the thermo-dynamic data important for the melting and crystallization properties, such as enthalpy of melting (ΔH_m), melting temperature (T_m), enthalpy of crystalliza-tion (ΔH_c) and crystallization temperature (T_c).

It can be seen from the results of the measurements that the melting point of the polyethylene terephthalate copolyester (Examples 1 through 3) rises and reaches the value for pure polyethylene terephthalate (Example 6) as the codiol content increases. The heat of fusion ΔH_m of the copolyesters and for com-parison, that of pure polyethylene terephthalate can be employed as a measure of the crystallinity. The crystallinity curve is analogous to that for the melting point. The crystallinity is at a maximum at a codiol content of 5 mol percent and at this point exceeds that of pure polyethylene terephthalate (Examples 2 and 6).

At a constant rate of cooling and under otherwise identical experimental condi-tions, the rate of crystallization is higher the earlier the polymer crystallizes out, that is to say the super-cooling: $\Delta T = T_m - T_c$ indicates when the rate of crystal-lization reaches its maximum under the cooling conditions used.

The super-cooling ΔT decreases as the codiol content increases and passes through a minimum at 5 mol percent of 3-methylpentane-2,4-diol and increases again from 7 mol percent. Accordingly, poly(ethylene glycol/3-methylpentane-2,4-diol)terephthalate has a maximum rate of crystallization at a codiol content of 5 mol percent of 3-methylpentane-2,4-diol.

It is known that the rate of crystallization is inversely proportional to the melt viscosity: the rate of crystallization is lower the higher the molecular weight (L. Mandelkern, *Crystallization of Polymers*, McGraw-Hill).

Poly(ethylene glycol/3-methylpentane-2,4-diol)terephthalate and polybutylene terephthalate have approximately the same rate of crystallization.

The copolyesters can be molded at mold temperatures between 110° and 150°C, preferably at about 120°C, and at injection pressures of 740 kp/cm^2 and follow-up pressures of about 380 kp/cm^2 and, under these conditions, the cycle time can be considerably shorter (namely 30 to 35 seconds) than in the case of con-ventional polyethylene terephthalates containing nucleating agents.

High Impact Strength Copolyesters

A copolyester having good impact strength is prepared by *R.W. Campbell and J.W. Cleary; U.S. Patent 4,107,150; August 15, 1978; assigned to Phillips Petroleum Company* utilizing terephthalic acid or the lower alkyl esters thereof and a mixture of 1,4-butanediol and 1,4-cyclohexanedimethanol as the sole monomers.

Examples: In a series of runs, dimethyl terephthalate was reacted with 1,4-butanediol to produce poly(tetramethylene terephthalate), a homopolymer out-side the scope of this process, or with 1,4-butanediol and 2,4-cyclohexanedi-methanol in varying ratios to produce copoly(tetramethylene/1,4-cyclohexylene-dimethylene terephthalate).

Each of the runs was conducted in a two-liter, stainless steel, stirred autoclave designed for polycondensation reactions. In each of the runs 5.0 mols of diol component, i.e., 1,4-butanediol plus 1,4-cyclohexanedimethanol, 2.0 mols of dimethyl terephthalate, and about 1.8 millimols of tetraisopropyl titanate were charged initially to the autoclave except that in Run 6 the 1,4-cyclohexanedimethanol was added after the mixture of other components had been heated to 220°C and the transesterification reaction was essentially completed. No 1,4-cyclohexanedimethanol was used in Run 1.

In each of Runs 1 through 5 the mixture of components was heated gradually at atmospheric pressure from about 160° to about 210°C during a period of about 1 hour, during which time methanol by-product was removed by distillation. The mixture was then maintained at about 210° to 220°C at atmospheric pressure for 1 hour, during which time additional methanol distilled. House vacuum (about 210 to about 330 mm Hg pressure) was then applied slowly over approximately 10 minutes as the temperature was raised to about 240° to 250°C, at which temperature and pressure the mixture was maintained for about 10 minutes. The pressure was reduced to 1 mm Hg or less and maintained there for about 30 minutes. With the temperature still about 240° to 250°C and the pressure at 1 mm Hg or less, a small stream of nitrogen was then bubbled through the melt for 3 to 4½ hours. The product was then removed from the autoclave.

Run 6 was conducted in essentially the same manner except that a mixture of the dimethyl terephthalate, 1,4-butanediol, and tetraisopropyl titanate was heated gradually at atmospheric pressure from about 160° to about 220°C, with distillation of methanol by-product over a period of about 1½ hours prior to the addition of the 1,4-cyclohexanedimethanol, followed by use of house vacuum, lower pressure, and a nitrogen stream as described above, except that the pressure rose to as much as 2.1 mm Hg during the bubbling of nitrogen through the melt.

Each of the resulting polymers was then evaluated. The copolyesters in Runs 2 through 6, all within the scope of this process, exhibited greater impact strength than did the homopolyester in Run 1. Additionally, the copolyesters in Runs 2 through 6 underwent less thermal degradation during molding than did the homopolyester in Run 1, as shown by the values for inherent viscosity (at 0.5 g polymer per 100 ml 3:2 by weight phenol/1,1,2,2-tetrachloroethane at 30°C) before and after molding each of the polymers. Furthermore, the copolyesters were more resistant to hydrolysis than was the homopolyester, as determined by a comparison of data in Runs 3 and 5 with data in Run 1 showing that the hydrolysis treatment resulted in a smaller reduction of inherent viscosity for the copolyesters than for the homopolyester as well as resulting in better weight retention of the copolyesters than of the homopolyesters.

Reduced Color Formation

According to *M.C. Baker; U.S. Patent 4,128,535; December 5, 1978; assigned to E.I. Du Pont de Nemours and Company* color formation in a polyester formed of a saturated acid, a glycol and 1,4-butanediol can be reduced if, during preparation of the polyester, a reaction mass is first formed of the acid and the glycol and the butanediol is then added to the reaction mass after the acid-glycol reaction is practically complete.

The polyester thus produced will have significantly less color formation than the same type of polyester conventionally prepared. For example, polyesters prepared according to this process will generally have APHA color values of about 100 or less, while conventionally prepared polyesters of the same type may have APHA color values as high as 400 to 500. Color is measured by the system developed by the American Public Health Association (APHA), using a standard colorimeter, a 40 mm x 20 mm cell and a number 42 blue filter. The colorimeter is calibrated according to ASTM D-1209-62, using a solution of platinum-cobalt APHA 500 color standard SOP-120 made by the Fisher Co.

Example: In the following example, all parts are by weight. A reactor was charged with 1,050 parts of ethylene glycol, and then purged with nitrogen. Adipic acid, 2,000 parts, was added to the glycol, with stirring, followed by the addition of 0.25 part of phosphoric acid in 2 parts of ethylene glycol.

The resulting reaction mass was heated to 240°C and held there, with continuous stirring, for 1 hour, during which period evolved volatiles were continuously withdrawn from the vessel.

1,4-butanediol, 600 parts, was then added to the reaction mass, with stirring, and the temperature of the mass was brought back to 240°C. Tetraisopropyl titanate, 0.02 part in 2 parts of ethylene glycol, was then added.

A vacuum of 10 mm of mercury was drawn on the vessel and the temperature of the reaction mass was held at 240°C for 4 hours, during which period evolved volatiles were again withdrawn from the vessel.

Heating was then discontinued and the vessel was filled with nitrogen. The resulting polyester had an APHA color value of 46. A polyester of the same composition, made according to a conventional process in which the acid, glycol and butanediol were charged to the reaction vessel simultaneously, had an APHA color value of 208.

Prevention of Color Degradation

W. Hewertson; U.S. Patent 4,082,724; April 4, 1978; assigned to Imperial Chemical Industries Limited, England describes a polyethylene terephthalate composition comprising highly polymeric polyethylene terephthalate and, in an amount sufficient to prevent color degradation of the polyethylene terephthalate, (1) a trihalide or tri(pseudohalide) of antimony, and (2) an organic oxo compound of phosphorus.

One method of estimating the effectiveness of the process is with reference to the luminance of the polymeric products, increase in the value of which reflects a reduction in greyness. Luminance is a measure of the proportion of the incident light reflected on examination of the polymer using a "Colormaster" which is the trade name for the differential colorimeter manufactured by Manufacturers Engineering and Equipment Corporation. The luminance may be measured on the as made polymer or on articles, e.g., filaments, fibers, films or molded articles, fabricated therefrom.

Example 1: 100 parts by weight of dimethyl terephthalate and 71 parts of ethylene glycol were added cold to a stainless steel vessel which had previously been purged with nitrogen and which was provided with heating means, a stirrer, an offtake for volatile material and a nitrogen inlet and outlet. The mixture was then melted at 120° to 130°C under nitrogen and an amount of manganese acetate tetrahydrate equivalent to 0.025 part of anhydrous manganese acetate was added. Gentle heating was then applied and the temperature was gradually raised to 220°C, and the methanol distilled off. When the theoretical amount of methanol had been collected, 0.015 part of phosphoric acid was added and the mixture was transferred under oxygen-free conditions to a stirred polymerization autoclave. 0.074 part of antimony trifluoride and 0.5 part of titanium dioxide were then added and the temperature raised to 235°C. Pressure was then reduced to 1 mm of mercury absolute while the temperature was raised to 290°C. The temperature was then maintained at 290°C until an adequate melt viscosity had been achieved, and the polymer then extruded and granulated. The polymeric product, intrinsic viscosity 0.71 (1 g polymer in 100 ml o-chlorophenol at 25°C), had a luminance value of 54.

Example 2: The process was repeated but using only 0.037 part of antimony trifluoride (i.e., about 0.028 wt % based on bis glycol ester assuming 100% conversion of the dimethyl terephthalate) and, in place of the phosphoric acid, 0.063 part of tetraisopropyl methylene diphosphonate The polymeric product, intrinsic viscosity 0.68, had a luminance value of 75. [The diphosphonate was prepared by the method described in the *Journal of the American Chemical Society* (1961) Volume 83, page 1722.]

Example 3: The process of Example 2 was repeated but using only 0.019 part of antimony trifluoride and, in place of the diphosphonate, 0.043 part of tetraphenyl methylene diphosphine dioxide. The polymeric product, intrinsic viscosity 0.71, had a luminance value of 67.

The tetraphenyl methylene diphosphine dioxide was prepared as follows: tetraphenyl methylene diphosphine (38.4 parts) in acetone (400 parts) was treated with 100 volume hydrogen peroxide (23 parts) in acetone (100 parts) at 0°C. The diphosphine dioxide so produced crystallized on evaporation of the solvent to about one-third of the original volume of solution. It was recrystallized from hot acetone and dried in a vacuum oven at 120°C for 3 hours to yield crystals having a melting point of 185° to 186°C; yield, 75 to 85%.

Improved Color While Retaining Tenacity

A polyester fiber must have a certain degree of tenacity retention to be commercially acceptable because the fiber must have a minimum tensile strength in order to be processed into a textile fabric and to be acceptable in typical textile uses, such as wearing apparel.

By the term "tenacity retention" is meant the retention of tenacity of a polyester fiber when the fiber is carried through the various processing steps, such as heat setting, carding, spinning, weaving, bleaching and the like, that occur between melt spinning and the manufacture of the finished polyester or polyester-cotton fabric.

The problem in achieving a combination of commercially acceptable blue-whiteness and commercially acceptable tenacity retention is a particularly difficult problem because an improvement in blue-whiteness often results in a reduction in tenacity retention and an improvement in tenacity retention often causes a reduction in blue-whiteness. For example, introduction of an oxidation stabilizer into the fiber increases the tenacity retention but decreases the blue-whiteness of the fiber.

R. Gilkey, S.D. Hilbert and T.H. Wicker, Jr.; U.S. Patent 4,049,621; Sept. 20, 1977; assigned to Eastman Kodak Company have developed a polyetherester fiber dyeable without a carrier that has commercially acceptable blue-whiteness and tenacity retention.

The textile fiber is broadly comprised of an admixture of:

(A) a polyetherester of
    (1) terephthalic acid, and
    (2) a diol component comprised of
        (a) ethylene glycol, and
        (b) from 6 to 12 wt %, based on the weight of the poly-etherester, of poly(oxyethylene)glycol having a molecular weight in the range of 200 to 600;
(B) from 50 to 200 weight ppm, based on the weight of the polyetherester, cobaltous aluminate;
(C) from 0.01 to 0.16 wt %, based on the weight of the polyetherester, of a stabilizer;
(D) a whitening agent which is
    (1) from 0.5 to 4.0 wt ppm, based on the weight of the poly-etherester, of a thermally stable organic compound characterized by having a high reflectance in the range of 400 to 500 nm and strong absorbance in the range of 550 to 650 nm, or
    (2) from 100 to 400 weight ppm, based on the weight of the polyetherester, of a thermally stable organic compound having a fluorescence emission spectrum in methylene chloride in the range of 425 to 445 nm.

A preferred antioxidant is pentaerythritol tetrakis[3-(3,5-di-tert-butyl)-4-hydroxy-phenyl] propionate which is sold commercially as Irganox 1010 by Geigy Chemical Company and corresponds to the structure

Among the other hindered phenols which are useful are 4,4'-butylidenebis(6-tert-butyl-m-cresol), 1,3,5-trimethyl-2,4,6-tris(3,5-di-tert-butyl-4-hydroxybenzyl)-benzene, tris(3,5-di-tert-butyl)4-hydroxyphenyl phosphate, and dioctadecyl 3,5-di-tert-butyl-4-hydroxybenzyl phosphonate.

Compounds suitable as the first whitening agent are described in "Color Index," The Society of Dyers and Colorists, 3rd Edition (1971). Examples of compounds are Indanthrene Brilliant Violet 3B (C.I. 60,005), Platinum Violet (C.I. 60,010), and Isoviolanthrone (C.I. 60,000).

In a preferred embodiment the whitening agent is C.I. Pigment Violet 23 and is described in British Patent No. 387,565.

Examples of compounds suitable as the second whitening agent are 2,2'-(vinylene-di-p-phenylene)bis(4,6-diphenyl-5-triazine) sold as Uvitex MES, 2,2'-(2,5-thio-phenediyl)bis-[5-(α,α-dimethylbenzyl)benzoxazole] sold as Uvitex 1980, 2,2'-vinylenedi-p-phenylenebisbenzoxazole sold as Eastman OB-1, and 7-(2H-naphtho-[1,2-d] triazol-2-yl)-3-phenylcoumarin sold as Leucopure EGM.

In a preferred embodiment the second whitening agent corresponds to the structure

This compound is well known in the art and is disclosed in U.S. Patent 3,260,715.

Good Melt-Forming Properties

Polyesters obtained by the ester-interchange reaction method have the disadvantage of poor melt-forming properties. In carrying out the ester-interchange reaction and the polycondensation reaction of the ester-interchange products, some metal compounds are used as catalysts to accelerate the reactions. These metal compounds remain in the obtained polyesters as insoluble and infusible foreign matter most of which have a diameter of more than about 20 μ, and cause some troubles in the melt-forming process of the polyesters. In manufacturing polyalkylene terephthalate filaments, e.g., the foreign matter builds up around the spinning orifices of a spinneret pack when a molten polyalkylene tereph-thalate is melt-spun through the spinneret pack, and thus makes it difficult to perform the smooth melt-spinning operation because of the resulting abrupt rise of the pressure in the spinneret pack and the breakage of spun filaments.

Y. Morimatsu, T. Tanaka, N. Okumura and T. Horiuchi; U.S. Patent 4,133,801; January 9, 1979; assigned to Teijin Limited, Japan have conducted continued and strenuous studies concerning the influence of impurities contained in the crude di-lower alkyl ester (1 to 4 carbon atoms) of terephthalic acid on the qualities and melt-spinning properties of the polyalkylene terephthalates. As a result of the strenuous studies, it has been found that if a di-lower alkyl ester of terephthalic acid and an alkylene glycol are reacted in the presence of a specific amount of an alkyl p-formylbenzoate, and di-alkyl isophthalate and p-toluic acid and/or a monoalkyl terephthalate, the abovementioned disadvantages can be substantially overcome.

Examples 1 and 2, Comparative Examples 1 through 4: A reaction vessel fitted with a stirrer and a rectifying column was charged with 100 parts dimethyl terephthalate which contains (A) methyl p-formylbenzoate, (B) dimethyl iso-phthalate, and (C) p-toluic acid and monomethyl terephthalate in the amount as shown in the table on the following page, 70 parts ethylene glycol and 0.063 part (0.069 mol percent based on dimethyl terephthalate) of calcium acetate monohydrate (an ester-interchange catalyst), following which an ester-interchange reaction was carried out by heating the mixture for 4 hours raising the temperaature from 150° to 240°C under normal atmospheric pressure while removing methanol formed. Thereafter, 0.029 parts phosphorous acid (a stabilizer), 0.04 part antimony trioxide (a polycondensation catalyst) and 0.5 part titanium dioxide (a delustering agent) were added to the ester-interchange product, following which a polycondensation reaction was carried out by heating the product for 4 hours at a temperature of 285°C under a reduced pressure of 0.5 to 1 mm Hg. The physical properties of the polyesters were as shown below, where L and b denote those of Hunter's Color Diagram in which L represents lightness. The greater the L value, the lighter is the color. The greater the absolute value of b, the deeper is the color shade. The preferred L value of the polyester is more than 65 and b value is less than 5.5.

Ex. No.	. . . . Compound (ppm)					Properties of the . . . Obtained Polyesters . . .			
	A*	B**	***	†	Total	η	SP (°C)	. . Color . . L	b
1	142	285	15	60	502	0.636	262.8	66.8	5.1
2	284	569	66	33	952	0.637	262.8	66.7	4.9
1††	3.8	3.2	0.5	10.3	17.8	0.637	263.0	68.3	5.5
2††	20	176	82	102	380	0.639	263.0	64.3	5.0
3††	65	1,572	305	284	2,226	0.639	262.0	61.0	6.1
4††	1,308	1,144	216	273	2,941	0.637	262.1	65.0	6.5

*Methyl p-formylbenzoate.
**Dimethylisophthalate.
***p-toluic acid.
†Monomethyl terephthalate.
††Comparative.

The polyesters according to Examples 1 and 2 had high softening point and good color as did the polyester in Comparative Example 1 which was obtained from the fiber grade dimethyl terephthalate. The pressure rise in the spinneret packs and the ratio of wrapping of the polyesters in Examples 1 and 2 were considerably less than the polyester in Comparative Example 1. The polyesters in Comparative Examples 2 through 4 did not have good color. In addition, the polyesters in Comparative Examples 1 through 3 had too much pressure rise in the spinneret packs and ratio of wrapping.

Melt-Processable Thermotropic Polyester

A wholly aromatic polyester is provided by *G.W. Calundann; U.S. Patent 4,130,545; December 19, 1978; assigned to Celanese Corporation* which, unlike the aromatic polyesters normally encountered in the prior art, is not intractable or difficultly tractable and readily undergoes melt processing with ease. The aromatic polyester of the process consists essentially of the recurring units (a) p-oxybenzoyl moiety, (b) m-oxybenzoyl moiety, (c) 2,6-dicarboxynaphthalene moiety, and (d) symmetrical dioxy aryl moiety, and is free of units which possess ring substitution. The resulting polyester exhibits a melting point below approximately 310°C, and preferably below 300°C. The ability of the wholly aromatic polyester readily to undergo melt processing can be attributed to its atypical inherent propensity to form a thermotropic melt phase at relatively low temperatures. The wholly aromatic polyester may be formed by a variety of procedures including a slurry polymerization technique described in the patent or a melt polymerization technique. The presence of the m-oxybenzoyl moiety in the wholly aromatic polyester has been found to render the polymer melt processable at even lower temperature than if this component were omitted.

Example: To a three-neck, round bottom flask equipped with a stirrer, nitrogen inlet tube, and a heating tape wrapped distillation head connected to a condenser are added the following: 58.54 g p-acetoxybenzoic acid (0.325 mol), 4.50 g m-acetoxybenzoic acid (0.025 mol), 16.21 g 2,6-naphthalene dicarboxylic acid (0.075 mol), and 14.56 g hydroquinone diacetate (0.075 mol).

This mixture is brought to a temperature of 250°C. At 250°C, the 2,6-naphthalene dicarboxylic acid is suspended as a finely divided solid in molten p-acetoxybenzoic acid, m-acetoxybenzoic acid and hydroquinone diacetate. The contents of the flask are stirred rapidly at 250°C under a slow stream of dry nitrogen for about 2 hours while acetic acid is distilled from the polymerization vessel. The polymerization suspension is then raised to a temperature of 280°C and is stirred at this temperature for 3 hours under a nitrogen flow while additional acetic acid is evolved.

About 80 ml of acetic acid is collected during these stages. The polymerization temperature is next increased to 320°C. The viscous polymer melt is held for 15 minutes at 320°C under a nitrogen flow and then subjected to a series of reduced pressure stages. The nitrogen is shut off and the pressure is reduced to about 300 mm of mercury for about 20 minutes, 210 mm for 15 minutes, 70 mm for 15 minutes and finally about 0.2 mm for 10 minutes. During these periods the polymer melt continues to increase in viscosity and is stirred more slowly while the remaining acetic acid is removed from the reaction vessel. The polymer melt is next allowed to cool to ambient temperature (i.e., about 25°C). Upon cooling, the polymer plug is finely ground in a Wiley Mill and dried in a forced air oven at 100°C for 50 to 60 minutes.

Approximately 54 g of polymer are obtained. The inherent viscosity of the polymer is approximately 2.4 as determined in pentafluorophenol solution of 0.1 wt % concentration at 60°C.

When the product is subjected to differential scanning calorimetry, it exhibits a large sharp endotherm at about 308°C (peak), which repeats at about 308°C on subsequent remelt scans. The polymer melt is thermotropic.

When the melt is cooled in a differential scanning calorimeter at a rate of –20°C per minute, a sharp polymer crystallization exotherm is observed at about 264°C (peak) indicating a rapid crystallization.

The resulting wholly aromatic polyester next is melt extruded to form oriented fibers directly from the melt or injection molded for forming three-dimensional shaped articles.

Good Moldability

T. Kodama, I. Sasaki and H. Mori; U.S. Patent 4,141,882; February 27, 1979; assigned to Mitsubishi Rayon Co., Ltd., Japan disclose a polyester composition obtained by blending, with a polyester comprising at least 80% by mol of ethylene terephthalate units, (1) an epoxy compound (A) having an isocyanuric acid ester construction or a cyanuric acid ester construction and at least one organic compound (B) capable of reacting with the epoxy group of the epoxy compound (A) and having one or more carboxyl, amino, isocyanate or hydroxyl groups or (2) a product of a melt reaction of the epoxy compound (A) with the organic compound (B), and then melt mixing and milling the blend. The polyester composition may comprise a fibrous filler and glass flakes, and has a high moldability and excellent physical properties.

Examples: Polyethylene terephthalate (PET) of an inherent viscosity of 0.72 obtained by the condensation of terephthalic acid and ethylene glycol was blended with prescribed amounts of triglycidyl isocyanurate (TGIC) and acrylic acid (Aa) in a tumbler for 3 minutes. The blend was melt mixed and milled in a nonvented extruder of a diameter of 30 mm (L/D = 25) at 275°C with a mean residence time of 5 minutes. The obtained pellets were again subjected to extrusion under the same condition and the melt index (MI) of each of the pellets obtained by the first and second extrusions was determined.

For comparison, melt indices of pellets were obtained as mentioned above but in this case, obtained by using TGIC and Aa in amounts outside the scope of this process.

When only TGIC is used, if the added amount is less than 0.0005 mol no heightening effect of the melt viscosity is obtained; while if the added amount is greater than 0.0005 mol the MI of the product is not uniform or the product becomes infusible.

When both TGIC and Aa are used, if TGIC is added in an amount less than 0.0005 mol or Aa is added in an amount greater than 2 mols, no heightening effect of the melt viscosity is obtained; while if TGIC is added in an amount greater than 0.02 mol, or Aa is added in an amount less than 0.2 mol the MI of the product is not uniform or the product becomes infusible.

If the added amount of TGIC is in a range between 0.0005 and 0.02 mol and that of Aa is in a range between 0.2 and 2.0 mols, the MI of the product is uniform and the product does not become infusible even after the second extrusion. In either the first extrusion or the second extrusion, the melt viscosity of the extruded strand remained high and stable during the continuous processing which occurred over a long period of time.

Thermoplastic Carbonate-Modified Copolyesters

According to *R. Binsack and H. Vernaleken; U.S. Patent 4,041,018; August 9, 1977; assigned to Bayer AG, Germany* carbonate-modified copolyesters which are suitable for the production of moldings by injection or for the production of pipes, films and fibers by extrusion, can be obtained by reacting aromatic dicarboxylic acids or their ester-forming derivatives and carbonic acid esters of dihydric aliphatic alcohols which have carbonic acid ester groupings in the molecule and hydroxyalkyl end groups.

The copolyesters are as a rule highly crystalline and have high melting points. The melting points of the carbonate-modified copolyesters are, surprisingly, substantially higher, for the same number of carbon atoms of the cocondensed carbonic acid hydroxyalkyl esters than those of corresponding homopolyesters of pure C–C diols. A further advantage of the copolyesters is their high toughness.

Example 1: *Preparation of Carbonic Acid Bis-(4-Hydroxybutyl Ester)* – 1,938 g (21.5 mols) of 1,4-butanediol and 1,181 g (10 mols) of carbonic acid diethyl ester are heated in the presence of 100 mg of calcium acetate, while the temperature slowly rises from about 100° to 180°C, in such a way that the ethanol split off distills off continuously, a packed column which gives sufficiently good separation being used. After about 8 hours, 912 g (19.8 mols) of ethanol have been split off. The unconverted diethyl carbonate and 1,4-butanediol is then distilled off through the column during which time the temperature of the reaction product is not allowed to rise above 180°C. This gives 2,042 g (9.9 mols) of carbonic acid bis-(4-hydroxybutyl ester) as the residue, characterized by an OH number of 556 mg of KOH/g (theory: 544) and a content of carbonate bonds of 21.0% of CO_2 (theory: 21.3%).

Example 2: 3.5 kg (18 mols) of dimethyl terephthalate and 3.72 kg (18 mols) of carbonic acid bis-(4-hydroxybutyl ester) are continuously heated, in the presence of 3.2 g of titanium tetrabutylate, to 230°C over the course of 3 hours. The pressure is then lowered to 1 mm Hg over the course of 60 minutes while at the same time raising the temperature to 245°C. After 4 hours, the polycondensation is terminated. The crystalline copolyester has a melting point of 180°C.

Improved Fiber- and Film-Forming Qualities

T. Miwa, S. Nakazawa, M. Itoga, K. Sano, I. Nakamura, T. Watanabe and Y. Shingu; U.S. Patent 4,067,855; January 10, 1978; assigned to Toray Industries, Inc., Japan describe polyester compositions suitable for fiber and film formation, containing from 0.2 to 3.0 wt % of fine particles of uniform size and possessing sufficient slippery tendency, appropriate opacity and crystallizing potential; the particles contain lithium and phosphorus as a lithium salt of a phosphoric acid ester constituent of short chain polyesters.

These polyester compositions have excellent frictional characteristics, opacity and adequate rate of crystallization compared with polyester compositions previously obtained. Excellent properties are exhibited when these polyester compositions are subjected to the fiber-forming, film-forming and molding processes.

The following results are to be expected when the polyester compositions are manufactured into fiber products through melt-spinning, hot-drawing and heat-treatment processes.

(1) Improved processability resulting from the increase in deformability of the polymer in spinning and drawing processes and of running speed of the fiber on the guide.

(2) Decreased breaking of threads during spinning and drawing (the effect is especially remarkable for a thread with modified cross section).

(3) Increased uniformity of deformation of the polymer in spinning and drawing gives rise to an improvement in uniformity of qualities.

(4) Improved processability in the processes of texturizing, twisting, weaving and knitting.

(5) Improved adhesiveness in the usage for rubber reinforcement.

Furthermore, the following excellent effects are also to be expected when the polyester compositions are melt-extruded into film and stretched biaxially, re-stretched in the longitudinal direction, heat-treated, slitted and wound up.

(1) Increase in the speed of film-production resulting from the improved deformability of the polymer.
(2) Uniformity of quality of the film is improved by increase in deform-ability of the polymer.
(3) Decrease of stretch tension in the transverse direction of the film re-duced the mechanical load of the tenter.
(4) Reduction of the friction between the film and blade of the slitter reduces the deformation of edges of the slitted film.
(5) The film can be wound up more uniformly giving the rise to a hard wound roll.
(6) Outstanding quality with an excellent opacity and slippery surface free from undesirable coarse particles and faults.

Example: 100 parts of dimethyl terephthalate and 70 parts of ethylene glycol are transesterified in the usual manner in the presence of 0.09 part of manganese acetate as catalyst.

Then 0.03 part of antimony trioxide, 0.20 part of lithium acetate and 0.15 part of trimethyl phosphate are added, then polymerized by the usual manner.

A product having an intrinsic viscosity of 0.63 (here and in the following, the intrinsic viscosity is measured at $25°C$ in o-chlorophenol) is obtained.

The polymer is pressed between two cover glasses on a hot plate at $290°C$ to a thin film, and observed in dark ground illumination under a polarization micro-scope. Many uniform particles, about 0.5μ in diameter are observed.

The maximum take-up speed in melt spinning, at an output of 30 g/min at $290°C$, of the polyester composition obtained in this example is 4,000 m/min without breaking of threads.

When the undrawn yarn from this polyester composition is drawn 3.5 times at $100°C$ into 100 pirns of 2 kg weight, no breakage of threads is observed.

The dynamic friction coefficient of this drawn yarn against a roll plated with chromium is 0.62.

The polyester composition obtained in this example is processed into biaxially oriented film of 25μ thickness and 3,000 m length under the usual film forming condition and is wound up (machine direction stretch ratio, 3.3; transverse di-rection stretch ratio, 3.4; heat set temperature, $215°C$; heat set time 13 sec) and then this rolled film is cut into 2 cm width by a knife edge.

The static friction coefficient between the above films is 0.40, the haze of the film is 32%, the maximum surface roughness by the stylus method is 0.2μ, and the surface of this film has many fine points of unevenness.

Wound forms of both unslitted and slitted film are excellent; it is especially noted that there is no deformation of edges on the slitted end of the wound film.

Comparative Example: By the same method as used in the above example except adding only antimony trioxide and lithium acetate after transesterification, a polyester is prepared and spinning tests and film-forming tests are carried out.

The maximum take-up speed of this polymer is 3,800 m/min, and breakage of threads is observed on two pirns, the dynamic friction coefficient of the yarn against the roll plated with chromium is 0.60, the static friction coefficient between 25 μ films is 0.38, the haze is 54%, and the maximum surface roughness is 0.3 μ.

The analytical data of the polyester composition with the differential scanning calorimeter indicates that the rate of crystallization is higher than that of the polymer obtained in the preceding example.

Low Molding Temperature with High Glass Transition Temperature

E.E. Paschke; U.S. Patent 4,035,342; July 12, 1977; assigned to Standard Oil Company (Indiana) discloses a new family of resins which are essentially linear copolyesters comprising units of a polyhydric alcohol component comprising at least one dihydric alcohol moiety and a dicarboxylate component wherein the dicarboxylate component comprises terephthalate moieties and 2,2',6,6'-tetramethylbiphenyl-4,4'-dicarboxylate moieties in a range of mol ratios of from 19:1 to 1:19.

The terephthalate polyesters have a relatively low molding temperature without reduction of the second order transition temperature of the polymer.

The copolyesters have a range of utility according to the mol ratios of the monomers. Copolyester films and molded parts based on ethylene glycol of mol ratios from 19:1 to 4:1 of ethylene terephthalate moieties and ethylene tetramethylbiphenyl dicarboxylate moieties can be used for hot-filled packaging applications because the glass transition temperature (T_g) of these films and molded parts range upward from about 78° to 95°C. Poly(ethylene terephthalate) film has a T_g of about 74°C (165°F) which restricts its use in hot-filling.

A series of copolyesters of mol ratios 19:1 to 3:2 of ethylene terephthalate moieties to ethylene tetramethylbiphenyl dicarboxylate moieties can be used in thermal applications where poly(ethylene 2,6-naphthalate) with a T_g of about 115°C has been required. The T_g of such a series, 19:1 to 3:2, ranges upward from about 78° to about 119°C. One such application where the mol ratio is 3 ethylene terephthalate moieties to 2 ethylene tetramethylbiphenyl dicarboxylate moieties, or 3:2, where the T_g is about 115°C, is in truck tire cord. Truck tires have not been fabricated from polyester cords since the heat buildup in the tires raises the temperature of the tire cord above 75°C and the tire cord loses strength as it stretches.

A series of copolyesters of mol ratios of 19:1 to 2:3 of ethylene terephthalate moieties to ethylene tetramethylbiphenyl dicarboxylate moieties can be used in thermal applications where polycarbonates with a T_g of about 145° to 148°C have been required. The T_g of such a series, 19:1 to 2:3, ranges upward from about 78° to about 145°C. A series of copolyesters of mol ratios of 19:1 to 1:4 of ethylene terephthalate moieties of ethylene tetramethylbiphenyl dicarboxylate moieties can be used in thermal applications where polyarylates with a T_g of about 173°C have been required. The T_g of such a series, 19:1 to 1:4, ranges upward from about 78° to about 169°C. One such application where the mol ratio is 1:4 is in molding medium to high-temperature electrical components with thermal properties engineered to desired requirements. Further, the copolyesters have good resistance to hydrocarbon and aromatic solvents, are colorless (white), have the ability to be drawn into fibers from the melt, and have a broad softening temperature above the glass transition temperature which allows molding and fabrication. In the following example parts and percentages are by weight unless otherwise indicated.

Example: This example illustrates the production of polyester containing 2,2'-6,6'-tetramethylbiphenyl-4,4'-dicarboxylate moieties, terephthalate moieties and glycol moieties. 3.1 g (0.016 mol) dimethyl terephthalate (DMT) mixed with 1.3 g (0.004 mol) dimethyl 2,2',6,6'-tetramethylbiphenyl-4,4'-dicarboxylate (M_2DMe), 2.8 g (0.044 mol) ethylene glycol, 0.05 g zinc acetate and 0.05 g calcium acetate were heated at 160°C for 120 minutes in a test tube equipped with a nitrogen bubbler and a side-arm. During the heating, nitrogen was passed slowly through the mixture.

After two hours the temperature was raised to 210°C and 0.05 ml antimony tris-butoxide was added. A partial vacuum was pulled on the mixture over a period of 10 to 15 minutes, using a vacuum pump attached to the side-arm. The temperature was raised to 275°C. When this temperature was reached, full vacuum (0.9 mm Hg) was applied and the reaction continued for 133 minutes. 3.8 g of low molecular weight polymer were obtained. It had an inherent viscosity (I.V.) of 0.29 dl/g, as determined in a 60/40 phenol-tetrachloroethane mixed solvent at 30°C.

Copolyesters of ethylene terephthalate (ET) and ethylene 2,2',6,6'-tetramethyl-biphenyl-4,4'-dicarboxylate (M_2D) in mol ratios 3:2, 2:3 and 1:4 were prepared in the same apparatus by the same procedure.

As the concentration of terephthalate moieties to tetramethylbiphenyl dicarboxylate moieties decreases from a molar ratio of 4:1 to 1:4, the molding temperature of the polymer increases. As the concentration of the tetramethylbiphenyl dicarboxylate moieties increases, the T_g of the polymer increases. A 19:1 mol ratio terephthalate to tetramethylbiphenyl dicarboxylate polyester having an I.V. of 0.63 dl/g has a T_g of 78°C and falls on the same line.

The 3:2 molar ratio ET/M_2D, based on 60% DMT and 40% M_2DMe, was further polymerized using a solid state polymerization procedure. A small sample of the 3:2 ET/M_2D copolyester was ground to No. 10 mesh using a laboratory grinder. The sample was then heated in a test tube at 210°C and 0.25 mm Hg vacuum for 16 hours. The resulting polymer had an inherent viscosity of 0.35 dl/g.

When the ethylene glycol in the 4:1 mol ratio terephthalate to tetramethylbi-phenyl dicarboxylate polyester was replaced with tetramethylene glycol and the polyester was prepared in the same way, a polyester was produced having an I.V. of 0.53 dl/g, a T_g of 52° to 54°C and a melting temperature of 173° to 179°C. In this case also the T_g of the copolyester falls on the linear line connecting the T_g points of homopolyesters of polytetramethylene terephthalate and polytetramethylene-2,2',6,6'-tetramethylbiphenyl-4,4'-dicarboxylate.

Improved Flexibility

D.R. Fagerburg and A.J. Cox; U.S. Patent 4,155,889; May 22, 1979; assigned to Eastman Kodak Company disclose a process for substantially increasing the flexibility of a polyester by annealing the polyester. The polyester is either a dimer acid or a poly(alkylene oxide)glycol modified poly(1,4-cyclohexylenedimethylene-1,4-cyclohexanedicarboxylate).

Example 1: *Random Polymer* — A random polyester of the process is prepared from 80 mol percent 1,4-cyclohexanedicarboxylic acid with 100% trans isomer content, 20 mol percent dimer acid, and 1,4-cyclohexanedimethanol.

Into a reaction flask equipped with a stirrer, nitrogen inlet, and outlet for volatile materials produced during the reaction is weighed the following: 80.0 g (0.40 mol) of dimethyl-1,4-cyclohexanedicarboxylate, 100% trans isomer, 56.6 g (0.10 mol) of Empol 1010 dimer acid (Emery Industries, Inc.), 124 g (0.60 mol, 20% excess) of a 70% solution in methanol of 1,4-cyclohexanedimethanol, 70% trans isomer, and 150 ppm. Ti metal as titanium tetraisopropoxide in n-butanol. The mixture is stirred under nitrogen and immersed in a metal bath held at 200°C. After 15 minutes of stirring, the bath temperature is raised to 225°C, held there for approximately 20 minutes, and raised to 290°C. Upon reaching 290°C, a vacuum is applied to a final pressure of 0.1 torr. The reaction is polycondensed for 3 hours and 15 minutes after which time the apparatus is repressurized with nitrogen and the polymer allowed to cool. The inherent viscosity of the polymer is 0.81.

A film is prepared by extrusion of the molten polyester. A portion of this film is annealed by placing the film in a hot air circulating oven for 1 hour at 120°C and exhibits a flexural modulus of 10,500 psi. The remaining portion of the film was not annealed and exhibited a flexural modulus of 14,200 psi.

Example 2: *Blocked Polymer* — A blocked polyester is prepared from 80 mol percent 1,4-cyclohexanedicarboxylic acid with 100% trans isomer content, 20 mol percent dimer acid, and 1,4-cyclohexanedimethanol.

Into a reaction flask equipment with a stirrer, nitrogen inlet, and outlet for volatile materials produced during the reaction is weighed the following: 64.0 g (0.320 mol) of dimethyl-1,4-cyclohexanedicarboxylate, 100% trans isomer, 82 g (0.40 mol) of a 70% solution of 1,4-cyclohexanedimethanol in methanol, and enough titanium tetraisopropoxide to give 150 ppm of titanium metal based on the final weight of the blocked polymer. The mixture is stirred under nitrogen at 200°C for 1 hour and the temperature raised to 290°C.

The vacuum is then applied to a final pressure of 0.1 torr. After one hour of polycondensation, the flask is repressurized with nitrogen, 54.4 g (0.08 mol) of a polyester of dimer acid and 1,4-cyclohexanedimethanol having an inherent viscosity of 0.4 is added and the polymer stirred for 10 minutes following which vacuum is reapplied and polycondensation continued for an additional 2 hours and 20 minutes. The blocked polymer is cooled under nitrogen and has an inherent viscosity of 0.75.

A film is prepared by extrusion of the molten polyester. A portion of this film is annealed by placing the film in a hot air circulating oven for 1 hour at 120°C and exhibits a flexural modulus of 8,000 psi. The remaining portion of the film was not annealed and exhibited a flexural modulus of 12,000 psi.

Improved Oxidation Resistance

Polyesters of 1,2-bis(4-hydroxyphenyl)ethane and aromatic dicarboxylic acids exhibiting improved oxidation resistance are obtained by *O.D. Deex and V.W. Weiss; U.S. Patent 4,115,357; September 19, 1978; assigned to Monsanto Company* by incorporating into the polyester units derived from a bis(hydroxyphenyl)sulfide. In addition, the polyesters possess a combination of strength, processability, high temperature performance, resistance to burning, crystallinity and solvent-resistance which makes them useful as molding resins, fibers and surface coatings.

Example 1: *Preparation of Poly[1,2-Bis(4-Hydroxyphenyl)Ethane] Isophthalate —* A charge consisting of 82 parts of isophthalic acid and 148 parts of 1,2-bis(4-acetoxyphenyl)ethane and 0.005 mol sodium carbonate per mol of 1,2-bis(4-acetoxyphenyl)ethane is placed in a reaction vessel equipped with a stirrer, condenser and receiver. The vessel is evacuated and purged with nitrogen 3 times. A nitrogen blanket is maintained in the reactor while it is heated to 250°C for about 3 hours during which period approximately 35 to 40 parts of acetic acid distills. Thereupon the vessel is evacuated to a pressure of about 200 torrs for about 5 minutes and the pressure is slowly reduced to about 1 torr while the temperature is increased to about 290°C. When the amperage on the stirrer motor increases by 0.02 to 0.04 amp, the stirrer motor is switched off and the vacuum is released with nitrogen.

From 95 to 97% of the theoretical amount of acetic acid is collected. The polymer is extruded from the vessel under slight nitrogen pressure and is reduced in a Thomas mill to a powder of particle size in the range of about 0.1 to 0.25 mm. The inherent viscosity is about 0.3. The powder is charged to a reaction vessel which is then purged with nitrogen. The pressure is reduced to 0.1 to 0.2 torr and the temperature is raised to about 10° to 15°C below the melting point of the polymer. Heating is continued for 12 hours. The vessel is cooled and the crystalline polymer is discharged. The inherent viscosity of the polymer is 0.90 at 30°C, at 0.5 g per 100 ml 60/40 by weight phenol/tetrachloroethane. The polymer melts at 279° to 290°C. The rate of crystallization is 2.56 min^{-1}.

Example 2: Example 1 is repeated with an equivalent amount of bis(4-acetoxyphenyl)sulfide substituted for 5 mol percent of the 1,2-bis(4-acetoxyphenyl)-ethane. The inherent viscosity of the polymer is 0.83. The polymer is crystalline.

Example 3: Example 1 is repeated with an equivalent amount of bis(4-acetoxyphenyl)sulfide substituted for 6 mol percent of the 1,2-bis(4-acetoxyphenyl)-ethane. The inherent viscosity of the polymer is 1.17. The polymer melts at 263° to 267°C; the T_g is 151°C. The crystallization rate is 1.78 min^{-1}.

Example 4: Example 1 is repeated with an equivalent amount of bis(4-acetoxyphenyl)sulfide substituted for 10 mol percent of the 1,2-bis(4-acetoxyphenyl)-ethane. The inherent viscosity of the polymer is 1.13. The polymer melts at 252° to 257°C; the T_g is 153°C. The crystallization rate is 2.13 min^{-1}.

Example 5: Example 1 is repeated with bis(4-acetoxyphenyl)sulfide substituted for the 1,2-bis(acetoxyphenyl)ethane. The polymer is amorphous.

Samples of the polymers are subjected to an accelerated high temperature oxidation test by the procedure set forth hereinabove. The data are presented in the following table and demonstrate the improved oxidation resistance of polymer Examples 3, 4 and 5 which are within the scope of this process.

Ex. No.	....Composition, Mol Ratio*....			Time to Develop Insolubility at 250°C (min)
	IA	BHPE	BHPS	
1	50	50	—	1.5
3	50	47	3	3.5
4	50	45	5	3.5
5	50	—	50	1.5

*IA is isophthalic acid; BHPE is 1,2-bis(4-hydroxyphenyl)ethane; and BHPS is bis(4-hydroxyphenyl)sulfide.

Sample of polymers of Examples 1, 2 and 5 are molded in thicknesses of 1.58 mm and 0.80 mm. The samples are then subjected to the UL-94 test and the average times of flame out in successive burns and the average time of afterglow is determined. The data of the following table show that the 1,2-bis(4-hydroxyphenyl)ethane polyesters of Examples 1 and 2 are superior in fire safety performance to the bis(4-hydroxyphenyl)sulfide polyester of Example 5. Furthermore, Example 2 prepared from 1,2-bis(4-hydroxyphenyl)ethane containing 5 mol percent bis(4-hydroxyphenyl)sulfide is superior in suppression of afterglow after the termination of flaming combustion in comparison with the polyester of 1,2-bis(4-hydroxyphenyl)ethane unmodified with bis(4-hydroxyphenyl)sulfide (Example 1).

Ex. No.	Composition ....Mol Ratio....			Sample Thickness (mm)	Average Flame Out Time* (sec)	Average Afterglow Time (sec)	UL-94 Designation
	IA	BHPE	BHPS				
1	50	50	—	0.80	6.0	23	V-O
2	50	47.5	2.5	1.58	2.0	2.5	V-O
				0.80	3.0	<1	V-O
5	50	—	50	1.58	**	—	—

*Second burns.
**Flaming drip.

Easily Removable Sizing

M.A. Lerman and J.C. Lark; U.S. Patent 4,145,461; March 20, 1979; assigned to Standard Oil Company (Indiana) provide a warp size for a variety of synthetic fibers which can be removed easily by scouring with mildly basic water solutions at 120°F within 30 minutes. This warp size does not require rigorous control over electrolyte balance, because the size is not precipitated or insolubilized by strong electrolytes, such as sodium hydroxide, hydrochloric acid, and sodium chloride or by highly concentrated solutions of relatively weak electrolytes.

It does not become so strongly coupled to a sized fabric which had been subjected to heat-setting prior to scouring that it is difficult to remove.

It has been found that the above objects can be achieved by the process of sizing with an aqueous composition comprising:

(A)   about 1 to 50 wt % of a low molecular weight water-dispersible polyester having a number average molecular weight in the range of about 850 to 2,500 and preferably 1,100 to 1,800 and an acid value in the range of about 40 to 85 mg of KOH/g of the polyester (mg/g) and preferably in the range of about 45 to 65 mg/g, which comprises:

(1)   a nonlinear polyester backbone having an acid value in the range of about 5 to 15 mg/g and preferably about 8 to 12 mg/g comprising the reaction product of (a) an aliphatic polyol wherein all the hydroxyl groups are preferably primary, (b) an aromatic dicarboxylic acid, and (c) a monocarboxylic acid, and

(2)   pendant polycarboxylic acid moieties attached to the backbone by ester linkages wherein each pendant moiety has at least one free carboxylic acid group;

(B)   about 0.5 to about 4 wt % and preferably about 1 to about 3 wt %, as based upon the weight of the polyester resin present, of a water-dispersible and water-stable chelate of titanium capable of interacting with the polyester so as to form a nontacky film; and

(C)   up to about 50 wt % of water as based upon the total composition.

The number average molecular weights were those found by the Vapor Phase Osmometry method. It has been found that in general polyesters of the warp sizing composition described above having a number average molecular weight in the range 850 to 2,500 gave rise to nontacky films having good to excellent adhesion.

Example 1: *Preparation of the Low Molecular Weight Polyester* — Into a 3-liter reaction vessel equipped with stirrer set at a rate of 5 rpm and means for gas sparging, add 1.275 mols of pelargonic acid and 2.48 mols of trimethylol propane. The reaction vessel was heated to 150°F under 0.3 scfm of nitrogen sparge.

When the reaction vessel has reached 150°F, 1.8 mols of isophthalic acid having a purity of 85% is added and the heating is continued to 340°F. When 340°F is reached, the heating is continued to 450°F over a 4 to 5 hour period.

The 450°F temperature is maintained until an acid value of 5 to 8 mg/g is obtained for the polymer in the reaction vessel. After an acid value of 5 to 8 mg/g is obtained, the reaction mass is cooled to 340°F, utilizing a water jacket.

0.52 mol of trimellitic acid anhydride is added to the reaction vessel. The temperature is maintained in the range of 340° to 350°F in order to maintain the presence of a free acid group and until an acid number of 62 to 65 mg/g is obtained, and then the reaction vessel was cooled to 280°F.

During this cooling process, isopropyl alcohol was added until 75% nonvolatile solution is obtained. To this isopropyl alcohol solution add slowly aqueous ammonia so as to completely neutralize all carboxyl groups and then continue adding aqueous ammonia until an excess of about 50% ammonia has been added. The resulting neutralized resin solution is to be stirred thoroughly.

Distilled water in small increments of roughly 5 g each are added with thorough stirring. It is to be noted that as distilled water is initially added the neutralized resin dispersion becomes very thick and translucent. However, as a 30% resin solids dispersion is approached, it begins to thin out. When about 50 g of water has been added then 0.3 g of Tyzor TE (1% on ester resin) is added with careful and thorough stirring. The final amount of distilled water and/or 28% aqueous ammonia to produce a 30% solution having an adjusted pH of about 8 to 9.5 is then added with stirring.

A 30% solution of ester resin with 1% Tyzor TE based thereon having a pH adjusted with 28% ammonia to 8 to 9.5 has a viscosity in the range 20 to 60 cp.

Example 2: *The Effectiveness of Desizing* — The completeness of removal by scouring of a sizing composition from a test fabric was determined as follows. A test fabric is immersed for 1 minute in one of several gently stirred scour baths consisting of 2 g/l of a surfactant such as Dextrajet Scour 99, and either 2 g/l or 10 g/l of soda ash, or about 2% sodium hydroxide, and enough soft water to provide a 60:1 liquor to goods ratio. The bath is at one of the following temperatures: 120°F, 130°F, 150°F, or 180°F.

After removal from scour bath and wringing off excess scour bath liquid by hand, a test fabric is immersed for 1 minute in a gently stirred soft water rinse bath having a 60:1 liquor to goods ratio. The rinse bath is at the temperature of the scour bath.

After removal from rinse bath and wringing off excess rinse bath liquid by hand, the test fabric is immersed for 1 minute in a mildly agitated dye bath with a pH of 6.0 to 7.0 adjusted by means of sodium hydroxide consisting of 1 g/l of Astrazon Blue liquid 50 (Verona) and enough soft water to provide a 60:1 liquor to goods ratio. The dye bath is at about 70°F.

After removal from dye bath, the test fabric is immersed for 1 minute in a mildly agitated rinse bath consisting of 10% acetic acid and 2 g/l Triton X-100 and enough soft water to provide a 60:1 liquor to goods ratio. The rinse bath is at room temperature.

After removal from rinse bath, the test fabric is rinsed for 1 minute under hot tap water (about 120°F) and air dried.

The depth of blue color over the total surface of the test fabric indicates the amount of size present. In all cases considered, substantially all of the size was removed.

Superior Lubricant

E.A. Weipert; U.S. Patent 4,169,062; September 25, 1979; assigned to Southern Sizing Co. describes a textile fiber lubricant, namely random copolymers of polyoxyethylene polyoxypropylene glycol monoester produced by the condensation reaction of an aliphatic fatty acid, or acids having from about 8 to about 22 carbons in the chain, with a mixture of ethylene oxide and propylene oxide, in the presence of an alkali catalyst. These fatty esters are water soluble, biodegradable and exhibit superior lubricating properties when applied to synthetic fibers. The esters have the empirical formula:

$$R-\overset{\overset{\displaystyle O}{\|}}{C}-O-(M)H$$

wherein R is an aliphatic chain having from about 7 to about 21 carbon atoms and M is a random mixture of oxyethylene $-CH_2CH_2O-$ and oxypropylene $-CH_2CH(CH_3)O-$ groups.

Examples of acids which are operable in the process include caprylic, pelargonic, capric, lauric, myristic, palmitic, stearic, oleic, linoleic, hydrogenated marine oil fatty acids, isostearic and mixtures thereof.

One advantage of the lubricant is the fact that it needs no emulsifier to produce a suitable lubricant. In other words, the random copolymers of polyoxyethylene polyoxypropylene glycol monoester are suitable for use, as such, or in aqueous solution and can be applied to both monofilament fibers and spun or staple synthetic fibers in the same manner as the prior art finishes are applied.

The lubricants appear to be stable at temperatures much higher than those temperatures which would be applied to fibers during the drying processes. Furthermore, the lubricants do not appear to readily be distilled by steam. This appears to be a major advance, e.g., for rayon treated with these lubricants over rayon treated with prior art lubricants which readily distill off of the rayon during the oven drying of the product, and thereby collect in the exhaust systems, causing severe fire hazards.

Indeed, these lubricants appear to remain on the fibers even after steam drying, thereby reducing the fire hazard, and reducing the cleanup time.

While heretofore the machinery utilized in the production of staple fibers tended to become extremely dirty, due to the loss of lubricant during the processing of the staple, the lubricants appear to have just the opposite effect. For example, when a lubricant of the process was employed on rayon fibers, it cleaned up the machinery through which the fibers passed, instead of causing an accumulation of "gunk" on such machinery.

The lubricants are not affected by the normal pH changes in the finish water. Of major economic importance when using the lubricants with rayon staple fibers is the fact that smaller amounts of heat are required to dry the fibers than with prior art lubricants. Thus, a mill can use its existing equipment for drying and by simply adding additional spinnerettes can feed more fiber poundage through this drying equipment, thereby increasing capacity by some 15 to 25%.

The lubricants, being esters, have a mild pleasant smell and, therefore, impart this smell to the mill during use. The bales of fibers treated with these lubricants appear to have a clean smell.

The evenness and uniformity of application of the lubricants to the fibers appear to be improved over prior art lubricants. When fibers treated with these lubricants are used, the finish solution in the mill remains clear and will remain stable without agitation.

The textile cards are able to run at a higher pounds per hour rate when staple fibers using these lubricants are processed in these cards.

The fly on drawing and especially on roving frames of synthetic fibers treated with these lubricants appears to be less.

The wet-out rate of fibers treated with these lubricants appears to be much faster than fibers treated with prior art lubricants which are water-insoluble. This is especially helpful on nonwoven fabrics, such as innerliners for disposable diapers.

The resiliency, openness and hand of staple fibers which utilize these lubricants appear to be excellent. With the addition of the good hand which is imparted by the lubricants, a mill should have a high rate of confidence in the finished fiber. The card web, roving package and yarn has a leaner and cleaner appearance.

The tackiness which is common in card laps from fibers treated with fatty acids is eliminated when fibers treated with these lubricants are processed.

BIORESORBABLE POLYMERS

Copolymer of Succinic and Oxalic Acids

Implantable surgical articles are provided by *J. Coquard, P. Sédivy, M. Ruaud and J. Verrier; U.S. Patent 4,032,993; July 5, 1977; assigned to Rhone-Poulenc Industries, France* which are at least partially bioresorbable and which consist at least partially of a copolyester of succinic acid and oxalic acid possessing receiving units of the general formulas:

$$-[O-\overset{\overset{\displaystyle O}{\|}}{C}-(CH_2)_2-\overset{\overset{\displaystyle O}{\|}}{C}-O-R]-$$

and

$$-[O-\overset{\overset{\displaystyle O}{\|}}{C}-\overset{\overset{\displaystyle O}{\|}}{C}-O-R]-$$

in which each R radical in the chain, which may be identical or different, represents a linear or branched alkylene radical possessing 2 to 6 carbon atoms, at least 2 of which carbon atoms form part of the polymer chain, a cyclo-alkylene radical possessing 5 to 8 carbon atoms, or a radical of the general formula

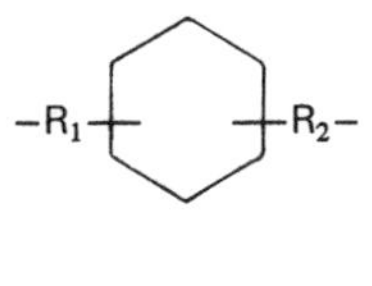

$$-R_1 -\hspace{-4pt}\left\langle\hexagon\right\rangle\hspace{-4pt}- R_2 -$$

in which each of R_1 and R_2, which may be identical or different, represents a methylene or ethylene radical. They are particularly suitable as sutures.

Example 1: *Preparation and Shaping of the Copolycondensate* — 159.42 g of succinic acid, 13.78 g of oxalic acid, 136.53 g of butane-1,4-diol, 3 g of p-toluenesulfonic acid and 1,200 cm^3 of toluene are introduced into a 1-liter, 3-necked flask equipped with a stirring system and a reflux condenser with a Florentine separator. The mixture is heated under reflux for 1 hour and 30 minutes so as to effect esterification of the oxalic and succinic acids, removing the water of the reaction by azeotropic distillation followed by decanting. The separator is then replaced by a Soxhlet apparatus packed with a 4 A dehydrating molecular sieve. The reaction mixture, which is kept under reflux, becomes increasingly viscous. After refluxing for 42 hours, the mass obtained is poured, with stirring, into 5 liters of methanol. The fibrous white polymer obtained is filtered off, washed with 3.5 liters of methanol, filtered off and then dried at ambient temperature in vacuo.

232 g of polymer are obtained, corresponding to a 90% yield. The reduced viscosity of the product, in the form of a solution in chlorophenol of concentration 2 g/l at 25°C, is 164 cm^3/g.

The copolyester is purified by dissolving it in 1,200 cm^3 of methylene chloride and then precipitating it by pouring into 5 liters of methanol; after drying in vacuo at ambient temperature, the copolyester has a reduced viscosity of 159 cm^3/g. Its melting point is 105°C and it is stable to heat under nitrogen up to 250°C.

Filaments of gauge 15 decitex are prepared from the above copolyester by melt spinning at 105°C followed by stretching on a plate at 77°C in a ratio of 6.5:1. These filaments possess the following mechanical properties (determined according to Standard Specification NF G 07-008, April 1961): tensile strength, 2.5 g/decitex; elongation at break, 49%; and tensile strength at a knot, 2.2 g/decitex.

Example 2: *Determination of the Local Tolerance of Implants in Rats* — A stretched monofilament of diameter 300 to 400 μ is produced using the copolyester prepared as described above and is sterilized by immersion for 1 hour in 70°GL (Gay-Lussac) strength ethyl alcohol.

1 cm long samples are implanted, by means of a semicurved needle, in rats (caesarean originated, barrier sustained) weighing 250 to 350 g; the implantation is carried out, on the one hand, in the paravertebral muscles and, on the other hand, under the skin of one of the sides of each animal. Four rats (two male and two female) each receive the two implants. Two months after the implantation, the animals are killed and the muscular implantation zone (implant and paravertebral muscular mass) and the subcutaneous implantation zone (skin, implant and underlying portion of muscle) are removed in each case. The following examinations are carried out on each sample removed:

(a)   Macroscopic examination after transverse section of one of the ends of the muscular mass (intramuscular implant) or detachment of the skin at one end (subcutaneous implant) of each sample; and

(b)   Histological examination after fixing the samples in Bouin liquid, inclusion in paraffin, cutting to 5 μ thickness and coloration using hematein/phloxine/saffron.

The following results were obtained:

Macroscopic Examination — Two months after the implantation, no tissue reaction which could be detected macroscopically was observed either with respect to the muscle or with respect to the subcutaneous tissue.

Histological Examination — Reaction of the Tissue: This reaction is restricted to a very discrete fibrous sheath around the implant and to the presence of a few multinucleate giant macrophage and histiocyte cells situated in contact with the implant itself.

No change which could relate to a possible toxic effect of the material was detected on the sections.

In short, the local tolerance in rats is very good from the macroscopic point of view and from the histological point of view.

Example 3: *Determination of the Bioresorbability "in Vitro"* — The bioresorbability of the copolyester obtained above is determined by measuring the variation in the reduced viscosity of a sample or polymer which is in the form of flakes and is incubated for varying periods of time is an enzyme extract. The test was carried out in the following way:

Preparation of the Enzyme Extract — "Fauves de Bourgogne" rabbits, of approximately 2.5 kg, are killed by means of chloroform. The muscles from the back and the thighs are removed and frozen at -20°C. They are then ground rapidly in a mincer. 300 g of ground product are dispersed in 500 ml of 0.2 M citrate buffer of pH 4.1 which has been cooled beforehand to +2°C. The suspension obtained is again ground for 1 minute (twice; grinder rotating at 2,000 rpm), with external cooling (mixture of ice and salt at approximately -13°C) so as to keep the temperature of the homogenized product below +10°C during the operation.

The homogenized product then undergoes an ultrasonic treatment (frequency: 20 kHz) for 2 minutes, taking the same cooling precautions. It is then centrifuged at 0°C for 20 minutes. The supernatant liquid, filtered through glass wool, forms the enzyme extract; its pH is approximately 4.5.

Reaction Between the Copolyester to be Investigated and the Enzyme Extract — A 100 mg sample of the copolyester to be investigated is incubated at 37°C with agitation (shaking table rotating at 80 rpm), in the following reaction medium: enzyme extract, 40 ml; streptomycin sulfate, 2 mg; sodium salt of penicillin G, 2 mg; and sodium nitride, 8 mg.

This reaction medium is renewed every day; during each renewal process, the samples are washed carefully with distilled water.

In parallel, other 100 mg samples of the product to be investigated are incubated in reaction media in which the enzyme extract is replaced by 0.2 M citrate buffer of pH 4.5.

The product to be investigated, which is in the form of flakes, is enclosed in a gauze bag for incubation purposes.

Examination of the Copolyester After Reaction — After 160 hours of incubation, the samples of the product to be investigated are recovered, washed several times with distilled water, drained on filter paper and dried under reduced pressure at 50°C.

The reduced specific viscosity of solutions of the product in 2-chloro-phenol at 25°C is then determined, as is that of the product before incubation. The following results were obtained:

Nature of the Incubation Medium	Reduced Specific Viscosity
None	143
Citrate buffer	130
Enzyme extract	114

It is found under these conditions that incubation in the active enzyme extract results in considerable degradation of the copolyester.

Reaction Between a Copolyester Yarn and the Enzyme Extract — The experiment described above is repeated, but is carried out on calibrated yarns instead of polyester flakes.

Examination of the Copolyester After Reaction — After 5 or 12 days of incubation, the samples of yarns are recovered, washed several times with distilled water, drained on filter paper and dried under reduced pressure at 50°C.

The reduced specific viscosity of solutions of the product in 2-chloro-phenol at 25°C is then determined. The following results are obtained.

Nature of the Incubation	Reduced Viscosity After Incubation for 5 Days	Reduced Viscosity After Incubation for 12 Days
Citrate buffer	134	109
Enzyme extract	131	104

The reduced viscosity of the control is 143.

Examination of the Yarns After Reaction — The breaking force of the yarns is measured on a tensometer of the Richard type. The following results, expressed in grams, are obtained.

Nature of the Incubation	Tensile Strength After Incubation for 5 Days	Tensile Strength After Incubation for 12 Days
Citrate buffer	537	423
Enzyme extract	476	430

A control has a tensile strength of 513 g.

Polyglycolic Acid plus Polyester of Diglycolic Acid and Unhindered Glycol

D.J. Casey and M. Epstein; U.S. Patent 4,118,470; October 3, 1978; assigned to American Cyanamid Company disclose bioabsorbable, hydrolyzable, polymeric reaction product of (1) a polyglycolic acid composition and (2) a polyester of diglycolic acid and an unhindered glycol and the process of preparing the same and the use of the polymeric reaction product as a sterile surgical element and as a device for the controlled continuous administration of a predetermined dosage of a drug to a living animal.

Example 1: Into a suitable reaction vessel preheated to 140°C, that has been purged with nitrogen gas, there is charged a mixture consisting of 160 parts of molten glycolide into which was placed 0.0032 part (0.002%) of stannous chloride dihydrate and 0.128 part (0.08%) of lauryl alcohol. The reaction vessel is heated to 218°C with constant stirring for 60 minutes after charging the molten glycolide to the reactor. The system is then placed under a 40 mm vacuum for about 30 minutes. The polyglycolic acid at this point has an inherent viscosity in hexafluoroacetone sesquihydrate of 0.94 when measured on a 0.5% solids solution at 30°C.

While maintaining the polyglycolic acid at about 218°C under a blanket of nitrogen, 40 parts of poly(1,4-butylene diglycolate), having an inherent viscosity in chloroform of 0.43 when measured on a 0.5% solids solution at 30°C, is added to the molten polyglycolic acid with constant stirring. Samples of the transesterified copolymer were removed from the reaction vessel at intervals of 40 minutes, 80 minutes and 120 minutes after the addition of the poly(butylene diglycolate). The samples all had inherent viscosities of about 0.68 to 0.69 when measured on a 0.5% solids solution in hexafluoroacetone sesquihydrate at 30°C and the samples when analyzed by nuclear magnetic resonance showed a content of 20.5 wt % of the poly(1,4-butylene diglycolate) and 79.5 wt % of the polyglycolic acid. The samples of the transesterified product were ground to a ten mesh size in a Wiley mill and dried in a vacuum oven at less than 1 mm of mercury and at 135°C for about 24 hours before being fabricated into a film and also into a fiber.

Example 2: Medicine, medication or other biologically active compositions including drugs, may be incorporated into a device comprising the polymeric reaction products by various techniques such as by solution methods, suspension methods or melt pressing.

For instance, 52 mg of pilocarpine hydrochloride was dissolved in 95% ethyl alcohol (0.5 ml) and 0.5 ml of the solution was added to a solution of the transesterification reaction product [52.8% poly(1,3-propylene diglycolate) by NMR] (0.95 g polymer dissolved in 3 ml of dioxane). The resulting solution was cast into a film and after drying in air and then under vacuum, the film was slightly hazy, strong and cold drawable.

As an alternative approach, finely ground pilocarpine hydrochloride (50 mg) was added to a solution of the transesterification reaction product, as in the preceding method, except that the hydrochloride was in suspension. The mixture was agitated until a good dispersion was obtained and the dispersion was then cast into a film and dried as before.

The film from the solution method set forth hereinabove was cut into large pieces and put between aluminum foils separated by 6 mil shims. The "sandwich" was pressed between chrome plated steel plates at a platen temperature of about 100°C for 30 seconds after preheating for 3 minutes. The sandwich was allowed to stand overnight in a desiccator to give the polyester transesterification reaction product time to crystallize.

Various other delivery devices may be manufactured from these transesterification compositions to administer drugs via a number of routes. For example, an intrauterine device for releasing an antifertility agent at a controlled rate for a prolonged period of time; a medical bandage for use in the continuous administration of controlled quantities of systemically active drugs over a prolonged period of time by absorption through the external body skin or mucosa; a strip which could be inserted between the gum and the cheek so that absorption of the medicament at a predetermined interval through the buccal mucosa into the bloodstream may take effect. Drugs could also be incorporated into fine particles of these polyester resin transesterification reaction products and subsequently a dispersion of these particles could be injected parenterally, subcutaneously, intramuscularly, etc., at which site the polymeric material would slowly biodegrade and release the drug over a prolonged period of time.

POLYLACTAMS

NYLON 6

Lactams of the general formula wherein R^1 is a C_{2-6} alkylene radical

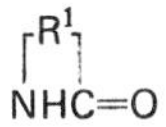

$$\begin{array}{c} \lceil R^1 \rceil \\ \mid \quad \mid \\ NHC=O \end{array}$$

are employed as starting materials for the preparation of high molecular weight polyamides. The homopolymer of ϵ-caprolactam is often referred to as nylon 6. These lactam polymers exhibit excellent tensile properties and can be formed into many products including textile fibers, films, castings and molded objects.

Pure, dry lactams, particularly ϵ-caprolactam, do not polymerize appreciably even at elevated temperatures; however, the polymerization can be activated using any one of a large number of materials as catalysts. Alkali metal salts of lactams are particularly effective catalysts, and are discussed by W.E. Hanford and R.M. Joyce in *Journal of Polymer Science*, Vol. 3, No. 2, pp 167 to 172 (1948). The polymerization proceeds rapidly at temperatures between 200° and 280°C and is conveniently carried out by adding the desired amount of alkali metal to the lactam to effect, first, salt formation followed by a polymerization which is catalyzed by the salt.

Reduced Coloration Due to Hexaorgano Distannane

A major shortcoming of the alkali metal salts of lactams is that they discolor relatively rapidly even under inert atmospheres, particularly at the elevated temperatures employed to liquify and polymerize lactams. The color imparted to the resultant polymer renders it unacceptable for many ends uses, particularly as textile fiber and film.

According to *J.W. Bouchoux and W.A. Larkin; U.S. Patent 4,108,828; August 22, 1978; assigned to M & T Chemicals Inc.*, hexaorgano distannanes of the general formula $R_3Sn-SnR_3$ significantly reduce or eliminate the discoloration which occurs when mixtures containing an alkali metal salt of a lactam are stored under an inert atmosphere for extended periods of time at temperatures between ambient and 100°C. R in the foregoing formula represents a monovalent hydrocarbon radical containing between 1 and 20 carbon atoms.

Example: The compositions employed to evaluate the stabilizers were prepared by transferring 50 cc samples of molten ε-caprolactam under a nitrogen atmosphere into containers which had previously been filled with nitrogen and charged with the desired amount of stabilizer. The containers were then placed in a water bath maintained at a temperature of 73°±2°C. Between 0.18 and 0.20 g of sodium metal was then placed in each of the containers.

The resultant hydrogen evolution was allowed to continue for between 3 and 5 min, as required, at which time the containers were sealed and the temperature of the water bath raised to 80°±5°C. The samples remained in the water bath for 24 hr, at which time they were removed and allowed to cool. The yellowness index of the resultant solid materials was measured using a Meeco Colormaster Differential Colorimeter. This instrument measures the percent green, red and blue light which is reflected from the sample. The yellowness index (YI) is calculated using the equation YI = (% red reflectance – % blue reflectance)/ % green reflectance.

Since the discoloration of the sodium salt in almost all instances produces a shade of yellow, the yellowness index provides a suitable means for determining the stability imparted by the various compounds tested. By definition the yellowness index is directly proportional to the intensity of this color.

The yellowness index values obtained using a variety of known stabilizers, including the hexaorgano distannanes are recorded in the following table together with the amount of stabilizer present in the test sample.

Stabilizer	Grams	Yellowness Index
None (control)	—	0.385, 0.385*
Hexamethyl distannane	0.2	0.160
Hexamethyl distannane	0.1	0.166
Hexabutyl distannane	0.2	0.193
Hexabutyl distannane	0.1	0.243
. . . . .Controls (prior art stabilizers)		
Butyl stannoic acid	0.2	0.300
2,4-Di-tert-butyl-p-cresol	0.2	0.304
Di-n-butyltin-S,S'-bis (lauryl mercaptide)	0.2	0.308
Bis(tri-n-butyltin) oxide	0.2	0.335
Tetrabutyltin	0.2	0.359
Stannous oxalate	0.2	0.373
Triphenyl phosphine	0.2	0.384
Triphenyl antimony	0.2	0.466

*Two trials.

The foregoing data demonstrate that the intensity of the yellow color which develops in compositions containing the sodium salt of ε-caprolactam and the hexaorgano distannanes is considerably less when compared against identical compositions containing known organotin and other stabilizers, even those which the prior art discloses as being effective stabilizers for polyamides.

Ozone-Resistant, Cationic Dyeable Nylon

An improved method of making nylon cationic dyeable is disclosed by *R.A. Lofquist and J.C. Haylock; U.S. Patent 4,083,893; April 11, 1978 and R.A. Lofquist; U.S. Patent 4,097,546; June 27, 1978; both assigned to Allied Chemical Corporation.* The soluble lithium, magnesium or calcium salt of a sulfonated polystyrene or polystyrene copolymer is added, to provide improved lightfastness and ozone resistance over conventional cationic dyeable nylon.

Example 1: A solution of the lithium salt of polystyrenesulfonic acid was prepared by dissolving 3.46 g of lithium carbonate in 57.5 g of a 30% aqueous solution of a 70,000 molecular weight polystyrenesulfonic acid which also contained 0.1 g of Dow Corning Antifoam 35 to reduce foaming. This entire solution was added to 1,520 g of caprolactam at 90°C. Manganese chloride (0.0576 g) and 0.1640 g of a 50% aqueous solution of hypophosphorous acid were added to serve as light stabilizers. The solution was homogeneous.

This solution was poured into a 3-liter agitated glass reactor equipped with a heating mantle, and a gas inlet and outlet to provide a nitrogen blanket over the molten mixture. 80 g of ε-aminocaproic acid was added as a polymerization initiator. The mixture was then heated over a period of about 1.5 hr to about 255°C. When the water flashed off, there was no phase separation.

At the end of 4.75 hr a polymer ribbon was extruded from the bottom of the reactor which was a pale yellow, without lumps, and of constant cross section. Unreacted caprolactam, about 10 wt % was removed by water extraction. The washed and dried polymer was submitted for analysis.

The formic acid relative viscosity was 65, with 72 equivalents of carboxyl and 31 equivalents of amine per 10^6 g of polymer. Sulfur analysis by x-ray fluorescence of the washed and dried polymer showed 2,150 ppm sulfur, or about 67 equivalents of sulfur per 10^6 g of polymer. A sample of the unwashed polymer contained about 2,400 ppm sulfur. The theoretical concentration of sulfur, based on the amount of polystyrenesulfonic acid salt added, was 2,240 ppm.

The polymer was submitted for spinning. The spinnerette had 14 holes each in the shape of a Y to get a yarn with a Y cross section. The spinning temperature was about 275°C. Pressure drop across the sand pack in the spin pot was about 5,900 psi.

The undrawn yarn had a total denier of 705 or an average of 50 denier per filament. The free fall yarn had a formic acid relative viscosity of 54, with 71 carboxyl equivalents and 24 amine equivalents per 10^6 g of polymer. Five ends of this yarn were gathered and drawn to 3.2 times the spun length, and then 2-plied to give a yarn of 2,260 total denier. This yarn had a tensile strength of 3.1 g per denier and an ultimate elongation of 45%. A control yarn (pure nylon 6) spun at the same time had a tensile strength of 3.3 g per denier and an ultimate elongation of 53%.

Example 2: A copolymer of 80 mol % styrene and 20 mol % methyl acrylate is made. The number average molecular weight is about 100,000. This polymer is sulfonated according to the technique described in U.S. Patent 3,072,618.

25.5 g of this polymer is neutralized with 4.6 g of lithium carbonate dissolved in water. This solution is added to 1,520 g of caprolactam at 90°C. Manganese chloride (0.576 g) and a 50% solution of hypophosphorous acid (0.1640 g) are added to serve as light stabilizers.

This solution is poured into a 3-liter agitated glass reactor equipped with a heating mantle, and a gas inlet and outlet to provide a nitrogen blanket over the molten mixture. 80 g of ϵ-aminocaproic acid is added as a polymerization initiator. The mixture is then heated over a period of about 1½ hr at about 255°C. When the water flashed off there is no phase separation.

At the end of 4¾ hr a polymer ribbon is extruded from the bottom of the reactor, which is pale yellow, without lumps and of constant cross section. Unreacted caprolactam, about 10 wt % is removed by water extraction. The washed and dried polymer is submitted for analysis.

The relative formic acid viscosity is 61, with 75 equivalents of carboxyl and 28 equivalents of amine per 10^6 g of polymer. Sulfur analysis by x-ray fluorescence of the washed and dried polymer would show 2,480 ppm sulfur, or about 67 equivalents of sulfur per 10^6 g of polymer. A sample of the unwashed polymer would contain about 2,550 ppm sulfur. The theoretical concentration of sulfur, based on the amount of polystyrene sulfonic acid salt added is 2,500 ppm.

A control yarn was made from a nylon 6 polymer having a formic acid relative viscosity of 46, about 90 carboxyl equivalents per 10^6 g, about 25 amine equivalents with about 81 sulfonate group equivalents, from sodium 5-sulfoisophthalate. It was spun on the same spinning equipment as described above, drawn, knitted into sleeves, and autoclaved at 270°F, that is, heat set.

Autoclaving consisted of putting the knitted sleeves into an autoclave, evacuating the chamber to 27" of vacuum and introducing steam to heat the chamber to 230°F. The chamber is held at 230°F for 5 min, and the pressure released. The chamber is then repressurized with steam to heat to 230°F. The pressure is held for 5 min and released. Then steam is introduced and the temperature is allowed to rise to 270°F. The pressure is released and then steam is put in again until the temperature is 270°F. It is held for 5 min and again the pressure released. Then it is repressurized to give a temperature of 270°F for 8 min. The pressure is then released and the yarns removed from the autoclave.

The yarns were dyed in separate dye baths to a moss green shade in a dye bath composed as follows: 0.3% owf (weight of the fiber) Sevron Yellow 8-GMF (du Pont); 0.25% owf Sevron Blue GCN (du Pont), C.I. Basic Blue 97; 2.0% owf Hipochem PND-11 (amine salt of an alcohol ester); 1.0% owf Hipochem CDL-60 (nonionic surfactant); and monosodium and/or disodium phosphate to adjust the pH of the dye bath to 7.0 ± 0.2.

The dyed sleeves were then exposed for 6, 12, and 18 hr in an atmosphere of about 20 parts per hundred million of ozone, at a temperature of 104°F, at a relative humidity of 95 to 100%. The results of ozone exposure are listed below.

| | . . . . . . Color Difference* . . . . . . | | |
	6 Hours	12 Hours	18 Hours
Control, polymer with 5-sulfoisophthalate	6.2	10.2	14.2
Polymer with lithium salt of polystyrene sul-			
fonic acid (Ex. 1)	0.8	2.2	2.5
Lithium salt of polystyrene copolymer (Ex. 2)	1.1	2.4	2.7
*Measured with a Hunterlab Color Difference Meter.			

The yarns with the lithium salts (Examples 1 and 2) and the yarn with 5-sulfo-isophthalic acid sodium salt were also dyed with 0.5% (owf) Astrazon Blue 5GL and exposed to xenon light in an Atlas Weatherometer for 60 hr. The former yarns took 40 hr to get a color break, while the latter broke in only 20 hr. By break or broke is meant a noticeable change in color or shade of the sample exposed compared to an adjacent unexposed portion.

Ultraviolet-Permeable Nylon

J.-P. Cornelis; U.S. Patent 4,012,557; March 15, 1977 describes a nylon-6 filament and a method for manufacturing the same to achieve a high degree of permeability to ultraviolet rays in fabric formed from such filament. The process consists of adding potassium or sodium bromide to the filament to render the same more pervious to ultraviolet rays, spinning the filament by extrusion, treating the spun filament with a solvent which superficially dissolves the filament surface with cooling after each of the steps to prevent the generation of spherulites. Fabrics made from such filaments are particularly suitable for making bathing costumes and beach clothes.

Example: A mass of nylon-6 polyamide, in powder form, is treated in a bath of 0.5% aqueous solution of potassium bromide and then spun by extrusion with intensive mixing in an extruding machine. The orifice of the extruder is oval-shaped. The extruded filament, at a temperature of about 200°C, is immediately cooled to room temperature by water or cooled air, or by passing the filament over cooling cylinders.

The thus cooled filament is then treated with a saturated aqueous solution of copper lactate at a temperature of about 55°C to an extent sufficient to dissolve the superficial shiny surface of the filament. Thereafter, the thus treated filament is again cooled in the manner abovedescribed. The resulting filament enjoys good resistance to weathering and to heat, while permitting the major part of the light to pass through yarn and fabric made therefrom.

Continuous Production with Initial Pressures Above Vapor Pressures of Reactants

H. Doerfel and C. Cordes; U.S. Patent 4,049,638; September 20, 1977; assigned to BASF AG, Germany provide a process for continuous production of polyamides by continuous transport of a mixture of one or more lactams and from 1 to 15% and preferably from 3 to 8% of water and optionally other polyamide-forming compounds such as the salts of dicarboxylic acids and diamines or aminocarboxylic acids through a number of reaction zones under polyamide-forming conditions, the mixture of starting materials being heated in a first reaction zone to temperatures of 210° to 330°C and preferably from 220° to 280°C, whereupon the polycondensation mixture is adiabatically vented in a further reaction zone and is then polymerized to completion in yet another reaction state to form high molecular weight polyamides. The process is characterized in the following manner.

(a) The mixture of starting materials is heated in the first re-action zone at pressures which are above the respective vapor pressures of the starting materials and prevent the formation of a vapor phase, the heating being effected for 5 min to 2 hr and preferably 10 min to 1 hr until a conversion of at least 70% and preferably 80% has been reached.

(b) The polycondensation mixture is vented in the second zone to pressures of 1 to 11 bars and preferably 1 to 6 bars and is then immediately heated in a third reaction zone, preferably together with the steam formed during adiabatic venting, with the application of heat and evaporation of the major portion of the water at the pressure obtained by venting or a lower pressure, the heating being carried out for less than 10 min and preferably less than 5 min at temperatures of 250° to 350°C and preferably 260° to 280°C.

(c) Thereafter, the polymerization mixture is separated from the steam in a fourth reaction zone and is polymerized to completion in a further reaction stage to form high molecular weight polyamides.

Example: A mixture of 100 kg of ε-caprolactam and 7 kg of water is heated in a heat exchanger at 280°C at a feed rate of 51 kg/hr and is pumped upwardly into a vertical reaction tube having a capacity of 45 liters. The reaction tube is maintained at a temperature of 280°C and the pressure in the tube is held at 40 bars. The product leaving at the top of the reaction tube is converted to an extent of 85%. The product is then passed through a cascade-connected tubular heat exchanger.

This heat exchanger is heated such that the temperature of the product at its outlet is from 288° to 290°C. On leaving the heat exchanger, the reaction mixture passes to a stirrer vessel heated at 288° to 290°C. In this vessel, the steam formed is separated off and a pressure of 1.02 bars is maintained. The rate of discharge of the melt from the tubular vessel is adjusted by means of a pump such that the vessel is always filled to an extent of 70 kg.

The pump-discharged product is passed downwardly through a vertical reaction tube. This reaction tube is provided with baffles such as to produce a plug flow as far as possible. Built-in heat exchangers cause the temperature of the molten material to fall from 288°C at the inlet of the reaction tube to 272°C at the outlet thereof. Following a residence time of 1.7 hr in the reaction tube, the product is discharged at the bottom and has a K-value of 70. The residual extract is 12%.

If the temperature of the melt passing through the reaction tube is lowered from 288°C at the inlet to 269°C at the outlet of the tube, there is produced, after a residence time in the reaction tube of 2.6 hr, a product having a K-value of 71.3, the residual extract being 11.4%. The K-value was determined by the method described by H. Fikentscher using 1% solutions in 96% sulfuric acid.

Concentration of Extract from Polymerization Process

C. Cordes and F. Zahradnik; U.S. Patent 4,053,457; October 11, 1977; assigned to BASF AG, Germany describe a process for the manufacture of polyamides from ε-caprolactam and/or other polyamide-forming starting compounds by polymerization and subsequent extraction of the polymer. The extract containing solvent, monomer and oligomers is concentrated in the absence of atmospheric oxygen, the surfaces which come into contact with the extract being made of materials which are inert under the conditions of the concentration process, and the concentrate, without further purification or separation, is polymerized by itself or together with other polyamide-forming starting compounds.

Example: Extract liquor containing 5.4 wt % of solids is taken from the extraction of a lactam-containing crude nylon-6 in an extraction made entirely of V2A steel and is concentrated to 50 wt % strength by distilling off water in the absence of atmospheric oxygen in a steam-heated kettle made of V2A steel. The steam used for heating is at 4 atm gauge pressure. The concentrate obtained is subsequently fed to a Sambay thin film evaporator made of V2A steel. The evaporator is heated with dry steam at 6 atm gauge pressure in the absence of atmospheric oxygen. After passing through the evaporator, a clean, colorless concentrate at 150°C, which contains less than 2% of water is obtained; this solidifies on a flaking roller to give pure white flakes.

4 parts of caprolactam and 1 part of water (Experiment A), or 4 parts of flakes and 1 part of water (Experiment B) are heated for 2 hr in an autoclave at 270°C and 20 atm gauge pressure. The autoclave is then let down and the mixture is post-condensed for 2 hr at 275°C under a slight stream of nitrogen. The amount of material extractable with hot water from the polymer obtained is 9.5% in Experiment A and 9.3% in Experiment B. The products obtained are extracted, dried and compounded with small amounts of seeding agents, on a twin-screw extruder, from which they are drawn off as strands which are chopped. The resulting granules are molded on an injection molding machine.

No difference in processing characteristics are observed between Experiments A and B. The intrinsic color of the moldings is assessed visually and their length contraction is measured. Furthermore, the energy of fracture required to destroy the moldings is determined by a standardized method. The results are summarized below.

	A	B
K value of the extracted material*	73.2	72.5
K value after compounding	72.4	72.9
Injection molding to give box-shaped moldings**		
Color of test specimen	ivory	ivory
Shrinkage in length of molding, %	0.43	0.43
Energy of fracture W_{50} required for definite destruction of the test specimen at 23°C,*** cm-kp	280	280

*Determined by the method of H. Fikentscher, *Cellulosechemie 13* (1932), 58, on a 1% strength solution in concentrated sulfuric acid.
**Box dimensions 120 x 60 x 41 mm; wall thickness 1 mm at 270°C material temperature.
***Determination analogous to DIN Draft 53,443, using box-shaped test specimens in place of discs.

Single Step Preparation of Block Copolymers

S.L. Nickol; U.S. Patents 4,044,071; August 23, 1977 and 4,045,511; August 30, 1977; both assigned to Suntech, Inc. describes a process for forming block copolymer which involves mixing the dry salt of a prepolyamide and a molten melt-spinnable polyamide. The mixture is heated to a temperature in the range of the melting point of the higher melting component of the mixture to below the amide-interchange temperature of a blend of the melt-spinnable polyamide and the homopolymer which would result from the polymerization of the salt. Mixing and heating is continued until substantially all of the salt and the polyamide are converted into a block copolymer. The latter can be used to make fibers.

Example 1: A salt having the following structure:

$$[\overset{+}{N}H_3(CH_2)_3-O-(CH_2)_2-O-(CH_2)_3\overset{+}{N}H_3]\ [\overset{-}{O}\overset{O}{\overset{\|}{C}}-(CH_2)_4-\overset{O}{\overset{\|}{C}}-\overset{-}{O}]$$

which can be referred to as a 30203-6 salt, was used as a component and was prepared in the following manner. First, 1,2-bis(β-cyanoethoxyethane), having the following structure: $NC-(CH_2)_2-O-(CH_2)_2-O-(CH_2)_2-CN$, was prepared. To prepare it a 5 liter double walled (for water cooling) glass reactor with a bottom drain and stopcock was charged with 930 g (15 mols) of ethylene glycol and 45.6 g of 40% aqueous KOH solution. Some 1,620 g (30.6 mols) of acrylonitrile ($NC-CH=CH_2$) were then added dropwise with stirring at such a rate that the temperature was kept below 35°C.

After the addition was completed the mixture was stirred an additional hour and allowed to stand overnight. The mixture was neutralized to a pH of 7 by addition of 6 molar HCl. After washing with a saturated NaCl solution three times, the product was separated from the aqueous layer, dried over $CaCl_2$ and passed through an Al_2O_3 column to insure that all basic materials had been removed. The yield was 90% of theoretical.

The next step was preparation of 4,7-dioxadecamethylenediamine:

$$NH_2(CH_2)_3-O-(CH_2)_2-O-(CH_2)_3-NH_2$$

Into an 800 ml hydrogenation reactor was charged 150 g of 1,2-bis(β-cyanoethoxyethane), 230 ml of dioxane and about 50 g Raney cobalt. After purging the air, the reactor was pressurized with hydrogen to 2,000 psi and heated to 110°C. As the hydrogen was consumed additional hydrogen was added until the pressure remained constant. Upon cooling, the pressure was released and the catalyst was filtered. The dioxane was removed by atmospheric distillation. The remaining mixture was distilled using a 3 ft spinning band distillation unit. The diamine distilled at 123° to 124°C and 3.75 mm Hg. About 98 g of 99.95% pure material were obtained. The resulting material can be referred to as a 30203 diamine.

The next step was the preparation of the salt. To a solution of 41.50 g of adipic acid dissolved in a mixture of 250 ml of isopropanol and 50 ml of ethanol were added, with stirring, 50 g of the 30203 diamine dissolved in 200 ml of isopropanol. An exothermic reaction occurred. Upon cooling, a polymer salt crystallized out of solution. The salt was collected on a Buchner funnel and subsequently recrystallized from a mixture of 400 ml of ethanol and 300 ml of isopropanol solution. The product, dried in vacuo overnight at 60°C, had a melting point of 127° to 128°C and the pH of a 1% solution was 6.9. 85 g (92% yield of theoretical) of the salt was obtained.

Example 2: The block copolymer was prepared in the following manner. A suitable container was purged with dry nitrogen and while under nitrogen 40 g of dry powdered nylon-6 were added to the container. The nylon-6 was a commercially available material. The container and its contents were heated to 245°C by a vapor bath. The nylon-6 used had onset melting point of 210°C. To the molten nylon-6 were added 17.1 g of the 30203-6 salt previously prepared. While the addition of the salt was made the container was kept under a positive pressure of nitrogen and during and after the addition the resulting mixture was constantly stirred. The container and its contents were maintained at a temperature of 245°C for 1 hr. After cooling the resulting polymer was analyzed as to its structure.

The method used to analyze the polymer structure involved the fractional precipitation of the polymer in formic acid. Generally the method was as follows: 1 g of dry polymer (copolymer; random or block or homopolymer) was weighed to the nearest tenth of a milligram. The 1 g sample was dissolved in a standardized formic acid (i.e., 90% formic acid).

The resulting solution was diluted with distilled water to a given percent formic acid, e.g., 55% formic. The solution was allowed to stand at ambient temperature for 3 hr and then filtered. The collected precipitate was then washed with water, dried and weighed to give the percent sample recovered at that particular formic acid concentration. A graph was then constructed by plotting the percent of the sample recovered vs formic acid concentration. The different polymers, i.e., random copolymer, block copolymer, homopolymer (e.g., nylon-6) each have different solubilities in formic acid. Thus each gave a different characteristic curve.

Almost all of the nylon-6 is recovered when the percent of formic acid is decreased to 55 to 56%. In contrast with the polymerized 30203-6 salt, none of the polyamide precipitates until the formic acid concentration is down to about 15%.

The precipitation of random 30203-6/6 copolymer does not occur until about a 45% of concentration of formic acid is reached. Furthermore, in contrast to the nylon-6, which has an almost perpendicular recovery line, the slope of the recovery curve for the random copolyamide is much more gradual.

The DSC (Differential Scanning Calorimeter) curves for the block copolymers prepared by the process, showed the absence of endothermic peaks corresponding to either the melting of 30203-6 salt or 30203-6 polymer. The block copolymer prepared by this process had endothermics which corresponded to those shown by 30203-6/6 block copolymers prepared by melt blending. The block copolymers had melting points more than 40°C above that observed for a random copolymer of the same overall composition. Thus, this comparison indicates that block polyamide copolymers can be prepared by this method.

Melt Blending Polyamide Process

R.M. Thompson and S.L. Nickol; U.S. Patent 4,045,512; August 30, 1977; assigned to Suntech, Inc. provide an improved melt blending process for preparing block copolymer of poly(4,7-dioxadecamethylene adipamide)-polycaprolactam, which is also known as N-30203-6/6, involving a continuation of the melt blending until the polymer is characterized in that the maximum amount of the block copolymer recovered from an aqueous formic acid solution containing the dissolved block copolymer exceeds about 90%. Fiber prepared from such a block copolymer does not fuse when scoured in boiling water.

Example 1: About 40 g of the polymer salt as prepared in Example 1 of the preceding patent (U.S. Patent 4,044,071) were charged to a heavy walled glass polymer D tube. Then the neck of the tube was constricted for sealing and purges of air by evacuating and filling with nitrogen 5 times. Finally the tube was heated in an aluminum block for 2 hr at 200°C. After cooling the tip of the tube was broken off and the remaining portion was bent over at a 45° angle by heating and then connected to a manifold and purged of air with nitrogen vacuum cycles. The tubes were heated at 222°C under nitrogen at atmospheric pressure for 6 hr using methyl salicylate vapor baths. On cooling, the tubes were broken and the polymer plug crushed to ⅛" size pieces.

Example 2: *Polymer Melt Blending* – Two different methods were used to melt blend the polyamides. Four of the samples, i.e., numbers 1 to 4, were made by feeding the dried polymers, i.e., N-30202-6 and N-6 to an extruder. The extruder melted the polymer and fed it to a static mixer where the block copolymer was formed as a result of the mixing and heating at a temperature of 260° to 282°C.

The two other samples, i.e., numbers 5 and 6, were made by charging suitable amounts of dried 30203-6 polymer and nylon-6 to a container having two openings in the rubber stopper. The openings were for a helical stirrer and a nitrogen inlet. The container was purged of air. Afterwards the nitrogen-filled container was heated using a suitable liquid-vapor bath. The mixture of the two polymers was agitated with the helical stirrer powered by an air motor for the required time. Before allowing the molten polymer to cool the stirrer was lifted to drain the polymer. In both methods after solidification the resulting copolymers were broken up and dried for spinning.

Example 3: *Polymer Spinning and Drawing* – After the melt blending the various N-30203-6/6 copolymers were spun into a fiber using a ram-extruder. The samples were spun through a spinneret having 7 orifices 12 mils in diameter and 24 mils in length. The dried samples, about 50 g, were changed to the extruder and allowed 25 min to melt and reach an equilibrium temperature, i.e., about 230°C. Then the samples were forced through screen filters, 40 mesh and 250 mesh stainless steel screens and the spinneret by a motor-driven ram.

The samples of N-30203-6/6 were melt spun at a suitable feed rate. The yarns passed from the spinneret through guides and were collected on paper tubes at a take-up speed of 138 ft/min. A spin finish, which is commercially available and which had been dissolved in heptane, was applied to the yarn as it passed over an applicator. The applicator was a nylon felt saturated with the spin finish and was attached to the yarn guide located about 4 ft below the spinneret. During the spinning the block temperature of the extruder was about 225°C while the ram pressure was about 400 psig.

Example 4: *Testing for Fiber Fusion* – The resulting fibers, i.e., samples 1 to 6 were then knitted into tubes and then placed in boiling water. After the knitted tubes were removed from the boiling water the tubes were deknitted to determine the relative amounts of fiber fusion. Samples 5 and 6 showed no fusion. Samples 1 to 4 showed amounts of fusion unacceptable in a commercial fabric.

Example 5: *Characterization of the Block Copolymer* – Portions of the samples 1 to 6 were tested as to their fractional precipitation in formic acid. Generally, the method was as follows. 1 g of dry copolymer, i.e., N-30203-6/6, was weighed to the nearest tenth of a milligram. The 1 g sample was dissolved in a standardized formic acid (i.e., 90% formic acid). The resulting solution was diluted with distilled water to a given percent formic acid, e.g. 55%. The solution was allowed to stand at ambient temperature for 3 hr and then filtered. The collected precipitate was then washed with water, dried and weighed to give the percent sample recovered at that particular formic acid concentration.

As the formic acid concentration at 50% copolymer recovery decreases, e.g., from 54.8 to 54.3%, the amount of filament fusion decreases. When the formic acid concentration decreased to 50.7%, no filament fusion is observed. Thus, it can be concluded that if the formic acid at 50% copolymer recovery is less than 50.7%, no filament fusion exists. In an equal manner it can be observed that once the maximum amount of copolymer recovered exceeds 90.6%, no further filament fusion is observed. It can also be concluded that if the maximum amount of copolymer recovered from the formic acid solution exceeds 90.6%, no filament fusion exists. Another conclusion is that when both characterizations exist, no filament fusion exists.

Poly(Dioxa-Amide) and Nylon 6 Block Copolymer

A copolymer formed by melt blending a melt spinnable polyamide, such as nylon-6, and a block of random poly(dioxa-amide), such as a copolymer prepared from the mixture of caprolactam and the salt of adipic acid and 4,7-dioxadecamethylene diamine, is disclosed by *R.M. Thompson and R.S. Stearns; U.S. Patent 4,113,794; September 12, 1978; assigned to Sun Ventures, Inc.*

The resulting copolymer, N-30203-6/6/6, has utility as a fiber. The fiber, for example, resulting from melt blending of nylon 6 and the aforementioned random poly(dioxa-amide) has moisture absorption characteristics similar to that of cotton. Furthermore, the resulting fiber still maintains the other desirable properties of the major constituent, for example, nylon 6.

Example 1: *Preparation of 50/50 Random Component* – About 27 g of caprolactam, 27 g of N-30203-6 salt, as prepared in the three preceding patents, 0.5 g of water and 0.1 g of acetic acid were charged to a heavy walled polymer tube. The latter was purged of air, sealed and then heated at 250°C for about 6 hr. Afterwards the tube was broken at the tip and a long glass tube inserted for a nitrogen purge. The filled tube was heated at 255°C in a biphenyl vapor bath for 2 hr. During the latter heating nitrogen was passed through the glass tube. After heating for 2 hr, the tube was broken and the random polymer removed. The softening point of the polymer was low so that it had to be cut by hand because machine cutting created too much heat. The polymer had an inherent viscosity of 0.78 in a meta-cresol solution at 100°C. The low softening point indicated that this component was amorphous.

Example 2: *Polymer Melt Blending* — Suitable amounts of dried random component and nylon 6 were charged to a large test tube having two openings in the rubber stopper. The openings were for a helical stirrer and a nitrogen inlet. The container was purged of air. Afterwards, the nitrogen filled container was heated using a suitable liquid-vapor bath. The mixture of the two polymers was agitated with a helical stirrer powered by an air motor for the required time. Before allowing the molten polymer to cool the stirrer was lifted to drain the polymer. After solidification the material was broken up and dried for spinning.

Example 3: *Polymer Spinning and Drawing* — After the aforementioned melt blending the polymer was charged to a micro spinning apparatus consisting of stainless steel tube ($\frac{3}{8}$" o.d. x 12") with a 0.037" capillary. The tube was heated with a vapor bath to the temperature consistent with the polymer. Generally about 245°C was used. Nitrogen was swept through the polymer until the polymer melted and sealed the capillary. After the polymer was completely melted and a uniform temperature had been reached (about 30 min) the nitrogen pressure was increased by about 30 to 50 psig (depending on polymer melt viscosity) to extrude the polymer.

The fiber as it left the tube was drawn on a series of rollers and wound up on a bobbin. The first roller or feed roll was traveling at 35 ft/min. The filament was wrapped 5 times around this. After crossing a hot pipe maintained at about 50°C, the filament was wrapped around the second roller or a draw roll (5 times) which speed varied depending on the draw ratio required (130 to 175 ft/min). Unlike commercial draw rolls, the fiber tended to abrade itself; that is the fiber coming off rubbed against the fiber coming on. This made high draw ratios difficult to obtain. The third roll had a removable bobbin and was driven at a slightly lower speed than the draw roller. The differences in speeds caused the fiber to be stretched.

Stretching or drawing orientates the molecules, i.e., places them in a single plane running in the same direction as the fiber. Draw ratios refer to the ratio of the speed of the second roller or draw roll to the speed of the first roller or feed roll. Thus, if the second roller was traveling at 175 ft/min and the first roller at 35 ft/min, the draw ratio is 5 (175/35). This difference in speeds of the rollers stretches the fiber.

Example 6: *Results of Tests and Comparative Runs* — Comparison runs indicate that as the amount of random N-30203-6 incorporated with nylon 6 increases the moisture regain at various relative humidities increases. Some comparison of runs indicates that increasing the proportion of N-30203-6 in the copolymer does not adversely influence the tenacity, elongation and initial modulus properties of nylon 6. Comparison indicates that as the amount of poly(dioxa-amide) incorporated with nylon 6 increases, the moisture regain approaches that of cotton.

Moisture regain refers to the amount of moisture a dried sample of fiber picks up in a constant relative humidity atmosphere. Measurement of this property was carried out using a series of humidity chambers made from desiccators containing suitable saturated salt solutions (i.e., $NaNO_2$ = 65%; $NaCl$ = 75%; KCl = 85% and Na_2SO_3 = 95%) at room temperature.

To determine moisture regain first a sample of the fiber was dried in a vacuum desiccator over P_2O_5. After a constant weight was obtained the sample was placed in one of the appropriate chambers. The chamber was then evacuated to speed up equilibrium. The fiber remained in the chamber until a constant weight was obtained. The increase in weight of the sample over the dried sample was the amount of moisture regained.

Poly(Dioxa-Arylamide) and Nylon 6 Block Copolymer

In a related patent, *R.M. Thompson; U.S. Patent 4,168,602; September 25, 1979; assigned to Sun Ventures, Inc.* discloses a block copolymer formed by melt blending a melt spinnable polyamide, such as nylon 6, and a poly(dioxa-arylamide), such as poly(4,7-dioxadecamethylene terephthalamide), also known as N-30203-T. The copolymer, which is known as N-30203-T/6, has superior moisture absorption and initial modulus characteristics than that of nylon 6. Furthermore, resulting fiber still substantially maintains the other desirable properties of the major constituent, for example, nylon 6.

Example 1: *Preparation and Polymerization of Poly(4,7-Dioxadecamethylene Terephthalamide)(N-30203-T)* — The 30203 diamine which has been previously described, 17.8 g dissolved in 50 ml of ethanol, was added with stirring to a heated slurry of 16.6 g of terephthalic acid in 200 ml of ethanol. Upon completion of this addition, 30 ml of water were added. After solution had occurred, the mixture was filtered hot. Then 10 ml of isopropanol were added. Upon cooling, the polymer salt precipitated. The salt was recrystallized from a mixture of 200 ml of ethanol and 400 ml of methanol. The yield was 21.3 g. A 1% aqueous solution of the recrystallized salt had a pH = 7.1. Its melting point was 221°C.

About 40 g of the polymer salt were charged to a heavy walled glass polymer D tube. Then the neck of the tube was constricted for sealing and purged of air by evacuating and filling with nitrogen 5 times. Finally the tube was heated in an aluminum block for 2 hr at 200°C. After cooling the tip of the tube was broken off and the remaining portion was bent over at a 45° angle by heating and then connected to a manifold and purged with nitrogen-vacuum cycles. The tubes were heated at 282°C under nitrogen at atmospheric pressure for 1 hr, then under a vacuum for 2 hr, using dimethyl phthalate vapor baths. On cooling, the tubes were broken and the polymer plug crushed to $\frac{1}{8}$" size pieces. Inherent viscosities ranged from 0.9 to 1.1 in meta-cresol solution at 100°F. The polymer was glossy white in color. Its melting point was 258°C.

Example 2: *Polymer Melt Blending* — Suitable amounts of dried N-30203-T polymer and nylon 6 were charged to a large test tube having two openings in the rubber stopper. The openings were for a helical stirrer and a nitrogen inlet. The container was purged of air. Afterwards the nitrogen filled container was heated using a suitable liquid-vapor bath. The mixture of the two polymers was agitated with the helical stirrer powered by an air motor for the required time. Before allowing the molten polymer to cool the stirrer was lifted to drain the polymer. After solidification the material was broken up and dried for spinning.

Example 3: *Polymer Spinning and Drawing* — After the aforementioned melt blending the polymer was charged to a micro spinning apparatus consisting of stainless steel tube ($\frac{3}{8}$" o.d. x 12"), with a 0.037" capillary. The tube was heated with a vapor bath to the temperature consistent with the polymer. Generally about 245°C was used. Nitrogen was swept through the polymer until the polymer melted and sealed the capillary. After the polymer was completely melted and a uniform temperature had been reached (about 30 min) the nitrogen pressure was increased by 30 to 50 psig (depending on polymer melt viscosity) to extrude the polymer.

The fiber, as it left the tube, was drawn on a series of rollers and wound up on a bobbin. The first roller or feed roll was traveling at 35 ft/min. The filament was wrapped 5 times around this. After crossing a hot pipe maintained at about 50°C, the filament was wrapped around the second roller or a draw roll (5 times) which speed varied depending on the draw ratio required (130 to 175 ft/min). Unlike commercial draw rolls, the fiber tended to abrade itself; that is the fiber coming off rubbed against fiber coming on. This made high draw ratios difficult to obtain. The third roll had a removable bobbin and was driven at a slightly lower speed than the draw roller.

Example 4: *Comparative Runs* — Comparison runs indicate that the inclusion of a substantial amount of N-30203-T into nylon 6 favorably influences moisture regain and initial modulus and does not adversely change other physical properties, i.e., tenacity, of the prepared fiber. The resulting N-30203-T/6 fiber has moisture regain approaching that of cotton.

Thermally Stable Nylon 6

S.D. Lazarus and J.H. Newland; U.S. Patent 4,061,708; December 6, 1977; assigned to Allied Chemical Corporation describe an improved polyamide characterized by resistance to thermal-oxidative degradation consisting essentially of a polyamide having a stabilizing amount of copper(II)dichlorobispyridine or copper(II)dichlorobisquinoline incorporated therein.

In a conventional melt-spinning process starting with polyamide polymer chips, e.g., polycaproamide chips, the stabilizing compound may be added to the polymer chips immediately prior to spinning.

In a continuous process, the stabilizing compound is preferably added to the process stream at a point where the polymer is molten by forming a stable dispersion comprising a liquid polyisobutene and the solid stabilizing compound, and continuously injecting the dispersion with mixing into the polymer melt after the final polymerization. Preferably, the liquid polyisobutene had a viscosity of 3 to 70 poises at 20°C, the solid stabilizing compound is ground to an average particle size of 2 μ or less in diameter, and the solid stabilizing compound is dispersed in the polyisobutene at a concentration of 10 to 60 wt % based on the total weight of the dispersion.

For use in Examples 1 to 4, a master batch of poly-ϵ-caproamide (nylon 6) chips is obtained from a large plant production batch. This polymer is prepared by the polymerization of ϵ-caprolactam without introduction of any other chemical component except water employed as a polymerization initiator. This polymer has a relative viscosity of 95.

Example 1: To 1,000 g of the nylon 6 polymer chips is added 0.216 g of copper(II)dichlorobispyridine. The chips and additive are blended to assure uniformity, melted in an extruder, and spun in the conventional manner. The tenacity of the resulting fiber is 8.3 g/den; thermal-oxidative stability is tested in accordance with Example 4, showing 93% retention of breaking strength.

Example 2: *Comparative Example* — Example 1 is repeated except that no additive is introduced. The tenacity of the resulting fiber is 8.3 g/den; thermal-oxidative stability is tested in accordance with Example 4, showing 28% retention of breaking strength.

Example 3: Example 1 is repeated except that 0.29 g of copper(II)dichlorobisquinoline is added. The tenacity of the resulting fiber is 8.4 g/den; thermal-oxidative stability is tested in accordance with Example 4, showing 99% retention of breaking strength.

Example 4: The fibers obtained in Examples 1 to 3 are twisted and plyed to form 2-ply cords of the type used in pneumatic tires. These cords are then evaluated for their thermal-oxidative stability.

The thermal-oxidative stability of the cords is measured in a test wherein the cords, mounted on metal spools, are exposed for 24 hr at 165°C in an oven having an air atmosphere. The cords are so wrapped and fastened on the metal spools that the cords are held to their original length during heat treatment. A measure of thermal stability is obtained by comparing the loss in breaking strength of the heat treated sample with a control consisting of similarly spool wound samples which are not placed in the air oven.

POLYPYRROLIDONES (NYLON 4)

The conventional polymerization of 2-pyrrolidone to poly-2-pyrrolidone is a bulk polymerization process not conveniently adapted to continuous processing. The process involves adding potassium hydroxide to 2-pyrrolidone, distilling off the water of reaction, adding carbon dioxide, sending the reaction mixture to a container or conveyor and reacting at about 50°C for several hours. The polymer is then recovered by cutting the solid product into manageable pieces and extracting the alkaline polymerization catalyst and excess monomer with water. This section deals primarily with improvements in this polymerization process.

Agitated Reaction Mixture with Nonsolvent and Insoluble Salt

R. Bacskai; U.S. Patent 4,017,465; April 12, 1977; assigned to Chevron Research Company has found that the problem of obtaining a polymer of 2-pyrrolidone in particulate form can be solved by a process adaptable to continuous operation.

The process advantageously includes agitating a reaction mixture comprising 2-pyrrolidone, alkaline polymerization catalyst, an inert nonsolvent in weight ratio to 2-pyrrolidone of 0.2-20:1, preferably 0.5-4:1, and an insoluble salt in weight ratio to 2-pyrrolidone of 1-10:1 at a temperature of from 18° to 100°C for 4 to 100 hr in order to form a particulate, or easily comminuted, poly-2-pyrrolidone.

Example 1: 200 g of 2-pyrrolidone (2.35 mols) distilled from 2% sulfuric acid solution was contacted with 15.5 g (0.24 mol) of potassium hydroxide pellets (86.2% KOH) in a stirred reactor and the mixture heated to incipient distillation over a period of 25 min under reduced pressure. The mixture contained 10 mol % potassium pyrrolidonate based on total monomer. The mixture was cooled to 30°C and a calibrated amount of CO_2 was introduced to produce carbonated potassium pyrrolidonate in 2-pyrrolidone solution containing 30 mol % carbonated potassium pyrrolidonate, based on total potassium pyrrolidonate. The solution was poured into a dried polyethylene bottle and placed in an oven at 50° to 54°C for 22 hr. The bottle, containing the hardened bulk-polymerized poly-2-pyrrolidone, was cut into four pieces and ground into granules. 148.58 g of the bottle's contents were washed with water until the filtrate had a pH of 6.5 and were dried in an oven at 70°C to constant weight. The yield was 60.5 g and the polymer had a viscosity on the Gardner scale of greater than Z.

Example 2 shows the use of a nonsolvent in an agitated reaction mixture without the addition of an insoluble salt.

Example 2: Carbonated alkaline polymerization catalyst in 2-pyrrolidone solution having 10 mol % potassium pyrrolidonate, based on total 2-pyrrolidone, and 30 mol % CO_2, based on total potassium pyrrolidonate, was prepared as in Example 1. 170.39 g of the carbonated catalyst solution and 130.08 g of purified hexane were charged into a stirred reaction vessel. After about 20 hr at about 60°±10°C the flask's contents were too hard to stir. After washing and drying as in Example 1, the yield was 47.01 g (29.4% conversion) and the viscosity of the polymer was rated Y on the Gardner viscosity scale.

Isolated Potassium Lactamate Catalyst

According to *R. Bacskai; U.S. Patent 4,125,523; November 14, 1978; assigned to Chevron Research Company*, an improved process for making a catalyst for the alkaline-catalyzed polymerization of 2-pyrrolidone includes the steps of contacting metallic potassium with a 5 to 7 membered-ring lactam and substantially isolating the product potassium lactamate. The isolated potassium lactamate is added to 2-pyrrolidone in amounts of 0.5 to 30 mol % based on total 2-pyrrolidone in order to effect polymerization at a temperature of 15° to 100°C for 4 to 100 hr.

The substantial isolation of the lactamate from this solution yields a satisfactorily effective catalyst if the alkali metal is potassium, but the same technique yields a less satisfactory catalyst if the alkali metal is sodium. Yet, other sodium-derived catalysts, such as sodium pyrrolidonate derived from sodium alkoxide, or from sodium hydroxide, are found to be satisfactorily effective catalysts for the polymerization of 2-pyrrolidone. No reason is presently known for the difficulty in obtaining as effective a catalyst from sodium metal as can be obtained from potassium metal by this process. This is especially difficult to understand in view of the greater reducing potential of potassium, since the unsuitability of the metallic alkali metals for catalyst preparation has been ascribed to reduction of the sensitive 2-pyrrolidone ring by these very strong metallic reducing agents.

Examples 1 through 5 deal with the isolation and/or use of sodium-derived catalysts for the polymerization of 2-pyrrolidone, and Examples 6 through 9 deal with the isolation and/or use of potassium-derived catalysts for the polymerization of 2-pyrrolidone.

All polymerizations were carried out for 22 hr at 50°C. Percent conversion is calculated as grams of polymer times 110/total monomer in grams. Viscosity is measured on the Gardner scale.

Polymer viscosity is measured at room temperature (about 25°C) on the Gardner viscosity scale using a Gardner-Holdt bubble viscometer. 0.5 g of polymer is dissolved in 10 ml of concentrated formic acid (90 wt % HCOOH, 10% water). The polymer solution is compared in viscosity to the Gardner Bubble Standards, e.g., Standard U corresponds to 6.27 stokes, Standard Z corresponds to 22.7 stokes [*Physical and Chemical Examination, Paints, Varnishes, Lacquers and Colors* by H.A. Gardner and G.G. Sward, 12 Ed., (1962)].

Example 1: This example is illustrative of the isolation of sodium pyrrolidonate and sodium pyrrolidonate/sodium N-carboxypyrrolidonate catalysts derived from sodium metal.

(1a) To a solution of 21.39 g of purified monomer in 125 ml of benzene was added 14.37 g of a 40% dispersion of sodium in mineral oil over a period of 1.5 hr at a temperature of 40° to 59°C. 2.06 g of monomer was additionally added. The mixture was diluted in 100 ml of benzene, filtered, and the solid product washed with two 100 ml portions of benzene, two 100 ml portions of hexane and two 50 ml portions of ether. The solid product, sodium pyrrolidonate, was dried at room temperature in a desiccator. The yield was 24.33 g.

(1b) To a solution of 21.37 g of purified monomer in 125 ml of benzene was added 14.37 g of a 40% dispersion of sodium in mineral oil over a period of 130 min at a temperature of 40° to 59°C. 2.12 g of additional monomer was added. The mixture was transferred to a distillation unit and 25 ml of benzene was added. After 25 ml of benzene was distilled off, the mixture was cooled to 30°C, and CO_2 bubbled in for 14 min. An exothermic reaction was observed in which the temperature of the mixture rose from 30° to 47°C. The mixture was diluted with 100 ml of benzene, filtered on filter paper and the solid washed with two 100 ml portions of benzene, two 100 ml portions of hexane and two 50 ml portions of ether. The product was dried in a desiccator (aspirator) and the yield was 31.72 g of a mixture of sodium pyrrolidonate and the adduct of CO_2 and sodium pyrrolidonate in a mol ratio of 5:4 respectively.

Example 2: This example is illustrative of the polymerization results obtained from isolated sodium-metal-derived catalysts.

(2a) Mixed in a polyethylene bottle were 45 g of purified monomer, 6.03 g of the isolated catalyst of Example (1b) (17.6 mmol of sodium pyrrolidonate-CO_2 adduct and 22.1 mmol of sodium pyrrolidonate) and 2.03 g of the isolated catalyst of Example (1b) (18.9 mmol of sodium pyrrolidonate). The catalysts were weighed out in a dry box. The polymerizate, consisting of about 90 mol % monomer, 3 mol % CO_2 adduct and 7 mol % sodium pyrrolidonate, was held at 50°C for 22 hr. The yield was 0.06 g of polymer, which is only 0.12% conversion of monomer.

(2b) A polymerizate was made up identically as in Example (2a) in a flask and heated to incipient distillation and cooled to 30°C before being poured into a polyethylene bottle. The polymerizate was held at 50°C for 22 hr. The yield was 1.88 g of polymer, corresponding to only 4.23% conversion of monomer. The polymer viscosity on the Gardner scale was less than A.

(2c) A polymerizate, made up identically as in Example (2a) but containing 55 g of purified monomer, was heated under reduced pressure to distill off 9.12 g

(16.4%) at a pot temperature of 188°C and a head temperature and pressure of 92°C per 2 mm. The polymerizate was poured into a polyethylene bottle and polymerized for 22 hr at 50°C. The yield was 3.84 g of polymer having a viscosity on the Gardner scale of less than A and corresponding to 8.38% conversion of monomer.

Example 3: This example illustrates that a satisfactory sodium-derived catalyst for the polymerization of 2-pyrrolidone can be obtained from a sodium compound.

(3a) 21.6 g of sodium methoxide, $NaOCH_3$ was dissolved in 200 ml of benzene and 20 ml distilled off. Distillation continued with the addition of 37.45 g of purified 2-pyrrolidone over a 17 min period. Distillation then continued until there was no longer methanol in the distillate. The total distillate was 151 ml. 180 ml of benzene was added and the mixture cooled to 30°C; CO_2 was bubbled in for 25 min. The exothermic reaction raised the temperature to 42°C. The solid product was filtered and washed with three 100 ml portions of benzene and two 100 ml portions of pentane and dried in a vacuum desiccator at room temperature overnight. The yield was 47.33 g of a mixture of sodium pyrrolidonate/sodium carboxypyrrolidonate in a molar ratio of 1:0.55 respectively, based on nitrogen analysis.

(3b) To 45 g of purified monomer was added 6.8 g of the isolated catalyst of Example (3a). The catalyst had been stored and weighed under dry conditions. The polymerizate was shaken and held at 50°C for 22 hr. The polymer yield was 19.37 g, corresponding to 40.8% monomer conversion and had a viscosity on the Gardner scale of X-Y.

(3c) The catalyst of Example (3a) was stored in a dry box for 1 wk. 6.8 g of the catalyst was then mixed with 45 g of purified monomer and polymerized as in Example (3b). The yield was 19.08 g of polymer having a Gardner viscosity of Y, corresponding to 39.6% conversion of monomer.

Example 4: This example shows that satisfactory catalyst for the polymerization of 2-pyrrolidone can be derived from sodium hydroxide by the conventional in situ preparation method.

50 g of purified monomer was contacted with 2.36 g of sodium hydroxide (98.2%) and heated at 106°C under reduced pressure for 29 min to dissolve the hydroxide. The clear solution was then heated to incipient distillation in 7 min at a pot temperature of 115°C and a head temperature of 80°C at 2.5 mm Hg. The mixture was cooled to 30°C and a calibrated amount of CO_2 was added to produce a 7:3 molar mixture of sodium pyrrolidonate and sodium pyrrolidonate-CO_2 adduct in 2-pyrrolidone solution. The solution was poured into a polyethylene flask and heated at 50°C for 22 hr. The polymer yield was 15.14 g, having a viscosity on the Gardner scale of U-V and corresponding to 35.26% conversion of monomer.

Example 5: The following results illustrate the poor polymer yields obtained from in situ sodium metal-derived catalysts.

(5a) 1.35 g of sodium was added to 60 g of purified monomer. The sodium was added in small portions which had been cut up under pentane. The sodium dis-

solved slowly. After 20 hr of stirring at room temperature, stirring was continued at 40°C in an oil bath for 4 hr to complete dissolution of the sodium.
A fine white precipitate was noted. 10.06 g of the solution was distilled off
under reduced pressure (pot temperature 105°C, head temperature and pressure 95°C/1 to 1.5 mm). The residue was cooled to 30°C and a calibrated
amount of CO_2 gas equivalent to 3 mol % based on total monomer was added.
The solution was heated at 50°C for 22 hr, yielding 5.61 g of polymer, which
represented 12.7% monomer conversion.

(5b) 1.38 g of sodium was added in small portions to 80 g of purified monomer at
40° to 46°C (oil bath) over 4.5 hr. Some precipitate was formed. After dissolution,
31.34 g (39%) of the solution was distilled off under reduced pressure (pot temperature 112°C, head temperature and pressure 104°C/2 mm). The residue was cooled
to 30°C and 3 mol % (based on total monomer) of CO_2 gas was added. The solution
was polymerized for 22 hr at 50°C, yielding 6.72 g of polymer having a viscosity
on the Gardner scale of D and corresponding to 16.15% monomer conversion.

(5c) 1.39 g of sodium was added to 50 g of purified monomer as in (5b) in 5.5
hr at 38° to 45°C, but there was no distillation. The same molar ratio of CO_2
was added and the solution polymerized for 22 hr at 50°C, yielding 0.15 g of
polymer, which corresponded to 0.35% conversion of monomer.

Example 6: The following results show the appreciable, but not substantial differences in polymer yields which are obtained from the use of, in situ, unisolated
potassium hydroxide-derived catalysts and isolated potassium hydroxide-derived
catalysts.

(6a) 200 g of purified monomer was contacted with 15.5 g of potassium hydroxide pellets (85.2% KOH) in a stirred reactor and the mixture heated to incipient distillation over a period of 16 min under reduced pressure (pot temperature 120°C, head temperature and pressure 95°C/1 mm). 3 mol % based on
total monomer of CO_2 was added and 130.94 g of the solution was polymerized
at 50°C for 22 hr, yielding 60.4 g of polymer having a viscosity on the Gardner
scale of Z_2 and corresponding to 49.19% conversion of monomer.

(6b) 110 g of purified monomer was contacted with 10 g of potassium hydroxide
pellets (85.2% KOH) and heated to incipient distillation under reduced pressure
for 13 min. The solution was cooled to 35°C and the vacuum replaced with nitrogen. 250 ml of dried benzene was added and CO_2 was bubbled through the
solution for 25 min. The precipitate was filtered, washed with benzene and
hexane, and dried in a vacuum desiccator. The yield was 25.86 g of potassium
pyrrolidonate-CO_2 adduct.

(6c) 48.5 g of purified monomer was contacted with 2.71 g of potassium hydroxide pellets (85.2% KOH) and the mixture heated to incipient distillation
under reduced pressure. After cooling to 30°C the solution was poured into a
polyethylene bottle containing 2.94 g of the isolated CO_2 adduct of (6b) and
shaken to dissolve. The polymerizate was heated for 22 hr at 50°C, yielding
24.67 g of a polymer having a viscosity on the Gardner scale of Z_1 and corresponding to 53.3% conversion of monomer.

Example 7: This example illustrates the unsatisfactory yield obtained from
potassium-metal-derived catalyst in the absence of a step achieving substantial
isolation of that catalyst.

To 50 g of purified monomer was added 2.29 g of potassium (weighed and cut under pentane) over a period of about 20 min. The potassium dissolved with some gas evolution, forming a colorless, cloudy solution. 3 mol % of CO_2 was added and then the solution was polymerized for 22 hr at 50°C. The yield was 6.99 g of polymer having a viscosity on the Gardner scale of B and corresponding to 15.1% monomer conversion.

The following examples show the substantial improvement in polymerization yield obtainable by substantial isolation of the potassium-metal-derived catalyst.

Example 8: This example illustrates the isolation of the potassium-metal-derived catalyst by precipitation from a nonsolvent for potassium pyrrolidonate.

(8a) To 20.68 g of purified monomer dissolved in 125 ml of dry benzene was added 9.47 g of potassium metal (cut under pentane) at 40°C over a period of 5.25 hr. The reaction was exothermic, and 25 ml of additional benzene was added to facilitate stirring. An additional 5.16 g of monomer was also added, making a 25 mol % excess of pyrrolidone. The precipitate was filtered, washed with benzene, hexane and ether and dried in a desiccator (aspirator) at room temperature. The yield was 31.33 g of a hygroscopic white powder, potassium pyrrolidonate.

(8b) 9.77 g of potassium metal was reacted with 24 mol % of excess monomer as in Example (8a). CO_2 was bubbled into the suspension of potassium pyrrolidonate in benzene and excess monomer at 30°C for 25 min. The solid precipitate was filtered, washed with two 100 ml portions of benzene, two 100 ml portions of hexane, and two 50 ml portions of ether, and dried in a vacuum desiccator. The yield was 35.76 g of a hygroscopic white powder. The molar composition of the product was 54.9% potassium pyrrolidonate and 45.1% of the CO_2 adduct of potassium pyrrolidonate. 1 g of the product contained 0.47 g (3.8 mmol) of potassium pyrrolidonate and 0.53 g (3.2 mmol) of the adduct.

Example 9: This example illustrates the satisfactory polymerization yields obtainable from an isolated catalyst derived from potassium metal.

(9a) 45 g of purified monomer, 2.47 g of the catalyst of Example (8a) and 5.52 g of the catalyst of Example (8b) were weighed into a polyethylene bottle in a dry box and shaken. The mixture was heated at 50°C for 22 hr, yielding 24.03 g of polymer having a viscosity on the Gardner scale of X-Y, corresponding to 49.26% monomer conversion.

(9b) In a separate experiment, the same polymerizate as in Example (9a) was heated to incipient distillation (pot temperature 115°C, head temperature and pressure 90°C/2.5 mm) for 9 min. It was cooled to 30°C and poured into a polyethylene bottle. The polymerizate was heated at 50°C for 22 hr, yielding 25.0 g of a polymer having a viscosity on the Gardner scale of X-Y and corresponding to 55.35% conversion of monomer.

(9c) In another experiment 55 g of a purified monomer was mixed in a flask under nitrogen with 2.47 g of the catalyst of Example (8a) and 5.22 g of the catalyst of Example (8b). 23.5% of the mixture was distilled off under vacuum (pot temperature 118°C, head temperature and pressure 100°C/3.5 mm). The solution was cooled to 30°C and poured into a polyethylene bottle. The poly-merizate was heated to 50°C for 22 hr, yielding 22.66 g of polymer having a vis-cosity on the Gardner scale of X-Y and corresponding to 56.93% conversion of monomer.

Inert Gas Removal of Alcohol in Catalyst Preparation

R. Bacskai; U.S. Patent 4,075,182; February 21, 1978; assigned to Chevron Re-search Company describes a process of preparing a catalyst for the polymerization of pyrrolidone which comprises contacting a C_{1-2} alkali metal alkoxide with an excess of 2-pyrrolidone to form a pyrrolidone solution comprising the catalyst and a C_{1-2} alcohol. An inert gas is passed through the solution at a temperature of 90° to 200°C. Unexpectedly, sufficient alcohol is removed from the catalyst solution by the inert gas treatment to permit substantial polymerization of mono-mer with this catalyst. Distillation is thus not required.

Example: A 500 ml 3-necked flask equipped with a magnetic stirring bar and a gas inlet tube was connected through a Claisen distillation head and water con-denser to a receiving flask and two dry-ice-cooled traps. The flask was charged with 50 g of 2-pyrrolidone and 3.18 g of sodium methoxide. Applying a vigor-ous air stream, the flask was heated to 100°C. The temperature inside the flask was maintained at 100°C for 60 min, in which time 1.89 g of liquid collected in the dry-ice traps. This liquid was identified, by vapor-phase chromatography, as methyl alcohol.

The reaction mixture was cooled to 30°C, and after removal of the dry-ice traps about 3 mol % CO_2, based on total 2-pyrrolidone, was added through a calibrated vacuum system. The contents of the flask were poured into a polyethylene bot-tle and heated at 50°C for 22 hr. After the usual grinding and washing with water, the poly-2-pyrrolidone was isolated in an amount corresponding to 42% conversion of monomer. The Gardner viscosity of the nylon-4 was Q-R. Es-sentially equivalent results are obtained by this process and by high vacuum distillation.

Alkali Metal Pyrrolidonate Catalyst

According to *R. House; U.S. Patent 4,100,145; July 11, 1978; assigned to Chevron Research Company*, a substantially anhydrous solid catalyst which effects the rapid polymerization of 2-pyrrolidone without run-away polymeriza-tion is made by reacting an alkali metal hydroxide with 2-pyrrolidone in equi-molar amounts to produce an aqueous product mixture containing alkali metal pyrrolidonate, and drying the mixture to obtain substantially anhydrous solid alkali metal pyrrolidonate catalyst. The solid catalyst is typically used with a polymerization activator, such as CO_2, and monomer to produce polypyrroli-done more rapidly than 2-pyrrolidone catalyst solutions obtained by contacting hydroxide with excess 2-pyrrolidone.

Example 1: 85.05 g (1 mol) of purified 2-pyrrolidone was contacted with 65.88 g (1 mol) of 85% KOH pellets for 30 min, at which time a soft, solid mass was observed to have formed. The solid was then ground in a mortar to insure complete mixing. The solid was dried for about 16 hr at 65°C under 1 to 3 mm Hg vacuum and was found to have lost 10% of its weight (theory 18.5% water). The partially dried solid was reground in a mortar to facilitate drying, and after 90 additional hours of drying under the same conditions it was found to have suffered a total weight loss of 19.8%. The recovered product weighed 98% of theory and had an alkalinity of 8.07 meq/g (theory 8.13 meq/g).

Example 2: A solid material composed of essentially equal molar amounts of CO_2 and potassium pyrrolidonate was prepared as described in U.S. Patent 3,721,652. 1.76 g (0.0105 mol) of this material was dissolved in 30.0 g (0.352 mol) of 2-pyrrolidone at 95°C. Then 3.04 g (0.0247 mol) of potassium pyrrolidonate, prepared as described in Example 1, was dissolved in the 2-pyrrolidone and the resulting solution divided into 4 portions which were allowed to polymerize at 50°C.

The polymerization of one sample portion was stopped after 2, 4, 8 and 22 hr, respectively. Each sample was worked up by washing the polymer with five 200 ml portions of water and then drying in a vacuum oven at 65°C for 16 hr. Conversion was determined from the weight of dry polymer based on the 2-pyrrolidone charged. The viscosity of the polymer was determined on a solution of 5.0 g of polymer in 100 ml of 88% formic acid using the Gardner-Holdt comparative viscosity method.

Example 3: *Comparative* – This polymerization was carried out using an in situ catalyst prepared in the usual way (U.S. Patent 3,721,652). A round-bottom flask, equipped with stirrer, gas inlet tube, thermometer and condenser was charged with 200 g (2.35 mols) of 2-pyrrolidone. To this was added 15.5 g (0.0235 mol) of 85% potassium hydroxide. The mixture was stirred until solution was complete and it was then heated under vacuum to remove water. When dry, the solution was cooled to about room temperature and 3.1 g (0.07 mol) of gaseous CO_2 was added to the solution. The resulting solution was divided into three parts which were allowed to polymerize for 4, 8, and 24 hr, respectively, at 50°C. The products were worked up essentially as described above. The results of the above polymerizations are summarized in the following table.

Catalyst	. . . . Mol Percent Conversion.			
	2 Hours	4 Hours	8 Hours	22 Hours
Example 2	18	30	44	72
Example 3	–	3	12	41

Pyrrolidonate plus Quaternary Ammonium Halide and Carbon Dioxide

In the process developed by *R. Bacskai; U.S. Patent 4,098,774; July 4, 1978; assigned to Chevron Research Company*, a catalyst for the polymerization of 2-pyrrolidone is made by contacting an alkali metal pyrrolidonate, certain quaternary ammonium halides and carbon dioxide in mol ratio of about 1:0.1-2:0.1-0.5, preferably in mol ratio of about 1:0.2-1.5:0.1-0.5, and most preferably in a mol ratio of about 1:1:0.3.

While poly-2-pyrrolidone of 300,000 weight average molecular weight is producible over a polymerization period of less than 24 hr by using a partially carbonated potassium pyrrolidonate catalyst, the catalyst of this process is capable of producing poly-2-pyrrolidone having a weight average molecular weight in excess of 1,000,000 under the same conditions.

This catalyst also achieves high yields and high conversion rates without diminution of molecular weight. The catalyst does not require an anhydrous source of alkali metal pyrrolidonate. The pyrrolidonate may be made by contacting 2-pyrrolidone with the hydroxide, rather than by contacting it with an alkali metal or alkali metal alkoxide.

Example 1: 100 g of 2-pyrrolidone (1.175 mols) was mixed with 1.55 g of 85.5% KOH pellets (0.0236 mol) to make a 2 mol % potassium pyrrolidonate solution which was dehydrated by heating to incipient distillation at 2 mm pressure for 11 min. To the dehydrated solution was added sufficient carbon dioxide to make a polymerizate containing 30 mol % CO_2 based on potassium. The carbonated pyrrolidonate solution was poured into two bottles, one of which was held at 50°C for 8 hr, and the other held at 50°C for 22 hr. After these time intervals, the contents of the bottles were chopped, washed with water, dried, weighed and subjected to viscosity measurement for molecular weight.

Example 2: This is the same as Example 1 except for the addition of 2.59 g (0.0236 mol) of tetramethyl ammonium chloride after carbonation by weighing the dried onium salt in a dry box and adding same under N_2 at room temperature to the polymerizate with stirring for 5 min. The remaining procedure was as in Example 1.

Examples 3 through 5: A 3-liter flask equipped with stirrer, thermometer, and distillation head, was charged with 1,000 ml of benzene and 108 g (2 mols) of sodium methoxide. The mixture was heated to the boiling point and 100 ml of benzene was distilled overhead. Then, while maintaining distillation, 187.24 g (2.2 mols) of pyrrolidone was added over 38 min. Distillation was continued until no more methanol came over. During this time, 1,200 ml of benzene was added, and the total overhead was 998 ml.

After cooling to 21°C, 24.5 g of CO_2 was bubbled into the slurry for 40 min. The precipitate was removed by filtration, washed with benzene and then pentane, and finally dried under a nitrogen atmosphere to give 238.67 g of a white solid. Analysis showed this to be a 7:3 (molar) ratio of sodium pyrrolidonate and a CO_2-sodium pyrrolidonate mixture.

Three flasks were charged with 24.26 g (0.285 mol) of 2-pyrrolidone and 1.8 g (0.015 mol) of the abovedescribed solid. The resulting mixture was heated at 100°C for 10 min to dissolve the solid. It was then cooled to room temperature before adding 1.64 g of tetramethyl ammonium chloride to the flask of Example 3; and 2.48 g of tetraethyl ammonium chloride to the flask of Example 4. Nothing was added to the flask of Example 5. The contents of the flasks were polymerized at 50°C for 22 hr. The results are shown below.

Example	Percent Conversion	$\overline{M}w \times 10^{-3}$
3	48.5	380
4	67.4	385
5	34.8	295

Example 6: 100 g (1.175 mols) of 2-pyrrolidone was mixed with 3.85 g of 85.5% KOH pellets (0.0588 mol) to form a 5 mol % potassium pyrrolidonate solution

which was dehydrated by heating to incipient distillation at 2 mm pressure for 10 min. To the dehydrated solution was added 30 mol % carbon dioxide based on potassium. Then 0.12 g of acetic anhydride (0.1 mol % based on total monomer) was added dropwise to the stirred polymerizate which was then polymerized for 8 hr, weighed and subjected to molecular weight determination. The addition of a small amount of acetic anhydride to this catalyst system produces a remarkable effect. 76% conversion is achievable after only 8 hr at 50°C, giving a product having a molecular weight of 175,000.

Pyrrolidonate plus Quaternary Ammonium Carboxylate and Carbon Dioxide

R. Bacskai; U.S. Patent 4,101,447; July 18, 1978; assigned to Chevron Research Company also produces a catalyst for the polymerization of 2-pyrrolidone by contacting an alkali metal pyrrolidonate, a quaternary ammonium carboxylate and carbon dioxide.

Examples: 200 g of purified 2-pyrrolidone (2.3 mols) was contacted with 7.7 g of KOH pellets (0.117 mol, 85.5% KOH) in a stirred reactor vessel and the mixture heated to incipient distillation under reduced pressure at a temperature of about 115°C. The mixture was cooled and a calibrated amount of carbon dioxide was introduced to produce a polymerizate containing 30 mol % carbon dioxide based on potassium. About 10 g of the polymerizate was poured into each of several successive polyethylene bottles, three of which contained 6 mmol of the dried onium salts shown in the table below. The bottles were shaken well and held at 50°C for 22 hr. The polymer was then removed, washed, dried and weighed.

Onium Salt	Mol Ratio K/Onium Salt	Percent Conversion	$\overline{M}w$ x 10^{-3}
None	—	37	405
$(CH_3)_4N^+OCOCH_3^-$	1.2	56	720
$(CH_3)_4N^+ClO_4^-$	1	31	400
$(CH_3)_4N^+BF_4^-$	1.1	35	415
$(CH_3)_4N^+PF_6^-$	1	33	405

The above table shows comparative polymerizations in the presence of 5 mol % potassium pyrrolidonate and 30 mol % carbon dioxide with and without tetramethyl ammonium acetate, polymerized 22 hr at 50°C. The tetraalkyl ammonium carboxylate in combination with carbon dioxide and potassium pyrrolidonate is found to be capable of producing polypyrrolidone of extremely high weight average molecular weight. All molecular weights are reported as the weight average molecular weight determined from the specific viscosity of 0.1 g of polymer/100 cc of m-cresol solution at 25°C.

Sulfur Dioxide as Chain Initiator

A method of forming white polymers of controlled molecular weight when polymerizing 2-pyrrolidone using an alkaline catalyst and sulfur dioxide as the chain initiator is disclosed by *A.C. Barnes and C.E. Barnes; U.S. Patent 4,105,645; August 8, 1978; assigned to Barson Corporation.* The temperature at which the polymerization is carried out is critical and is carefully controlled. The SO_2

which must also be controlled within limits is diluted with an inert gas to diminish its activity and make it easier to measure the small amounts involved. The viscosity and thermal stability of the polymers formed are as good as or better than those formed when CO_2 is used as the initiator.

Examples: 100 g of 2-pyrrolidone purified by recrystallization was added to a 500 ml 3-necked flask equipped with a gas inlet tube, a thermometer for measuring pot temperature and a distillation head bearing a thermometer for measuring vapor temperature and a receiver having a vacuum connection. To this was then added 3.3 g (0.05 mol) of potassium hydroxide (85% assay). The system was then swept with nitrogen, evacuated to a pressure of 10 to 15 mm of mercury, and the liquid warmed gently to help form the potassium salt of 2-pyrrolidone.

The water formed as the result of this reaction was removed by further heating, and to insure dryness, 15 g of pyrrolidone was distilled over, leaving 85 g or 1 mol of pyrrolidone. The water-white contents of the flask was then cooled to about 25°C and the SO_2 diluted with nitrogen was admitted through the gas inlet in amounts of 0.0003 to 0.02 mol.

After the addition of the SO_2, the syrupy liquid was poured into a polyethylene polymerization bottle, capped and then placed in a polymerization oven at the temperatures of 25° to 50°C for 30 to 48 hr. At the end of the polymerization time the color of the polymer cake was noted and the polymer was further treated with water in a Waring blender to reduce the particle size so as to facilitate the removal of alkali and unpolymerized monomer. The polymer was then further washed several times with water and the color of the wet polymer noted. The polymer in each example was then dried and both the percentage of conversion and the viscosity determined.

The viscosity measurements were made using a 5.0% solution of the polymer in 85% formic acid. When the polymer had completely dissolved, the solution was poured into an empty Gardner bubble viscometer tube for comparison at 25°C with the bubble flow of Gardner No. VG-7375 standard bubble tubes. Some of the higher viscosities required the use of the Gardner VG-7380 series of tubes, calibrated in stokes.

At a polymerization temperature of 50°C as disclosed in the prior art, white polymers are not produced except at concentrations of SO_2 which are so low that the yield is uneconomical and even then the molecular weight (viscosity) of the product is unsatisfactory. Polymerization at room temperature likewise results in conversions which are so low as to be impractical.

Not only is the whiteness of the polymer formed dependent on the temperature at which the polymerization is carried out, but so is the viscosity. The viscosity was observed to drop from a value of 9.0 to 3.0 stokes (s) as the polymerization temperature was raised from 30° to 50°C. Similarly, the viscosity of the polymer formed dropped from 10.0 to 3.0 s as the temperature was raised to 50°C.

As observed in the prior art, the concentration of SO_2 employed also affects the viscosity of the polymer formed, but it is also apparent that this factor is not nearly as critical as is the polymerization temperature. It was found that at 50°C a five-fold increase in the SO_2 concentration caused a drop in viscosity from 4.4 to 1.4 s. At 45°C it required a ten-fold increase in the SO_2 concentration to effect a similar drop in viscosity. At a polymerization temperature of 35°C a ten-fold increase in the SO_2 concentration resulted in only a relatively minor change in viscosity from 6.3 to 5.0 s. Therefore, it may be concluded that the temperature at which the polymerization is carried out is a much more critical factor in determining both viscosity and whiteness than is the SO_2 concentration. It may also be concluded that in order to insure uniform maintenance of viscosities from batch to batch, the polymerization temperature must be very accurately controlled.

Sulfonic Acid Derivatives as Anionic Polymerization Promotors

R. Bacskai; U.S. Patent 4,042,573; August 16, 1977; assigned to Chevron Research Company describes a process of polymerizing pyrrolidone in the presence of an alkaline polymerization catalyst and a polymerization promoter which is a derivative of a sulfonic acid selected from among sulfonic acid anhydride, N-sulfonyl-2-pyrrolidone and sulfonyl halide.

Pyrrolidone polymerizes under anionic polymerization conditions preferably in the presence of a polymerization activator such as carbon dioxide. The sulfonic acid derivatives used in this process are promoters of pyrrolidone polymerization.

Example 1: A 500 ml flask equipped with a thermometer-stirrer, a gas inlet tube and a side-arm take-off with condenser was charged with 199 g (2.35 mols) of 2-pyrrolidone and 15.5 g (0.235 mol) of 85% KOH. After dissolution was complete, the contents of the flask were heated to 110°C under 1 mm Hg pressure to remove the by-product water.

The solution was then cooled to 25°C and CO_2 was added until the solution contained 3 mol % CO_2, based on monomer. The resulting solution was stirred for 2 min and then poured into 4 polyethylene bottles, each holding approximately 50 g. The bottles were placed in an oven at 50°C. A bottle was removed and worked up after 4, 8, 12 and 22 hr, respectively. The work-up consisted of cutting off the plastic bottle, chopping the polymer into small pieces and extracting 3 times with water.

After drying, the polymer was weighed and its viscosity determined. Monomer conversions were 6, 14.4, 23.7 and 48.3% at 4, 8, 12 and 22 hr, respectively. Gardener viscosities were M, U-V, X-Y and Z_1, respectively, for the same samples.

Example 2: Example 1 was repeated except that trifluoromethane sulfonic acid anhydride (TFSA), 133 mg (0.02 mol % based on monomer), was added after the CO_2 was bubbled in. Other experiments were carried out with 233 mg (0.035 mol %); 342 mg (0.05 mol %); 684 mg (0.10 mol %); 1.33 g (0.20 mol %); and 1.0 g (0.15 mol %) of TSFA. The results are given in the table below as percent conversion and viscosity after 22 hr at 50°C.

TFSA (mol percent)	Percent Conversion	Gardner Viscosity
0	48	Z_1
0.02	58	$Z-Z_1$
0.035	59	Y-Z
0.05	60	X

(continued)

TFSA (mol percent)	Percent Conversion	Gardner Viscosity
0.10	61	W
0.15	59	V
0.20	60	I-J

With TFSA, the molecular weight is still high at the 0.1 mol % concentration. These results indicate that TFSA can be used to promote polymerization to 60% conversion or higher at 22 hr and 50°C without a concomitant substantial loss in molecular weight.

Anionic Polymerization Activated by Crown Ethers

In the process described by *J.P. Collman; U.S. Patent 4,073,778; February 14, 1978; assigned to Chevron Research Company*, the anionic polymerization of lactam is activated by the presence of crown ethers which are macrocyclic polyethers having the ability to form stable complexes with salts.

The macrocyclic polyethers encompass those cyclic compounds consisting of 4 to 10 $-O-X-$ units wherein O is oxygen and X in each of the units is independently selected from:

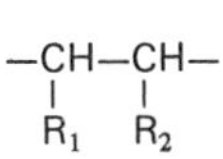

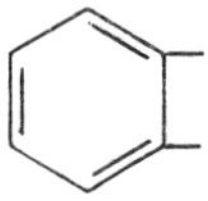

$$-CH-CH- \atop \quad R_1 \quad R_2 \qquad\qquad \text{or} \qquad -CH-CH-CH- \atop \quad\quad R_3 \quad R_4 \quad R_5$$

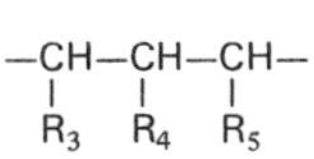

wherein R_1, R_2, R_3, R_4 and R_5 are radicals independently selected from hydrogen, or C_{1-4} alkyl, or R_1 and R_2 are linked together to form a cyclohexylene group. Preferably, R_1 through R_5 are hydrogen or C_{1-4} alkyl groups. Most preferably they are hydrogen.

These macrocyclic polyesters are most simply defined as [N]-crown-M wherein N, representing the number of atoms in the polyether ring, is 12 to 30 and M, representing the number of oxygen atoms in the polyether ring, is 4 to 10. Each oxygen atom in the ring is separated from its adjoining oxygen atoms by two or three carbon atoms. Examples of these polyethers under several common name designations are given as follows:

[12]-Crown-4

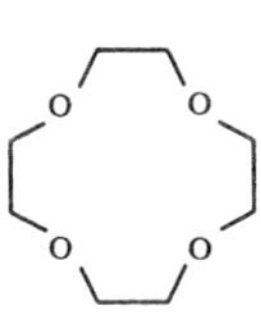

(1) 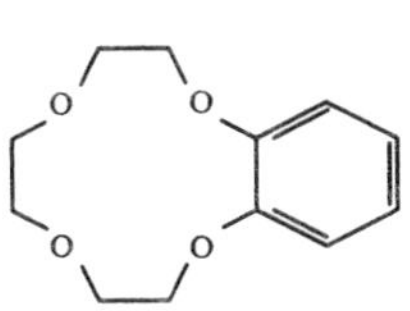   (2) 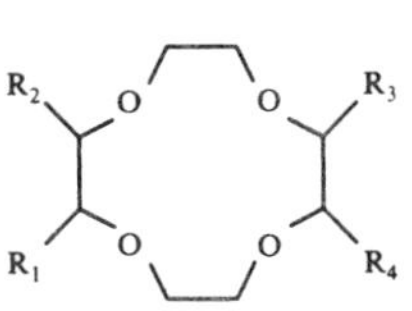   (3)

[14]-Crown-4   [18]-Crown-6

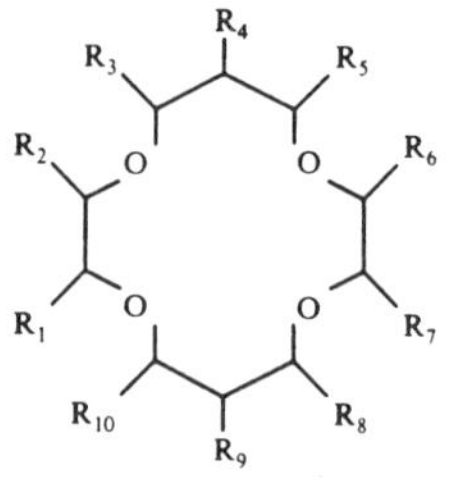

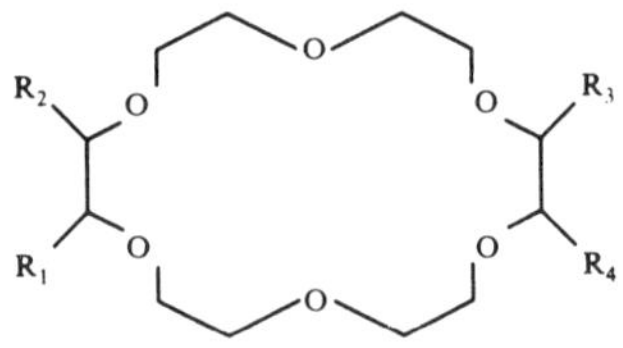

(4)   (5)

The macrocyclic polyethers are synthesized by known methods. A recent reference is G.W. Gokel et al, *J. Org. Chem.*, Vol. 19, p 2445 (1974). In general, a linear polyether having halide and alkoxide functions at opposite ends yields the macrocyclic polyether by ring closure; or two linear polyethers, each terminally functionalized as above, can be joined to form a ring. Catechol is used to introduce phenylene groups into the polyether. Saturated ethers are prepared from the corresponding aromatic ones by catalytical hydrogenation.

The polymerization process is applicable to the polymerization of lactams under anionic polymerization conditions, i.e., alkaline catalyzed polymerization. It is specifically applicable to the polymerization of 5 to 7 membered-ring lactams, principally the polymerization of epsilon-caprolactam to polycaprolactam (nylon-6) and the polymerization of 2-pyrrolidone to poly-2-pyrrolidone (nylon-4).

The process is specifically preferred for the polymerization of 2-pyrrolidone to form a polymeric carbonamide of high molecular weight in good yield in a reasonably short polymerization time. The polymer is capable of being melt-spun into filaments having substantial orientation along the filamentary axis, high-tensile strength and other properties suitable for making into textiles. It can be made into shaped articles and film by melt-molding or extrusion.

Example: A flask equipped with distillation head was charged with 200 g (2.35 mols) of 2-pyrrolidone and 3.1 g (0.047 mol) of 85% KOH. The flask was then heated to 115°C at 2.5 mm for 13 min to remove water. To this 2 mol % catalyst in 2-pyrrolidone solution was added sufficient carbon dioxide to provide 0.3 mol carbon dioxide per mol of potassium. The resulting solution was added to polyethylene bottles in about 12 g portions. Some of the bottles contained crown ethers in the amount, and of the type shown in the table below. Each bottle was shaken well and placed in an oven for 8 to 22 hr. At the end of that time, the bottle was cooled, cut open and the product ground up. Unreacted monomer was extracted with water. The crown ether was extracted with chloroform. After drying, the polymer was weighed and its viscosity determined. The results follow.

Effect of Crown Ethers on Nylon Polymerization*

. . Crown Ether, mol %. . .		Percent Conversion	$\overline{M}w \times 10^{-3}$
None	—	2.4	175
[18]-crown-6	1.5	28.9	365
[18]-crown-6	1.5	10.4	225

*Pyrrolidone with 2 mol % KOH and 0.6 mol % CO_2 with mol ratio
K:crown of 1:0.77. Polymerization at 50°C for 8 hours.

Anionic Copolymerization of 2-Pyrrolidone with Caprolactam

Although the alkaline-catalyzed copolymerization of 2-pyrrolidone with higher lactams is known, the carbon dioxide process (U.S. Patent 3,721,652), which rapidly produces the solid, white, melt-spinnable, high-molecular-weight homopolymer (nylon-4), is inhibited by the presence of higher lactams. The process developed by R. Bacskai; U.S. Patents 4,101,531; July 18, 1978; and 4,107,154; August 15, 1978; both assigned to Chevron Research Company provides a copolymer of 2-pyrrolidone and C_{5-6} lactams having all the aforementioned favorable properties of the carbon dioxide-produced homopolymer, and in addition, having greater thermal stability than the homopolymer.

In the preferred embodiment, 2-pyrrolidone and a C_{5-6} lactam are contacted with an alkaline polymerization catalyst, an auxiliary catalyst, and a copolymerization promoter, as well as carbon dioxide. The carbon dioxide, as has heretofore been taught, is believed to form an adduct with the alkaline polymerization catalyst. The lactams finding use in this process are 2-pyrrolidone, caprolactam and piperidone, and preferably 2-pyrrolidone and caprolactam.

The alkaline polymerization catalyst is a lactamate salt, such as pyrrolidonate or C_{5-6} lactamate, of the alkali or alkaline earth metals. The alkali metal lactamates, particularly alkali metal pyrrolidonate, are preferred. Potassium pyrrolidonate is a particularly preferred catalyst.

In a preferred embodiment, potassium hydroxide is reacted with excess 2-pyrrolidone and the coproduct of the reaction, water, is removed to give a substantially anhydrous catalyst solution in 2-pyrrolidone. The alkaline polymerization catalyst is normally used in amounts of 1 to 20 mol %, based on total lactam (2-pyrrolidone + C_{5-6} lactam) in the polymerizate, including the lactam salt itself. Preferably 3 to 10 mol % of an alkali metal pyrrolidonate catalyst is used.

The auxiliary catalyst is, generally, a quaternary Group V-A "onium" salt, more particularly the ammonium or phosphonium salt. A preferred class of onium salts is the halides and carboxylates, such as quaternary ammonium chloride, phosphonium bromide and ammonium acetate. The onium salts of this process contain lower alkyl, lower alkylaryl, and/or lower aralkyl groups. Tetraalkyl quaternary ammonium halide is the preferred auxiliary catalyst.

The auxiliary catalyst should be substantially soluble under the alkaline conditions of copolymerization. The auxiliary catalyst is normally used in amount corresponding to 0.1 to 2 mols auxiliary catalyst to each mol of alkaline catalyst (1:0.1-2 mol ratio), preferably in mol ratio of 1:0.2-1.5, and most preferably in a mol ratio of about 1:1.

Example 1: 100 g (1.175 mols) of 2-pyrrolidone was mixed with 3.85 g (0.0587 mol) of 85.5% potassium hydroxide to give a 5 mol % potassium pyrrolidonate solution. This solution was dehydrated by heating to incipient distillation at 2 mm pressure for 10 min. Sufficient carbon dioxide was added to this dehydrated solution to make a polymerizate containing 30 mol % carbon dioxide based on potassium. A 24 g portion of the carbonated pyrrolidonate solution was poured into a bottle containing 1.70 g (15 mmol) of caprolactam and 1.64 g (15 mmol) of tetramethylammonium chloride. After thoroughly mixing the contents of the bottle, it was held at 60°C for 22 hr. At the end of this time, the solid polymer was chopped into small pieces, washed with water and dried to give 8.40 g of polymer. Analysis of this polymer by NMR showed it to contain only homopoly-2-pyrrolidone without any copolymer.

Example 2: This polymerization was carried out in essentially the same manner as Example 1, except that the polymerization bottle contained 0.20 g (0.0016 mol) of N-acetylpyrrolidone in addition to caprolactam and tetramethylammonium chloride. The final yield was 11.64 g of copolymer which contained 0.5 mol % of caprolactam in the 2-pyrrolidone/caprolactam copolymer.

At least some homopolymerization of 2-pyrrolidone occurs at 70°C in the presence of alkaline polymerization catalyst and carbon dioxide (4% conversion), but the polymerization is completely inhibited by the addition of 5 mol % caprolactam. The use of the auxiliary catalyst and copolymerization promoter of the process prevents the complete inhibition of the homopolymerization under the same conditions. The presence of the auxiliary catalyst and the copolymerization promoter produces copolymerization.

Substantial conversion to copolymers of high molecular weight, containing substantial amounts of C_6 lactam (up to 17 mol %), are possible with the choice of higher than normal polymerization temperatures and larger amounts of C_6 lactam comonomer. The temperature effect is opposite to that observed in the homopolymerization of 2-pyrrolidone, where increasing the polymerization temperature over 40°C produces less and less conversion to polypyrrolidone as the polymerization temperature rises, all other polymerization variables being the same.

In a thermal stability test (a thermogravimetric analysis), a sample of polymer or copolymer is held under flowing nitrogen on the continuously heated pin of a microbalance which constantly monitors the sample weight. The temperature of the sample is programmed to increase at 10°C/10 min, starting from room temperature.

From the thermogravimetric analysis, the temperature at which the sample had suffered a 50% loss of the initial weight was recorded. The higher this temperature (T_{50}), the more thermally stable the copolymer or polymer is. T_{50} of 300 to 301°C were found for the copolymers containing as little as 3 mol % caprolactam, vs 279 to 280°C for the homopolymer. Differences of about 20°C in T_{50} are very significant, since the polymer decomposes rapidly at temperatures higher than, but close to, its melting point. This makes melt-spinning to synthetic fibers very difficult. The increase in T_{50} of 20°C is expected to effect a significant improvement in the melt-spinnability of the copolymer.

Polypyrrolidone Prepared with Inert Drying Agent

S.F. Pusztaszeri; U.S. Patent 4,104,260; August 1, 1978 describes a process for polymerizing 2-pyrrolidone in high yield to produce a high molecular weight polymer having the advantageous physical properties of synthetic polymers, such as strength and wear-resistance, and the advantageous chemical properties of cotton, such as water-absorptivity and release, dye-receptivity and similar ironing capability. The process is a hydrogen transfer which is catalyzed by an anion and bone dry carbon dioxide, nitrogen dioxide or sulphur dioxide gas and pushed to high yield with drying agents. Pure anhydrous 2-pyrrolidone is first reacted with an analytical grade alkali metal or its hydroxide or bicarbonate as an anion source to form a mixture comprising 2-pyrrolidone and the metal salt of 2-pyrrolidone; the mixture is then treated with dry CO_2, NO_2 or SO_2 gas as a catalyst to open the ring of the lactam salt; a drying agent is then added; and finally the hydrogen transfer polymerization is accomplished under controlled heating conditions to produce a high yield of polymer.

Examples 1 and 2 illustrate the criticality of the amount of alkali metal compound used to react with the pure 2-pyrrolidone monomer in the first step of this process.

Example 1: 300 g (3.53 mols) of pure, substantially anhydrous 2-pyrrolidone, purified according to a purification process using potassium hydroxide as reactant, were mixed with 15 g (0.27 mol) of analytical grade potassium hydroxide (to keep the water content in the KOH constant) in a 3-necked flask. The mixture was heated up slowly under a reduced pressure and in the presence of an inert atmosphere (nitrogen). At 60°C the potassium hydroxide was dissolved in the 2-pyrrolidone at 30 mm mercury pressure. At 86°C dehydration began and the vacuum was 30 mm mercury. At 102°C the 2-pyrrolidone began to distill and the vacuum dropped to 20 mm mercury, evidencing the completion of dehydration. However, this does not insure anhydrous conditions since pyrrolidone hydrate will not release all the bound water under the effects of heat alone.

The temperature was raised to 107°C under 20 mm mercury pressure and approximately 10% of the 2-pyrrolidone was distilled off. At this point sufficient 2-pyrrolidone had been distilled to insure the best anhydrous conditions possible and heating was discontinued and the solution mixture cooled down to 38°C.

Next, bone dry carbon dioxide gas (most economical) was added to the solution in the flask by bubbling the gas into the solution over a 5 min period. During the CO_2 addition, the temperature of the solution increased from 38° to 48°C, evidencing the dissociation energy released by the rupture of the 2-pyrrolidone salt ring. The weight gain of the solution was 1.8 g.

After the CO_2 addition, the cloudy, milky solution mixture was poured into a bottle which contained approximately 5%, based upon the weight of the solution, of anhydrous potassium sulfate. The bottle was sealed and put into a constant temperature incubator maintained at 50±0.5°C for 48 hr to polymerize.

The white polymer obtained was ground to chips and washed to a neutral pH with water and dried in an electrically heated oven at 105°C. The weight of the pure polymer was 224 g, equaling a yield of 93.5% based upon the weight of the 2-pyrrolidone and its salt which had been subjected to CO_2 treatment. The viscosity of the polymer in solution at room temperature (about 25°C), 0.05% solution in formic acid, was 8.84 s.

Example 2: To illustrate the criticality of the starting ratio of potassium hydroxide to 2-pyrrolidone, the above procedure was repeated exactly except that only 7.5 g (0.13 mol) of analytical grade potassium hydroxide was used together with 300 g of the purified 2-pyrrolidone. During the CO_2 addition, the temperature of the solution increased from 38°C up to only 44°C, evidencing less dissociation energy, and after heating for 48 hr at the constant temperature of 50±0.5°C. in the presence of about 5% drying agent, the weight of ground, washed and dried polymer was only 90.6 g equaling a yield of 37.8% based upon the weight of 2-pyrrolidone and its salt which had been subjected to CO_2 treatment. The viscosity of the polymer in 0.5 formic acid solution at room temperature was 2.75 s.

Example 3: To illustrate the criticality of anhydrous conditions to the polymerization the following work was carried out.

Example 1 was repeated using 250 g of the purified 2-pyrrolidone and 12.5 g of analytical grade potassium hydroxide to produce a solution which, after the CO_2 treatment, weighed 215 g and was divided into four substantially equal portions, A, B, C and D. No water was added to portion A. One drop of water (0.05 ml) was added to portion B, two drops of water (0.10 ml) were added to portion C, and three drops of water (0.15 ml) were added to portion D.

The four portions were placed in separate bottles without any drying agent, sealed and heated at $50\pm0.5°C$ in a constant temperature oven for 48 hr. Based upon the weight of the starting portions, portion A underwent a polymer conversion of only 54.6% and had a viscosity of 6.28 s, indicating that the solution mixture was not anhydrous after the CO_2 reaction; portion B underwent a polymer conversion of only 36.6% and had a viscosity of 5.06 s; portion C underwent a polymer conversion of only 2.1% and had a viscosity of 0.26 s; and portion D did not polymerize.

The results are in sharp contrast to the results obtained when water is excluded from the system and a drying agent is included to tie up any water formed during the reaction, in which event the polymer conversion is over 90% (based upon the weight of the 2-pyrrolidone and its salt prior to CO_2, H_2O or SO_2 treatment and the viscosity is at least 8 s and preferably as high as 12 to 14 s as measured at room temperature as a 0.5% solution in formic acid. The higher the viscosity the greater the molecular weight of the polymer.

Washing Polypyrrolidone with 2-Pyrrolidone

K. Katsumoto and R.A. Wuopio; U.S. Patent 4,013,626; March 22, 1977; assigned to Chevron Research Company describe an improved process for the production of poly-2-pyrrolidone in which the solid product is washed with monomer to remove the catalyst and the monomer wash, containing catalyst and monomer, is recycled for further polymerization reaction.

Figure 2.1 is a flow diagram illustrating an embodiment of the process in which polypyrrolidone is produced by bulk polymerization. In this process crude 2-pyrrolidone in line **1** is combined with recycle 2-pyrrolidone in line **3**, and charged to a monomer purification zone **2**.

The 2-pyrrolidone is topped in zone **2**, low boiling impurities are removed through line **4** and high boiling impurities through line **5**. The purified 2-pyrrolidone then passes through lines **6** and **14** into the separation and catalyst extraction zone **16**, except for a small fraction of the monomer (1 to 20 wt %) which passes through line **7** into the make-up catalyst preparation zone **10**. In zone **10** potassium hydroxide is added through line **8** and the catalyst-containing solution is then passed through line **28** into the water separation zone **12** wherein water is removed by distillation at reduced pressure through line **11**.

Figure 2.1:   Flow Chart for Bulk Polymerization with Recycle of Monomer Wash

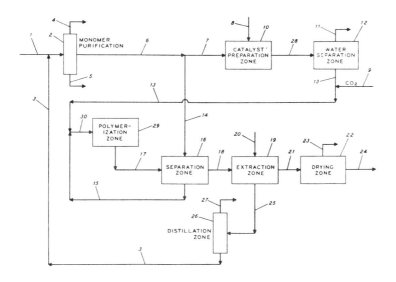

Source: U.S. Patent 4,013,626

To the dried make-up catalyst solution is added carbon dioxide via line **9**. The make-up catalyst solution in line **13** is then combined with the catalyst-containing monomer stream from line **15**, and the combined streams are charged to the polymerization zone **29** via line **30**.

The polymerization zone comprises one or more reactors, e.g., tanks, stirred reactors, ball mills, Brabenders, etc., wherein the temperature is maintained within the range of 18° to 100°C. The period of polymerization is from 4 to 100 hr. The crude reaction mixture containing solid product, unreacted monomer and active catalyst, passes to the separation and catalyst extraction zone **16** via line **17**. The separation and catalyst extraction zone comprises a vessel, e.g., a counter-current extractor, wherein the fresh monomer extracts the active catalyst from the solid product, forming a catalyst-containing monomer stream which may be recycled back to the polymerization zone via lines **15** and **30**.

The separated solid product from which substantial amounts of catalyst have been extracted, passes via line **18** into an extraction zone **19** wherein monomer and catalyst residues are removed by extraction with water introduced through line **20**. The polymer then passes through line **21** into a drying zone **22** where residual water is removed through line **23**. The product polymer is recovered at **24**. The water extracts of the polymer pass through line **25** into a distillation zone **26** where the water is stripped off through line **27**. The dry monomer is recycled through line **3** back to the monomer purification zone **2**.

The success of the process of this embodiment depends on the continued activity of the catalyst in the polymerization reaction, and the extractability of the catalyst from the solid product. The demonstration of continued catalyst activity is shown by the following examples wherein polymerization of fresh monomer takes place after contact with the unextracted solid reaction product.

Example 1: Into a ball mill reactor consisting of a horizontally mounted rotatable vessel of 1 liter capacity containing ceramic cylinders 12 mm in diameter and 12 mm in height and having a total volume of 430 ml was charged 200 ml of heptane and 99.2 g of a 2-pyrrolidone solution containing 7 mol % potassium pyrrolidonate and 3 mol % of an adduct of carbon dioxide and potassium pyrrolidonate, based on total monomer.

The reactor was rotated for 22 hr at 48°C. At the end of this time the reactor was opened and a second batch of monomer consisting of a mixture of 100 g of 2-pyrrolidone and 200 ml of heptane was added. Rotation of the ball mill reactor was continued for an additional 22 hr at 48°C. At the end of this additional time the mixture comprising heptane, unreacted monomer, catalyst and particulate solid product was filtered to separate the product. The product was then washed with water to remove unreacted monomer and dried. The product poly-2-pyrrolidone weighed 105.3 g, corresponding to a yield of 54.5% based on total 2-pyrrolidone charged.

A control polymerization was carried out for purposes of comparison by charging 52.7 g of a 2-pyrrolidone solution containing 7 mol % potassium pyrrolidonate and 3 mol % of the CO_2 adduct of potassium pyrrolidonate to a polyethylene bottle and reacting the same at 48°C for 44 hr (U.S. Patent 3,721,652), giving a poly-2-pyrrolidone yield of 60.5%. Therefore, the second batch of monomer charged to the ball mill reactor was polymerized to essentially the same extent as the first batch of 2-pyrrolidone, showing that the catalyst retained its full activity after the first 22 hr polymerization. The extractability of the catalyst from the solid product is shown in the following example.

Example 2: Into a ball mill reactor of Example 1 was charged 200 ml of heptane and 47.2 g of 2-pyrrolidone solution containing 7 mol % potassium pyrrolidonate and 3 mol % of an adduct of carbon dioxide and potassium pyrrolidonate. The reactor was rotated at 48°C for 22 hr. At the end of this time the solid product was separated by successive filtrations. The yield was 50.1%.

The product was divided into 2 equal parts. The first fraction, 20.3 g, was stirred with 100 g of 2-pyrrolidone in a Waring blender for 30 min. The second fraction of 20.2 g was stirred in a Waring blender for 30 min with 100 g of water. Each product fraction was then separated from its extractant by filtration and analyzed for potassium.

It was found that the 2-pyrrolidone had extracted about 23% of the catalyst from the product fraction whereas the water had extracted about 97% of the catalyst from the product fraction. The experiment shows that a significant fraction of the catalyst can be extracted from the polymeric product by a single 30 min contacting with monomer. Multiple or continuous 2-pyrrolidone extractions by the aforementioned techniques permit recovery of most of the catalyst from the polymer for use in recycle and provide a polymeric product substantially free of catalyst.

Thermal Stabilization with Epoxides

Polypyrrolidone is melt-spun into filaments by extrusion from multihole spinnerets. In melt-spinning, the polymer composition is extruded in a molten condition at a melt temperature which is generally greater than 270°C. The extrusion must be carried out with care because of the tendency of the polymer to thermally degrade and revert to monomer. Degradation produces an unacceptable extrudate containing foam or bubbles. If extrusion is attempted at an appreciably lower temperature to avoid thermal decomposition, fibers of lower tensile strength are produced. Consequently, in order to melt extrude polypyrrolidone efficiently, one may either seek to increase the thermal stability of the polymer, or to improve the extrudability of the polymeric composition.

According to *P.H. Parker; U.S. Patent 4,071,486; January 31, 1978; assigned to Chevron Research Company*, alkyl epoxides are thermal stabilization additives for normally solid polypyrrolidone at melt temperatures.

The alkyl epoxides are alkyl compounds whose principal functionality is one or more oxirane or oxetane rings, preferably an alkyl epoxide of the general formula:

$$RCH\!\!-\!\!CHR'$$
$$\diagdown_{\;O\;}\diagup$$

wherein R is hydrogen, alkyl or epoxyalkyl and R' is alkyl, polyalkylene oxide, epoxyalkyl, epoxypolyalkyl oxide, alkyl ester, epoxyalkyl ester, hydroxyalkyl, or hydroxyalkyl ester. The alkyl epoxide preferably contains less than 60 carbon atoms. Epoxidized esters of C_{16-20} fatty acids and C_{1-10} alkanols are a preferred class of alkyl epoxide.

The melt extrusion of normally solid poly-2-pyrrolidone is improved by the inclusion of a thermal stabilizing amount, or an extrusion-assisting amount of the alkyl epoxide. Such amounts of alkyl epoxide are minor amounts based on polypyrrolidone, ranging from 0.1 to 10-15 wt %, preferably about 0.1 to 1 wt %. The melt extrusion may be improved either by an appreciable lowering of the melt extrusion temperature due to the addition of the alkyl epoxide, or by reduction in the rate of monomer formation at the melt temperature, with the production of fibers of good tensile strength. The improvement is evidenced by a continuous extrusion of the filamentary poly-2-pyrrolidone composition at melt temperatures in the range of 260° to 280°C and preferably less than 270° to 275°C, without breaks, dripping, foam or bubbles, using ordinary commercial spinning equipment known to the synthetic textile art.

The normally solid poly-2-pyrrolidone is polypyrrolidone having a weight average molecular weight in excess of about 5,000 and preferably in excess of about 50,000. The alkyl epoxide is normally added to the solid polypyrrolidone by coating pellets of the polymer with the alkyl epoxide before extrusion, or by pelletizing the polypyrrolidone resin with added alkyl epoxide, but any convenient method may be used.

Thermal stabilization is determined, among other methods, by the measurement of weight loss by the polypyrrolidone polymer on a Mettler FP-1 hot stage at 269°C over a period of 5 min, with and without the presence of the alkyl epoxide. The monomer produced by this heat treatment is completely removed by extraction with water. The difference in weight between the starting polymer and the thermally treated dry extracted polymer is the weight loss. The alkyl epoxide produces an average decrease in weight loss of more than 10 wt % in this test.

Aromatic epoxides or unepoxidized related compounds, are generally found to produce no decrease in weight loss or an actual increase in weight loss in the thermal stabilization test.

Thermal Stabilization with Dialkylaminobenzaldehyde

According to *R. House; U.S. Patent 4,083,827; April 11, 1978; assigned to Chevron Research Company*, the dialkylaminobenzaldehydes are thermal stabilization additives for solid polypyrrolidone at melt temperatures.

The dialkylaminobenzaldehydes are p-, o- or m-dialkylaminobenzaldehydes. The preferred dialkylaminobenzaldehydes are di(C_{1-5} alkyl)aminobenzaldehydes and dimethylaminobenzaldehyde is most preferred. Thermal stabilization of polypyrrolidone is determined as in the preceding patent. The results are shown in the following table.

% p-Dimethylamino- benzaldehyde	Weight Loss, %	Effect on Weight Loss, %
0.0	14.5	−
0.2	21.9	+51
1.0	10.9	−25
5.0	12.2	−16

Densification of Polypyrrolidone

The continuous bulk polymerization of 2-pyrrolidone under agitation produces a finely comminuted product which is washed with water to remove both unreacted monomer and the alkaline catalyst. The polypyrrolidone product is then in the form of a wet powder containing about 50 wt % water, based on polypyrrolidone.

This powder is normally dried, but it has too low a bulk density for economical commercial processing, e.g., it is difficult to feed into an extruder for melt-spinning into filaments. Consequently, the dry powder is melt-extruded to form dense pellets which are suitable for feeding into extruders in melt-spinning processes. In the process of melt-extruding the polypyrrolidone to form dense pellets, its weight average molecular weight is degraded from a high initial value of about 200,000 or more to a value of 35,000 to 100,000 or less, because of the tendency of nylon-4 to decompose into its monomer upon melting.

K. Katsumoto and E.L. Nimer; U.S. Patent 4,130,521; December 19, 1978; assigned to Chevron Research Company provide a method of treating polypyrrolidone powder by compressing the powder having a water content of 5 to 60 wt %, based on polypyrrolidone, under pressure and temperature sufficient to produce solid polypyrrolidone having a dry bulk density suitable for feeding an extruder in a melt-spinning process.

The maximum temperature of the polypyrrolidone during compression is always well below its melting point; consequently, degradation of the polymer to monomer does not occur rapidly. The monomer content of the dried product of this process is normally less than about 0.2 wt %, which is appreciably better than that obtainable from melt-extrusion pelletizing of the powder without further treatment.

The temperature selected for the polypyrrolidone during compression is inversely proportional to its water content. At the lower water contents employed, about 10%, higher temperatures of about 150°C or greater are preferred. At the higher water contents employed, about 40%, lower temperatures of 100°C or less may be utilized, i.e., even as low as about 50°C.

Generally, maximum temperatures of about 75° to 225°C, preferably 100° to 200°C, and pressures of 50 to 300 psi are preferred for compression densification of polypyrrolidone in this process. The moisture content of the compressed polymer is generally in the range of 5 to 40 wt % based on polypyrrolidone. This moisture is removed by drying at temperatures of 100° to 150°C, and optionally, under reduced pressures of 1 to 5 torrs.

Example 1: Polypyrrolidone powder of 170,000 weight average molecular weight, containing 22 wt % of water, was compacted by forcing the powder through the holes of a die of a kibble mill. The size of the holes was $\frac{1}{8}$" in diameter by $1\frac{1}{2}$" long. The powder was heated by mechanical mixing, fused into a hardened cylindrical shape and cut into $\frac{3}{16}$" lengths at the die holes by a doctor blade. Kibble temperatures were about 80°C as measured by a thermometer in the receiver. A smooth surfaced kibble with irregular ends and 19% water content was thus formed. The bulk density of the kibble was 37 lb/ft^3 after drying to less than 0.1 wt % water. The molecular weight of the kibble was the same as that of the polymer powder fed to the mill.

Example 2: Polypyrrolidone powder of 290,000 weight average molecular weight (10 to 200 mesh) containing 37 wt % water (based on polypyrrolidone) was subjected to compaction by passing it through a 2 roll mill. Roll surface temperature was about 148° to 154°C. The polymer was compressed into dense sheets approximately $\frac{1}{16}$ to $\frac{3}{32}$" thick which had 19% water content. The sheets were chopped into approximately $\frac{1}{8}$" x $\frac{1}{8}$" granules and dried. The particle density of these granules was approximately 62 lb/ft^3 and the average bulk density was about 32 lb/ft^3. The molecular weight of the granules was identical to that of the starting polymer powder.

OTHER POLYLACTAMS

Decrease in Melting Temperature due to Lithium Halide Additive

S. Russo, E. Bianchi, A. Ciferri, B. Valenti and G. Bonta; U.S. Patent 4,092,301; May 30, 1978; assigned to Consiglio Nazionale Delle Ricerche, Italy describe a process by which transformations of linear polyamides are performed at temperatures below the melting and decomposition temperatures of pure polyamides, wherein anionic polymerization of dry lactams with 4 to 12 carbon atoms is performed in the presence of one or more lithium halides, in amounts from 1 to 10 wt % based on the weight of polymer.

Among the lactams susceptible to polymerization in the presence of the additive, with production of a low-temperature melting polymer, there are ring compounds with 4 to 12 carbon atoms. Particularly interesting is the case of pyrrolidone whose polymer (nylon 4) in the pure state decomposes at the melting temperature, a fact which strongly reduces its processability.

Another advantage of lactam polymerization in the presence of lithium halide is found when preparing composites with glass beads or glass fibers; in this case the process enables the production of perfectly homogeneous samples with optimal adhesion between glass and polymer. It is known that the mixing of glass filler with a preformed polymer does not normally produce a good, homogeneous distribution of the filler.

Example 1: Lithium chloride (LiCl) is added to 250 g of dry ϵ-caprolactam with a percentage ranging from 1 to 4 wt % of the monomer; with some stirring, the mixture is heated to 100° to 140°C to insure the complete solubilization of the salt. The mixture is then cooled to 90°C and metallic Na is added, enough to give 0.4 to 0.6 mol % of sodium caprolactam (the catalyst). The temperature is then increased to 100° to 170°C and the activator, 0.4 to 0.6 mol % of N-acetyl caprolactam, is added; the polymerization reaction takes place in a few minutes, with yields very near to 100%.

From intrinsic viscosity measurements of the polymer in m-cresol at 25°C, it can be seen that the molecular weight of polycaprolactam is not particularly affected by the presence of LiCl. Thermodynamic melting temperatures T_m, measured by a differential scanning calorimeter range from 192°C at 4% LiCl to 230°C at 1% LiCl, compared to 240°C with no additive.

It is to be noted that the decrease in T_m is not permanent; hot water washing will in fact restore T_m to the standard value of the conventional polymer. Melt viscosity data, measured at 250°C and extrapolated to zero gradient, show a viscosity increase due to LiCl.

Example 2: Polycaprolactam with 4% of LiCl in weight, prepared as in Example 1 but with a molecular weight of 14,000 (by suitably adjusting catalyst and activator concentration) has been used to prepare fibers by extrusion of the polymer melt at 195°C through a capillary rheometer. Fibers of 1 mm in diameter can be extruded at a temperature which is 45°C below the melting point of the pure polyamide.

Films of 2 mm of thickness were prepared by compression molding at 150 kg/cm^2 for 40 sec at 200°C. LiCl was then washed away with boiling water and the torsional modulus was measured on a Clash-Berg apparatus. The torsional modulus at 10 sec and at room temperature was 1.6×10^4 kg/cm^2. A sample of the same polymer prepared without LiCl required a temperature of 245°C for the compression molding (i.e., 5°C above its melting temperature) and the corresponding modulus was 1.1×10^{-4} kg/cm^2, smaller than that of the polyamide-LiCl sample.

Example 3: Polymerization is performed as in Example 1 but with the addition of 20 wt % of glass beads (40 to 50 μ in diameter). Adequate stirring results in a homogeneous distribution of the filler into the mass polymerizing at 100° to 170°C; the same is true for the final polymer, as shown by microscopy observation.

Mechanical properties of these samples were studied by an Instron instrument; elongation at break of mixtures with 2% LiCl was decreased by a factor of 3, whereas samples without LiCl had an increase in elongation of a factor of 4, resulting from the addition of glass beads. These results show the role of LiCl in increasing the adherence between polymer and glass.

Transition Metal Oxides as Catalysts

V.I. Serenkov, J.S. Deev, E.A. Ryabov, E.V. Gorbunova, V.S. Tikhomirov, M.K. Dobrokhotova and A.V. Berezovsky; U.S. Patent 4,124,468; November 7, 1978 describe a process for the production of polyamides, which comprise polymerizing lactams in the presence of oxides of transition metals. The process is carried out at a temperature from 200° to 350°C in air or in an inert gaseous medium. The process may likewise be conducted with the reactants being exposed to an ionizing radiation, e.g., gamma-radiation, which boosts the polymerization rate by 5 or more times. The end product shows good mechanical, electrical and thermophysical properties.

The proposed process for lactam polymerization in the presence of transition metal oxides offers the following advantages over the hydrolytic process used for industrial purposes:

(a) The proposed process takes less time;

(b) The polymer does not undergo destruction in processing;

(c) The process proceeds in a single step, materially simplifying the flow chart and also permitting a continuous procedure;

(d) No high-pressure equipment is required;

(e) The product is superior in performance to the polymer produced hydrolytically and anionically;

(f) At a higher level of the oxides, viz above 5 wt %, the oxide serves as a filler; thus, the polymer is filled right in the course of polymerization;

(g) The process gives already colored polymers; thus, if the oxide is TiO_2 the product is white, while in the case of Cr_2O_3 the polymer is green;

(h) The polymer shows excellent wear resistance, being superior in this respect by a factor of 5 to the products of the prior art processes;

(i) The process produces no wastes, emissions and other waste products;

(j) Owing to the simpler procedures and higher throughput, the proposed process features improved technical and economic characteristics.

As distinct from the product of the prior art anionic technique, the polymer produced in this process (the polymerization times being compatible) lends itself to all known kinds of processing without any noticeable adverse effect on its molecular weight or mechanical properties.

Example 1: A steel reactor equipped with a stirrer and heater is charged with 300 g of dodecalactam and 15 g of titanium dioxide (5 wt %).

The reactor is evacuated to a residual pressure of 10^{-3} torr. The reaction is carried out at a temperature of 300°C for 4 hr with continuous stirring, after which the reactor is cooled down, and the polymer is discharged therefrom with the aid of a screw conveyor. The degree of conversion is 98%; the molecular weight is 43,000; and the color of the polymer is white.

Example 2: A steel reactor equipped with a stirrer and a heater and having an inlet port made of aluminum foil is charged with 200 g of dodecalactam and 10 g of titanium dioxide (5 wt %), purged with nitrogen to remove all air therefrom and closed. The reaction is carried on at atmospheric pressure in an atmosphere of nitrogen at a temperature of 300°C for 15 min, the reaction mixture being exposed to fast electrons of energy 5 MeV, mean current density 1 μA/cm^2 and dose rate 200,000 rad/sec. Then the reactor is cooled down and the product polymer discharged therefrom. The degree of conversion is 98%; the molecular weight is 23,000; and the color of the polymer is white.

Example 3: A steel reactor equipped with a stirrer and a heater is charged with 200 g of dodecalactam and 20 g of chromium oxide (10 wt %), purged with argon and closed. The reaction is carried on at atmospheric pressure in an atmosphere of argon at a temperature of 320°C for 45 min with continuous stirring, the reaction mixture being exposed to gamma-radiation of dose rate 180 rad/sec. Then the reactor is cooled down and the product polymer discharged. The degree of conversion is 98.5%; the molecular weight is 30,000; and the color of the polymer is green.

Cationic Polymerization Initiated by Selected Anhydrous Salts

Most of the polymers of lactams are prepared by hydrolytic polymerization. Hydrolytic polymerization is usually carried out at an elevated pressure and requires high reaction temperatures and long polymerization times. Processes which utilize anionic catalysts to promote polymerization have become popular. Processes which utilize anionic catalysts proceed more rapidly than hydrolytically catalyzed processes. However, the processes are very sensitive to the presence of small amounts of impurities which react with the catalyst system.

The anionic polymer is somewhat less thermally stable due to the presence of residues of the strongly basic initiator. Anionic polymerization of lactams with high melting points, e.g., dodecanolactam, is technologically difficult because of the high polymerization rates achieved on mixing of the components.

R. Puffr and J. Sebenda; U.S. Patent 4,111,869; September 5, 1978; assigned to Ceskoslovenska akademie ved, Czechoslovakia describe a method for the cationic polymerization of lactams either substituted or unsubstituted on the lactam ring, which contains 6 to 16 carbon atoms, in the presence of 0.05 to 10 mol % based on the lactam of an anhydrous salt of Cu, Zn, B, Al, Ti, Si, V, Sb, Bi, Cr, Mo, W, Mn, Fe, Co, Ni and an inorganic or organic anion such as phosphate, pyrophosphate, phosphite, sulfate, halide, carboxylate and sulfonate in the presence of 0.05 to 5 mol % of a protic organic or inorganic acid such as pyrophosphoric, phosphorous, hydrohalic, carboxylic and sulfonic. These additives can be added to lactam below 180°C and the polymerization is carried out at 150° to 320°C for 0.1 to 200 hr. The above-specified salts, also their mixtures, can be used. The polymerization can be carried out with mixtures containing two or more lactams. The process can be carried out in a solvent and in the presence of various anhydrous inorganic fillers, such as glass fibers, ground minerals, powdered metals, metal oxides, phosphates, sulfates or sulfides or graphite. The process can also be effectively conducted in the presence of organic dyes, plasticizers and emulsifiers. If the polymerization is performed with the salts of cations which are spontaneously reduced by the lactam during polymerization, finely dispersed metals are formed and remain suspended in the polymer.

Example 1: Lauric acid in an amount of 0.032 g and 0.063 g of WCl_6 were dissolved in 6.25 g of dry dodecanolactam at 170°C under a nitrogen atmosphere. The mixture was heated to 260°C for 4 hr. The resulting light-blue polymer contained 1.5% of portions extractable with boiling benzene and had an intrinsic viscosity of 1.3 (tricresol at 25°C).

Example 2: Lauric acid in an amount of 0.0285 and 0.0425 g of $SbCl_5$ were dissolved in 5.6 g of dodecanolactam at 170°C. The mixture was heated at 260°C under a nitrogen atmosphere. During heating at 260°C, metallic Sb was gradually formed and remained evenly dispersed due to the high viscosity of the medium. After 20 hr of heating, a polymer was obtained which had an intrinsic viscosity 0.7 (tricresol 20°C) and contained 1.5% of material extractable with boiling benzene.

Example 3: Titanium tetrachloride in the amount of 0.35 g was dissolved in a solution of 30 g of capryllactam and 0.06 g of 1,3,5-tricarboxybenzoic acid in 62 ml of N-methylpyrrolidone (dried over a molecular sieve). After heating at 240°C for 20 hr a viscous solution of nylon 8 (molecular weight 10,000) in N-methylpyrrolidone was obtained.

Example 4: A dry melt of capryllactam containing 7 g tungsten hexachloride and 3.6 g of sebacic acid in 1 kg of mixture was held in a container at 90°C under a nitrogen atmosphere. The mixture was pumped by a gear-pump at a rate of 10 kg/hr into a reactor heated to 265°C. It polymerized for 10 min and was extruded through a nozzle. The cooled polymer strip was cut into granules.

Continuous Production of Polylaurolactam

G.A. Enenshtein, A.V. Berezovsky, M.K. Dobrokhotova, E.K. Lyadysheva, L.A. Nosova, S.N. Nurmukhomedov, E.L. Tarasova, S.S. Gusakov, L.D. Pertsov and I.A. Ebel; U.S. Patent 4,077,946; March 7, 1978 provide a process of producing polylaurolactam comprising the steps of hydrolytic polymerization of laurolactam performed in the presence of water, orthophosphoric and dicarboxylic acids at a temperature from 280° to 310°C under a pressure from 70 to 120 atm, leading to the formation of reaction mixture containing prepolymerizate and water, the mixture being a homophase system; removing water from the mixture by throttling the mixture down to a pressure of 0.1 to 0.005 atm under isothermal conditions at a flow rate of 20 to 90 m/sec.

Subsequent polymerization of the prepolymerizate is accomplished in a layer at a temperature of 280° to 300°C, which is 0.5 to 2.00 mm thick. The process makes it possible to obtain a high quality end product by simple technology.

Carrying out of the continuous process under the operating conditions reduces the time of polylaurolactam production at least tenfold as compared to the industrial periodic process and makes it possible to obtain the polymer with required quality characteristics. As compared to the continuous prior art process the production time is reduced at least 4 to 6 fold.

Example: The starting mixture, i.e., a laurolactam melt, containing 10 wt % of water, 0.2 wt % of orthophosphoric acid, and 0.2 wt % of adipic acid with respect to the weight of lactam are fed for prepolymerization into a first stage reactor in amounts of 3 kg/hr with the aid of a batching device, such as a piston or gear-type pump. In the reactor the reaction mixture is heated up to 290°C under a pressure of 96 atm. The reactor is a tube 12 m in length and 12 mm in inner diameter. When the content of low molecular weight products in the prepolymerizate as determined by its extraction in ethanol attains 5.5 to 5 wt %, the reaction mixture containing the prepolymerizate and water is throttled down to a pressure of 0.013 atm at a temperature equal to 290°C.

The flow rate is 50 m/sec and the number of local resistances of a narrow slit-type which ensures the required conditions is 360. The water and unreacted monomer evaporated under these conditions are removed from the reaction mixture and do not participate in the process. The resulting prepolymerizate is delivered into a thin-film apparatus where the prepolymerizate is finally polymerized in a 1.7 to 1.8 mm layer at a temperature 290°C under 0.013 atm. The residence time is 20 min. The end product contains 1.6 wt % of low molecular products and has the specific viscosity of 0.74 units as measured in a 0.5% m-cresol solution at 25°C.

Lactam/Polyol/Polyacyl Lactam Terpolymers

Lactam/polyol/polyacyl lactam terpolymers having both ester linkages and amide linkages between the monomeric segments of the terpolymer are disclosed by *R.M. Hedrick and J.D. Gabbert; U.S. Patent 4,031,164; June 21, 1977; assigned to Monsanto Company* .

The lactam/polyol/polyacyl lactam terpolymer of this process has the general formula:

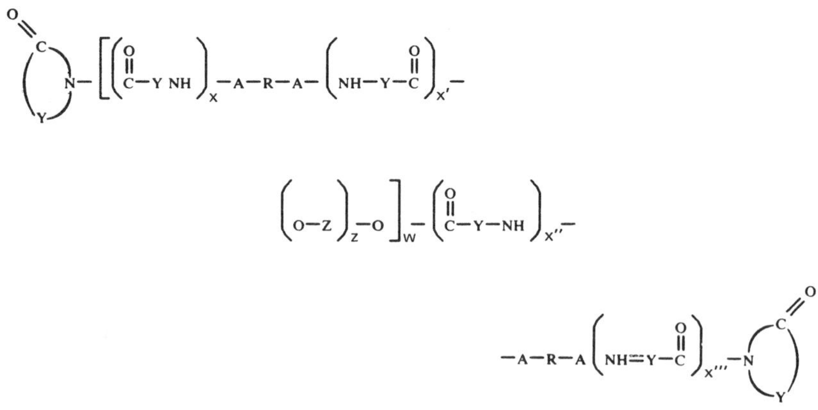

wherein $(O-Z)_z$ is a polyol segment and Z is a hydrocarbon or substituted hydrocarbon group, the group being alkylene, arylene, alkylene carbonyl, arylene carbonyl, and mixtures thereof; A is a carbonyl group; R is a polyvalent hydrocarbon group; Y is an alkylene or substituted alkylene having from 3 to 14 carbon atoms; x, x', x'' and x''' are integers and the total number of x's is equal to 2w + 2; and z and w are integers equal to one or more.

Example 1: A quantity of 2,100 g of ε-caprolactam is heated under a vacuum to remove water. The heating is continued until 100 g of the caprolactam is removed from the reaction vessel. The caprolactam is then allowed to cool, after which 725 ml of pyridine is added. The mixture is then allowed to cool further to about 40°C, after which time 609 g of terephthaloyl chloride is added at a rate sufficient to keep the pot temperature at about 80° to 90°C.

The mixture is then heated to 135° to 140°C for 2 hr. After that time, the resultant solution is poured into 14 liters of ice water to precipitate the product, terephthaloyl bis-caprolactam. The precipitate is filtered and washed three times with cold water and once with methanol. Then the powder is dried at 50°C.

Example 2: A crosslinked terpolymer was prepared using the quantities of ingredients listed below. The catalyst, bromomagnesium caprolactam (prepared by reaction of ethyl magnesium bromide with caprolactam) was used in a concentration of 5 mmol/mol of caprolactam.

Voranol 2000	60.0 g
Trimesoyl tris-caprolactam	11.0 g
Caprolactam	138.0 g
Flectol H	1.0 g
BMC*	15.5 ml

* 0.4 molar bromomagnesium caprolactam in molten caprolactam.

The Voranol 2000, trimesoyl tris-caprolactam and Flectol H were heated with stirring under vacuum to dry the mixture by distilling 25 g caprolactam. The solution was cooled to 100°C and cast into a 100°C mold by means of a metering pump. Catalyst was injected into the mixture by means of a second metering pump. Mixing of the reactant streams was accomplished by means of a Kenics static mixer. The reaction mixture set into a firm gel within 15 sec after casting. The mold was then heated to 160°C and maintained at this temperature for 30 min after which time the mold was opened and the sample removed. The tensile properties of the terpolymer are as follows:

Tensile strength	6,320 psi
Elongation	295 %
Tensile modulus	76,000 psi

Example 3: A terpolymer containing ethylene glycol as the polyol segment was prepared using the quantities of ingredients specified. The catalyst, ethyl magnesium bromide, was used in a concentration of 8 mmol/mol of caprolactam.

Caprolactam	155.0 g
IBC*	59.8 g
Flectol H	1.0 g

(continued)

Ethylene glycol	10.0 g
Ethyl magnesium bromide**	4.5 ml

  *Isophthaloyl bis-caprolactam
**Catalyst – 3 molar solution in diethyl ether

The caprolactam, isophthaloyl bis-caprolactam, and Flectol H were mixed together and dried by heating under vacuum to distill 25 g caprolactam. The mixture was cooled to 75°C, and ethylene glycol was added. Ethyl magnesium bromide solution was added and diethyl ether and ethane evolved from the catalyst was removed by evacuation of the reactor to 2 mm pressure for 2 min. The resultant mixture was cast into a vertical mold having dimensions of 10" x 10" x ⅛" and which had been preheated to 100°C. The mold consisted of 2 Teflon coated plates separated by a ⅛" Teflon spacer. After casting, the mold was then heated to 160°C and held at that temperature for 30 min, after which time it was opened and the sample removed. The tensile properties of the terpolymer are as follows:

Tensile strength at yield	10,300 psi
Elongation at yield	10 %
Tensile strength at break	9,100 psi
Elongation at break	320 %
Tensile modulus	155,000 psi

Ester-Terminated Terpolymers

R.M. Hedrick and J.D. Gabbert; U.S. Patent 4,034,015; July 5, 1977; assigned to Monsanto Company discloses similar lactam/polyol/polyacyl lactam or lactam/polyol/acyl polylactam terpolymers having up to 100% ester end group termination.

Example: A charge of 325 g Voranol 2000 (polyoxypropylene polyol having a molecular weight of about 2,000), 118 g caprolactam, 60.9 g isophthaloyl-bis-caprolactam, and 2.5 g Flectol H (polymerized 1,2-dehydro-2,2,4-trimethylquinoline) stabilizer was introduced into a reaction chamber equipped with stirrer, thermal controller, nitrogen inlet, and vacuum distilling head. The mixture was heated under vacuum ($<$1 mm) to remove moisture by distillation of 50 g of caprolactam.

The vacuum was released to nitrogen atmosphere and the solution cooled to 100°C. Dissolved nitrogen was removed under vacuum followed by the addition of 2.3 ml decyl alcohol and mixing of the resulting solution by stirring under nitrogen. The solution was catalyzed and cast into a stainless steel vertical sheep press maintained at 100°C by means of gear pumps. The solution was delivered into the press at a rate of 876 ml/min. The catalyst solution (0.4 molar bromomagnesium caprolactam) was injected into the solution stream by a second gear pump at a rate of 103 ml/min resulting in a catalyst concentration of 12 mmol catalyst per mol of caprolactam. After casting the press was heated to 160°C.

Seven lactam/polyol/polyacyl lactam terpolymers were prepared and cast using this process, however, the alcohol and alcohol concentration were varied as indicated in the following table.

Run No.	Percent Molar Equivalence Alcohol of Excess Imide*	Tensile Strength Fail (psi)	Elongation at Break	Modulus psi x 10^{-3}	230°C Melt Index
. . . . Effect of Decyl Alcohol					
1	90	1,840	1,100	4.3	54
2	65	2,280	1,200	3.0	37
3	0	1,640	800	1.9	0.09
. . . . Effect of 2-Octanol					
4	90	2,200	793	1.3	37.2
5	50	1,360	667	5.0	16.2
6	40	1,790	923	4.4	2.4
7	0	1,580	828	0.79	0.14

*Based on molar equivalence of polyol hydroxyl.

Deep-Dyeing Polycaprolactams

E.Radlmann, H.-G. Gelhaar and G. Nischk; U.S. Patent 4,104,324; August 1, 1978; assigned to Bayer AG, Germany disclose polycaprolactam filaments and fibers capable of being deeply dyed with acid dyes to give dye finishes with extreme fastness to light, characterized by the fact that they consist of physical mixtures of unmodified polycaprolactam and a basic polycaprolactam containing structural units corresponding to the formula:

$$
\begin{array}{c}
\text{H} \qquad\quad \text{H} \qquad\quad \text{H} \ \ \text{O} \ \ \ \text{O} \\
| \qquad\qquad | \qquad\qquad | \ \ \| \ \ \ \| \\
(-N-CH_2-CH_2-N-CH_2-CH_2-N)_{\overline{x}}C-R-C-
\end{array}
$$

in which R represents an alkylene radical with 4 to 12 carbon atoms and x is an integer from 1 to 5, have relative solution viscosities of from 2.9 to 3.5 (as measured on solutions of 1 g of nonextracted polyamide in 100 ml of cresol at 25°C) and contain free amino groups in a concentration from 60 to 180 millivals per kilogram of polymer mixture.

Example 1: *Production of Basic Polycaprolactam Concentrate* – In a polycondensation apparatus a mixture of 33,900 pbw of ϵ-caprolactam, 5.47 pbw of adipic acid, 4.056 pbw of diethylene triamine and 0.2 pbw of hydrazine sulfate was heated under nitrogen while stirring to a temperature of 190°C. After 3.5 hr, most of the water had been split off. The pressure was then reduced to 3 mm and the temperature was increased to 210°C. After another 3 hr under these conditions, the condensation reaction had ceased. The viscous melt was then spun off onto a drying belt and the strands obtained were granulated. The homogeneous, pale yellow colored granulate may be characterized as follows: relative viscosity η_{rel} is 2.70, water content is 0.01%, nitrogen content (amino groups) is 1,058 millivals per kilogram, and the softening range is 168° to 172°C.

Example 2: *Production of Basically Modified Polycaprolactam Filaments* – The basic concentrate described in Example 1 was melted in a Barmag three-zone extruder (l = 12 d, d = 30 mm). The nominal temperatures were adjusted to 230°, 240° and 250°C so that the temperatures in the product amounted to about 210°, 240° and 250°C. The melt pressure in the extruder was 25 bars. The concentrate is delivered by an extruder to a gear pump which pumps it into a master melt of polycaprolactam (with a terminal amino group content of 32 millivals per kilogram and a relative viscosity η_{rel} of 3.2). After adjustment of the delivered quantity of 5.7 wt %, its drive was firmly coupled with that of a pump delivering the master melt. This ensured the delivery of a constant amount of concentrate.

The two melts were mixed over a period of 3 min in a dynamic Wiemann mixer at a nominal temperature of $270° \pm 10°C$. The resulting homogeneous mixture was then delivered to the spinning beam and spun into filaments which, after texturing, had a denier of dtex 950 f 42. The amino group content of these filaments was determined by analysis at 91 millivals per kilogram. Using the acid dye Telon Echtgelb 4 GL (corresponds to Supranol 4 GL), color index, acid yellow, the filaments can be dyed in a deep color with a stability to light of 5 to 6. The filament material had a relative viscosity η_{rel} of 3.14.

OTHER POLYAMIDES

CATALYSTS AND ACID-ACCEPTORS

Lithium or Calcium Halide with Improved Solvent System

Aromatic polyamides having high polymerization degree are described by *N. Yamazaki and F. Higashi; U.S. Patents 4,118,374; October 3, 1978; and 4,045,417; August 30, 1977; both assigned to Sumitomo Chemical Company, Limited, Japan.* These are produced by polycondensing an arylene containing aminocarboxylic acid, or a dicarboxylic acid with an aromatic diamine in the presence of a phosphorus compound, a tertiary amine and a metal halide using N-methylpyrrolidone and/or dimethylacetamide as a solvent.

The improved process is characterized by using as a solvent for the liquid phase reaction N-methylpyrrolidone and/or dimethylacetamide and adding either lithium halide or calcium halide to the reaction system.

The phosphorus compound includes so-called phosphites or phosphonites, for example, triphenyl phosphite, tri(o-, m- or p-methylphenyl) phosphite, tri(o-, m- or p-chlorophenyl) phosphite, diphenyl phosphite, ethyldiphenyl phosphonite, n-butyl diphenyl phosphonite, phenyl diphenyl phosphonite, and the like.

Example 1: A mixture of 2.74 g (0.02 mol) of p-aminobenzoic acid, 2 g of LiCl, 6.21 g (0.02 mol) of triphenyl phosphite, 10 ml of pyridine and 40 ml of N-methylpyrrolidone was heated at 100°C for 6 hours with stirring under nitrogen atmosphere. After the reaction, the reaction solution was poured into methanol and the deposited polymer was filtered off.

The polymer was ground, washed with methanol, filtered and dried. The yield was 100%. The inherent viscosity of a 0.5% solution at 30°C in concentrated H_2SO_4 was 1.27.

Example 2: The process of Example 1 was repeated except for changing the solvent. The results are shown in the following table.

Solvent	Volume Ratio (ml/ml)	. Polymer Produced .	
		Yield (%)	η_{inh}
N-methylpyrrolidone/ pyridine	20/30	100	1.21
N-methylpyrrolidone/ pyridine	30/20	100	1.52
N-methylpyrrolidone/ pyridine	40/10	100	1.27
N-methylpyrrolidone/ pyridine	45/5	100	1.26
Dimethylacetamide/pyridine	10/10	99	0.71
Pyridine*	50	100	0.21
Dimethylformamide/ pyridine*	40/10	11	0.08

*Comparative example.

Alkali Metal Salt of an Alcohol

According to *B.K. Onder; U.S. Patents 4,061,622; December 6, 1977; and 4,061,623; December 6, 1977; both assigned to The Upjohn Company* the use of certain catalysts provide for a process for the preparation of soluble polyimides, polyamides, and polyamideimides. The catalysts are alkali metal salts of formula MOR, wherein R represents alkyl or aryl and M represents an akali metal.

The process comprises reacting organic diisocyanates with polycarboxylic compounds consisting of tetracarboxylic acids or the intramolecular dianhydrides thereof, tricarboxylic acids or the monoanhydrides thereof, dicarboxylic acids, and mixtures thereof, in the presence of the catalysts. The polymers are obtained in solution at low reaction temperatures and short reaction times thereby avoiding side-reactions which otherwise would be detrimental to polymer molecular weight and ultimate polymer properties.

Example 1: A dry 500 ml resin flask equipped with a stirrer, condenser, thermometer, nitrogen inlet tube, and addition funnel was charged with 64.4 g (0.2 mol) of commercial grade (97.44% anhydride) 3,3',4,4'-benzophenonetetracarboxylic acid dianhydride (BTDA) and 0.05 g (0.000915 mol) of sodium methoxide catalyst. The flask contents were dissolved in 234 g of dry dimethylformamide (distilled over calcium hydride).

The temperature of the contents was raised to 80°C and, during constant stirring under nitrogen, a solution consisting of 10.0 g (0.04 mol) of 4,4'-methylenebis-(phenylisocyanate) (MDI) and 28.0 g (0.16 mol) of 2,4-toluenediisocyanate (TDI) dissolved in 30 g of dry dimethylformamide (DMF) was added dropwise over 4.5 hours. At the end of this time, infrared analysis of a sample of the viscous solution revealed only a trace amount of unreacted isocyanate (−NCO) and anhydride groups.

An additional 2 mol percent excess [0.0008 mol (0.2 g) of MDI and 0.0032 mol (0.56 g) of TDI] of a mixture of the diisocyanates dissolved in 30 g DMF was added over 1 hour. Approximately 6.5 hours from the beginning of the polymerization, IR analysis revealed no unreacted −NCO or anhydride. The DMF solution, having an inherent viscosity (0.5% at 29.1°C) = 0.41, consisting of approximately 25% by weight of copolyimide, was characterized by a structure wherein approximately 80% of the recurring copolyimide units had the formula

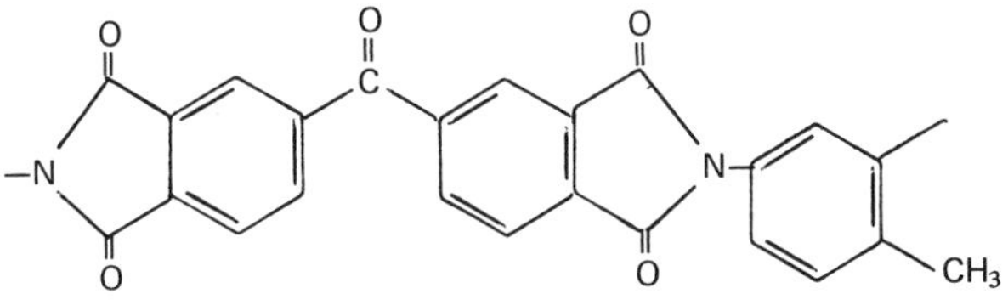

and the remaining 20% of the recurring units had the formula

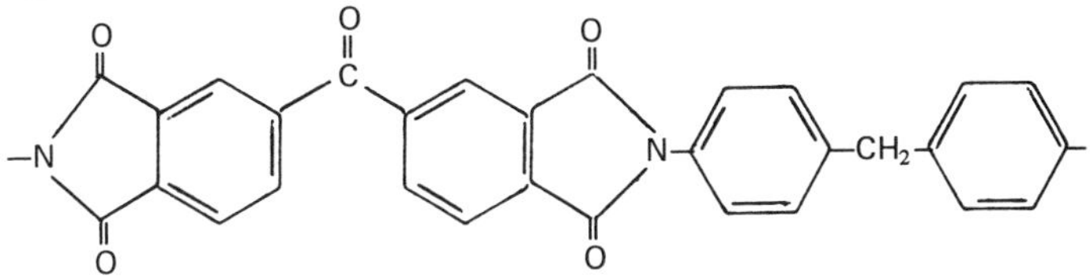

Example 2: *(Comparative)* — Using the procedure and reactants set forth in Example 1 except for the fact that no catalyst was used, the polymerization described therein was repeated. At the reaction temperature of 80°C after 8.75 hours, IR analysis showed an appreciable quantity of –NCO and anhydride groups remaining. Further, the solution which was 25% by weight in solids was quite turbid which was a result of the preferential reaction of the more reactive MDI to form the homopolyimide which is known to be insoluble, thereby leaving at least a portion of the TDI unreacted.

Example 3: *(Comparative)* — The following example is a polymerization reaction carried out in the presence of a known catalyst for the reaction of an isocyanate with an anhydride (see J. Drapier, et al, *Tetrahedron Letters* No. 6, 419–422, 1973) but not a catalyst according to the process.

Using the procedure and reactants set forth in Example 1, except that the DMF was replaced by 175 g of dry distilled N-methylpyrrolidone (NMP), the quantities of reactants were reduced by one-half, and 0.05 g (0.00015 mol) of dicobalt octacarbonyl was employed as the catalyst. After a 7 hour reaction period at 80°C, strong bands in the IR absorption spectrum for –NCO and anhydride groups showed the polymerization was proceeding only at a slow rate.

As in Example 2, the turbidity of the polymerization solution was an indication of the preferential formation of the insoluble MDI based polyimide. The dicobalt octacarbonyl did not catalyze the copolymerization process.

Heterocyclic Tertiary Amine Acid Acceptor in Selected Reaction Media

P.W. Morgan; U.S. Patent 4,025,494; May 24, 1977; assigned to E.I. Du Pont de Nemours and Company describes a process for the preparation of high molecular weight poly(1,4-benzamide) from a p-aminobenzoyl halide salt in selected reaction media and in the presence of a heterocyclic tertiary amine acid acceptor or ethyl diisopropylamine.

The monomers preferred for use are p-aminobenzoyl bromide hydrobromide, p-aminobenzoyl chloride hydrobromide, and p-aminobenzoyl chloride hydrochloride. These monomers may be prepared by known procedures and are stored under anhydrous conditions prior to being used.

Not all tertiary amines are operable in the process. Heterocyclic tertiary amine acid acceptors are useful and preferred among them are pyridine; 2,6-lutidine; 2,4-lutidine; 2,5-lutidine; 3,5-lutidine; 2,3-lutidine; with pyridine being most preferred. Ethyl diisopropylamine is also suitable. These acceptors form salts with the hydrogen chloride and/or hydrogen bromide by-products of the polymerization and are used in sufficient quantity to react with all the acidic by-product generated.

Not all liquid media are operable for the production of the high molecular weight polymers. The liquids useful in the process to provide a medium for the polymerization reaction include chloro-, bromo-, and nitrobenzene, and orthodichlorobenzene; n-pentane, n-hexane, n-heptane and cyclohexane, acetonitrile, propionitrile, and benzonitrile; acetone, methyl ethyl ketone and ethyl acetate, diethyl ether, ethylene glycol dimethyl ether, and diethylene glycol dimethyl ether; methylene chloride, chloroform, carbon tetrachloride, chlorobromethane, 1-chloro-2-bromoethane; ethylene dichloride, 1,1,2-trichloroethane; and unsymtrichlorotrifluoroethane (Freon-113). The aliphatic hydrocarbons are most preferred.

The following examples are illustrative of the preferred embodiments. In these examples, inherent viscosity (η_{inh}) has been determined in accordance with the following equation

$$\eta_{inh} = \ln(\eta_{rel})/C$$

wherein (η_{rel}) represents the relative viscosity, C represents a concentration of 0.5 g of the polymer in 100 ml of the solvent. The relative viscosity (η_{rel}) is determined by dividing the flow time in a capillary viscometer of a dilute solution of the polymer by the flow time for the pure solvent. The dilute solution used herein for determining (η_{rel}) is of the concentration expressed by C above; flow times are determined at 30°C, using concentrated (95 to 98%) sulfuric acid as a solvent.

Example 1: This example illustrates preparation of poly(1,4-benzamide), using chlorobenzene as the reaction medium and pyridine as the acid acceptor. To a mixture of 50 ml chlorobenzene and 6.4 g (0.033 mol) of p-aminobenzoyl chloride hydrochloride, stirred under nitrogen and cooled in a bath of solid carbon dioxide, is added 10 ml (added within 2 to 4 sec. time) of pyridine. The cooling bath is removed in 15 minutes; stirring is continued for 2 hours. The reaction mixture is combined with water, and the solid poly(1,4-benzamide) product isolated, washed well with water, and dried in vacuo. There is obtained 3.87 g polymer, η_{inh} = 1.38.

Example 2: This example illustrates preparation of high molecular weight poly(1,4-benzamide) in n-hexane. In an ice-cooled 1 liter resin kettle equipped with a disc-stirrer are placed 250 ml of n-hexane and 32 g (0.17 mol) of p-aminobenzoyl chloride hydrochloride. To these stirred ingredients, under nitrogen, is added 75 ml of pyridine. The cooling bath is removed in 15 minutes and the reaction mixture stirred for 2 hours before being allowed to stand overnight at room temperature. The solid product is then filtered off, washed well with water, and dried to yield poly(1,4-benzamide), 19.6 g, η_{inh} = 4.5.

Sodium Carbonate Hydrate as Acid Acceptor

T. Noma, H. Fujie and S. Ozawa; U.S. Patent 4,009,154; February 22, 1977; assigned to Teijin Limited, Japan describe a process for preparing aromatic polyamides which comprises contacting (1) a solution or dispersion in a polar, nonbasic and inert organic liquid medium of at least one starting material selected from the group consisting of

(a) an aromatic aminocarboxylic acid halide hydrohalogenide,

(b) a mixture of an aromatic dicarboxylic acid halide and an aromatic diamine dihydrohalogenide, and

(c) a low molecular weight aromatic polyamide having an inherent viscosity of no greater than 0.2 as measured on a solution of 0.5 g of the polyamide in 100 ml of concentrated sulfuric acid at 30°C, with

(2) an aqueous slurry consisting of a sodium carbonate hydrate.

Example 1: 48.1 g of m-aminobenzoyl chloride hydrochloride was added to 250 ml of tetrahydrofuran of 3 mg/100 ml water content, after which this mixture was cooled to –25°C with stirring to prepare a liquid dispersion (1). Separately, 53 g of anhydrous sodium carbonate was dissolved in 250 ml of water at room temperature, following which this mixture was cooled to 3°C with stirring to precipitate sodium carbonate hydrate crystals and thus prepare an aqueous slurry (2) of a solid phase concentration of 35 wt %.

The foregoing liquid dispersion (1) and aqueous slurry (2) were then mixed for 2 minutes in a domestic mixer with vigorous agitation followed by dilution with 200 ml of water, whereupon the resulting polymer was precipitated as a white powder. The polymer obtained after filtration, water-washing and drying, had an inherent viscosity of 2.41, and the yield was 99%.

Example 2: *(Comparative)* — Example 1 was repeated but using instead of the aqueous slurry (2) an aqueous solution obtained by dissolving 69 g of potassium carbonate in 250 ml of water and thereafter cooling to 3°C. The resulting polymer was submitted to the same aftertreatment as Example 1. When the so obtained polymer was measured for its inherent viscosity, it was only 1.67.

Example 3: *(Comparative)* — Example 1 was repeated but using instead of the aqueous slurry (2) a 16°C aqueous solution obtained by dissolving 53 g of anhydrous sodium carbonate in 250 ml of water. The inherent viscosity of the resulting polymer obtained after having been given the same aftertreatment as in Example 1 was 1.18.

MANUFACTURING PROCESSES

Continuous Manufacture of Polyamides

F. Mertes, H. Doerfel, E. Heil and C. Cordes; U.S. Patent 4,060,517; Nov. 29,

1977; assigned to BASF Aktiengesellschaft, Germany describe a process for the
continuous manufacture of polyamides by continuously conveying the aqueous
solution of a salt of essentially equivalent amounts of a diamine, or of a mixture
of several diamines, and of a dicarboxylic acid, or of a mixture of several dicarboxylic acids, or of mixtures of such salts with lactams and/or aminocarboxylic
acids, through several reaction zones under polyamide-forming conditions, wherein the mixture of starting materials is heated, in a first reaction zone, to from
200° to 300°C, preferably 220° to 280°C, at pressures which are above the corresponding saturation vapor pressure of water and prevent the formation of a
vapor phase. The pressure action on the polycondensation mixture is released
in a second reaction zone and condensation of the mixture, to form high molecular weight polyamides, is then completed in further reaction stages.

Figure 3.1 shows a flow diagram illustrating a preferred embodiment of the
subject process, the individual elements of the drawing being conventional.

Figure 3.1: Continuous Manufacture of Polyamides

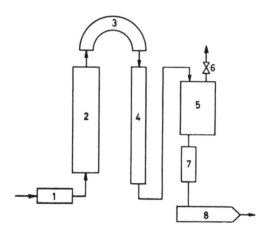

Source: U.S. Patent 4,060,517

The process is carried out by pumping the aqueous solution of the salts of diamines and dicarboxylic acids, or the aqueous solutions of the starting materials
for corresponding copolyamides, continuously, under pressure, from a stock
vessel via a heat exchanger **1** into the precondensation reactor **2**, which is completely filled with liquid and through which the material preferably flows upwards, to avoid forming gas cushions. The precondensation reactor is operated,
e.g., at 280°C and 80 atm gage, with residence times of 45 min.

The precondensate, which has been polycondensed to the extent of more than
80%, is adiabatically released through a valve into a distributor tube **3** kept at
from 10 to 20 atm gage. While being released, the precondensate and water
vapor cool to from 180° to 210°C and then pass into a tube bundle evaporator

4 in which the bulk of the water is evaporated and at the same time the precondensate is heated to from 220° to 330°C, preferably from 250° to 300°C. After evaporation, the pressure settles to a value of from 8 to 15 atm gage.

In a downstream separating vessel **5**, the water vapor escapes through an orifice at the top, via a pressure-regulated valve **6**. The polymer melt collects in the lower part of the separating vessel and remains there, e.g., for 30 min at from 165° to 270°C. The walls of the separating vessel are preferably kept at from 2° to 10°C below the temperature of the polycondensate which enters from the heat exchanger **4**, to avoid the formation of vapor bubbles and foam in the separating vessel.

The polycondensate is continuously released adiabatically through a second pressure-release valve in the bottom of the separating vessel and is then heated in a subsequent heat exchanger **7**, together with the water vapor formed, to from 270° to 290°C, passed into the post-condensation apparatus **8**, separated therein from the water vapor, and post-condensed at from 270° to 290°C under atmospheric pressure or subatmospheric pressure, using times of from 25 to 40 min.

Example: A solution of 60 kg of the salt of adipic acid and hexamethylenediamine, 600 g of hexamethylenediamine and 40 kg of water is kept under autogenic pressure at 110°C in the absence of air. 10 kg/hr of this solution are continuously brought to 80 atm pressure and dumped through a tube **1**, 1 m long and 25 mm diameter, which is heated to 290°C, whereby the solution is brought to 280°C.

The polycondensation mixture then passes upward through a vertical pressure vessel **2**, 2 m long and 70 mm internal diameter, in which the polycondensation mixture remains for an average of 45 min at 280°C and 80 atm gage. It is then released, through an orifice at the highest point of the pressure vessel, via a pressure-regulated needle valve, into a pipe arc **3** which is heated to counteract heat losses and from there into a vertical tube **4**, 2.5 m long and of 25 mm diameter, which is heated to 290°C. As soon as the installation is in operation, the precautionary heating of the pipe arc can be switched off.

Virtually regardless of the heating, the water vapor saturation temperature corresponding to the pressure to which the mixture is being released, in this case 180°C and 10 atm gage, establishes itself in the pipe arc without any interference with the polycondensation through premature solidification of nylon precondensate. In the downstream heat exchanger, which is heated to 290°C, the polycondensate which has been released adiabatically, and the water vapor, are heated to 275°C, with evaporation of the bulk of the water present, and the material then passes into an elongated separating vessel **5**, of about 10 liters capacity, the walls of which are kept at 270°C.

Water vapor excapes through an orifice in the upper part of the separating vessel, via a pressure-regulated valve **6** which maintains the release pressure. The polymer melt remains in the lower part of the separating vessel for 15 min and is then released, via a second pressure-release valve into a pipeline **17** which is heated to 290°C and is under atmospheric pressure, through which the material, together with the water vapor formed, is conveyed to a type ZSK 53 twin-screw extruder **8**.

In the intake zone, the water vapor which accompanies the polycondensate is separated off and escapes through a vent pipe. The polycondensate is conveyed through the twin-screw extruder at from 280° to 285°C and atmospheric pressure in the course of 20 min, during which little back-mixing occurs, and is extruded as a strand. This is granulated and dried, giving colorless polyamide granules of K value 71. The end product can be used to injection-mold articles having excellent mechanical properties, or to spin filaments.

Comparative Example: The process is carried out as above except that 9 kg/hr of a 50% strength aqueous solution of the salt obtained from equivalent amounts of adipic acid and 2,2-di(4-aminocyclohexyl)propane are used. In addition, the process is modified in the following respects: the residence time of the precondensate in the separating vessel, in which the water vapor and polycondensate are separated, is restricted to a few minutes, and the post-condensation in the twin-screw extruder is carried out at 50 mm Hg instead of atmospheric pressure. The end product is a glass-clear polyamide of K value 60, which has a melt viscosity of about 60,000 poise and can be injection-molded to give glass-clear articles.

Because of the high melt viscosity of the end product it is not possible to manufacture the same polyamide by batchwise polycondensation in an autoclave; instead only a relatively low molecular weight precondensate can be produced batchwise and has to be converted to an end product of sufficiently high molecular weight by heating in the solid state. This process is uneconomical compared to the continuous process and its second stage results in yellowing and oxidative degradation of the product because small traces of oxygen cannot be totally excluded from the granules being heated.

Vapor-Phase Preparation of Aromatic Polyamides

Vapor-phase condensation polymerization of aromatic polyamides is carried out by *H. Shin; U.S. Patent 4,009,153; February 22, 1977; assigned to E.I. Du Pont de Nemours and Co.* by vaporizing polyamide-forming monomers, diluting with an inert gas, and reacting the monomers in a reaction zone heated from 150° to 500°C, from 0.01 to 5 seconds or longer. The polyamide is preferably deposited on a removable inorganic or organic substrate maintained within the zone.

Aromatic polyamides can be prepared from polyamide-forming monomers. By that term is meant the pairs of monomers that form alternating units (e.g., units in the polymer corresponding to diamines and diacids) into polymers where the units are joined by amide linkages. Examples are aromatic diamines with derivatives of aromatic diacids which will react with these diamines, such as aromatic diacid halides.

Representative monomers are described in Kwolek et al., U.S. Patent 3,063,966, and Hill et al., U.S. Patent 3,006,899, the teachings of which are incorporated herein by reference. Particularly preferred diamines are para-phenylenediamine (PPD), meta-phenylenediamine (MPD), and mixtures thereof. Particularly preferred diacid halides are terephthaloyl chloride (TCl), isophthaloyl chloride (ICl), and mixtures thereof. Especially preferred polymers are the condensation products of PPD with TCl and MPD with TCl or ICl.

As depicted in Figure 3.2, vapors of two different monomers (A and B) together with hot inert gas are fed to a mixer (such as a jet mixer, a simple short tube or a combination) and then to the reactor inlet. Additional inert gas is optionally applied to the reactor. The reactor effluent stream (consisting of some polymer, possibly oligomers and by-products such as HCl) is conducted through a quenching chamber where the stream is cooled by a flow of relatively cold inert gas. Other cooling means including water sprays could be used. The cooled stream is then led through a separator such as a combination of a cyclone separator and filters to remove solid material. The filtered stream is then passed through a water scrubber to remove hydrogen halide and vented to the atmosphere or recycled.

Figure 3.2: Vapor-Phase Preparation of Aromatic Polyamides

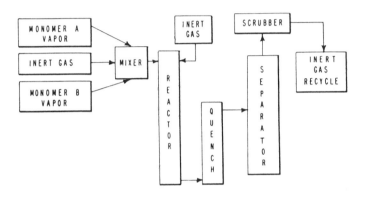

Source: U.S. Patent 4,009,153

Inherent viscosity (I.V.) is defined by the equation:

$$\text{I.V.} = \ln(\eta_{rel})/c$$

where c is the concentration (0.5 g of polymer in 100 ml of solvent) of the polymer solution and η_{rel} (relative viscosity) is the ratio between the flow times of the polymer solution and the solvent (concentrated sulfuric acid of 95 to 98% H_2SO_4) as measured at 30°C in a capillary viscosimeter. Inherent viscosities determined on crude polymer are designed I.V.i and are generally lower by 0.1 I.V. unit than the I.V. on the purified polymer although occasional samples will have an I.V. value as much as 1 unit lower.

Branching value (B) is defined as η_B/η_{BO} where η_B is the bulk viscosity at 60°C of a 2 to 4.5 wt % solution of a sample polymer in 100% sulfuric acid and η_{BO} is the bulk viscosity at 60°C of a solution of a control polymer in 100% sulfuric acid at the same concentration as used for the sample polymer. The control polymer of the same I.V. as the sample polymer is prepared in a solvent at low temperatures (cf U.S. Patent 3,063,966) which affords no branching. The concentrations of the solution are selected to give convenient measurements with the Brookfield Viscometer (Model HBT) used as follows: I.V. <2.5, 4.53%, I.V. 2.5 to 5.0, 4.75%, I.V. 5.1 to 6, 3.07% and I.V. above 6.0, 2.01%.

Tenacity (Ten), elongation (E) and initial modulus (Mi) are obtained from breaking a single filament (gage length of 1.0") or a multifilament yarn having 3 turns per inch twist (gage length 10") on an Instron tester at a rate of extension of 10%/min. Gas flows are expressed in the example as scfm (standard cubic feet per minute) and slm (standard liters per minute), both referring to volumes at 0°C and 1 atmosphere of pressure. All gases and vapors are at atmospheric pressure in the example. Monomer flow rates are expressed as gram-mols/minute (mols/min). The gas residence time in the reactor is calculated assuming that the entire volume of monomer vapors and inert gas pass through the reactor at the average temperature of the reaction.

Example: Para-phenylenediamine (PPD) vapor and terephthaloyl chloride (TCl) vapor both at 325°C and at flow rates of 0.359 and 0.357 mols/min, respectively, are mixed with 2.5 scfm (70.8 slm) of N_2 gas at 400°C in a mixing jet heated to 470°C by electrical heaters.

The reaction mixture is conducted through a pipe 7" long of 0.375" i.d., heated to 450°C by a heating tape, to the reactor. The reactor consists of 24.5" long nickel pipe of 2" i.d. in a vertical position containing a removable inside lining of nickel foil, 2 openings on opposite sides of the top of the pipe wall for the reaction mixture entrance and additional N_2 gas [2.1 scfm (59.5 slm)] at 315°C, a top closure containing a lead to the thermocouple in the reactor, a bottom removable plug and an exit in the lower end of the wall of the pipe. The exterior of the reactor wall is held at 290°C by heating tapes.

The flow rates give a gas residence time in the reactor of about 0.22 second. The reactor exit connects to a vertical quenching tube where 11.7 scfm (331 slm) of room temperature N_2 gas are added. The quench tube contains a reamer downstream of the quench gas entrance and means for removing deposited polymer powder from the wall of the quench tube. The quench tube is connected with a cyclone separator and a bag filter (where polymer powder is removed) and then through a water scrubber (to remove HCl vapors) to the atmosphere.

Before the run is started, the equipment is preheated to near operating conditions by passing the heated nitrogen streams at their required flow rates and temperatures through the reactor and mixing jet. At the start of each run, both monomer vapor streams (which have previously been established, but not allowed into the mixing jet) are introduced simultaneously into the mixing jet. The temperature of the reacting mixture (containing 11 mol % of the two monomers) is found to vary from 350° to 370°C at the reactor inlet, and from 320° to 370°C at the reactor outlet.

After 5 minutes of operating time, the monomer flows are terminated, the plug at the bottom of the reactor is removed followed by the nickel foil liner and the solid polymer deposited thereon. The nickel foil liner and attached polymer is allowed to cool to room temperature in air. The polymer (approximately 30 g) is then peeled from the nickel foil and ground in a Wiley Mill until it passes through a 60-mesh screen.

The polymer is washed 3 times in a blender, with methanol which contains 3.0% HCl, to remove any unreacted monomer. The polymer is then washed 2 times with methanol and then several times with distilled water until the pH of wash water is above 6.5. The polymer is dried in a vacuum oven. The I.V. of the extracted polymer is 4.3 with a B value of 2.1. This polymer is spun into filaments using the technique of U.S. Patent 3,767,756 to give a yarn with Ten of 18 g/den, E of 3.3% and Mi of 500 g/den for the 600 denier yarn.

Additional polymer (about 550 g) is recovered from the spaces between the reactor and the scrubber. This polymer typically has an I.V. of between 0.6 and 1.6.

Preparation of Polyamide from Diacid-Rich and Diamine-Rich Components

Prior art processes for preparing high molecular weight polyamides, whether starting with a dicarboxylic acid (also called diacid) and a diamine or a diester or a diacyl chloride and a diamine, usually were carried out in solution in an organic solvent or solvent mixture, water, or a solvent-water system. The most widely practiced industrial process starts with a dicarboxylic acid and a diamine, which are combined together in a stoichiometric ratio in solution in a large amount of water, that must eventually be evaporated and removed. The solids concentration in such a solution is normally only about 50%. It can be seen that these prior art processes require large amounts of energy and, in addition, call for large capital investment because of the size and complexity of equipment, which must handle large volumes of liquids per amount of polyamide produced.

An ideal process for making high molecular weight polyamides should be carried out in the liquid state because of the ease of metering and conveying the reactants, in the absence of organic solvents or large amounts of added water, and at sufficiently low temperatures to avoid substantial thermal degradation.

J.W. Sprauer; U.S. Patent 4,131,712; December 26, 1978; assigned to E.I. Du Pont de Nemours and Company describes a process for making a high molecular weight polyamide, wherein the diacid-rich component and a diamine-rich component are prepared separately in nonstoichiometric proportions, each of these components melting below the melting temperature of the polyamide product, and preferably below 200°C; and then the diacid-rich component and the diamine-rich component are contacted in liquid state at high enough temperature to prevent solidification, and in proportions such that the total amounts of diacid and diamine, whether combined or not, are as much as possible stoichiometric. A major utility of this process is in the production of nylon 66, a known fiber-forming polyamide and engineering plastic.

Example 1: *Preparation of Low Melting Acid-Rich Component* — (A) 1 mol of adipic acid and one-half mol of hexamethylenediamine (containing 13.4% water) were mixed and cautiously melted in a glass laboratory resin kettle equipped with heater, stirrer, thermometer, nitrogen purge, and condensate receiver. When the free water (9.0 g) had distilled and the temperature reached 171°C, a homogeneous clear melt was attained. In order to form a carboxylic oligomer, the melt was cautiously heated to 211°C as water formed by dehydration of the components distilled, until distillation practically ceased. The product was chilled and solidified and tested by differential thermal analysis, showing multiple melting peaks of which the highest was 174°C.

(B) In a resin kettle, as in (A) above, 1.5 mols of adipic acid and 0.5 mol of anhydrous hexamethylenediamine were cautiously mixed and melted to a clear melt at about 180°C. Water was slowly distilled until distillation practically ceased; and the clear melt was chilled, solidified, and tested by differential thermal analysis, showing a single melting peak at 123°C. The product was crushed and twice leached with boiling methanol. The methanol-insoluble fraction was tested by differential thermal analysis showing twin melting peaks at 167° and 179°C. This confirmed eutectic depression of oligomeric (methanol-insoluble) melting points by excess adipic acid (methanol-soluble, normal melting point 152°C).

Example 2: *Preparation of Low Melting Amine-Rich Component* — As in 1(B) above, 0.80 mol of anhydrous hexamethylenediamine, and 0.20 mol of adipic acid were mixed, melted and heated until distillation practically ceased at 215°C. The chilled solid product showed by differential thermal analysis a maxiumum melting peak at 166°C. The product was crushed and extracted with water and then with methanol. The methanol-insoluble fraction was tested by differential thermal analysis showing twin melting peaks at 197° and 210°C, confirming depression of oligomeric (methanol-insoluble) melting points by excess hexamethylenediamine (water- and methanol-soluble, normal melting point about 40°C).

Example 3: *Mixtures of Hexamethylenediammonium Adipate with Excess Diacid or Excess Diamine* — (A) Equal weights of adipic acid and purified hexamethylenediammonium adipate (a plant-run nylon 66 intermediate) were intimately mixed and melted together by heating at a programmed rate in an encapsulated differential thermal analysis sample cup. First heating program was terminated at 155°C. The sample was cooled and remelted, showing melting peaks at 103° and 127°C before melting program terminated at 195°C. This shows that eutectic depression of melting points occurs not only between adipic acid and oligomer but also between adipic acid and nylon salt.

(B) In the same way as in (A) above, one pbw of hexamethylenediamine and 0.75 pbw of hexamethylenediammonium adipate were intimately mixed and melted together by heating at a programmed rate in an encapsulated differential thermal analysis cup. First heating program was terminated at 175°C; the sample was cooled and remelted showing peaks at 35° and 156°C. This experiment shows eutectic depression of melting points between hexamethylenediamine and nylon salt.

Example 4: *Preparation of High Molecular Weight Polyamide* — To a resin kettle as in Example 1(A) were charged 0.50 mol hexamethylenediamine, 0.02 mol acetic acid (to regulate molecular weight of the product), and 1.00 mol adipic acid. The charge was cautiously mixed and melted to a clear melt at 182°C. The temperature was slowly raised to 250°C while simultaneously distilling water and adding further 0.25 mol of hexamethylenediamine. The temperature was gradually increased to 285°C, while simultaneously distilling water and adding the last 0.25 mol of hexamethylenediamine. The resulting clear melt was discharged and solidified, then tested by differential thermal analysis, showing a melting peak of 254°C. This examples demonstrates the preparation of nylon 66 with anhydrous feed materials at atmospheric pressure.

Polyamide Gel Removal

Polyamides are often used in the form of nylon fibers, coatings or molded objects. The usual process for the manufacture of many polyamides involves heating the reactants under reduced pressure so as to remove water of reaction and force the amidation reaction to completion. The molten polymer so produced is then often maintained at elevated temperatures while it is mixed with additives.

Extended exposure to these elevated processing temperatures causes crosslinking of a part of the polyamide. Crosslinks convert the polymer into a gel which is insoluble in molten polymer and in all polymer solvents. Gel collects on the walls of the processing equipment, requiring periodic shut down of the equipment for its removal. Various techniques have been used for the removal of this gel from the process equipment. However, none of these techniques has proved entirely satisfactory, and the burning of deposits from isolated equipment parts and mechanical techniques has remained as the general industrial practice for removing gel.

T.A. Wood; U.S. Patent 4,070,425; January 24, 1978; assigned to E.I. Du Pont de Nemours and Company provides a process whereby aliphatic polyamide gel is broken down into fragments which can be readily flushed from process equipment without its being dismantled.

The process comprises contacting an aliphatic polyamide gel at a temperature of from 100° to 300°C and for a period of from 15 minutes to 12 hours, with at least one compound having at least one alcoholic or phenolic hydroxyl group, the compound being liquid at the contact temperature and having:

a. an infrared absorption characterized by a wavelength difference of at least 0.2 μ in hydroxyl group stretching frequency when measured in carbon tetrachloride and in tetrahydrofuran;

b. a normal boiling point greater than 100°C; and

c. not more than one aliphatic carbon-carbon double bond per molecule.

A wide variety of known liquids satisfy the requirements of the process. Particularly satisfactory compositions have been found to include 1,1-bis(trifluoromethyl)heptadecanol, 2-hydroxy-α,α-bis(trifluoromethyl)benzyl alcohol, and 3,4-dichloro-α,α-bis(trifluoromethyl)benzyl alcohol of which the last has been found to be particularly effective. Accordingly, it is preferred that the hydroxyl-containing liquid comprise a major portion of at least one of these compounds, that is, at least 50% by weight.

Example: Process vessels and associated piping coated with nylon 66 and its gel were partially filled with 3,4-dichloro-α,α-bis(trifluoromethyl)benzyl alcohol fluid. The amount of fluid introduced was sufficient to maintain the dissolved nylon concentration below 10% by weight. The fluid was recirculated through the system while being heated to 200°±10°C and held at temperature for 4 hours while continually spraying the exposed vessel surface with the fluid. The particulated gel was filtered out of the recirculating fluid. Pressure in the system was adjusted with nitrogen to achieve adequate pressure to the recirculating

pump at the desired recirculation rate. The vessel was drained while recirculating and spraying. The cycle was repeated with a fresh charge of the processing
fluid.

Phosphorus Compound to Reduce Ammonia Formation

An improvement in the process for preparing random copolyamides by the melt
polymerization of a mixture of hexamethylenediammonium terephthalate (6TA)
and hexamethylenediammonium isophthalate (6IA) is described by *R.D. Chapman,
D.A. Holmer and G.A. Mortimer; U.S. Patent 4,113,708; September 12, 1978;
assigned to Monsanto Company.* The improvement comprises accomplishing the
melt polymerization in the presence of a small amount of a certain phosphorus
compound, such as benzene phosphinic acid, whereby ammonia formation during
polymerization is reduced.

The phosphorus compound has the formula

$$\text{(I)} \quad Q-\overset{\displaystyle \overset{O}{\|}}{\underset{\displaystyle |}{P}}-Q \qquad \text{or} \qquad \text{(II)} \quad Q-\underset{\displaystyle |}{P}-Q$$
$$\quad\quad\quad Q \qquad\qquad\qquad\qquad\qquad Q$$

where Q is a hydrogen atom or a radical selected from the group consisting of
R, OR, OM and OH, wherein R is a hydrocarbon radical, such as phenyl or a
C_1 to C_4 alkyl, and M is an alkali metal atom such as Na, K, or Li, with the proviso that each phosphorus compound has less than 2 (i.e., 0 or 1) OH radicals
and the phosphorus atom has a valence of less than +5, as determined by assign-
ing a valence of (i) +1 to hydrogen atoms, (ii) –1 to R, OR, OM and OH radicals and (iii) –2 to =O radicals.

Representative such phosphorus compounds include benzene phosphinic acid,
sodium hypophosphite, sodium phosphite and triphenylphospite. If desired, a
mixture of formula (I) phosphorus compounds and/or a mixture of formula (II)
phosphorus compounds may be used.

Example: Eight 6TA/6IA copolymers containing 58 mol % 6TA units and 42
mol % 6IA units were prepared by the following 3 cycle melt polymerization
procedure. An aqueous solution consisting of 87 g of 6TA salt, 63 g of 6IA
salt, an antifoaming agent, from 0 to 2,000 ppm, based on the weight of the
resulting polymer of benzene phosphinic acid, and 100 ml of water (deionized)
was charged to a stainless steel autoclave. The exact amount of benzene phosphinic acid used in each of the eight runs is given in the table below.

After thoroughly purging the autoclave contents with nitrogen, the autoclave
was pressurized to 250 psig, 1.72 MPa, with nitrogen. In the first cycle the
autoclave was heated to 218°±2°C over a period of 32 minutes with stirring
while maintaining the pressure constant at 1.72 MPa. In the second cycle the
temperature was increased to 300°C over a period of 64±5 minutes while maintaining the pressure at 1.72 MPa. In the third cycle the pressure was reduced
to atmospheric pressure over a period of 25 minutes while increasing the temperature to 325°±3°C.

A 14 hole spinneret was then attached to the base of the autoclave and the
polymer was extruded through the spinneret by application of 200±35 psig

(1.38±0.24 mPa) nitrogen to form a yarn. Each yarn was drawn 4.9±0.3 times while passing the yarn over a heated pin (80° to 90°C) and/or through a heated furnace (140° to 200°C) to provide a drawn yarn having a denier of about 80. The amount of ammonia formed during each melt polymerization is given in the table in terms of mol % ammonia, based on the mols of total salts used. The intrinsic viscosity of each copolymer and the tenacity of the yarn prepared therefrom are also given.

Benzene Phosphinic Acid (ppm)	NH₃ (mol %)	...Yarn Tenacity...		Intrinsic Viscosity*
		(g/den)	(MJ/kg)	
None	2.5	2.4	0.21	0.88
500	1.2	3.1	0.27	0.81
1,000	0.9	3.9	0.34	0.81
1,000	1.2	3.8	0.34	0.81
1,000	1.1	4.2	0.37	0.82
1,000	1.0	4.1	0.36	0.85
1,500	0.9	4.2	0.37	0.87
2,000	1.0	4.0	0.35	0.85

*10 x m^3/kg, in 96% H_2SO_4 at 25°C.

When the above melt polymerization procedure was repeated using one of the following phosphorus compounds instead of benzene phosphinic acid, no significant reduction in the amount of ammonia formed during the polymerization was observed:

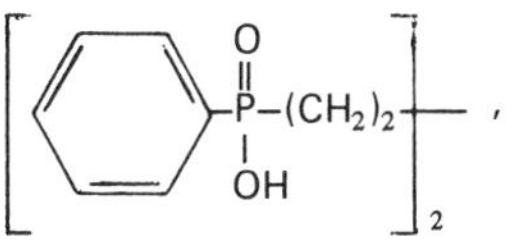

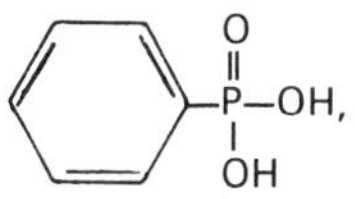

H_3PO_4,

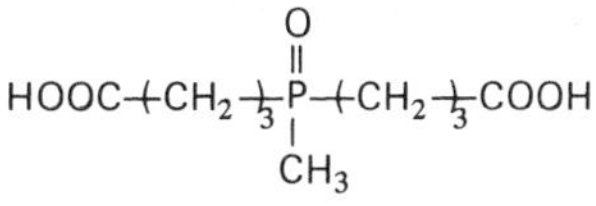

THERMALLY STABLE POLYMERS

Nitrile-Containing Polymers

Aromatic nitrile-containing polyester, polyamide, and silicone polymers are described by *W.J.H. Chang; U.S. Patent 4,021,415; May 3, 1977.* These nitrile-containing polymers are heat resistant and are used as fibers and resins and to prepare fiber-reinforced composites exhibiting outstanding mechanical strength at room and particularly elevated temperatures. Styrene-butadiene and styrene-acrylonitrile copolymers are described wherein at least some of the phenyl groups contain nitrile substituents. Composites having various shapes such as tires are described utilizing these nitrile-containing copolymers reinforced with fibers either containing aromatic nitrile groups or coated with resins containing nitrile substituted aromatic groups.

Example 1: This example illustrates the preparation of a nitrile-bearing aromatic polyamide by the melt polymerization process. In an open reaction vessel, equal molar amounts of the dimethyl ester of 5-cyanoisophthalic acid and 4,4'-diaminodiphenylmethane are heated to melt at a temperature above 250°C and maintained at this temperature to effect polymerization to a molecular weight of over 10,000. The nitrile-bearing polyamide prepared in this manner is a high temperature resistant polymer.

Example 2: A mixture comprising 50% by weight of the polymer of Example 1 and about 0.1% based on the weight of the polymer of zinc chloride in a solvent such as dimethylformamide is prepared. This liquid mixture is impregnated into fibers prepared from the polymer of Example 1, which fibers are in sheet form. The solvent is removed by heating the impregnated sheet at a temperature sufficient to volatilize the solvent. The impregnated sheet is placed in a mold and a pressure of about 500 psi is applied while the temperature of the mold is gradually raised to 300°C over a period of 4 hours. The composite is then post cured at a temperature of 350°C for 3 hours. The composite prepared by this manner exhibits high mechanical strength at room and elevated temperatures.

Example 3: A graft copolymer is prepared by adding styrene and acrylonitrile monomers to a latex of polybutadiene with agitation while maintaining the mixture at about 50°C. Potassium persulfate catalyst is then added and the reaction is allowed to continue until the monomers have been copolymerized. The emulsion is coagulated, washed and dried to give a white powder. This powder can be used as formed or blended with additional styrene-acrylonitrile copolymer as desired.

The above prepared grafted copolymer is iodinated and the iodine substituent converted to a nitrile substituent as described above.

The nitrile-containing grafted copolymer is dissolved in a suitable solvent and about 0.1% of zinc chloride, based on the weight of the copolymer, is added. The liquid mixture is then poured into a mold around fibers arranged in fabric form. In this example, the fibers are glass fibers coated with an aromatic nitrile-containing silicone polymer of the type described above.

The solvent is removed by heating the mixture at a temperature sufficient to volatilize the solvent whereupon a pressure of about 500 psi is applied on the mold while the temperature is gradually raised to about 300°C over a period of 4 hours. The composite then is post cured at 350°C for 3 hours. This composite exhibits high mechanical strength at room and elevated temperatures.

Example 4: The procedure of Example 3 is repeated except that the fibers of Example 2 are utilized in lieu of the coated glass fibers.

The fiber-polymer composites described above may be utilized in applications where thermal stability and high strength are desirable. Of course, the polymer fibers themselves, being high temperature resistant, may be utilized in a wide variety of applications and the preparation of textile materials. Examples of the application of the composites of the process include ovenware, appliance parts, auto parts, electrical laminates, electric switch gears, insulators, heavy duty circuit breakers, aircraft structural parts, rocket motor cases, pressure bottles, encapsulation of electronic components, grinding wheels and other temperature mechanical parts, jet engine components, tires, soles of footwear, etc.

Polyamides Made from Cycloaliphatic Diamines

Cycloaliphatic diamines have been prepared by *L.T.C. Lee; U.S. Patent 4,024,185; May 17, 1977; assigned to Allied Chemical Corporation* which consist of two un-equivalent amine groups. The diamines are prepared by reduction of aromatic dinitro compounds. The polyamides made from these diamines are useful in forming fibers and films having excellent high temperature properties.

The unsymmetrical cycloaliphatic diamines have the formula

$$H_2N-A-CH_2-\underset{\underset{CH_3}{|}}{\overset{\overset{CH_3}{|}}{C}}-NH_2$$

wherein A is a cycloaliphatic moiety having 6 to 15 carbon atoms, such as trans-1,4 cyclohexyl-cis-1,4-cyclohexyl-bis(p-cyclohexyl)methylene and bis(p-cyclo-hexyl)isopropylene.

Polyamides are prepared from the diamines by polymerization with a diacyl halide having the formula

$$\underset{XC-R-CX}{\overset{O\quad\ O}{\overset{\|\quad\ \|}{}}}$$

wherein R is a divalent organic radical and X represents a halogen, preferably chlorine or bromine, more preferably chlorine. The polymerization is conveniently carried out in accordance with conventional interfacial or solution poly-merization techniques.

Thus, the polyamides of this process have recurring units of the formula

$$-NH-A-CH_2-\underset{\underset{CH_3}{|}}{\overset{\overset{CH_3}{|}}{C}}-NH-\overset{\overset{O}{\|}}{C}-R-\overset{\overset{O}{\|}}{C}-$$

wherein A and R are as defined above.

The polyamides are characterized by having excellent high temperature properties. In particular, the polyamides have high glass transition temperatures and yet are melt processable. In general, a high glass transition temperature is desirable in that it reflects the resistance of the polymer to deformation at temperatures up to the glass transition temperature. For example, fibers of a polymer having a high glass transition temperature have greater resiliency and resistance to wrinkling.

A notable feature of the polyamides is their high glass transition temperatures coupled with their comparatively low melting points. A low melting point is desirable for ease in melt processing. Hence, a high ratio of glass transition temperature to melting point, T_g/T_m (the temperatures being expressed in degree Kelvin), is especially desirable. The polyamides are characterized by having exceptionally high T_g/T_m ratios. The ratios are normally above 0.74.

Example 1: *Preparation of 2-(4-Aminophenyl)-1,1-Dimethylethylamine* — 10 g of 2-(4-nitrophenyl)-1,1-dimethyl-1-nitroethane, prepared from the reaction of p-nitrobenzyl chloride and 2-nitropropane catalyzed by lithium hydride, dissolved in dioxane, was hydrogenated in the presence of 0.8 g of 10% ruthenium on carbon at 120°C and 1,600 psig for 7 hours. The reaction mixture was filtered and the filtrate was fractionally distilled to produce 6.7 g of trans-2-(4-aminocyclohexyl)-1,1-dimethylethylamine (A) boiling point 78°C at 0.2 mm Hg, and 1.5 g of cis-2-(4-aminocyclohexyl)-1,1-dimethylethylamine (B), boiling point 69°C at 0.2 mm Hg.

Example 2: *Preparation of Polyamide from trans-2-(4-Aminocyclohexyl)-1,1-Dimethylethylamine and Adipyl Chloride by Interfacial Method* — A polyamide was prepared from 100% trans-2-(4-aminocyclohexyl)-1,1-dimethylethylamine by polymerization with adipyl chloride in accordance with the interfacial polymerization technique.

The polyamides were prepared in accordance with the following typical method. A solution of 2.96 g of adipyl chloride in 60 ml of 1,2-dichloroethane was added to a vigorously stirred solution of 2.76 g of 2-(4-aminocyclohexyl)-1,1-dimethylethylamine. A polymer formed immediately and after 5 minutes the reaction mixture was poured into ice water. The polymer, which was recovered by filtration, washed and dried, weighed 4.7 g (85% yield). The properties of the polyamide are T_g, 92°C; T_m, 209°C; T_g/T_m, 76; reduced viscosity, 0.19; and crystallinity, 33%.

Soluble Polyamide-Imides

It is well known that aromatic polyamide-imides are tough, thermally stable and flame resistant. The existing polyamide-imide polymers, however, have found only limited commercial acceptance due to certain undesirable properties. For example, in coating applications it is highly desirable for the polymer to be soluble in relatively nonpolar volatile solvents so that solvent removal may be easily accomplished.

Most high molecular weight polyamide-imides, however, are soluble or only marginally soluble only in highly polar nonvolatile, expensive solvents. Furthermore, these solutions generally are not stable.

J.H. Bateman and D.A. Gordon; U.S. Patent 4,026,876; May 31, 1977; assigned to Ciba-Geigy Corporation have found that polyamide-imides with phenylindane diamines incorporated into the polymer backbone are soluble in high concentration in nonpolar organic solvents.

The polyamide-imides are characterized by a recurring unit having the following structural formula:

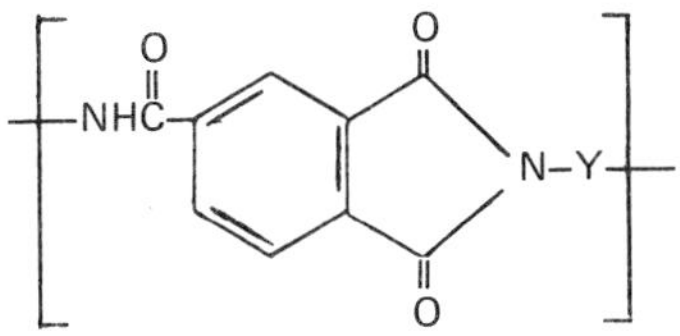

wherein Y is a divalent organic radical selected from carbocyclic-aromatic, aliphatic, araliphatic, cycloaliphatic, and heterocyclic radicals, combinations of these, and radicals with heteroatom containing bridging groups where the heteroatom in the bridge is oxygen, sulfur, nitrogen, silicon or phosphorus, provided that, out of the total number of polyamide-imide recurring units, 1 to 100%, preferably 10 to 100% of such units, have Y equal to a phenylindane radical of the structural formula:

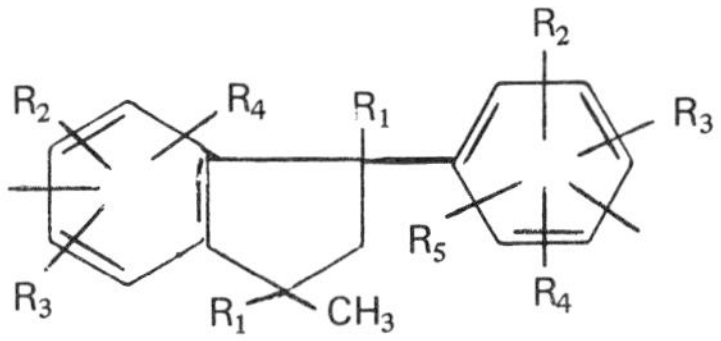

wherein R_1 is hydrogen or lower alkyl, preferably lower alkyl of 1 to 5 carbon atoms, and R_2, R_3, R_4 and R_5 are independently hydrogen, halogen or lower alkyl, preferably, lower alkyl of 1 to 4 carbon atoms, and either aromatic ring of the unsymmetrical phenylindane radical may be bonded to an amide or imide nitrogen.

Example 1: *5(6)-Amino-1-(4'-Aminophenyl)-1,3,3-Trimethylindane* — (A) Preparation of 5,4'-Dinitro- and 6,4'-Dinitro-1,3,3-Trimethyl-1-Phenylindane Isomers: To a solution of 236 g (1.0 mol) 1,3,3-trimethyl-1-phenylindane (α-methylstyrene dimer) in 750 ml chloroform at 2° to 8°C was added a previously mixed solution of 396 ml sulfuric acid and 132 ml nitric acid dropwise over a 2.5 hour period. The two phase reaction mixture was allowed to stir an additional 4 hours at 5°C. The chloroform phase was isolated and washed with aqueous sodium bicarbonate until neutral and then with distilled water.

A light yellow oil was obtained after drying and stripping the chloroform solution. Two triturations in hexane at room temperature afforded 295 g light yellow powder, melting point 109° to 125°C. This material was shown to be a mixture of the 5,4'-dinitro- and 6,4'-dinitro-1,3,3-trimethyl-1-phenylindane isomers by NMR analysis.

Analysis–Calculated for $C_{18}H_{18}N_2O_4$: C, 66.25%; H, 5.55%; N, 8.58%. Found: C, 66.13%; H, 5.50%; N, 8.42%.

(B) Preparation of 5(6)-Amino-1-(4'Aminophenyl)-1,3,3-Trimethylindane: A mixture of 250 g (0.767 mol) of the dinitro isomer and 250 g (4.60 gram-atoms) reduced in iron powder in 1 liter 50% aqueous ethanol was brought to reflux and a previously prepared solution of 60 ml concentrated hydrochloric acid in 400 ml 50% aqueous ethanol was added over a 1 hour period. Reflux was continued an additional 3 hours, the reaction cooled to 50°C and 50 ml concentrated hydrochloric acid added. The reaction mixture was filtered. The filtrate was made basic with 20% NaOH and extracted with ether, dried and stripped under vacuum to afford 145 g (71%) of a clear brown glassy solid, melting point 47° to 54°C. NMR analysis indicated the product was 62% 6-amino- and 38% 5-amino-1-(4'-aminophenyl)-1,3,3-trimethylindane.

Analysis—Calculated for $C_{18}H_{22}N_2$: C, 81.18%; H, 8.32%; N, 10.52%. Found: C, 81.27%; H, 8.20%; N, 10.48%.

Example 2: *Preparation of Polyamide-imide from 5(6)-Amino-1-(4'-Amino-1-(4'-Aminophenyl)-1,3,3-Trimethylindane and Trimellitic Anhydride Acid Chloride* — Solution Method: To a 500 ml flask fitted with mechanical stirrer and nitrogen inlet was added 20.20 g (0.0758 mol) 5(6)-amino-1-(4'-aminophenyl)-1,3,3-trimethylindane and 260 ml N,N-dimethylacetamide. The reaction mixture was stirred until solution was achieved and 7.73 g (0.0763 mol) triethylamine was added.

The reaction mixture was then cooled using an ice bath and 15.97 g (0.0758 mol) trimellitic anhydride acid chloride was added all at once. The ice bath was removed after the trimellitic anhydride acid chloride had dissolved. Reaction was continued for 3 hours at room temperature during which time triethylamine hydrochloride precipitated as a crystalline solid. The polyamic acid was precipitated by adding the reaction mixture to rapidly stirring cold water. The polyamic acid was then recovered by filtration, washed thoroughly and dried at 50°C at 50 mm Hg vacuum. The polyamic acid, a white powdery solid, had an inherent viscosity of 0.22 (0.5% in N-methyl-2-pyrrolidone).

The polyamide-imide was prepared by treating 25 g of the polyamic acid, dissolved in 150 ml N-methyl-2-pyrrolidone, with 30 g (0.3 mol) acetic anhydride and 3 g (0.04 mol) pyridine at room temperature for 18 hours. The polyamide-imide was precipitated into water, recovered by filtration, washed thoroughly and dried at 80°C per 50 mm Hg vacuum. The polyamide-imide, a yellow powdery solid, had an inherent viscosity of 0.29 (0.5% in N-methyl-2-pyrrolidone at 25°C).

Example 3: By essentially following the procedure described in Example 2, soluble polyamide-imides are obtained by reacting equivalent amounts of trimellic anhydride acid chloride and each of the following aromatic diamine mixtures:

(a) 50 mol % 5(6)-amino-1(4'-aminophenyl)-1,3,3-trimethylindane (PIDA) and 50 mol % 4,4'-methylenebis(o-chloroaniline);

(b) 50 mol % PIDA and 50 mol % 4,4'-sulfonyldianiline;

(c) 50 mol % PIDA and 50 mol % m-phenylenediamine; and

(d) 50 mol % PIDA and 50 mol % p-phenylenediamine.

Copolyamides Prepared from 3,4'-Diphenylene Type Diamines

S. Ozawa, Y. Nakagawa, K. Matsuda, T. Nishihara and H. Yunoki; U.S. Patent 4,075,172; February 21, 1978; assigned to Teijin Limited, Japan describe a fiber- or film-forming high-molecular-weight aromatic copolyamide consisting essentially of (1-A) a diamine recurring unit of the formula:

$$-HN-\langle\text{phenylene}\rangle-Y_1-\langle\text{naphthylene}\rangle-NH-$$

such as 3,4'-diaminodiphenyl ether or 3,4'-diaminodiphenyl sulfone, (1-B) a diamine recurring unit of the formula $-HN-Ar_1-NH-$ such as p-phenylene diamine, and (2) a dicarboxylic acid recurring unit of the formula $-OC-Ar_2-CO-$ such as terephthaloyl dichloride, wherein Y_1 is $-O-$, $-S-$, $-SO_2-$, $-C(O)-$, $-NH-$, $-CH_2-$ and $(CH_3)_2C=$, and Ar_1 and Ar_2 are the same or different and are phenylene, naphthylene or biphylene with bond chains extending in a coaxial or parallel direction. The total amount of recurring units (1-A) and (1-B) are substantially equimolar to recurring unit (2), and the proportion of recurring unit (1-A) is 7.5 to 40 mol % of the entire recurring units.

The copolyamide is well soluble in solvents to give solutions having good flowability and stability that can afford shaped articles with superior thermal stability, fire retardancy and mechanical properties.

In the following examples, the inherent viscosities (η_{inh}) of the polymers were determined at 30°C for a solution of 0.5 g of polymer dissolved in 100 ml of concentrated sulfuric acid after isolating the polymers from the reaction mixture. The weight loss beginning temperature is a value obtained from a differential thermal analysis curve measured by a thermal analyzer (Thermoflex) at a temperature raising rate of 10°C/min using 8.0 mg of a sample.

Example 1: This example illustrates an aromatic copolyamide obtained by polymerizing 10 mol % of 3,4'-diaminodiphenyl ether, 40 mol % of para-phenylenediamine and 50 mol % of terephthaloyl dichloride.

1.201 g (0.006 mol) of 3,4'-diaminodiphenyl ether and 2,595 g (0.024 mol) of para-phenylenediamine were dissolved in 150 g of N-methylpyrrolidone-2 containing 1.0% by weight of calcium chloride in a stream of dry nitrogen. The solution was cooled to 0°C and with vigorous stirring, 6.091 g (0.030 mol) of a powder of terephthaloyl dichloride was added rapidly. The monomers were reacted at 35°C for 1 hour.

Then, 1.68 g of calcium oxide was added to neutralize the by-product hydrochloric acid, and at the same time, 3.00 g of calcium chloride was added. The mixture was stirred at 70°C for 2 hours. The polymer concentration of the resulting solution was 4.7% by weight, and the concentration of calcium chloride was 4.8% by weight. The polymer had an inherent viscosity of 1.95.

The polymer solution was filtered and deaerated, and then spun into a verticaltype aqueous coagulating bath containing 50% by weight of calcium chloride and maintained at 75°C at a linear extrusion speed of 5.5 m/min through a spinning nozzle having 5 orifices each with a diameter of 0.2 mm.

The as-spun filaments were passed through the coagulating bath over a distance of about 1 m and then wound up at a rate of 6.2 m/min. The filaments were then passed through a water-washing bath at 80°C over a distance of 5 m and through a water-washing bath at 95°C over a distance of 6 m.

The filaments were then dried by bringing them into contact with a drying roller at 110°C over a distance of 3 m and a drying roller at 200°C over a distance of 5 m. The dried filaments were drawn to 8.5 times the original length in a heated cell through which a nitrogen gas at 510°C was flowing at a rate of 3 l/min.

The resulting drawn filaments had a monofilament denier size of 0.95 den, a tensile strength of 21.5 g/den, an elongation of 4.5%, and a Young's modulus of 590 g/den.

Example 2: The procedure of Example 1 was followed using 15 mol % 3,4'-diaminodiphenyl ether, 35% p-phenylenediamine, and 50% terephthaloyl dichloride.

The weight loss beginning temperature of the polymer obtained by the polymerization was 410°C, and the weight loss beginning temperature of the filaments obtained by spinning and drawing was 464°C.

Crosslinkable Mixtures of Oligomers or Polymers and Diamines

R. Darms, V. Kvita and G. Greber; U.S. Patent 4,115,231; September 19, 1978; assigned to Ciba-Geigy Corporation describe crosslinkable mixtures of oligomers or polymers, containing terminal imidyl groups, and diamines, for example mixtures of a polyamide oligomer or polyamide-acid oligomer, or terminal polyamide polymer or polyamide-acid polymer, having maleimidylphthalic acid groups, and a diamine.

These mixtures and the prepolymers obtainable therefrom are soluble in customary organic solvents, and in general, they can also be processed from the melt. The crosslinkable products obtainable therefrom are distinguished in particular by their good thermo-oxidative stability.

The crosslinkable mixtures consist of at least one oligomer or polymer of the formula Ia or Ib

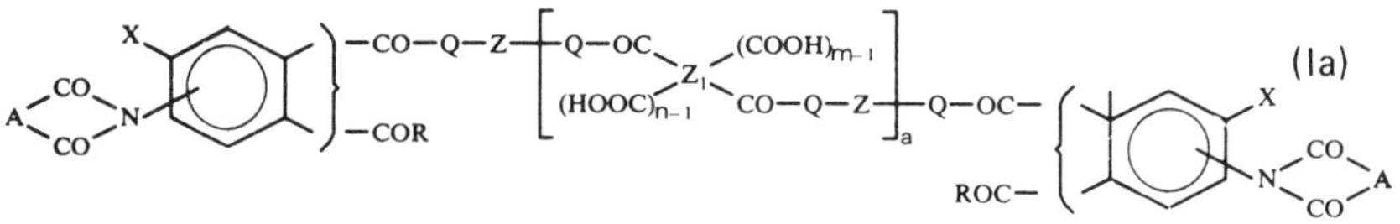

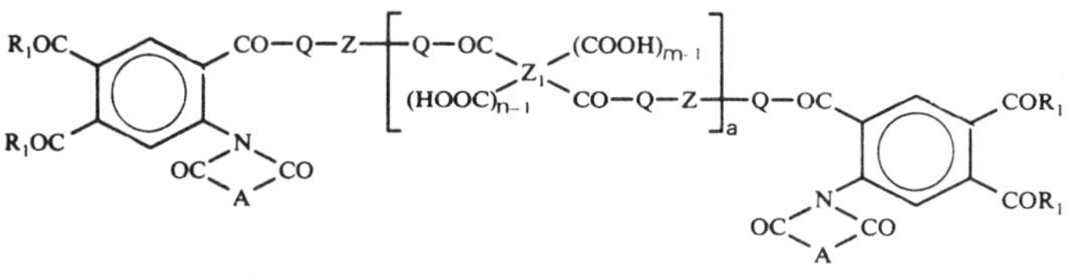

or of a corresponding cyclized derivative, and of at least one diamine of the formula II

$$HQ-Z_2-QH \qquad (II)$$

whereby the molar ratio of oligomer or polymer of the formula Ia and/or Ib, or of a corresponding cyclized derivative, to diamine of the formula II is between 1.2:1 and 50:1, and wherein the X's independently of one another represent hydrogen or, if the radical

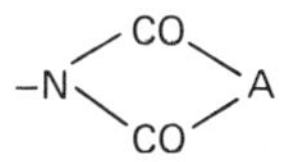

$$-N\underset{CO}{\overset{CO}{<}}A$$

is in the 4 position of the benzene ring, also $-COR_2$. The R's and R_2's independently of one another represent a hydroxyl group, an unsubstituted or substituted phenoxy group, an alkoxy group having 1 to 18 carbon atoms or an $-O^-M^+$ group. The R_1's independently of one another represent a hydroxyl group, an unsubstituted or substituted phenoxy group, an alkoxy group having 1 to 18 carbon atoms, an $-O^-M^+$ group, or two adjacent R_1 together represent the $-O-$ grouping. The A's independently of one another represent a radical of the formula

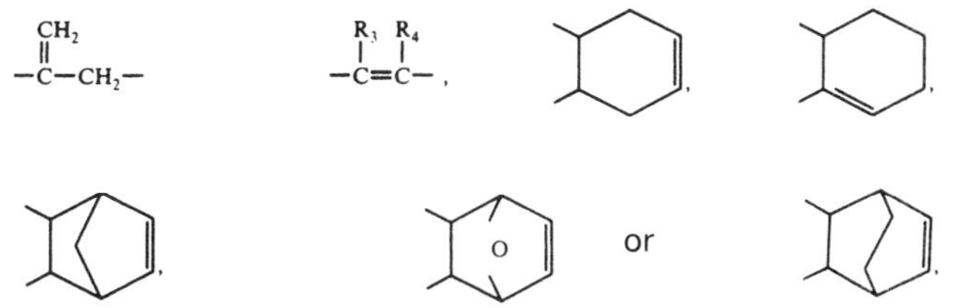

$$-\overset{\overset{\displaystyle CH_2}{\|}}{C}-CH_2- \qquad \underset{}{-\overset{R_1}{\underset{}{C}}=\overset{R_4}{\underset{}{C}}-} ,$$

R_3 and R_4 independently of one another represent hydrogen, chlorine, bromine or methyl. M^+ represents an alkali metal cation, a trialkylammonium cation having 3 to 24, especially 3 to 12 carbon atoms, or a quaternary ammonium cation. A represents a number from 1 to 100, preferably 1 to 60, and particularly a number from 1 to 10. m and n independently of one another represents the number 1 or 2. The Z's and Z_2's independently of one another represent an aliphatic radical having at least 2 carbon atoms, a cycloaliphatic, carbocyclic-aromatic or heterocyclic aromatic radical. The Q's represent $-NH-$, or $-Q-Z-Q-$ or $-Q-Z_2-Q-$ represent the grouping

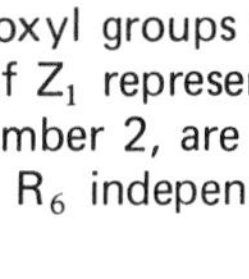

$$-N\underset{R_6}{\overset{R_5}{<}}N-$$

The Z_1's represent an aliphatic radical having at least two carbon atoms, or a cycloaliphatic, carbocyclic-aromatic or heterocyclic-aromatic radical, in which the carbonamide groups and carboxyl groups are bound to different carbon atoms, and the carboxyl groups, if Z_1 represents a cyclic radical and at least one of m and n represents the number 2, are each in the o-position with respect to a carbonamide group. R_5 and R_6 independently of one another represent hydrogen, methyl or phenyl.

The molar ratio of the oligomers or polymers of the formula Ia and/or Ib or of the corresponding cyclized derivatives, to the diamines of the formula II is preferably between 1.3:1 and 10:1.

The oligomers or polymers of the formula Ia or Ib can be obtained by a process wherein (a) a compound of the formula IIIa:

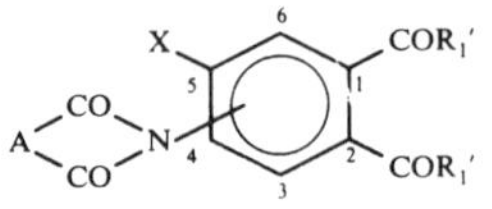

(IIIa)

or (b) a compound of the formula IIIb:

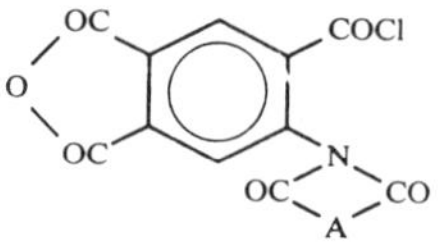

(IIIb)

or a mixture of two different compounds of the formula IIIa or IIIb, wherein in respect of A and X that applies which has been stated under the formula Ia and Ib, and the R_1's independently of one another represent a hydroxyl group, an unsubstituted or substituted phenoxy group, an alkoxy group having 1 to 18 carbon atoms, or the two R_1's together represent the $-O-$ grouping, is reacted, in the molar ratio of 2:1, with an oligomer or polymer of the formula IV:

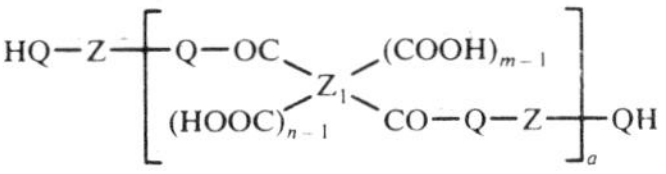

$$HQ-Z\left[Q-OC\underset{(HOOC)_{n-1}}{\overset{(COOH)_{m-1}}{Z_1}}CO-Q-Z\right]_a QH \qquad (IV)$$

or with a corresponding cyclized derivative, wherein in respect of a, m, n, Z_1, Z and Q or $-Q-Z-Z-$ that applies which has been stated under the formulas Ia and Ib.

By heating of the crosslinkable mixtures at temperatures of between 100° and 300°C, preferably 100° and 200°C, it is possible to produce imidized or cyclized, and optionally partially crosslinked, prepolymers. The prepolymers still contain crosslinkable groups, are soluble in the usual organic solvents, such as N,N-dimethylformamide, N,N-dimethylacetamide and N-methylpyrrolidone, and can also be processed in general from the melt.

The prepolymers can be converted in a second stage, by known methods, into fully crosslinked products, which are insoluble in the customary organic solvents. Alternatively, it is also possible to process and to crosslink directly the mixtures of compounds of the formula Ia and/or Ib and diamines. The crosslinking can be performed in a known manner, chemically or thermally, or under the influence of electromagnetic waves, especially light.

Chemical crosslinking is generally performed at temperatures of 50° to 250°C, in the presence of radical initiators, e.g., peroxides or azo compounds. Thermal crosslinking is performed advantageously by heating at temperatures of 350°, preferably 150° to 250°C, and also radical initiators may concomitantly be used. Crosslinking under the influence of electromagnetic waves can be performed, e.g., by irradiation with x-rays or with UV light, and optionally in the presence of sensitizers, such as benzene, 1,4-diacetylbenzene, phenol, benzonitrile, acetophenone, benzophenone, benzaldehyde, diisopropyl ketone and fluorene.

Example: In a nitrogen atmosphere in a sulfonating flask, 15.86 g (0.08 mol) of 4,4'-diaminodiphenylmethane is dissolved in 180 ml of anhydrous dimethylacetamide (DMA), and the solution is cooled from –15° to –20°C. With stirring, 12.63 g (0.06 mol) of trimellitic acid anhydride chloride in solid form is added portionwise in such a manner that the temperature of the reaction solution does not exceed –15°C. The solution is then stirred at –15°C for 30 minutes and at 20° to 25°C for one hour. 6.07 g (0.06 mol) of triethylamine in 50 ml of DMA is then added dropwise, whereupon a fine precipitate is formed.

The reaction mixture is stirred for a further hour at 20° to 25°C, and an addition is subsequently made portionwise at 15° to 20°C of 9.72 g (0.04 mol) of a 1:1 mixture of 3- and 4-maleimidylphthalic acid anhydride. After one hour's stirring at 20° to 25°C, the precipitated triethylamine hydrochloride is separated by filtration. To the filtered reaction solution is then added dropwise with stirring, 160 ml of acetic acid anhydride and stirring is continued at 20° to 25°C for 16 hours. The reaction solution is subsequently poured onto water; the precipitated product is separated, washed several times with water and dried in a vacuum drying chamber for 16 hours at 80°C/100 torr and for 16 hours at 80°C/10⁻²
torr. 32 g of a crosslinkable oligoamide-imide in the form of a yellow powder is obtained.

12.84 g (0.0075 mol) of the above oligoamide-imide and 0.89 g (0.0045 mol) of 4,4'-diaminodiphenylmethane are dissolved in 85 ml of anhydrous DMA, and the solution is heated in a nitrogen atmosphere, with stirring, at 130°C for 2½ hours. The solution obtained after cooling to about 25°C is poured into ice water; the precipitated product is filtered off, washed with water and dried for 20 hours at 80°C/100 torr and 80°C/10⁻¹ torr, respectively.

For processing by the compression process, the prepolymer obtained is introduced into a compression mold, preheated to 240°C for circular discs, and molded at this temperature for 35 minutes under a pressure of 450 kp/cm². Annealing for 16 hours at 240°C yields solid molded specimens having good electrical properties.

Bis(2-Methyl-2-Cyanopropyl) Aromatics

According to *A.H. Frazer, J.F. Harris, Jr. and E.L. Martin; U.S. Patent 4,130,579; December 19, 1978; assigned to E.I. Du Pont de Nemours and Company* aromatic-aliphatic dinitriles of the formula

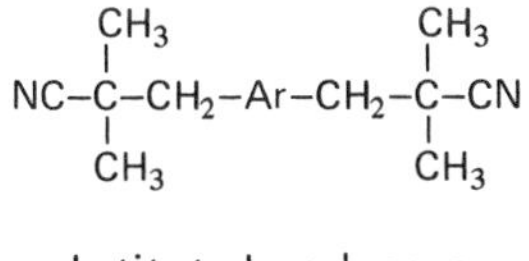

$$\underset{\underset{CH_3}{|}}{\overset{\overset{CH_3}{|}}{NC-C}}-CH_2-Ar-CH_2-\underset{\underset{CH_3}{|}}{\overset{\overset{CH_3}{|}}{C}}-CN$$

in which Ar is an arylene or substituted arylene, are useful in the preparation of the corresponding amines which are intermediates for the preparation of thermally stable, rigid, polyamides.

Example 1: *1,4-Bis(2-Methyl-2-Cyanopropyl)Benzene* –

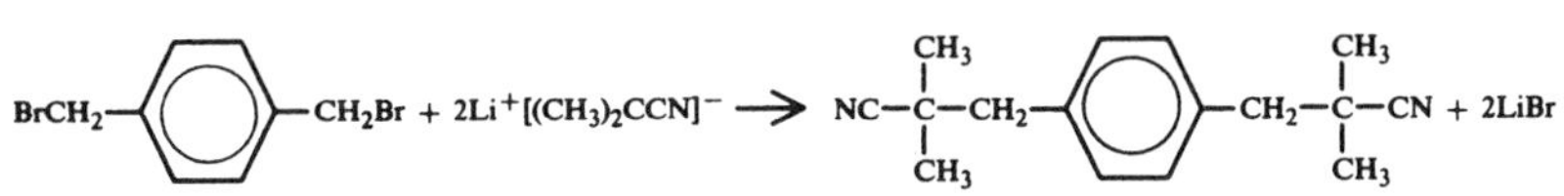

In a 2 liter flask, equipped with a magnetic stirrer, a reflux condenser capped with a nitrogen bubbler, a dropping funnel, and a syringe adapter, was placed 900 ml of anhydrous tetrahydrofuran (THF) and 42 ml (30.32 g, 0.30 M) of diisopropylamine (via syringe). The stirred mixture was cooled in a dry ice-acetone bath, and then 138.6 ml of 2.17 N (0.30 M) n-butyllithium in hexane was added via syringe.

After the mixture had stirred for 1 hour, a solution of 20.52 g (0.297 M) of freshly distilled isobutyronitrile in 60 ml of anhydrous THF was added in 20 minutes. Following an additional 1 hour and 7 minutes of stirring at dry ice temperature, a solution of 39.57 g (0.150 M) of α,α'-dibromo-p-xylene in 450 ml of anhydrous THF was added in 1 hour and 23 minutes.

The mixture was stirred at dry ice temperature for 2¼ hours and then overnight as the cooling bath warmed to room temperature. Stirring was continued for 4 days at room temperature. The suspended white solid was removed by filtration, rinsed on the filter with THF and dried: wt = 13.74 g; MP = 193° to 195°C. The filtrate was distilled on the water pump to remove the solvent, and the residue, a mixture of brown oil and solid, was stirred with 100 ml of methanol which dissolved the brown oil.

Filtration of the mixture, rinsing of the solid on the filter with methanol, and drying of the solid under nitrogen gave an additional 15.7 g of crude 1,4-bis(2-methyl-2-cyanopropyl)benzene melting at 192° to 194.5°C (total yield = 82%). Dissolving of this material in refluxing acetone (28.5 ml/g), filtration of the hot solution through a coarse sintered glass funnel to remove some insoluble material, and cooling of the filtrate at 8° to 10°C gave the product as colorless needles melting at 194° to 195°C.

Analysis—Calculated for $C_{16}H_{20}N_2$: C, 79.95%; H, 8.39%; N, 11.66%. Found: C, 79.79%, 79.96%; H, 8.21%, 8.37%; N, 11.83%, 11.67%.

Example 2: *Preparation of 1,4-Bis(2,2-Dimethyl-3-Aminopropyl)Benzene* —

In a 2 liter flask, equipped with a magnetic stirrer, a reflux condenser capped with a nitrogen bubbler, and an additional funnel, was placed 7.50 g (0.0312 M) of 1,4-bis(2-methyl-2-cyanopropyl)benzene and 300 ml of reagent grade toluene which had been passed through acid alumina under nitrogen directly into the reaction vessel. With stirring at room temperature, 107 ml of a 24.1% solution (0.150 M) of diisobutylaluminum hydride in toluene was added from the addition funnel in 28 minutes. The mixture was then refluxed for 16 hours.

After the mixture had been cooled in an ice water bath, a solution of 6 ml of water in 30 ml of methanol was added dropwise with stirring. This was followed by the dropwise addition of a solution of 30 ml of water in 60 ml of methanol. The mixture was stirred vigorously for 1 hour while being cooled in the ice water bath, and then for an additional hour at room temperature.

The mixture was filtered under nitrogen, the solid was washed thoroughly on the filter with toluene, and the combined filtrate and rinsings were distilled on the water pump. The resulting residue crystallized on cooling to room temperature. Further drying on the oil pump gave 5.28 g (68%) of crude 1,4-bis(2,2-dimethyl-3-aminopropyl)benzene melting at 53° to 56°C to a cloudy melt. Distillation of this material through a small Vigreux still gave the product as a colorless liquid boiling at 131° to 132°C/0.60 mm. The solidified material melted to a clear melt at 53.5° to 54.75°C.

Analysis—Calculated for $C_{16}H_{28}N_2$: C, 77.36%; H, 11.36%; N, 11.28%. Found: C, 77.68%, 77.07%, 77.15%; H, 11.44%, 11.30%, 11.27%; N, 11.04%, 11.14%.

The infrared spectrum contains bands at 2.93, 3.00 and 6.15 μ ($-NH_2$), 3.28 μ (shoulder) ($=CH$), 3.38 and 3.48 μ (saturated CH), 6.59 and 6.77 μ (aromatic $C=C$), 7.21 and 7.33 μ (gemdimethyl), and 11.86 μ (p-disubstituted aromatic).

Example 3: *Preparation of a Polyamide from 1,4-Bis(2,2-Dimethyl-3-Aminopropyl)Benzene and Sebacyl Chloride —*

$$H_2NCH_2-\overset{\overset{\displaystyle CH_3}{|}}{\underset{\underset{\displaystyle CH_3}{|}}{C}}-CH_2-\langle \bigcirc \rangle -CH_2-\overset{\overset{\displaystyle CH_3}{|}}{\underset{\underset{\displaystyle CH_3}{|}}{C}}-CH_2NH_2 + Cl\overset{\overset{\displaystyle O}{\|}}{C}(CH_2)_8\overset{\overset{\displaystyle O}{\|}}{C}Cl + 2(C_2H_5)_3N \xrightarrow{HCCl_3}$$

$$\left[-CH_2-\overset{\overset{\displaystyle CH_3}{|}}{\underset{\underset{\displaystyle CH_3}{|}}{C}}-CH_2-\langle \bigcirc \rangle -CH_2-\overset{\overset{\displaystyle CH_3}{|}}{\underset{\underset{\displaystyle CH_3}{|}}{C}}-CH_2\overset{\overset{\displaystyle H}{|}}{N}-\overset{\overset{\displaystyle O}{\|}}{C}-(CH_2)_8-\overset{\overset{\displaystyle O}{\|}}{C}-\overset{\overset{\displaystyle H}{|}}{N}-\right]_n + 2(C_2H_5)_3N \cdot HCl$$

In a 3 liter flask, equipped with a paddle stirrer, a reflux condenser, and a nitrogen bubbler, was placed 25.00 g of 1,5-bis(2,2-dimethyl-3-aminopropyl)benzene, 31.0 ml of triethylamine, and 350 ml of chloroform which had been passed through basic alumina under nitrogen directly into the reaction flask. With vigorous stirring at room temperature, 24.07 g of freshly distilled sebacyl chloride in 100 ml of purified chloroform was added all at once. The mixture was stirred for 45 minutes and then 1,500 ml of hexane was added to precipitate the polymer.

After 15 minutes of stirring, the mixture was allowed to stand overnight. With stirring, a solution of 150 ml of concentrated hydrochloric acid in 600 ml of water was added. The coagulated polymer was filtered, rinsed on the filter with water, and then washed in a blender once with 600 ml of water, once with 600 ml of acetone and 3 times with 600 ml of water. The isolated polymer was dried overnight in a vacuum oven at 70°C. There was thus obtained 32.6 g (78%) of product: inherent viscosity (0.05% in m-cresol at 25°C) = 1.32.

A clear, tough, colorless film was pressed at 180°C and 500 lb pressure from a portion of this polymer. Another portion of the polymer was melt spun through a spinnerette (0.020" x 0.04") at 248° to 270°C to give filaments which, after cold drawing, had strengths of 1.5 g/den.

The product of another experiment, on 1/10 the scale of that just described, was further characterized by elemental analysis and infrared spectroscopy.

Analysis—Calculated for $(C_{26}H_{42}N_2O_2)_n$: C, 75.31%; H, 10.21%; N, 6.76%. Found: C, 75.20%, 75.66%; H, 10.90%, 10.89%; N, 6.94%, 6.95%.

The infrared spectrum contained bands at 3.03 μ (—NH), 3.42 and 3.48 μ (saturated CH), 6.08 and 6.45 μ (amide I and II bands), 6.60 μ (aromatic C=C), and 7.30 and 7.32 μ (gemdimethyl).

FLAME-RESISTANT POLYMERS

Polyamide-Melamine-Cyanurate Composition

A flame-retarding polyamide composition having excellent mechanical and flame retardant properties is disclosed by *T. Tsutsumi, N. Kato, M. Morikawa, T. Aoyama and K. Nagai; U.S. Patent 4,001,177; January 4, 1977; assigned to Toray Industries, Inc., Japan.* The composition comprises polyamide, about 3 to 40% by wt of melamine on the basis of the polyamide, 0.5 to 20% by wt on the basis of the polyamide, and 2 to 75% by wt on the basis of melamine of a cyanurate or isocyanurate compound represented by a compound selected from the group consisting of the following formulas (I) and (II):

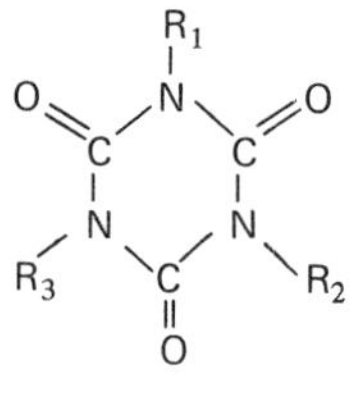

(I) 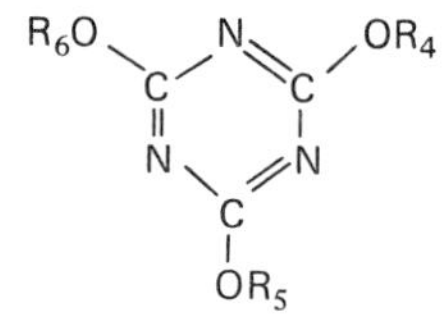   (II)

wherein R_1, R_2, R_3, R_4, R_5 and R_6 are the same or different radicals selected from the group consisting of hydrogen, alkyl, alicyclic and aromatic radicals having from 1 to 10 carbon atoms, and from such radicals having a substituent selected from the group consisting of —OH, —COOR' and —CONH$_2$, wherein R' is aliphatic alkyl group having from 1 to 5 carbon atoms.

Example: 10 pbw of finely pulverized melamine (average particle size 5 μ) and 3 pbw of tris(2-hydroxyethyl)isocyanurate were blended with 100 pbw of polyhexamethylene adipamide (nylon 66) having a relative viscosity (ηrel) (as measured on a solution of 1 g of polyamide in 100 ml of sulfuric acid) in a mixer. The blend was extruded through a 65 mm diameter screw extruder at a temperature of 285°C. The extruded strand was allowed to cool, cut into chips, and dried at 95°C for 15 hours under vacuum. Using chips thus obtained, test specimens were prepared by injection molding.

The flammability of these white opaque test specimens was tested in accordance with the method specified in UL Subject 94, and it was found that polyamide composition was classified as 94 V-O. The polyamide composition was also tested for tensile strength (ASTM D-638-56T) and impact strength Izod (ASTM D-256-56), and the following results were obtained: tensile strength, 735 kg/cm^2; elongation, 25%; and impact strength Izod notch, 4.3 kg-cm/cm.

The composition of this example has good flame retarding properties together with excellent mechanical properties.

Comparative Example: For purpose of comparison, polyamide compositions were prepared by blending 10 pbw consisting solely of melamine, or 10 pbw of melamine together with 3 pbw of tetrachlorophthalic anhydride. The compositions were formed and tested by the same method as above. In each case, the extruded strands obtained from the above procedure had many protuberances, and as shown by this fact, the degree of dispersion of the additives was not good.

In addition, the combustion behavior of the material was unstable due to the coagulation of melamine powder though the samples were ranked as 94 V-O. The results of the tests showed that these compositions were fragile and did not have good mechanical properties.

Naphthalene Dicarboxylic Acid, Halogenated Dicarboxylic Acid, and Aliphatic Amine

S. Kawase, T. Shima and K. Moriyama; U.S. Patent 4,042,571; August 16, 1977; assigned to Teijin Limited, Japan describe a process for preparing a fire-retardant polyamide, which comprises polymerizing a carboxylic acid component consisting essentially of at least 1 naphthalene dicarboxylic acid selected from the group consisting of naphthalene-2,7-dicarboxylic acid and naphthalene-2,6-dicarboxylic acid and/or its amide-forming derivative, and a halogen substituted-carboxylic acid and/or its amide-forming derivative, and an aliphatic diamine, the amounts of the naphthalene dicarboxylic acid and/or its derivative being at least 40 mol % of the total acid component, and the amount of the halogen-substituted carboxylic acid being such that the halogen atom content of the polyamide is at least 0.13% by wt.

The following examples illustrate the process in greater detail. In these examples, all parts are by weight. The evaluation of fire retardancy was performed by an oxygen index (LOI) according to JIS-K-720101972 (corresponding to ASTM D-2863), the number of ignitions, and UL 94 (September 1973). The number of ignitions and UL values were measured as follows.

Number of Ignitions: A sample of fiber is dried at 50°C for 24 hours, and then stored for 2 hours in a desiccator. 1 g of the fiber is packed uniformly over a distance of 100 mm in a coil having an i.d. of 10 mm and a length of 150 mm and made of a stainless steel wire (coil pitch 2 mm). The coil is inclined at an angle of 45°, and the fiber is ignited by bringing a butane gas burner close to the lower end of the coil. When the fire extinguishes, the fiber is again ignited by the burner. This is repeated, and the number of ignitions required until the entire sample has been burned out is measured. The larger number of ignitions shows better fire retardancy.

UL 94/September 1973 (Underwriters' Laboratories): Test pieces of 3 different thicknesses with a size of 5" x 0.5" x ($\frac{1}{4}$, $\frac{1}{8}$, $\frac{1}{16}$") are prepared, and maintained for 48 hours at 23°C, and a RH of 50%. The test pieces so pretreated are suspended perpendicularly, and the flame of a Bunsen burner is applied to the center of the lower end of each test piece for 10 seconds. It is then withdrawn, and the flaming time is recorded.

When the flaming time after the contact of the flame is within 30 seconds and the total flaming time after 10 replicates is within 250 seconds with respect to 5 test pieces as 1 group, and if the molten test piece that has fallen off causes fire in a surgical cotton placed under the test piece, the evaluation is 94 V-1. If the molten test piece that has fallen off does not cause fire, the evaluation is 94 V-2.

$\eta_{sp/c}$ is the reduced viscosity measured at 35°C using concentrated sulfuric acid as a solvent in a concentration of 0.4 g/100 ml.

Example 1: Equimolar proportions of naphthalene-2,7-dicarboxylic acid and hexamethylene diamine were dissolved in water to form a salt. Then, ethyl alcohol was added to form hexamethylene diammonium naphthalene-2,7-dicarboxylate as a white powder which did not contain water of crystallization.

A synthetic linear polyamide was prepared by melt-polymerization of 299 parts (0.9 mol) of hexamethylene diammonium naphthalene-2,7-dicarboxylate, 30 parts (0.1 mol) of hexamethylene diammonium isophthalate (monohydrate), 4.0 parts of m-bromobenzoic acid (2.0 mol % based on the hexamethylene diammonium naphthalene-2,7-dicarboxylate), and 0.42 parts of phosphorus acid (stabilizer) in the following manner.

An autoclave equipped with a stirrer was charged with the above compounds, and after purging with nitrogen, was heated at 290°C in a sealed condition. The pressure inside the autoclave increased with time, but after 3½ hours, the pressure stayed constant at 12.7 kg/cm². Then, the pressure inside the autoclave was released to the atmospheric pressure while raising the temperature to 310°C over the course of 1½ hours. After the pressure reduction, heating was continued with stirring at 310°C for 1 hour while passing nitrogen gas through the autoclave, thereby to complete the polymerization.

The polymerization product was withdrawn from the bottom of the autoclave in the form of ribbon, and cooled with water to form a transparent polyamide. The polymer was dried, and its physical properties were measured. It had a Vicat softening point of 163°C, a glass transition temperature of 138°C, an $[\eta_{sp/c}]$ of 0.789, and a bromine content of 0.50% by wt.

The polymer was dried, and then compression molded at a molding temperature of 300°C using a mold on a hot press to form a rectangular rod with a size of 3 mm x 6.5 mm x 127 mm. The oxygen index (LOI) measured of this rod was 29.0%. The UL 94 value of this copolymer was 94 V-2. The polymer was stable to boiling and methanol.

Example 2: 334 parts (1 mol) of hexamethylene diammonium naphthalene-2,7-dicarboxylate, 4.0 parts (0.02 mol) of p-bromobenzoic acid, and 0.41 part of a phosphorus acid were polymerized in the same way as in Example 1 to form a polymer having a melting point of 289°C, an $[\eta_{sp/c}]$ of 0.967 and a bromine content of 0.52% by wt.

The polymer was dried, and melt spun at 310°C by an extruder-type melt-spinning machine. The filaments were drawn to 4.0 times the original length while being passed over a hot plate held at 210°C to form a yarn (24 filaments per 71.8 den). The yarn had a tenacity of 4.1 g/den, an elongation of 12.2% and a Young's modulus of 980 kg/mm². The number of ignitions was tested, and the average was 9.0.

For comparison, the above procedure was repeated except that 2.4 parts (0.02 mol) of benzoic acid was used instead of p-bromobenzoic acid. The average number of ignitions was 6.0.

Bispiperazido Phosphorus Polyamides

J.C. Hermans; U.S. Patent 4,098,768; July 4, 1978; assigned to SA Texaco Belgium NV, Belgium provides polymers containing repeating units of the formula

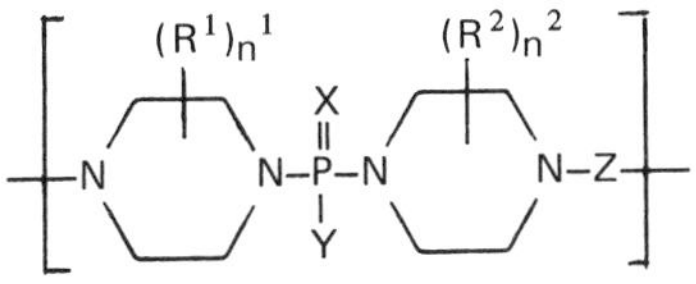

wherein X is absent or represents an oxygen or sulfur atom, or a group of the formula $=N-R$, in which R represents a hydrogen, an aliphatic, cycloaliphatic or aromatic group, or a heterocyclic group; Y represents an aliphatic, cycloaliphatic or aromatic hydrocarbon group or a heterocyclic group; a group of the formula $-NR_2$ in which each group R has the meaning given above, or the two groups R, together with the nitrogen atom to which they are attached, represent a N-containing heterocyclic ring; a group of the formula $-OR$ in which R has the meaning given above; or a group of the formula

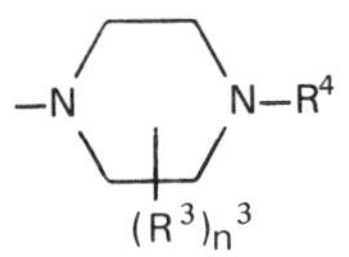

R^1, R^2 and R^3, which may be the same or different, each represents a substituent on the piperazine ring.

R^4 represents a hydrogen atom or a substituted or unsubstituted aliphatic, cycloaliphatic, or heterocyclic radical or heterocyclic group, an acyl group, a sulfonyl group; or a substituted or unsubstituted carbamoyl group; n^1, n^2 and n^3, which may be the same or different, each represents 0 or an integer; and Z represents a group of the formula $-CO-CO-$; an organic dicarboxylic group of the formula $-CO-R^5-CO-$, in which R^5 represents a substituted or unsubstituted aliphatic, cycloaliphatic or aromatic hydrocarbon radical, or a heterocyclic radical, or a group of the formula $-CO-Ar^1-Q-Ar^2-CO-$, in which Ar^1 and Ar^2, which may be the same or different each represents an arylene group. Q represents a group of the formula $-O-$, $-S-$, $-SO-$, $-SO_2-$,

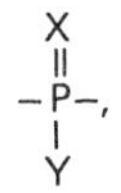

$-NR^6-$ (wherein R^6 represents an aliphatic, cycloaliphatic or aromatic radical or heterocyclic radical), $-SiR_2^6-$, or is a divalent aliphatic or cycloaliphatic group, a group of the formula:

$$
\begin{array}{cc}
X & \quad X \qquad X \\
\| & \quad \| \qquad \| \\
-P- & \ \ or \ \ -P-R^5-P- \\
| & \quad | \qquad \ | \\
Y & \quad Y \qquad Y
\end{array}
$$

in which X, Y and R^5 have the meanings given above; a group of the formula $-SO_2-$, $-SO-$, $-SO_2-R^5-SO_2-$, $-SO-R^5-SO-$ or $-SO_2-R^5-SO-$ in which R^5 has the meaning given above; or a group of the formula

$$
\begin{array}{ccccc}
R^6 & R^6 & R^6 & OR^6 & OH \\
| & | & | & | & | \\
-Si-, & -Si-R^5-Si-, & -Si- & or & -Si- \\
| & | & | & | & | \\
R^6 & R^6 & R^6 & OR^6 & OH
\end{array}
$$

or a group of the formula

$$
\begin{array}{ccccc}
X & X & X & & X \\
\| & \| & \| & & \| \\
-P-R^5-CO-, & -P-R^5-SO-, & -P-R^5-SO_2-, & -OC-R^5-SO_2-, & -P-R^5-SiR_2^6- \ \ or \ \ OC-R^5-SiR_2^6- \\
| & | & | & & | \\
Y & Y & Y & & Y
\end{array}
$$

but wherein Y cannot be dimethylamino when X represents oxygen and n^1 and n^2 are both zero.

Example 1: *Homogeneous Polycondensations* — Using chloroform ($CHCl_3$) or dimethylformamide (DMF) as solvent, such polycondensations have been carried out according to the following standard procedure. To 0.005 mol of the bis-piperazidophosphorus compound and 0.010 to 0.012 mol of triethylamine, dissolved in 25 ml of dry solvent, is added, in one portion at room temperature with gentle stirring, 0.005 mol of the diacid chloride dissolved in 25 ml of dry solvent. Usually a temperature rise from 20° to 35°C within 5 minutes was observed, and the mixture was stirred for 15 to 60 minutes.

The reaction mixture was worked up as follows: (a) If DMF was used as a solvent, the reaction mixture was added dropwise with stirring to 600 ml of water. If the polyamide did not precipitate, dry ice was added to provoke precipitation. The precipitated polyamide was filtered off, washed with water, and dried in vacuum over P_2O_5 at 40° to 70°C/1 mm.

(b) If $CHCl_3$ was used as a solvent, the reaction mixture was precipitated with petroleum ether or n-hexane (600 ml) and the polymer was filtered off, washed with water to remove triethylamine hydrochloride and dried in vacuum over P_2O_5 at 40° to 70°C/1 mm. Sometimes the polyamide became a sticky mass on washing with water. When this happened, the polyamide was dissolved in chloroform (30 to 50 ml), ethanol and benzene were added and the water and solvents were distilled off under reduced pressure at 60° to 100°C. The remaining polyamide was redissolved in chloroform (30 ml), and precipitated with petroleum ether or n-hexane (500 to 600 ml) or, if precipitation was not possible, the polyamide was isolated by freeze drying from a chloroform solution.

Examples 2 and 3: *Heterogeneous (Interfacial) Polycondensations* — The systems $CHCl_3/H_2O$ and CCl_4/H_2O were used according to following standard procedure. To a solution of 0.005 mol of the bispiperazidophosphorus compound and 0.012 mol of potassium hydroxide in 40 ml of water, was added in one portion, at room temperature and with vigorous stirring, a solution of 0.005 mol of the diacid chloride in 40 ml of chloroform or carbon tetrachloride, and the mixture was stirred for 15 to 30 minutes.

When the reaction was carried out using 0.10 mol of the bispiperazidophosphorus compound and 0.1 mol of the diacid chloride in 600 to 750 ml of solvent and 200 to 270 ml of water, the temperature of the reaction mixture rose from $20°$ to about $35°C$ within 5 minutes, due to the heat evolved.

In the mixture CCl_4/H_2O, the polyamide precipitated as it was formed, and formed a thick swollen lump, but in the system $CHCl_3/H_2O$, the polymer remained dissolved. In general there was no considerable difference between the results (yield MW) obtained in the two solvent systems. An emulsion was, however, often formed when using the $CHCl_3/H_2O$ system, necessitating the use of different methods for isolating the polyamides from the various reaction mixtures.

From a CCl_4/H_2O Mixture – The precipitated polyamide was filtered off, cut into small pieces, washed with water and dissolved in 60 to 100 ml of chloroform. This solution was washed twice with 70 to 100 ml of water, to remove inorganic and organic potassium salts, and concentrated to about 30 ml in vacuum, and the polyamide was isolated by precipitation with 600 ml of petroleum ether or n-hexane. Then it was filtered off and dried over P_2O_5 in vacuum (1 mm) at $40°$ to $70°C$.

From a $CHCl_3/H_2O$ Mixture – When no emulsion was formed, or when separation between the chloroform and the water layer still occurred, the organic layer was separated, washed twice with 70 to 100 ml of water to remove potassium salts, and concentrated in vacuum to about 30 ml, and the polyamide was precipitated by pouring this solution into 600 ml of petroleum ether, filtered off and dried.

Sometimes an additional amount of polymer could be recovered by extracting the water layer and the wash water with chloroform, and by pouring the water-washed and concentrated chloroform solution into hexane.

When the separation of the layers was too difficult, or when an emulsion was formed, the chloroform was stripped off in a thin film evaporator $60°$ to $90°C/$ 20 mm and the remaining water was decanted from the precipitated polymer, which was rinsed with 100 ml of water and redissolved in 60 to 100 ml chloroform. This solution was water-washed and concentrated, and the polyamide was precipitated with 600 ml of n-hexane, filtered off, and dried.

If during purification of the polyamide by washing a chloroform solution with water, an emulsion was again formed, the chloroform was distilled off at reduced pressure, the supernatant water was decanted from the precipitated polymer, the rest of the water was removed as an azeotrope with some added alcohol and benzene. The remaining polyamide was further treated as above.

If precipitation of the polymer was not possible, the polyamide was isolated by freeze drying from chloroform solution. The ratio of organic solvent:water can be varied over a wide range, e.g., 1:1 to 3:1, with no substantial variation in the results obtained.

The polyamides had good solubility in chloroform and dimethyl formamide; some were soluble in methanol and ethanol, and a few in dioxane, benzene and acetone. The polyamides were not soluble in carbon tetrachloride, petroleum ether, ether, or paraffinic oils. The polyamides show, in general, melting temperatures in the range of 150° to 280°C, and start to decompose at about 300°C. The polyamides derived from sebacoyl chloride have lower melting and decomposition temperatures than derivatives of terephthaloyl chloride.

Film Forming Properties of the Polyamides — Films cast from chloroform solutions of the polyamides, in general, have a good appearance. They are colorless, flexible and nonburning or self-extinguishing when ignited in a Bunsen flame. Polyamide films dissolve in 10 N sulfuric acid within a few hours, but have good resistance to 0.1 N sulfuric acid in which they turn hazy and show only small changes in weight when immersed for 6 to 14 days. The polyamide films are resistant to caustic solutions; when immersed in 10 N NaOH and in 0.1 N KOH at room temperature the film remains unchanged or turns hazy; in some instances, small changes in weight are observed.

The polyamide films seem unchanged on aging; no changes in appearance or solubility in chloroform were observed on film strips which were exposed to daylight and air for several months.

Example 4: To 0.005 mol (1.552 g) of phenylbispiperazidophosphate and 0.01 mol (1.4 ml) of triethylamine in 25 ml of chloroform (free of ethanol) was added, in one portion at room temperature with stirring, 0.005 mol (1.015 g) of terephthaloyl chloride dissolved in 25 ml of chloroform. The reaction temperature rose to 35°C within 5 minutes, and the mixture was stirred for 15 minutes. From it the polyamide was precipitated by pouring the solution into 600 ml of vigorously stirred n-hexane. The polyamide was filtered off, thoroughly washed with water and dried over P_2O_5 at 70°C in vacuum (1 mm). Yield: 2.092 grams (95%); $[\eta]$ = 0.425 dl/g; η_{red} = 0.46 dl/g ($CHCl_3$; 0.40 g/dl; 25°C). Melting range 258° to 270°C; decomposition temperature 350°C.

The polyamide could be cast from a chloroform solution to give a colorless, flexible film, which was nonburning.

Example 5: To a solution of 0.005 mol (1.552 g) of phenylbispiperazidophosphate and 0.010 mol (1.4 ml) of triethylamine in 35 ml of chloroform was added, in one portion with stirring, 0.005 mol (1.066 ml) of sebacoyl chloride in 45 ml of chloroform. The temperature rose from 22° to 32°C and the mixture was stirred for 15 minutes, during which it remained a clear solution. The mixture was concentrated in vacuum and poured into 600 ml petroleum ether. The polyamide precipitated as a white powder and was filtered off. On washing with water it formed a sticky mass which was dried in vacuum over P_2O_5.

The polymer was redissolved in chloroform and isolated by freeze drying in the form of a white foam. Yield: 2.142 g (89.9%); $[\eta]$ = 0.133 dl/g; η_{red} = 0.145 dl/g ($CHCl_3$; 0.4 g/dl; 25°C). Melting range 70° to 80°C; decomposition temperature 240° to 250°C.

When cast from a chloroform solution, the polymer formed a colorless flexible film which was nonburning.

Tetrakis(Hydroxymethyl)Phosphonium Compound with Melamine or Guanidine Compound

The process developed by *G.M. Moulds; U.S. Patent 4,112,016; September 5, 1978; assigned to E.I. Du Pont de Nemours and Company* relates to fibers, filaments, yarns and fabrics of selected polyamides wherein the fibers contain the reaction product of absorbed reactants comprising a tetrakis(hydroxymethyl)phosphonium compound and a member of the group consisting of melamine-formaldehyde condensate, melamine-formaldehyde condensate and magnesium ammonium phosphate hexahydrate, guanidine magnesium phosphate, and guanidine phosphate, in such amount as to provide between 0.4 and 3.0% by wt of phosphorus.

The suitable aromatic polyamides employed in this process are selected from the group of poly(p-phenylene terephthalamide), poly(chloro-p-phenylene terephthalamide), poly(p-phenylene chloroterephthalamide), poly(p-benzamide). They may be prepared by the low temperature solution polymerization procedures reported in, e.g., Kwolek et al., U.S. Patent 3,063,966 and Kwolek U.S. Patent 3,671,542, and should have an inherent viscosity within the range of 1.5 to 5.5, preferably between 2.5 and 3.5.

Polyamides which have been prepared, isolated, washed, and dried by the procedures shown in the above-cited references are combined with concentrated sulfuric acid to form spinning dopes, some of which are anisotropic, e.g., as described and characterized in the previously mentioned Kwolek U.S. Patent 3,671,542. Preferably, the concentration of the sulfuric acid used for dope preparation is within the range of $100\pm0.5\%$, and most preferably, is about $100\pm0.2\%$, by wt H_2SO_4. These dopes contain from 3 to 15% by wt polymer solids.

The dopes are prepared by combining and mixing (mechanical agitation) appropriate quantities of dry polyamide and sulfuric acid at 70° to 95°C for a time sufficient to form the desired dope, e.g., from 1 to 4 hours. Appropriate dyes can be dissolved in the spinning dope to color the product fiber. Vacuum (e.g., 25 to 30" of Hg) is applied during the mixing operation in order to reduce the occlusion of air in the spinning dope. If stored, the dope is heated to about 65°C prior to being spun.

For spinning, the warm dope is filtered through fine mesh stainless steel screens, and then wet spun through a corrosion-resistant, fine capillary spinneret into a coagulation bath maintained within the range of 70° to 95°C. The bath may be water or dilute sulfuric acid, preferably 5 to 10% by wt sulfuric acid. After traveling horizontally about 10 to 12" in the coagulation bath, the freshly formed filaments are carried by driven rolls into a series of baths which remove residual spinning solvent, and neutralize traces of residual sulfuric acid. The threadline is then led into a tank of pure water for a final rinse. If desired, a suitable textile finishing agent may be added to the last tank in order to facilitate textile processing, e.g., of the staple prepared from the fiber.

During the spinning operation a spin stretch factor, i.e., SSF (Kwolek U.S. Patent 3,600,350), of from 0.5 to 2.5 is experienced by the threadline. The washed fibers emerge from the baths with speeds from 30 to 120 yd/min. Essentially no drawing occurs in the extraction baths. The wet tow (very porous) from the final wash may be piddled with fluted rolls into small tow cans and stored in this manner while wet and free of tension.

Fibers produced by the abovedescribed processes (but dried) exhibit tenacities within the range of 1.5 to 7.0 g/den, elongations within the range of 8 to 50%, and initial moduli within the range of 10 to 250 g/den. Preferably the compositions of this process in fiber form exhibit the following filament tensile properties: tenacity in g/den (T), 4 to 6; elongation in percent (E), 9 to 14; and initial modulus in g/den (Mi), 70 to 140. Preferably, the product of the spinning operation exhibits a denier per filament value of about 1.25±0.15.

Absorption Process: Into the wet tow from the aforementioned spinning operation is absorbed a tetrakis(hydroxymethyl)phosphonium compound and a member of the group consisting of (a) melamine-formaldehyde condensate (e.g., Aerotex resin UM); (b) melamine-formaldehyde condensate (e.g., Aerotex resin UM) and magnesium ammonium phosphate hexahydrate ($MgNH_4PO_4 \cdot 6H_2O$); (c) guanidine magnesium phosphate; and (d) guanidine phosphate. Melamine-formaldehyde condensate is preferred for the treatment.

Suitable tetrakis(hydroxymethyl)phosphonium compounds include the chloride (THPC), the so-called hydroxide (THPOH), the oxalate (THPOX), the phosphate, the acetate, etc., and mixtures of these compounds. Although the tetrakis(hydroxymethyl)phosphonium moiety of each of these compounds is sufficiently reactive to form a suitable insoluble reaction product with any of the coreactants (a) through (b) above, the preferred tetrakis(hydroxymethyl)-phosphonium compounds are THPOH and THPOX.

The weight ratio of tetrakis(hydroxymethyl)phosphonium compound to its coreactant (a) through (d) will range from 2:1-5:1 with a 5:1 ratio preferred. In the case of coreactant (b) the magnesium ammonium phosphate hexahydrate is present in amounts up to 30% by wt of (b). As will be known to those skilled in the art, small amounts of a catalyst such as ammonium chloride may be used to promote the crosslinking of the melamine-formaldehyde condensate. Vircol 82 contains 11.3% phosphorus, 2 hydroxyl groups per phosphorus atom and a hydroxyl number of about 205.

Absorption into the fiber may proceed as follows. The previously described porous, wet, as-spun tow (containing about 5 parts of water per part of polymer) is piddled into a stock dryer and treated with an aqueous solution containing, e.g., based on weight of the wet fiber, 1% THPC, 0.2% Aerotex UM, and 0.007% ammonium chloride crosslinking agent. The tow and treating solution are heated together for 45 minutes at 80°C, then cured for 1¼ to 2 hours in 30 to 35 psi steam. The tow is then given a neutralizing scour, e.g., with a solution of sodium carbonate and Duponol D surface active agent (sodium salt of mixed long chain alcohol sulfates), after which it is dried at temperatures up to 150°C, preferably at 130° to 150°C for 1 hour. Alternatively, the THPC could be neutralized before being added to the fiber.

The phosphorus content of the treated textile structure of this process is within the range of 0.4 to 3.0% by wt, preferably 0.8 to 1.3% by wt.

Product Utility: The fabrics exhibit excellent thermal and flame-retardant performance. As previously described herein, and illustrated further in the examples which follow, their use in textile applications is especially preferred. Protective clothing prepared from these compositions can be advantageously worn, e.g., by workers who handle flammable liquids and gases in industrial plants, by firemen, by aviators, and by others who may be exposed to the hazards of catastrophic fires wherein escape from total immersion in flames becomes necessary. In these situations, being insulated from an extremely high heat flux for even a few seconds is vital.

These fabrics perform very well when exposed to high heat fluxes, e.g., one of 2.6 calories/cm^2/sec, a flux representative of that produced by a jet fuel fire. Under such conditions, these fabrics effectively resist breaking open, an event which would destroy the air barrier existent between the fabric and the skin of the wearer who would be directly exposed to the convective and radiant energy of the heat source. They exhibit low shrinkage when exposed to high temperatures, thus maintaining between fabric and wearer the insulating air layer which reduces the rate of heat transfer to the wearer.

Strength retention during exposure to flames is excellent for these fabrics, an important feature for a wearer who may be climbing or running to escape a fire. When the fabrics are no longer exposed to open flame but are still subjected to a rather high level of radiant energy, they support minimal after-flaming (i.e., 0 to 5 seconds), a very important quality. The limiting oxygen index value (LOI) for the compositions of this process is high, e.g., 0.40.

In addition to these excellent thermal and flame-retardant characteristics, the fabrics exhibit other desirable qualities. These fabrics may be obtained in a range of colors, an important aesthetic consideration. They exhibit good hydrolytic stability, e.g., as evidenced by a high level of retention of fabric breaking strength after 7 hours at the boil in 10% sulfuric acid. Fabrics exhibit excellent resistance to both pilling and abrasion and display desirable wash- and light-fastness properties.

Testing Apparatus: An apparatus by which the heat- and flame-resistant characteristics of fabrics (e.g., those commonly used for military and protective clothing purposes) prepared from the fibers of this process may be evaluated is shown in Figure 3.3.

The heat flux is supplied by combined radiant and convective sources. The radiant energy is supplied by 9 quartz infrared tubes 1 (e.g., G.E. T-3, 500 W each) to which a total of up to 45 A current is supplied from a power supply not shown. These tubes are located within a transite box 2, ¼" thick, whose top is a water cooled ⅜ to ⁷/₁₆" thick stainless steel jacket. Radiant energy from the quartz tubes is directed upward toward the fabric sample through a 4" x 4" opening in the top of the box.

Convective energy is supplied by 2 Meker burners 3 positioned (on opposite sides) over the top of the transite box, each at an angle of about 45° from horizontal. The tops of the Meker burners are separated from each other by a distance of about 5". In order to insure a constant gas flow rate, gas is fed to the burners through a flow meter from the fuel supply. The gas flow to these burners can be shut off by a toggle switch. The test fabric sample 4 held in holder 5 can be brought into horizontal position above the heat flux provided by the tubes and burners by means of a carriage, not shown.

Figure 3.3: Fabric Testing Apparatus

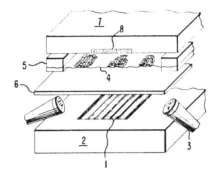

Source: U.S. Patent 4,112,016

When the sample is in this position, it is about 2¼" above the tops of the burners and about 3¾" above the infrared tubes. A 4" x 4" area of the fabric test sample is exposed to the heat flux unless otherwise indicated.

Located in a fixed position above the tubes and burners but below the test position plane of the fabric sample is a movable, water-cooled steel shutter **6**. When located in the closed position, i.e., directly above the heat flux, the shutter insulates the fabric test sample from the heat flux. When the shutter is removed from above the heat flux, the open position, the fabric sample is exposed to the heat flux. The duration of the fabric exposure to the heat flux can be controlled by movement of the shutter into or out of closed position.

The top member of the apparatus shown is an insulating (Marinite) block **7** containing a copper slug calorimeter **8** whose output is fed to an appropriate recording apparatus, not shown, by which the temperature rise (°F) experienced by the calorimeter can be recorded on chart paper. The distance between the calorimeter and the top surface of the fabric sample is ¼".

In each of the following test procedures there is employed the apparatus previously described. Unless otherwise indicated, the heat flux in each test is a combination of radiant and convective energy in about a 50-50 ratio; the total heat flux to which each fabric sample is subjected is 2.6 calories/cm^2/sec. In each test the quartz tubes and Meker burners are at operating temperatures and the shutter is in the closed position prior to exposure of the fabric sample which has been placed on the carriage in the test position.

Heat Transfer Test: A fabric sample is held taut in the holder, the Marinite block containing the calorimeter is clamped tightly to the top of the holder **5**, and the shutter **6** is opened and closed by use of a timer, not shown, in order to expose the sample to the heat flux for a predetermined interval (e.g., 1 sec, 2.5 sec, 4 sec). The temperature rise (ΔT, °F) experienced by calorimeter is recorded. The ΔT which is measured and recorded during the exposure time plus the first 10 sec after the shutter closes is taken as the heat transfer index. Between each successive test the Marinite block is unclamped from the holder **5** to allow the calorimeter to cool to room temperature prior to being used again.

After-Flame Test: For this test horizontal sample holder **5** is replaced with another sample holder consisting of an inclined U-shaped metal plate whose base lies in the plane previously occupied by the sample holder, whose parallel legs (each 4" long) are inclined in a downward direction at an angle of about 45°, such that the open end of the U points generally toward quartz infrared tubes **1**, and oriented such that the vertical plane which passes through the 2 Meker burners **3** is a perpendicular bisector of the base of the U and also passes through the center of the (inclined) 2" wide aperture defined by the legs of the U. A 3" wide strip of test fabric is mounted, spanning the legs of the U, one edge of the strip being flush with the end of each leg.

With shutter **6** open, quartz tubes **1** and burners **3** operating, the test is commenced by dropping the holder and sample into position, and allowing the burner flames to impinge on both fabric surfaces for a given interval, e.g., 3 to 5 seconds. The Meker burners are then turned off, and the time required until any flames present on the fabric surface extinguish in the presence of the radiant energy supplied by the quartz tubes is measured. The latter time interval is reported as the after-flame time.

Fiber Systems: Spin dopes were made by dissolving 60 lb of dry poly(p-phenylene terephthalamide) and optionally, i.e., where "sage green" fibers are called for 0.48 lb of Ponsol Khaki 2G and 0.12 lb of Ponsol Brilliant Violet 4R N dry, crude grade in 440 lb of sulfuric acid at 100.0±0.1% concentration to give 12% polymer solids. The solution was agitated in a Ross Planetary mixer at 70°±5°C for 3 to 4 hours with application of 26 to 27" of vacuum.

Upon completion of mixing, solution was transferred to a storage tank jacketed to 60°±1°C, heated to 65°C in transfer lines and filtered at 65°C through 325 x 2300 mesh stainless steel screens. The wet spinning was carried out with spinnerets with about 1,500 holes (0.0012" diameter, each) with a capillary length to diameter ratio of 1:1. The spin bath was maintained at 80°±2°C and contained 8 to 10% H_2SO_4 by adjusting the flow of pure water to compensate for the acid added from the spin dope.

The 1.25±0.15 denier per filament product was spun at 60 ypm with a spin stretch factor of about 1.1. The threadline of about 1,900 den was washed by passing through 3 tanks of essentially pure water, a neutralizing tank containing sodium carbonate solution, and a final rinse tank with pure water. The increase in the speed of the rolls through the extraction-neutralization process was 3.4%.

The threadline is piddled with a gear piddler into small tow cans. The wet tow (17 lb dwb) is placed in a stock dryer of 20 lb capacity. To this is added a room temperature solution comprising 300 lb of water, 4 lb of THPC, 0.8 lb of Aerotex UM, and 12.7 g of ammonium chloride.

The contents of the dyer are heated to 80°C and the solution therein is recirculated for 45 minutes. The solution is then pumped out and the tow is treated with steam (35 psi) for 90 minutes.

The tow is then scoured for a few minutes with a solution of 0.8 lb of sodium carbonate and 0.2 lb of Duponol D in 40 gallons of fresh water. After this solution is removed, the tow is then rinsed with fresh water. The tow is removed and dried at 145°C for 1 hour.

The fibers exhibit the following tensile properties (measured on bundles): T, 4.9; E, 12.8; and Mi, 96 with a denier per filament of 1.25±0.15. The dry tow is then crimped in a mechanical crimper and cut into 1½" staple. The dry, dyed fiber contains 1% by wt phosphorus (based on total weight of the dry fiber). The staple is then spun into yarn from which is prepared test fabrics.

Example 1: This example illustrates that a fabric of this process retains its desirable low heat-transfer and low after-flaming characteristics even after repeated commercial launderings.

A fabric sample (plain weave) is prepared from the fiber system identified herein. This fabric sample is subjected to 10 commercial launderings. Separate portions of the laundered fabric sample are then subjected to 1, 2.5 and 4 second exposures, respectively, to the standard 2.6 calories/cm^2/second combined heat flux by means of the heat-transfer procedure and apparatus described in the testing apparatus section above. Data is obtained for an unlaundered control, also. The results are shown below.

	... Temperature Rise (°F) After Exposure of ...		
	1 Second	2.5 Seconds	4 Seconds
Fabric samples	9.5	26.5	42
Control	–	27	–

Two separate portions of the abovedescribed commercially laundered fabric exhibit after-flame times of 3.4 and 3.6 seconds, respectively, when tested by the After-Flame Test procedure described herein (3 second exposure). An unwashed control fabric exhibits an after-flame time of 3.2 seconds.

Example 2: This example illustrates that a fabric of this process retains its desirable low heat-transfer and low after-flaming characteristics even after repeated home launderings.

A fabric sample (plain weave) is prepared from the same fiber system as in Example 1. This fabric sample is subjected to 30 home laundering operations. Separate portions of the laundered fabric sample are then evaluated for heat transfer and after-flaming as in Example 1 herein. The results are shown below.

	... Temperature Rise (°F) After Exposure of ...		
	1 Second	2.5 Seconds	4 Seconds
Fabric samples	9	25.5	40

Two separate portions of the abovedescribed laundered fabric each exhibit after-flame times of 3 seconds when tested as described in Example 1 herein.

Nonblooming Nylon

R.C. Nametz and P.H. Burleigh; U.S. Patent 4,141,880; February 27, 1979; assigned to Velsicol Chemical Corporation describe a flame retarded, nonblooming nylon composition with excellent thermal stability. This composition is comprised of from 5 to 35% by weight of a condensation product derived from brominated phenol by the displacement of bromine from the phenol wherein

(a) the phenol is selected from the group consisting of tribromophenol, tetrabromophenol, pentabromophenol, and mixtures thereof;

(b) the condensation product has a repeating structural unit of the formula

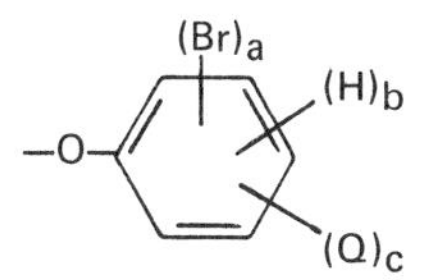

wherein a is an integer of from 0 to 4, b is an integer of from 0 to 2, c is an integer of from 1 to 5, a plus b plus c equal 5, Q is a monovalent bond from a carbon atom in the aromatic nucleus of the repeating structural unit to an oxygen atom bonded to an aromatic nucleus, and the polymeric units containing the repeating structural unit comprise at least 80% by weight of the product;

(c) the condensation product contains from 17 to 31% by weight of elemental carbon, from 0 to 1.0% by weight of elemental hydrogen, from 3 to 8% by weight of elemental oxygen, and at least 60% by weight of elemental bromine; and

(d) the condensation product has a molecular weight of at least 750, and one or more polymeric units containing at least four aromatic nuclei per unit comprise at least about 80% by weight of the product.

When the condensation product described above is incorporated into polyester, the resulting composition has a thermal stability which is mediocre to poor. However, when this condensation product is incorporated into nylon, the resulting composition exhibits an excellent thermal stability.

Example 1: 2,000 ml of water, 164 g of sodium hydroxide, 10.7 g of Emulsifier 334 (an aryl polyether emulsifier, Milliken Chemical Corporation), 0.7 g of dodecyl sodium sulfate, and 1,324 g of 2,4,6-tribromophenol were charged to a 5-liter flask fitted with mechanical stirring, a thermometer, and a reflux condenser. The reaction mixture was first heated to 100°C and maintained at that temperature for 1 minute; then it was cooled to a temperature of 33°C. To this mixture was charged 133 ml of toluene and 20 g of benzoyl peroxide. An exothermic reaction occurred, and the reaction temperature was then maintained at 55°C for 0.5 hour. Thereafter, 25 g of sodium hydroxide were added to the reaction mixture.

The reaction mixture was then filtered, the filter cake was washed with 15 liters of water, and the filter cake was dried to give 932 g of product.

Zytel nylon 6,6 chip was dried for 17 hours at a temperature of 79°C. There-after it was used in Examples 2 and 3.

Example 2: 300 g of the dried nylon chip, 240 g of Dechlorane 515 (a poly-chlorinated cycloaliphatic flame retardant), and 60 g of antimony trioxide were dry-mixed and compounded in a high shear C.W. Brabender mixer (Model R6) for one minute at a temperature of 224°C. Thereafter, the concentrate so pre-pared was mixed with an additional 600 g of the nylon chip and the mixture so formed was fed into a Newbury HI-30RS 30-ton injection molding machine with compounding mixer. A stock temperature of 475°F was used, and test specimens were prepared.

The specimens contained 20.0% by weight of Dechlorane 515 and 5.0% by weight of antimony trioxide.

The samples were tested for flammability in accordance with Underwriter's Lab-oratory Subject No. 94 test (UL Tests for Flammability of Plastic Materials, UL 94, February 1, 1974). Self-extinguishing properties were measured using this test which is carried out on test specimens 6 x ½ x ⅛ inch. In this test, the test specimen was supported from the upper end, with the longest dimension vertical, by a clamp on a ring stand so that the lower end of the specimen was ⅜ inch above the top of the burner tube. The burner was then placed remote from the sample, ignited, and adjusted to produce a blue flame ¾ inch in height. The test flame was placed centrally under the lower end of the test specimen and allowed to remain for 10 seconds.

The test flame was then withdrawn, and the duration of flaming or glowing com-bustion of the specimen was noted. If flaming or glowing combustion of the specimen ceased within 30 seconds after removal of the test flame, the test flame was again placed under the specimen for 10 seconds immediately after flaming or glowing combustion of the specimen stopped. The test flame was again with-drawn, and the duration of flaming or glowing combustion of the specimen was noted. If the specimen dripped flaming particles or droplets while burning in this test, these drippings were allowed to fall onto a horizontal layer of cotton fibers (untreated surgical cotton) placed one foot below the test specimen.

Significantly flaming particles were considered to be those capable of igniting the cotton fibers. The duration of flaming or glowing combustion of vertical speci-mens after application of the test flame (average of three specimens with six flame applications) should exceed 25 seconds (maximum not more than 30 sec-onds) and the portion of the specimen outside the clamp should not be com-pletely burned in the test.

Materials which complied with the above requirements and did not drip any flam-ing particles or droplets during the burning test were classified as V-1. Materials which complied with the above requirement but dripped flaming particles or drop-lets which burned briefly during the test were classified as V-2. A V-0 rating was given to materials wherein the duration of flaming or glowing combustion aver-aged less than 5 seconds under the conditions specified above.

The samples from Example 2 had ratings of V-1 at ⅛ inch and V-2 at 1/16 inch.

Example 3: In substantial accordance with the procedure described in Example 2, test specimens comprised of 18.0% by weight of the product of Example 1 and 3.6% by weight of antimony trioxide were prepared. 216 g of the product of Example 1, 340.8 g of the Zytel nylon 6,6 chip of Example 2, and 43.2 g of antimony trioxide were used to form the concentrate. Thereafter, 600 additional g of the nylon 6,6 of Example 2 were mixed with the concentrate and injection molded into test specimens in accordance with the procedure at a stock temperature of 480°F.

These specimens were tested in accordance with the procedures described in Example 2. Furthermore, they were also tested for migration (blooming) by being subjected to a temperature of 100°C for 100 hours and then being visually observed to determine whether any flame retardant migrated to the surface.

The nylon composition of this example had a V-0 rating at $\frac{1}{8}$ inch and a V-2 at $\frac{1}{16}$ inch.

Unlike many prior art flame retarded nylon compositions, the composition of this process exhibits good thermal stability, good heat distortion temperature, excellent tensile strength and Izod impact strength and very good migration resistance.

Sulfonic Acid or Sulfonate Group Attached to Aromatic Ring

High performance wholly aromatic polyamide fibers are provided by *R.S. Jones, Jr., M. Tan and E.W. Choe; U.S. Patents 4,162,346; July 24, 1979 and 4,075,269; February 21, 1978; both assigned to Celanese Corporation.* The fibers have chain-extending bonds from the aromatic nuclei thereof which are coaxial or parallel and oppositely directed, and have a sulfur content of about 0.5 to 10% by weight (e.g., 0.5 to 3% by weight) as sulfonic acid and/or sulfonate groups attached to the aromatic nuclei. Not only do the fibers exhibit highly satisfactory tenacity (at least 15 grams per denier), elongation (at least 1.5%), and initial modulus (at least 400 g per denier), but they exhibit an increased resistance to burning when compared to conventional unsubstituted aromatic polyamides. The fibers also exhibit particularly good resistance to organic solvents.

Nonlimiting examples of aromatic diamines useful in preparing the aromatic polyamides include

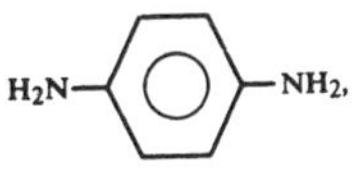

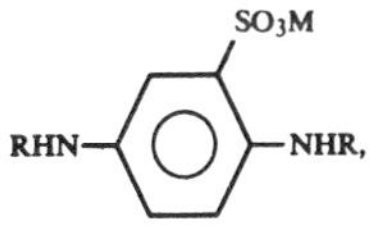

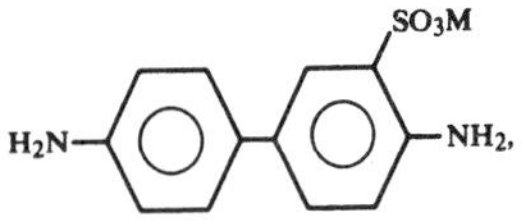

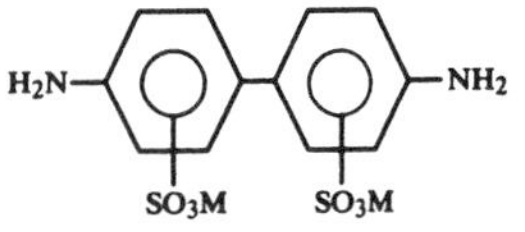

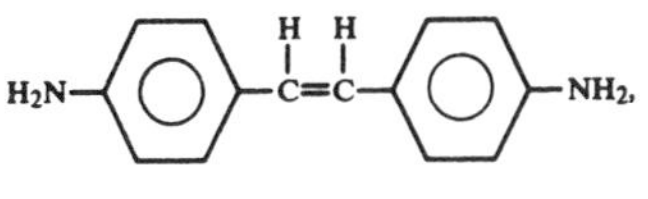

etc., where R = H, $-CH_3$ or $-C_6H_5$, and M = H, Li, Na, K, Ca, Ba, or $-CH_3$.

Nonlimiting examples of aromatic dicarboxylic acids or their derivatives useful in preparing the aromatic polyamides include:

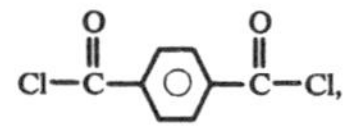

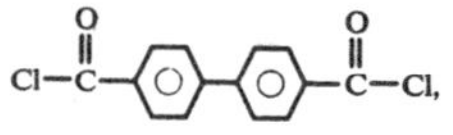

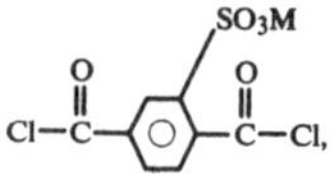

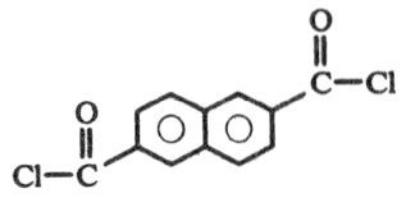

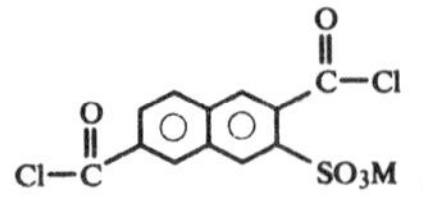

etc, where M = H, Li, Na, K, Ca, Ba, or $-CH_3$. Particularly useful monomers are

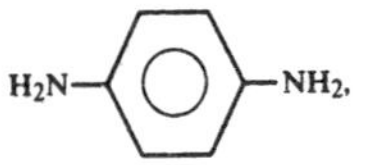

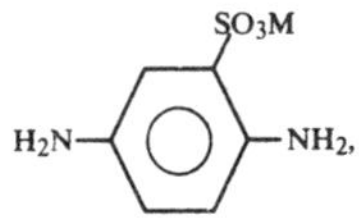

Example: A spinning solution was selected containing sulfonated poly-p-phenylene terephthalamide having a sulfur content of 1.7% by weight as sulfonic acid groups attached to the aromatic nuclei. The spinning solution contained 20% by weight of the sulfonated poly-p-phenylene terephthalamide dissolved in 100% concentrated sulfuric acid.

The spinning solution initially was prepared from poly-p-phenylene terephthalamide having an inherent viscosity of 4.8 by dissolving it in 100% concentrated sulfuric acid which was heated at 90°C for 10½ hours and at 95°C for 5¼ hours. The inherent viscosity of the resulting spinning solution of sulfonated poly-p-phenylene terephthalamide was 2.7.

The spinning solution while at a temperature of about 86°C was extruded at a rate of 40 m/min employing a ten-hole spinneret, each hole being 50 microns in diameter, through a ½-inch air gap and into a water bath maintained at 4°C.

The fibers were passed to a take-up unit, the speed of the take-up roll of which was adjusted so that the rate of wind-up was three times greater than the theoretical jet speed. The latter was calculated from the volumetric through rate of polymer solution in the extrusion cyclinder and from the number and size of the spinneret orifices.

The as-spun fibers having an inherent viscosity of 2.7 were thereafter washed in a base solution and then washed with water to remove residual sulfuric acid. Both steps were carried out batchwise. The washed yarns were then air dried.

The dried fibers were thereafter passed through a tube heated to a temperature of 250°, 300°, 350°, 400°, 450°, or 500°C for a residence time of about 6 seconds while blanketed with nitrogen and under a tension such that they were elongated 1.005 times (0.5%) their as-spun length.

The fibers produced after the thermal treatment possess outstanding tensile properties. In particular the fibers possess a tenacity of at least 15 grams per denier (e.g., 17 to 28 and typically greater than about 18 grams per denier; an elongation of at least 1.5% and typically greater than about 2% (e.g., 2 to 3.5%); an initial modulus of at least 400 grams per denier (e.g., 600 to 1,200 grams per denier) and typically higher than about 700 grams per denier; and an I.V. of about 2.0 to 4.0. The large increase in the physical properties, upon heat treatment of the as-spun sulfonated aromatic polyamide fibers is considered to be unexpected. The increase in tenacity is considered to be particularly unique.

IMPROVED COLORATION AND DYEABILITY

Anionic Dye-Leveling Agents

According to *W.T. Holfeld and A.J. Strohmaier; U.S. Patent 4,030,880; June 21, 1977; assigned to E.I. DuPont de Nemours and Company*, synthetic polycarbonamide filaments having more uniform acid dyeability are prepared in a melt-spinning and drawing process by applying a dye-leveling agent to the filaments in a spin finish prior to drawing. The filaments containing the finish are drawn to increase their molecular orientation. Dye uniformity is improved with washfast, rate-sensitive acid dyes. The filaments may be processed conventionally prior to dyeing.

The dye-leving agent is selected from the group consisting of:

(a) aromatic sulfonic acids and their condensation products;

(b) sulfated derivatives of C_{16-22} unsaturated fatty acids and their alkyl esters; and

(c) alkali metal and ammonium salts of (a) and (b).

Preferred aromatic sulfonic acids are p-(C_{8-18}-alkyl)-diphenyl ether disulfonic acids; particularly the p-dodecyl derivative. Preferred sulfate derivatives are sulfates of fatty esters of C_{1-8} alcohols; particularly butyl oleate. The agents should contain at least 0.5, and preferably at least 1.0, meq/g of the sulfonic or sulfate groups. The preferred salts in each instance are the sodium salts.

Effective agents include sodium dodecyl diphenyl ether disulfonate and sulfated butyl oleate.

Example: Poly(hexamethylene adipamide) flake, prepared by a conventional autoclave process, is melted under vacuum processing conditions to increase the molecular weight and spun into 2,450 denier, 128 filament yarn. The yarn has relative viscosity of about 70, measured at 25°C on an 8.4% by weight solution in 90% formic acid, an amine-end level of 40, and a carboxyl-end level of 55, both expressed as equivalents per million grams of polymer.

Yarns are spun and drawn at 2.7 and 3.0 draw ratios (feed roll speeds 856 and 771 ypm respectively) using a draw roll speed of 2,311 ypm. Finish application is by a standard rotating finish roll contacting the yarn prior to the draw zone. After the draw zone, the yarn is heated on two chest rolls operating at 215°C, then fed to a bulking jet using 240°C air at 120 psig (844 kg/cm²). The yarn is wound up at 2,077 ypm (1,899 m/min) at a nominal 425 g tension.

The finish compositions are comprised of 4% dye leveling agent (active ingredient basis), 16% of a nonionic, ethylene oxide-propylene oxide condensation product lubricant, and 80% water. A control yarn finish is similar except that the leveling agent is deleted. Finish level on yarn is about 0.6 to 1.0% (nonaqueous components) giving a retarding agent level of 0.12 to 0.20% on weight of fiber. To increase dyeability differences, draw ratios of 2.7 and 3.0 are run. The draw ratio is changed by changing the speed of the hot chest rolls.

Leveling agents used were Alkanol ND (sodium salt of dodecyl diphenyl oxide disulfonic acid), Univadine MC (condensation products of aromatic sulfonic acids) and Nylomine DN (80% sulfated butyl oleate, 20% ethoxylated tridecyl alcohol). Prior art finishes were (1) primarily paraffinic hydrocarbons emulsified in water with sulfated oleyl triglycerides and soaps, and (2) primarily coconut oil emulsified with sulfated oleyl triglycerides, nonionic detergents and soaps.

The yarn samples are dyed with Anthraquinone Milling Blue BL (C.I. Acid Blue 122), a typical rate-sensitive, large molecule, acid dye.

Results of these experiments illustrate that the application of a leveling agent via the spin finish dramatically reduces dye variability caused by the differences in draw ratio.

Change in dyeability ranged from 8.2 to 17.9 when the leveling agents were used, compared to 22.3 to 32.9 with prior art finishes. Dye values are calculated from reflectance readings.

Sulfopolyester Incorporated into Polyamide

As a result of various studies on improvements in a process for preparing a polyamide having excellent receptivity or affinity for basic dyes and yet providing a polyamide whose fiber properties such as spinnability, tensile strength, elongation and the like were effected to a lesser extent, T. Kusunose, M. Ikeda, K. Kitamura, T. Shima and H. Henmi; U.S. Patent 4,083,894; April 11, 1978; assigned to Asahi Kasei Kogyo KK, Japan found that polyamides which satisfy the above requirements can be obtained by polymerizing a polyamide-forming compound selected from the group consisting of ω-amino carboxylic acids, ω-lactams and an equimolar mixture of a diamine and a dicarboxylic acid and modifying the polyamides formed in the polymerization. The polyamides are modified by incorporating at least one compound of the formula

wherein M represents an alkali metal or an alkaline earth metal, n is an integer of 2 to 4; and m is an integer of 2 or more, before, during or after the polymerization. In the following examples all parts, percents, ratios and the like are by weight.

Example: *(a) Preparation of Polyethylene 5-Sodium Sulfoisophthalate* — 100 parts of dimethyl 5-sodium sulfoisophthalate and 110 parts of ethylene glycol were charged into a reaction vessel, and the mixture was kept at a temperature of 197°C for 5 hours under a nitrogen stream to effect the transesterification reaction while distilling off most of the methanol formed. The temperature of the reaction mixture was then increased to 230°C, and any excess of ethylene glycol was distilled off under reduced pressure. The mixture was then polymerized under a reduced pressure of 1.5 mm Hg at a temperature of 240°C for 2 hours. The resulting polymer was white in color and had a melting point of 236°C.

The average degree of polymerization of the resulting polymer was determined using the following procedure. That is, 5 ml of an acetylation reagent consisting of acetic anhydride and pyridine (1:3 by volume) was added to 1 g of the resulting polymer, and the mixture was heated on a steam bath for 45 minutes. The resulting sample solution was poured into 10 ml of water and, after allowing the mixture to cool, 10 ml of n-butanol was added thereto. After addition of an indicator (a mixture of a 0.1% aqueous solution of cresol red and a 0.1% aqueous solution of thymol blue, 1:3 by volume), the mixture was titrated with a standard 0.5 N ethanolic sodium hydroxide solution. The degree of polymerization (Pn) was calculated using the following equation

$$Pn = \frac{4 \times S}{M \times (V_1 - V_2)}$$

wherein S designates the weight of the sample (mg), M designates the molecular weight of the polyester recurring unit, V_2 designates the amount of the 0.5 N ethanolic sodium hydroxide solution required for the titration of the test sample (ml), and V_1 designates the amount of the 0.5 N ethanolic sodium hydroxide solution required for the titration of a blank sample. The degree of polymerization obtained in this example was found to be 13.

(b) Polymerization and Results of Spinning and Dyeing — 116 parts of hexamethylene diammonium adipate, 73 parts of water and 2.0 parts of polyethylene 5-sodium sulfoisophthalate having a sulfonate group content of 1.5 mol % per polyamide recurring unit were charged into an autoclave, and the temperature of the mixture was gradually increased from 230° to 280°C over a period of 1.5 hours while maintaining the mixture under a nitrogen atmosphere at a pressure of 17.5 kg/cm². The pressure was then reduced to atmospheric pressure while maintaining the temperature at 280°C, and the reaction mixture was kept under a nitrogen stream for an additional hour to complete the polymerization. The resulting polymer was found to have a chip yield of 91.4%, an η_r of 2.52 and an η_m of 375.

The sulfuric acid relative viscosity (η_r) was determined with an Ostwald viscometer using a 1.0% solution of a polyamide in 95.5% sulfuric acid as a test sample. The η_r value is calculated by dividing the falling time of the sample by the falling time of the solvent (95.5% sulfuric acid).

The melt viscosity (η_m) was determined in Shimazu Type Flow Tester Model 301. The determination was carried out by melting, at a temperature of 290°C, a polyamide chip which had been dried at 70°C for 24 hours under reduced pressure, and the melt viscosity is expressed in poises.

The chip yield was determined by dividing the amount of chips actually obtained by the theoretical amount of chips. The lower the value, the larger the amount of chips remaining unrecovered in the polymerization tank used for the polycondensation.

The spinning yield designates the number of spindles which could be wound without breaking of the yarn relative to the total number of spindles wound when the melt spun fibers are stretched, twisted and wound as a 3 kg pirn.

The dye adsorption was determined by the following procedure using a basic dye Cathilon Blue 5 GH (Hodogaya Chemical Co., Ltd.). A 1-g sample of the polyamide fibers was dyed in a dye bath having a dye concentration of 1% owf, a pH of 4 and a bath ratio of 1:100 at a temperature of $100°C$ for 1 hour. The sample was determined from the bath, and the absorbance of the dye bath was determined using a Hirama Type photoelectric colorimeter at a wavelength of 610 mμ.

The resulting polymer was then melt spun at a temperature of $294°C$ and the spun fibers were drawn to a draw ratio of 3.5. The spinning yield of the spun fibers was found to be 98.5% and the fibers had a tensile strength of 4.43 g/d and an elongation of 34.5%. The fibers obtained above exhibited a dye absorption of 98.5%.

Example 2: 116 parts of hexamethylene diammonium adipate, 73 parts of water and 0.15 parts of acetic acid as a viscosity-stabilizing agent were polymerized in the same manner as described in Example 1 to obtain a polymer having a chip yield of 91.5%, an η_r of 2.51 and an η_m of 377.

The resulting polymer was then spun and subsequently drawn in the same manner as described above. The spinning yield of the spun fibers was 98.0%. The dye absorption of the fibers with respect to Cathilon Blue 5 GH was found to be 10.1%.

Amino-Poly(Imidazoline-Amide) Compositions

D.E. Peerman, D.G. Swan and H.G. Kanten; U.S. Patent 4,049,598; September 20, 1977; assigned to General Mills Chemicals, Inc. describe amino-poly(imidazoline-amide) compositions which are useful for occlusion in polyolefins to render them dyeable with acid textile dyes. Polyolefins are useful in textile fibers or filaments. The compositions are certain azelaic acid, polyamide-imidazoline products of a mixture of diethylene triamine and a diamine. Optionally, fatty compounds such as a dimeric fat acid or monomeric fatty acid or another, different aliphatic dicarboxylic acid may be present.

The products are employed with polymers of α-olefins to provide dyeability thereto and are employed in an amount of from 0.1 to 10% and more preferably 1 to 5% by weight based on the amount of a α-olefin polymer. Generally, all normally solid polymers of the lower (2 to 4 carbon atoms) α-olefins are useful. Copolymers of the monomers abovementioned may be employed as well as copolymers of these monomers and another comonomer such as hexene-1, decene-1 or butadiene. Particularly useful is polypropylene. The product of this process may be combined with the polypropylene in various ways such as blending, melting, extruding, milling, or other procedures.

In evaluating the products of the process with polypropylene as a dyeability additive, the crushed resin is dry blended with the polypropylene (Amoco grade 10-5013 or 11-5013) and coextruded at 450°F to effect the first stage of melt blending and then pelletized. The second stage of blending comes in the extruder which feeds the fiber die, the extruder being fed with pellets from the preliminary extrusion. The fibers emerging from the machine are drawn for greater strength, run through a water bath to cool, are dried and collected on spindles. Dyeing tests are conducted by treating the fibers with acid dyes of various hues in boiling water for two hours. A typical dye is Acid Nylon Red GLM (Allied Chemical). The dyeing uniformity and depth of color are evaluated by comparison with standard dyed fibers.

Processability tests include observation of fiber continuity and uniformity. If the additive does not blend well or is incompatible with the polypropylene the dye absorbs nonuniformly, frequently in a streaked or spotty manner. Fiber or film uniformity is measured with a micrometer and continuity involves fiber breakage records. More than two fiber breaks per hour of drawing is considered unsatisfactory.

Example: 10 eq % stearic acid, 40 eq % azelaic acid, 25 eq % hexamethylene diamine (HMDA), and 25 eq % diethylene triamine (DETA) were used in this example. Ten drops of phosphoric acid were also added as a catalyst to ensure complete reaction and 15 drops of an antifoam reagent (DC Antifoam, 1%) were added to minimize foaming.

All of the HMDA and 94% of the azelaic acid, comprising Part B, were heated with stirring to 240°C over a period of one hour and maintained there an hour. The reaction was then cooled to 200°C and discharged.

All of the stearic acid and DETA and 6% of the azelaic acid, comprising Part A, were heated with stirring 1¼ hours to 260°C and maintained there for 30 minutes. It was then cooled to 200°C and the pulverized Part B was added. In 35 minutes it was heated to 250°C, in an additional hour to 265°C and maintained there for 1½ hours. The product was cooled to 200°C and discharged.

The product had a measured total amine number of 79 and a tertiary amine number of 51, providing a ratio of tertiary to total of 0.65. The product had a R&B melting point of 175°C. The product was evaluated for dyeability and processability and found to meet fully the requirements therefor.

Copolycondensation of Monomers and Pigments

The process developed by *P.Y.E. Gangneux; U.S. Patent 4,002,591; January 11, 1977; assigned to Produits Chimiques Ugine Kuhlmann, France* for the coloration of a linear polymer, especially polyamide, comprises effecting the polycondensation of the monomer in the presence of a pigment of the polycyclic polycarboximide series containing two primary amine or haloformyl functional groups.

When the dyestuffs have only one function capable of reacting with an amino or carboxy group of the polymer during its formation, the fixation of the dyestuff may stop the process of polycondensation on the end of the chain. When the dyestuffs have three or more groups capable of reacting, the polycondensation gives rise to a network system which may be polydimensional.

Accordingly a pigment of the polycyclic polycarboximide series containing two primary amine or haloformyl functional groups is used.

Of these pigments those of the following general formula may be particularly mentioned

$$Z-(R_2)_n-N \underset{\substack{\diagup\diagdown \\ C \\ \| \\ O}}{\overset{\substack{O \quad O \\ \| \quad \| \\ C \quad C \\ \diagup\diagdown\diagup\diagdown}}{\quad R \quad}} \underset{\substack{C \\ \| \\ O}}{\overset{C}{\diagup\diagdown}} N-(R_2)_n-Z$$

wherein R represents the tetravalent residue of a substituted or unsubstituted mono- or polycyclic aromatic hydrocarbon or the tetravalent residue of a substituted or unsubstituted perylenone, n represents 0 or 1, R_2 represents a substituted or unsubstituted aliphatic, aromatic or heterocyclic radical, and Z represents a primary amine group, a haloformyl group, or a radical of the general formula: $X-R_1-CONH-$ wherein X represents a primary amine or haloformyl group and R_1 represents a substituted or unsubstituted aliphatic, aromatic or heterocyclic radical.

Examples of substituents of R are halogen atoms or nitro, hydroxy, or alkoxy groups and examples of substituents of R_2 are halogen atoms or nitro, alkyl or alkoxy groups.

In order to carry out the process, the monomers and dyestuffs are mixed and the copolycondensation is effected by the usual processes for the preparation of polymers. The proportions of the dyestuff may vary from 0.1% (i.e., 1/1,000) to 2% of the weight of the monomer. In the following examples all parts are by weight.

Example 1: 49.95 parts of hexamethylene-ammonium adipate are intimately mixed with 0.05 part of an equimolecular mixture of N,N'-bis(4-amino-phenyl)-3,4-9,10-bis(dicarboximide)perylene and adipic acid. The composition is introduced into an autoclave, which is closed and purged with nitrogen. The temperature is raised to 180°C over a period of one hour and maintained at this temperature for ½ hour. After decompression, the autoclave is scavenged with nitrogen and heated to 280°C and maintained at this temperature for 2½ hours. The autoclave is decompressed and allowed to cool while being maintained under an atmosphere of nitrogen. The polymer thus obtained has an intrinsic viscosity in formic acid of 1.42 (solution of 1 g of polymer in 100 cc of 75% formic acid was measured at 25°C). It is impossible to extract the dyestuff from the polymer.

Example 2: 200 parts of water, 1.6 parts of sodium hydroxide, 0.5 part of sodium lauryl sulfate, 2.26 parts of hexamethylenediamine and 0.02 part of N,N'-bis(4-amino-2-nitro or 3-phenyl)- 1,8-4,3-bis(dicarboximide)naphthalene are introduced with vigorous stirring into an apparatus provided with a stirring device. Vigorous stirring is effected at the ambient temperature for 2 minutes, then a solution of 3.66 parts of adipoyl dichloride in 100 parts of tetrachloroethylene is introduced over a period of 15 seconds, and the mixture is maintained at the ambient temperature for 2½ minutes while stirring. The product is filtered off and washed to remove mineral products. A copolyamide is thus obtained from which it is impossible to separate the constituents and which gives yellow filaments.

Example 3: One operates as in Example 2, but 220 parts of water, 0.82 part of sodium hydroxide, 0.1 part of sodium lauryl sulfate, 1.16 parts of hexamethylenediamine, and 0.01 part of N,N'-bis-aminoperylene-bis-3,4-9,10-dicarboximide) are used. The organic phase consists of 2.39 parts of sebacoyl dichloride in 150 parts of tetrachloroethylene. A copolyamide of rose color is obtained.

Polyamides Colored with Acid Azobenzene Dyestuffs

According to *D. Razavi; U.S. Patent 4,063,881; December 20, 1977; assigned to Produits Chimiques Ugine Kuhlmann, France* polyamide fibers are colored with dyestuffs of the formula:

(1)

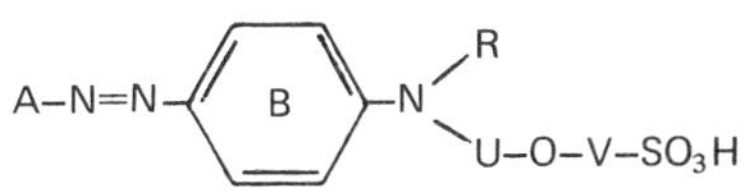

$$A-N=N-\underset{B}{\bigcirc}-N\overset{R}{\underset{U-O-V-SO_3H}{}}$$

wherein A represents the residue of a carbocyclic diazotizable amine, R represents an alkyl group, U represents a linear or branched alkylene group of 1 to 5 carbon atoms, V represents an arylene group, and A, B, R and V are unsubstituted or substituted with non-water-solubilizing substituents, and a process for their preparation which comprises diazotizing an amine of the formula $A-NH_2$ and coupling the obtained diazo derivative with a coupling component of the formula:

(2)

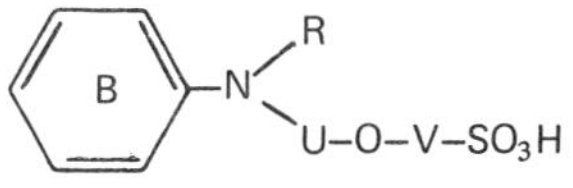

$$\underset{B}{\bigcirc}-N\overset{R}{\underset{U-O-V-SO_3H}{}}$$

wherein A, B, R, U and V have the same significance as above. The polyamide fibers exhibit excellent tinctorial yield, very remarkable brilliance, good fastness to washing and to light and good covering of bars. In the following examples the parts indicated are by weight unless otherwise stated.

Example 1: 13 parts of p-chloraniline dissolved in 100 parts of water and 25 parts of hydrochloric acid are diazotized by means of 7 parts of sodium nitrite at 0°C and coupled at 0° to 5°C with 34.5 parts of the sodium salt of N-ethyl N-(2'-p-sulfophenoxy-ethyl) aniline dissolved in 200 parts of water and 10 parts of concentrated hydrochloric acid. After one hour of stirring at 0° to 5°C, the pH is adjusted to approximately 3 by means of sodium acetate, the mixture is stirred overnight at ordinary room temperature, heated to 50°C for a ¼ hour, cooled to ordinary room temperature, filtered and dried at 50°C. 50 parts of a brown-red powder are obtained which dyes nylon fibers in a yellow-gold shade which is fast to light and washing.

Example 2: 13 parts of p-chloraniline are diazotized as in Example 1 and coupled with 36 parts of the sodium salt of N-ethyl N-(2'-p-sulfophenoxy-ethyl) m-toluidine dissolved in 150 parts of water and 10 parts concentrated hydrochloric acid. The coupling of the dye is completed as in Example 1 and, after drying of the dye, a brown-red powder is obtained which dyes nylon in a golden-yellow shade, which is slightly more orange than that of Example 1 and also fast to light and washing.

The sodium salt of N-ethyl N-(2'-p-sulfophenoxy-ethyl) m-toluidine used for the manufacture of this dye may be prepared by the following process.

A mixture of 453 parts of the β-bromoethyl ether of phenol and 608 parts of N-(monoethyl) m-toluidine is heated over 1½ hours to 95° to 100°C, then over 7 hours to 115° to 120°C and then a solution of 110 parts of sodium hydroxide dissolved in 225 parts of water is added; the mixture is decanted and the upper layer is separated. This is distilled in vacuum and the excess ethyl m-toluidine is recovered followed by N-ethyl N-(2'-phenoxy ethyl) m-toluidine which distills in a vacuum of 1.5 mm Hg at 172° to 173°C (uncorrected). 531 parts of this product are obtained which solidifies after cooling. For the sulfonation 196 parts of the amine so prepared are added at between 50° and 60°C to 174 parts by volume of 66°Bé sulfuric acid.

This mixture is then heated to 90°C, maintained for ½ hour at that temperature, poured into 2,000 parts of water, neutralized with approximately 540 parts by volume of a 40% solution of sodium hydroxide, cooled to 30°C and precipitated by the addition of 700 parts of sodium chloride. The solid is filtered at that temperature, washed with a solution of 30% sodium chloride and dried at 50°C. A white powder is obtained which contains a little sodium chloride.

Example 3: 65 parts by volume of 66°Bé sulfuric acid are added to a suspension of 27.5 parts of 6-ethoxy-2-amino-benzothiazole 91.4% in 80 parts of water. The mixture is cooled down to 0°C and in 30 minutes 25 parts by volume of 2 N sodium nitrite are added. Stirring is kept up for 1½ hours at between 0° and 5°C. 1 part of sulfamic acid is added and then gradually 25 parts of the sodium salt of N-ethyl N-(2'-p-sulfophenoxy ethyl) m-toluidine prepared as in Example 2.

Stirring is continued for 1 hour at 0°C then overnight at ordinary room temperature. The mixture is poured onto ice water, stirred for 1 hour, the pH is adjusted to 8 by means of sodium hydroxide solution, filtered and dried at 50°C. 30 parts of a violet powder are obtained which dyes polyamide fabrics such as nylon fabric in a red shade which is fast to washing and to light.

Transparent Copolyamides

K. Moriyama, S. Kawase and T. Shima; U.S. Patent 4,012,365; March 15, 1977; assigned to Teijin Limited, Japan provide polyamides which are transparent while retaining their high melting points and softening points and excellent solvent resistance, especially excellent resistance to alcohols.

The fiber-forming polyamides are prepared by polycondensing naphthalene-2,7-dicarboxylic acid and a straight-chain aliphatic diamine containing 4 to 13 carbon atoms and 15 to 50 mol % of another carboxylic acid and/or another diamine as a copolymerizing component. These polyamides can be melt-shaped into fibers and films having a high Young's modulus.

In the following example LOI (limiting oxygen index), resistance to ethanol, specific viscosity and cloudiness were measured by the following methods:

(1)   LOI—measured in accordance with the method of ASTM D-2863.

(2) Resistance to ethanol—the appearance of a test sample was tested in accordance with the method of ASTM D-543-67, and evaluated on the following scale: 0, no change in appearance after immersion for 7 days; x, swelled, became cloudy, or dissolved after immersion for 7 days.

(3) Specific viscosity—400 mg of a polymer was dissolved in 100 cc of concentrated sulfuric acid at 80°C over the course of 1 hour. The solution was cooled, and then the viscosity was measured at 35°C.

(4) Cloudiness—measured in accordance with the method of ASTM D-1003-61.

Example: Equimolar proportions of hexamethylene diamine and naphthalene-2,7-dicarboxylic acid were dissolved in water to form a salt, and then methyl alcohol was poured into the solution to form a white powder of hexamethylene diammonium naphthalene-2,7-carboxylate (abbreviated as 6-N salt). The salt did not contain water of crystallization.

The 6-N salt and hexamethylene diammonium isophthalate (abbreviated as 6-I salt) were charged in the proportions of 0.6:0.40 mol and polymerized.

An autoclave equipped with a stirrer was charged with 1 mol in total of the 6-N salt and the 6-I salt, 0.005 mol of phosphorous acid (stabilizer) and 0.01 mol of sebacic acid (for adjusting the degree of polymerization), and after being purged with nitrogen, sealed. Then, the internal pressure of the autoclave was reduced to normal atmospheric pressure over the course of 1 hour, during which time the temperature was raised to 290°C. Purified nitrogen gas was passed through the autoclave at a rate of 1 l/min, and the contents were heated at 290°C for 1 hour with stirring to complete the polymerization. The resulting polymer was cooled and solidified in water. Test results of this and a comparative example similarly prepared but omitting the 6-I are shown below.

	Example	Comparative Example
Vicat softening point, °C	154	158
Specific viscosity	1.06	0.95
Second order transition point, °C	129	131
Crystallinity	none	crystalline
Resistance to ethanol	0	*

*The polymer was already cloudy before testing because of the crystallization of the polymer.

Polyester-Amides of Superior Whiteness

Moldable polyester-amides have been developed by *M. Ducarre; U.S. Patent 4,116,943; September 26, 1978; assigned to Rhone-Poulenc-Textile, France* which are characterized in that they are derived from (1) an aromatic organic diacid whose carboxyl groups are directly attached to an aromatic ring in the para-position, (2) a primary diol of the general formula $HO-CH_2-R-CH_2-OH$, in which R represents either a linear aliphatic chain having at least 4 carbon atoms, optionally branched with alkyl or aryl groups, or an aromatic ring, and (3) a salt of an organic diacid and a primary diamine of aliphatic character. The polymers have a degree of whiteness greater than or equal to 80% and a luminosity greater than or equal to 35%, as defined in standard specification ASTM E 308-66.

They have a number of terminal $-NH_2$ groups, per ton of polymer, equal to or greater than 30, preferably greater than 40. Preferably, they have a melting point equal to or greater than 180°C, in order to be usable in the field of synthetic textiles. In the examples below all parts are expressed by weight.

Example 1: 1,494 parts of terephthalic acid, 2,124 parts of 1,6-hexanediol, 846 parts of the salt obtained from terephthalic acid and hexamethylenediamine, and 0.6 part of triethanolamine orthotitanate are introduced simultaneously into an autoclave equipped with the usual means for heating and regulation.

The temperature is raised progressively to 235°C and at the end of 1 hour 30 minutes, the full theoretical quantity of water formed is collected.

A further 0.6 part of triethanolamine orthotitanate is added and the pressure is progressively reduced to 0.5 mm Hg over a period of 1 hour, while the temperature is raised to 267°C, which removes the excess hexanediol. The polycondensation is carried out over a period of 3 hours at 267°C under 0.5 mm Hg. A white polymer is obtained which is extruded in the form of strands and then ground into granules. Its properties are as follows:

Ratio of ester units/amide units	75/25
Intrinsic viscosity*	0.79
Viscosity in the molten state at 260°C	1,540 poises
Number of $-COOH$ end groups per ton of polymer	54
Number of $-NH_2$ end groups per ton of polymer	59
Softening point	216°C
Degree of whiteness**	85%
Luminosity	51.4%
Dominant wavelength**	573.6 nm

*Measured on a 0.5% strength solution in a 60/40 mixture of phenol/tetra-chloroethane, at 25°C.
**As defined in standard specification ASTM E 308-66.

The polyester-amide thus prepared is melt-spun through a spinneret comprising 7 orifices of 0.34 mm diameter.

The polymer, which melted at 235°C, is spun at 250°C. The filaments are then stretched to a ratio of 2.95, over a finger at 40°C and a plate at 85°C, at a speed of 150 m/min.

The yarns thus obtained have the following characteristics: gauge per strand 2.22 dtex, elongation 16.7%, tenacity 25 g/tex and modulus of elasticity 350 g/tex.

The stretched yarns, when examined by means of x-rays, show a high degree of crystallinity.

Example 2: The same raw materials are used as in Example 1 except that the 1,6-hexanediol is replaced by neopentyl glycol (as proposed in French Patent 2,193,845).

An autoclave identical to that used in Example 1 is charged with 498 parts of terephthalic acid, 624 parts of neopentyl glycol (2,2-dimethyl-propanediol), 282 parts of the salt derived from terephthalic acid and hexamethylenediamine, 0.19

part of triethanolamine orthotitanate, 0.27 part of phosphorous acid, and 0.30 part of triphenyl phosphite.

The phosphorous acid and the triphenyl phosphite are used in French Patent 2,193,845, in order to avoid the coloration which can be produced by the neopentyl glycol. The method of working is exactly the same as in Example 1.

A highly colored polymer with intrinsic viscosity 0.83 (measured as above) is obtained with other properties as follows: degree of whiteness 40%, luminosity 21.3%, dominant wavelength 579.6 nm and softening point 140°C.

Given the amorphous character of the polymer, the melting point cannot be determined with precision, and only the softening point can be measured (with a penetrometer).

From this experiment it is clearly evident that the substitution of neopentyl glycol for 1,6-hexanediol results in a yellow coloration and too low a softening point, which characteristics rule out any use in the field of textiles.

IMPROVEMENT IN OTHER PROPERTIES

Light Resistance, Flexibility, and Wear Resistance

T. Konomi, S. Endo, M. Yamaguchi and K. Katsuo; U.S. Patent 4,072,664; February 7, 1978; assigned to Toyobo Co., Ltd., Japan have developed aromatic polyamides which are improved in light resistance, flexibility and wear resistance while retaining excellent strength, rigidity and heat resistance inherent to conventional all-aromatic polyamides.

An N,N'-bis(p-aminobenzoyl)ethylenediamine unit is introduced in a polyamide chain mainly consisting of a unit derived from terephthalic acid and a unit derived from p-phenylenediamine.

Example 1: *Preparation of N,N'-Bis(p-Nitrobenzoyl)Ethylenediamine* — In a 2-liter flask equipped with a reflux condenser, a thermometer, a stirrer and an inlet for reagents, the atmosphere in the flask being replaced by dry nitrogen to obtain an anhydrous state, dried benzene (1,000 ml), ethylenediamine (30 g; 0.5 mol) and triethylamine (111 g; 1.1 mol) are charged, and the resultant uniform solution is cooled at 5° to 10°C. A solution of p-nitrobenzoyl chloride (185 g; 1.0 mol) in benzene (500 ml) is added thereto in 60 minutes with vigorous stirring.

After stirring is continued for 2 more hours, the reaction mixture is filtered, and the collected precipitate is washed with methanol, water and methanol in order in a mixer and then dried. Recrystallization from dimethylacetamide (1.5 liters) affords white needles (143 g) melting at 269° to 271°C. Yield, 80%.

This substance is identified to be N,N'-bis(p-nitrobenzoyl)ethylenediamine by the IR absorption spectrum and the elementary analysis (N: found, 15.68%; calculated, 15.64%).

Example 2: In a 300-ml volume flask equipped with a stirrer, a thermometer, an inlet for reagents and an inlet for dry nitrogen, the atmosphere in the flask being replaced by dry nitrogen to obtain anhydrous state, hexamethylphos-phoramide (23 ml) and N-methyl-2-pyrrolidone (46 ml) are charged, and N,N'-bis(p-aminobenzoyl)ethylenediamine (2.983 g; 0.01 mol) and p-phenylenedia-mine (1.081 g; 0.01 mol) are portionwise added thereto under heating. The resultant uniform solution is cooled to 2°C and fine powders of terephthalyl di-chloride (4.060 g; 0.02 mol) are added thereto all at once while stirring whereby the polymerization reaction takes place immediately. Stirring is continued for about 10 minutes, and the reaction mixture is allowed to stand overnight. The product is washed with water and crushed into powders to obtain a polyamide with a logarithmic viscosity of 2.903.

Example 3: The same procedure as in Example 2 is repeated but using N,N'-bis(p-aminobenzoyl)ethylenediamine (0.895 g; 0.003 mol), p-phenylenediamine (1.838 g; 0.017 mol), terephthalyl dichloride (4.060 g; 0.02 mol), hexamethyl-phosphoramide (19 ml) and N-methyl-2-pyrrolidone (38 ml) to obtain a poly-amide showing a logarithmic viscosity of 3.756.

Example 4: A polyamide obtained as in Example 3 (logarithmic viscosity, 4.16) is dissolved in 100% sulfuric acid while stirring under nitrogen stream at 85°C in 5 hours, and the resultant dope (concentration of polyamide, 20% by weight; viscosity of dope, 1,360 poises) is, after filtration and defoaming, extruded through a spinneret (hole diameter, 0.08 mm; number of holes, 7) at a rate of 1.6 ml/min by a sending gear pump under a pressure of nitrogen of about 4 kg/cm^2 by way of a layer of air (thickness, about 7 mm) into a coagulating bath of water of 3°C and taken up at a rate of 200 m/min. The spinnability is extremely good. The taken-up product on a bobbin is immersed in water overnight and dried at 80°C for 4 hours in a hot air drier.

For comparison, a dope (concentration of polyamide, 20% by weight) prepared from poly(p-phenyleneterephthalamide), logarithmic viscosity, 5.07, is extruded through a spinneret (hole diameter, 0.08 mm; number of holes, 10) by way of a layer of air into a coagulating bath of water of 2°C and taken up at a rate of 200 m/min to obtain a fiber.

As shown in Table 1, the fiber prepared from the polyamide of the process pos-sesses excellent physical properties and shows a high knot strength.

Table 1

	Example 4	Poly(p-Phenylene-terephthalamide)
Fineness, d	4.02	2.14
Dry strength, g/d	20.26	23.4
Elongation, %	4.8	3.6
Initial modulus, g/d	453	640
Knot strength, g/d	6.54	4.96

Example 5: The same procedure as in Example 4 is repeated but adopting a taking-up rate of 180 m/min to obtain a fiber showing a fineness of 6.01 d, a dry strength of 20.67 g/d, an elongation of 5.2%, an initial modulus of 420 g/d and a knot strength of 6.73 g/d. The fiber is subjected to a heat resistance test, and exhibits a heat resistance almost equal to the fiber prepared from poly-(p-phenyleneterephthalamide).

Example 6: The fiber used in Example 5 is subjected to a light resistance test. As shown in Table 2, the fiber prepared from the polyamide of the process is superior to the fiber prepared from poly(p-phenyleneterephthalamide) in the strength maintaining degree after irradiation of 100 hours by a fade meter.

Table 2

	Example 5	Poly(p-Phenylene-terephthalamide)
Dry strength, g/d	69.1	50.1
Elongation, %	70.2	54.3
Initial modulus, g/d	93.2	91

Good Melt Processability and High Impact and Flexural Strengths

B.K. Onder; U.S. Patent 4,072,665; February 7, 1978; assigned to The Upjohn Company describes copolyamides having the recurring unit

$$-\overset{O}{\underset{||}{C}}-R-\overset{O}{\underset{||}{C}}-NH-Ar-NH-$$

wherein R, in 60 to 85% of the recurring units, is $+CH_2\frac{1}{x}$ wherein x is an integer from 7 to 12 inclusive, and in 15 to 40% of the recurring units, R is m-phenylene; and Ar is an arylene radical. The copolyamides are characterized by ease of melt-processing such as in molding, extruding, and injection molding, while at the same time possessing good high temperature stability, and being further characterized by unexpectedly high impact and flexural strengths.

Example: A 2,000-ml resin flask equipped with a mechanical stirrer, addition funnel, thermometer, a nitrogen inlet tube (entering below the surface of the reaction mixture), and a reflux condenser equipped with a gas outlet tube, was thoroughly dried by heating with an electric heat gun while under a constant flow of dry N_2 gas.

After cooling to room temperature, 158.63 g (0.8 mol) of 4,4'-methylenedianiline (freshly distilled, BP 200°C/1.5 mm Hg) was charged to the flask along with 900 g of dimethylacetamide (DMAc), distilled from calcium hydride and stored over 3 A molecular sieves, and the mixture stirred to form a slightly colored solution. It was cooled to −20°C by a dry ice-acetone bath.

The addition funnel was charged with a mixture of 126.07 g (0.56 mol) of azelaoyl chloride, 48.73 g (0.24 mol) of isophthaloyl chloride, 2.75 g (0.01 mol) of palmitoyl chloride (used as a chain capping agent), and 75 g of dimethylacetamide. The contents of the addition funnel were added dropwise during continual stirring and cooling, over a period of 11 minutes whereupon the temperature rose from −20 to 0°C. After about 10 minutes, the solution turned milky due to the formation of the DMAc·HCl salt. A rinse of 125 g of DMAc was added from the addition funnel to yield a solution of about 20% solids. The reaction mixture was stirred overnight and allowed to return to room temperature.

The polymer solution was neutralized by slowly pouring it into 3 gal of water in which 169.6 g (1.6 mol) of Na_2CO_3 (anhydrous) had been dissolved. The polymer precipitated as rope-like strands. It was pulverized in a Waring blender with water, filtered, washed again by stirring with 3 gal of water containing 106 g (1.0 mol) of Na_2CO_3, filtered, and washed twice with fresh water, or until the wash water was neutral to pH test paper. The polymer was filtered in a coarse porosity funnel and dried initially at 100° to 105°C overnight in a circulating oven. Final drying was accomplished at 140° to 145°C under vacuum of about 0.1 mm for 8 hours. The copolyamide had an η = 1.01 at 30°C (0.5% in m-cresol) and corresponded to the following structure

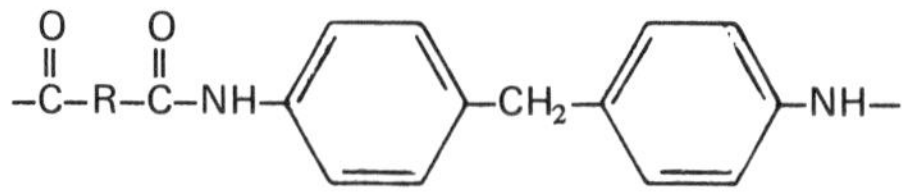

wherein R in about 70% of the recurring units is $+CH_2\frac{}{}_7$ and in about 30% of the recurring units is

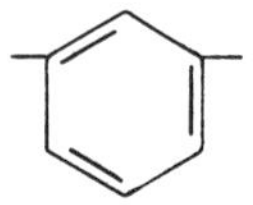

The powdered copolyamide was dry-blended with 1% by weight of N,N'-bis[3-(3',5'-di-tert-butyl-4'-hydroxyphenyl)propionyl]hexamethylenediamine (Irganox-1098, an antioxidant), then extruded into smooth rods using a Brabender Plasti-Corder (Brabender Instruments Inc.). Extrusion conditions were as follows, screw speed, 40 rpm; die size, ¼ diameter; temperature conditions: zone 1, 270°C; zone 2, 270°C; zone 3, 260°C; and zone 4, 250°C.

Test bars were prepared by placing a 5-inch length of the extruded rod (approximately 10 g) in an ASTM ½ x 5 inch bar mold preheated to 180°C and compression molding the sample at 180°C and 4,000 psi. The following physical properties represent the average values derived from the measurements of five different test bars.

Tensile strength at yield (psi)	11,400
Tensile modulus (psi)	167,000
Elongation (%)	21.9
Izod impact strength (ft-lb/in notch)[*]	3.19
Heat deflection temperature at 264 psi[**]	140°C

[*]ASTM Test Method D256-56
[**]ASTM Test Method D648-56

Good Melt Processability with Versatility of Fabrication

Copolyamides having the recurring unit

are disclosed by *P.S. Andrews, W.J. Farrissey, Jr., B.K. Onder and J.N. Tilley;*

U.S. Patent 4,065,441; December 27, 1977; assigned to The Upjohn Company wherein x, in 50 to 85% of the recurring units, is an integer from 6 to 10 inclusive, and, in 15 to 50% of the recurring units is 4, and Ar is an arylene radical. The copolyamides are characterized by ease of melt-processing such as in molding, extruding, and injection molding, while at the same time possessing good physical properties, and being further characterized by their ease of conversion from the amorphous to the crystalline state which, in turn, gives rise to a high degree of versatility in their fabrication.

Example 1: A 1,000-ml resin flask was equipped with a stainless steel motor driven stirrer, a nitrogen inlet tube, and a side-arm distillation condenser leading to a 250-ml collection flask. The system was flamed out under vacuum and flushed several times with dry N_2. The flask was charged with a preblended mixture consisting of 176.46 g (0.9375 mol) of azelaic acid (recrystallized three times from chloroform), 47.50 g (0.325 mol) of adipic acid (99%+ purity), 247.50 g (1.25 mols) of 4,4'-methylenedianiline (recrystallized two times from hot benzene under N_2), and 2.13 g of N,N'-bis[3-(3',5'-di-tert-butyl-4'-hydroxyphenyl)propionyl]hexamethylenediamine (Irganox-1098, an antioxidant). The mol ratio of dicarboxylic acids to the diamine represented a 1% excess.

The resin flask was lowered into an oil bath at 250°C and stirring of its contents begun at a low rate of about 4 rpm, under N_2 while at the same time the oil bath temperature was slowly raised. After about 13 minutes, the reactants were completely melted and beginning to react. Thereafter the temperature of the bath was controlled to about 288°C and stirring increased to 20 rpm. Over 2½ hours reactant viscosity increased which was reflected in the stirrer rpm dropping to 10. Vacuum was applied (1.3 mm Hg) with a nitrogen bleed while liquid condensate collected in the collection flask. Vacuum and stirring were maintained for 1¾ hours at a bath temperature of about 290°C.

The vacuum was released and the very viscous reaction mixture maintained under N_2 while the oil bath was allowed to cool slowly. At about 210°C, crystallization of the flask contents began. The collection flask contained 42.25 g of water (theory, 45.0 g). When the temperature had cooled to about 150°C, the resin flask was removed from the oil bath, cooled, and the contents provided 413 g of product. It was ground to a particle size of about 2 mm in a Wiley mill and was characterized by an $\eta = 0.75$ at 30°C (0.5% in m-cresol); and a Tg = 110°C followed by a double melting at 236° and 252°C as determined by DTA (differential thermal analysis) using a DuPont 900 Thermal Analyzer instrument, under N_2 at a rate of 20°/min, and corresponded to the following structure

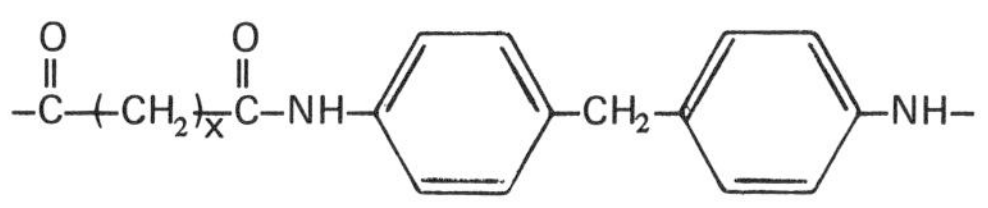

wherein x in 75% of the recurring units is 7 and in the remaining 25% is 4.

The powdered copolyamide was then extruded into smooth rods using a Brabender Plasti-Corder (Brabender Instruments, Inc.). Extrusion conditions were as follows: screw speed, 40 rpm; die size, ¼ inch diameter; and reverse temperature conditions which ranged from 270°C in the first zone to 220°C in the extruder nozzle. The clear transparent amorphous polymer was then simply molded under one of the three different sets of conditions set forth below to provide: (a) amorphous, (b) amorphous-annealed, and (c) crystalline copolyamides in accordance with the process.

Example 2: *Amorphous Polymer* — A series of test bars were prepared by placing a 5-inch length of the extruded rod (about 10 g) in an ASTM ½ x 5 inch bar mold preheated to 150°C and compression molding the sample at 150° to 155°C under about 4,000 psi and thereafter allowing the mold to cool to about 90°C over a period of about 1 hour to 1½ hours, or force cooling, by using a fan and cooling to the same temperature over about 45 to 50 minutes. The bars were demolded and were characterized by the average physical properties set in the table below.

The heat deflection data (HDT) was determined on four different test bars starting with the first molded at 150°C in accordance with the conditions set forth above for obtaining amorphous polymer. The second, third, and fourth bars were obtained by taking three bars already molded at 150°C and remolding them at 160°, 170°, and 180°C, respectively. The increase in HDT from 124° to 145°C is due to the copolyamide being converted from an amorphous polymer in the first bar, to a crystalline material in the fourth bar.

Example 3: *Amorphous-Annealed Polymer* — Test bars prepared in accordance with the procedure described in Example 2 above were annealed by storage in an oven at 102°C over a period of 24 hours.

Example 4: *Crystalline Polymer* — Test bars prepared in accordance with the procedure described in Example 2 above were remolded in the ASTM ½ x 5 inch mold at 180°C and 4,000 psi. They were allowed to cool and after 1 hour and 10 minutes, at about 110°C, were demolded and were now completely opaque whereas originally they were clear to translucent.

	. . . . . . Example Number		
Properties	2	3	4
Tensile strength at yield, psi	3,770	4,040	1,970
Tensile modulus, psi	138,900	160,800	200,000
Elongation, %	3.02	3.15	1.1
Izod unnotched impact strength, ft-lb/in*	3.24	3.2	1.4
Gehman Tg, °C**	120	–	207
DTA Tg, °C***	115	–	–
Heat deflection temperature, °C, at 264 psi†			
Bar molded at 150°C	124	–	–
Bar molded at 160°C	120	–	–
Bar molded at 170°C	144	–	–
Bar molded at 180°C	145	–	–

   *ASTM Test Method D256-56.
  **Glass transition temperature determined using the procedure of ASTM D 1053-58T
      on a modified Gehman Torsion Stiffness Tester, fitted with a heavy duty furnace
      to allow operation up to 500°C.
 ***Glass transition temperature determined by differential analysis using a Dupont
      900 Thermal Analyzer DTA.
 †ASTM Test Method D648-56.

Heat and Chemical Resistance with Improved Solubility

*T. Konomi, K. Yukimatsu, K. Katsuo and M. Yamaguchi; U.S. Patent 4,011,203;
March 8, 1977; assigned to Toyobo Co., Ltd., Japan* have developed a process
for producing aromatic polyamides having excellent heat resistance, toughness
and chemical resistance and further having improved solubility, molding properties and light resistance, which comprises polycondensing a mixed diamine consisting of piperazine and p-phenylenediamine with a substantially equimolar
amount of terephthaloyl dihalide in at least one amide type polar solvent (e.g.,
hexamethylphosphoramide, N-methyl-2-pyrrolidone, alone or a mixture thereof),
where the piperazine is contained in a ratio of 10 to 35% by mol on the basis
of the whole mixed diamine component, and fibers produced from the aromatic
polyamides having high tenacity, flexing characteristics and wear resistance and
further having improved elongation ratio, knot tensile strength, fatigue resistance
and fibrillation resistance.

Example 1: A 1-liter glass-made reactor provided with a stirrer, an inlet for introducing dry nitrogen and an outlet thereof, an inlet for adding terephthaloyl
dihalide and a thermometer is substituted with dry nitrogen and thereby is made
anhydrous. The reactor is charged with powdery p-phenylenediamine (4.596 g;
0.0425 mol), anhydrous piperazine (0.646 g; 0.0075 mol), hexamethylphosphoramide (50 ml) and N-methyl-2-pyrrolidone (90 ml) while passing through a small
amount of dry nitrogen. The mixture is stirred at room temperature to give a
homogeneous diamine solution.

After cooling the diamine solution on an ice bath to 4°C, to the solution is added
under stirring at once fine powdery terephthaloyl dichloride (10.152 g; 0.05 mol)
by using a funnel with a wide neck and then the remaining terephthaloyl dichloride adhered to the funnel is rapidly flowed down with N-methyl-2-pyrrolidone
(10 ml).

After the addition of terephthaloyl dichloride, the temperature of the reaction
mixture is immediately raised, and after 5 minutes, a clear, viscous and homogeneous dope is obtained. The reaction mixture is stirred and washed with a high-
speed mixer together with a large amount of water to give a fine powdery polymer, which is separated by filtration, washed with water and further with alcohol
and then dried at 100°C for 16 hours in a vacuum.

The copolymer thus obtained starts to decompose at 460°C in an atmosphere of
argon and has an inherent viscosity of 3.08.

Example 2: To the copolymer obtained in the same manner as in Example 1
is added 98% concentrated sulfuric acid and the mixture is stirred at room temperature for 3 hours to give a dope having a concentration of the polymer of
12% by weight and a viscosity of about 1,800 poises. After deaerating, the dope
is filtered by two filters having 400 mesh. The dope is extruded through a spinneret having 50 orifices under a pressure of nitrogen gas of 3.5 kg/cm^2 into a
solidifying medium (water) of 20°C, washed water at the second bath and then
wound up on bobbins.

On the other hand, in the same manner as described above, poly(p-phenyleneterephthalamide) having an inherent viscosity of 3.01 is spun.

The characteristics of the filaments are shown below:

	Copolymer	Poly(p-Phenylene-terephthalamide)
Denier	4.43	4.46
Tenacity in dry state, g/d	3.81	4.57
Elongation ratio, %	12.8	10.4
Initial modulus, g/d	171.3	193.2
Knot tensile strength	1.41	1.01

Example 3: The copolymer having an inherent viscosity of 3.09 obtained in the same manner as in Example 1 is dissolved in 100% concentrated sulfuric acid at 85°C to give a dope having a concentration of the polymer of 20% by weight and a viscosity of 1,100 poises. The dope is extruded through a spinneret having 10 orifices of 0.08 mm in diameter at a speed of 1.6 m/min into a solidifying medium (cold water) of 5°C via an air zone of about 7 mm in width and then wound up on bobbins at various speeds. The filaments wound on the bobbins are dipped in water overnight and dried at 80°C for 4 hours.

In the same manner as described above, poly(p-phenyleneterephthalamide) having an inherent viscosity of 5.07 is extruded into a solidifying medium (cold water) of 2°C and wound up on bobbins at a speed of 200 m/min.

Example 4: On the product of Example 3 (winding speed 150 m/min), the heat resistance and light resistance are tested. The fibers obtained from the copolymer have a similar heat resistance to that of the conventional wholly aromatic polyamide at high temperatures and are superior in the light resistance in comparison with the latter.

Example 5: On the fibers obtained in Example 3, the wear resistance and twist resistance are tested. The fibers obtained from the copolymer are superior in comparison with the conventional wholly aromatic polyamide.

Good Mechanical Properties, Thermal Stability, and Fire Retardancy

Y. Nakagawa, T. Noma and H. Mera; U.S. Patent 4,018,735; April 19, 1977; assigned to Teijin Limited, Japan describe an anisotropic aromatic polyamide dope comprising more than 10% by weight of an aromatic polyamide and a solvent for the polyamide, the polyamide consisting essentially of, at least 5 mol %, but less than 35 mol %, based on its entire units, of an only partially p-oriented structural unit expressed by the following formula

$$-\text{HN} \overset{\displaystyle N}{\underset{\displaystyle X}{\bigcirc\!\!\!\!\bigcirc}}\!\!-\!C-Ar^1-NH-$$

wherein Ar^1 is a divalent aromatic ring whose chain-extending bonds are coaxial or parallel, X is a member selected from the group consisting of O, S and NH, and the NH group bonded to the benzene ring of the above benzoxazole, benzothiazole or benzimidazole ring is meta or para to the carbon atom of the benzene ring.

The resulting polyamide dope has superior stability, flowability and spinnability, and can be formed into fibers or films having good mechanical properties, thermal stability and fire retardancy, by being extruded, for example, through a spinning orifice or a slit.

Example 1: *Preparation of Polymer* — 2.25 g (0.01 mol) of 5-amino-2-(p-amino-phenyl) benzoxazole and 1.08 g (0.01 mol) of p-phenylenediamine were dissolved in 60 ml of dry N-methylpyrrolidone and 30 ml of dry hexamethylphosphonamide in a stream of nitrogen, and the solution was cooled to -20°C. This solution was stirred vigorously, and 4.04 g (0.02 mol) of powdery terephthaloyl chloride was added at a time to the stirred solution. The reaction mixture became clear, then gradually became highly viscous, and in about 15 minutes assumed a clear jelly-like form. 400 ml of N-methylpyrrolidone was added to the jelly-like reaction mixture, and the mixture was stirred at room temperature for about 6 hours to form a flowable solution which was then reprecipitated from about 2 liters of water.

The precipitate was filtered, washed with water and dried to afford a polymer of the following structure having an inherent viscosity of 5.85.

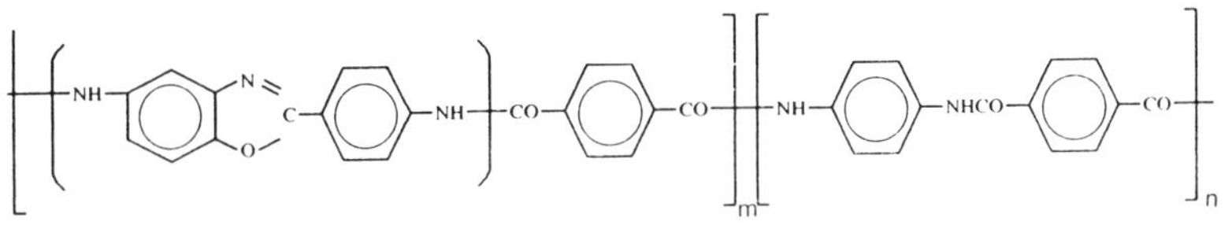

(m/n = 50/50)

Preparation of a Dope and Its Properties — The aromatic polyamidebenzoxazole powder (with an inherent viscosity of 5.85) and 100% sulfuric acid were mixed with vigorous stirring at 0°C so that the polymer concentration of the resulting dope ranged from 12 to 25%. The viscosity of the resulting dope at 40°C was measured by a falling-ball method. The dope was observed under a polarizing microscope, with all samples found to be anisotropic.

After storing each of these 12 to 25% solutions at 40°C for 5 hours, the polymer was collected by reprecipitation, and its inherent viscosity was measured. All of the polymers collected were found to have an inherent viscosity of 5.82 to 5.87, showing no reduction in the degree of polymerization.

Example 2: 2.0 g of a powder of poly(p-phenylene terephthalamide) with an inherent viscosity of 5.41 was vigorously mixed with 80 g of 100% sulfuric acid at 0°C, and the mixture was heated to 40°C. The mixture was still nonuniform. When this mixture was heated to 90°C, its flowability and spinnability became superior, and it became somewhat optically anisotropic. However, the viscosity of the dope gradually decreased. After storing the dope for 5 hours at 90°C, the polymer was collected by reprecipitation. The polymer collected had an inherent viscosity of 3.58.

Low Temperature Moldability

According to *S. Minami, H. Kakida and J. Nakauchi; U.S. Patent 4,081,430; March 28, 1978; assigned to Mitsubishi Rayon Co., Ltd., Japan*, an aromatic polyamide crystalline complex capable of being molded at low temperatures is produced by contacting an aromatic polyamide in which at least 75 mol %

of repeating unit comprises m-phenyleneisophthalamide with at least one member selected from the group consisting of hexamethylphosphoric amide, N-methyl-2-pyrrolidone, ϵ-caprolactam and 2-pyrrolidone to form a crystalline complex and then removing the hexamethylphosphoric amide, N-methyl-2-pyrrolidone, ϵ-caprolactam and 2-pyrrolidone which are not contained in the crystal.

Example: Poly-m-phenyleneisophthalamide having a relative viscosity of 3.2 was dry spun and then the resultant fibers were drawn to 4.0 times in boiling water. Furthermore, they were additionally drawn to 1.1 times to obtain drawn fibers (size 2 d, strength 5.0 g/d and elongation 28%). These fibers were cut to 51 mm and a web was formed in carding process.

Separately, the poly-m-phenyleneisophthalamide powder was immersed in hexamethylphosphoric amide (HMPA) at HMPA/polymer = 2 (weight ratio) at 50°C for one hour. Thereafter, the product was filtered off and dried in vacuum at 80°C for 5 hours to obtain powder of crystalline complex of the poly-m-phenyleneisophthalamide and HMPA, which had a melting point of 120°C measured from the differential scanning calorimetry curve and an average particle size of 0.2 mm.

The powder of crystalline complex was uniformly added on the web and this was subjected to calender treatment at 100 kg/cm and 140°C, then treated with steam of 1 kg/cm³ gauge for 3 minutes and thereafter dried. The thus obtained nonwoven fabric which had the properties as shown in the table below had smoothness and self-extinguishing properties, and its strength and elongation did not change even after treatment at 250°C for 5 hours.

Weight of Complex Compound (%)	Basis Weight (g/m²)	Tensile Strength (kg/50 mm)	Elongation (%)
5	27	5.0	20
10	30	7.2	16
30	33	10.1	10
50	37	20.3	7

Chemical Resistance, Low Water Absorption, and Thermal Stability

G. Corrado; U.S. Patent 4,115,370; September 19, 1978; assigned to Snia Viscosa Societa' Nazionale Industria Applicazioni Viscosa SpA, Italy describes polyester-amide polymeric materials containing double bonds which are capable of cross-linking with vinyl or allyl monomers and to which the concurrent presence of ester and amide groups imparts particular characteristics.

Some properties are:

(a) high elongation at break under traction, often higher than that of polyesters having equal bending elasticity modulus,

(b) resilience coupled with a good elongation under traction,

(c) low density,

(d) general chemical resistance which is noteworthy even for highly flexible resins,

(e)   low water absorption, and

(f)   linear thermal dilatation coefficients which are high with respect to the values found for polyester resins.

If the above characteristics are taken into consideration it becomes evident that the uses to which the polyesteramides may be destined are numerous. This is particularly true in view of the fact that the resins cover a wide range of mechanical characteristics going from rigid resins through semirigid to flexible resins. The good chemical resistance coupled with the low water absorption and the resilience suggests the use of some of the resins in question in the field of varnishes and gel coats. The resilience and the thermal behavior direct some of these polyesteramide resins in particular to uses in the encapsulation of electric circuits and their support plates, this especially considering their good electrical characteristics.

The polyesteramide resins having particular resilience, chemical resistance and low water absorption may further be found excellent for the production of laminates and other structural products. Similarly the resilience, the chemical resistance and the low water absorption coupled to a good flexibility make the polyesteramides suitable for mixing with bisphenol or isophthalic polyester resins as particularly interesting agents for making the resins flexible. The rigid polyesteramide resins, on the other hand, may find use in mixtures with solid monomers of the type of diacetoacrylamide and of triallylcyanurate for the purpose of obtaining particularly interesting resins usable for hot molding.

The advantages resulting from the low density of the hardened polyesteramides should also be taken into consideration. This feature permits the obtaining of lighter products than those obtained from polyesters when they are used in particular cases, such as the manufacture of floating hulls.

The unsaturated linear polyesteramide is characterized in that it comprises, in its linear chain, units which in turn comprise a base sequence including a radical of a COOH-terminated saturated polyamide, a glycol radical, at least a radical of an α,β-unsaturated bicarboxylic acid, and once again a glycol radical.

Example: In a reaction flask provided with a mechanical stirrer, a descending cooler and a draw pipe, a mixture of 348 g of fumaric acid and 637 g of diethylene glycol with the addition of a small amount (about 0.1%) of an esterification catalyst, is slowly heated to 180°C in an inert gas atmosphere and while distilling the reaction water. After about three hours of heating at the same temperature (180°C), when the acidity has reached the value zero, the reaction is stopped. 220 ppm of hydroquinone are added and a di-diethylene-glycol-fumarate is thus obtained which may be stored for a long time.

In a reaction vessel provided with a stirrer, a descending cooler and a draw pipe, a mixture of 181 g of Empol 1014 very low chain acids (Unilever) and 25 g of 2,4,4-trimethyl-1,6-diaminohexane, with the addition of 0.1% of a catalyst, are slowly heated to 160°C in an inert gas atmosphere and while distilling the reaction water. After about two hours at 160°C, when the amine number has reached the value zero, 47 g of the di-diethylene-glycol-fumarate prepared above are slowly added at that temperature, under stirring and in an inert gas atmosphere. The temperature is brought to 200°C and the mixture is left under such conditions until, after about three hours, the acid number has reached the value 20. In the last half hour of reaction a vacuum of 0.2 mm Hg has been applied.

An unsaturated polyesteramide having a reddish color, which is very viscous and has an acid number 20 and amine number 0, is thus obtained.

Dimensionally Stable Copolyamides

According to a process developed by *J.S. Ridgway; U.S. Patent 4,024,114; May 17, 1977; assigned to Monsanto Company* polyamides derived from cyclohexanebis(ethylamine) have exceptional stability of modulus to heat and moisture, i.e., dimensional stability. Particularly useful polyamides of this class are those prepared from trans-1,4-cyclohexanebis(ethylamine) and unbranched saturated aliphatic dicarboxylic acids.

The polyamides are characterized in that they contain repeating units of the structure

$$-NHCH_2CH_2 \!-\!\!\!\bigcirc\!\!\!-\! CH_2CH_2NH-$$

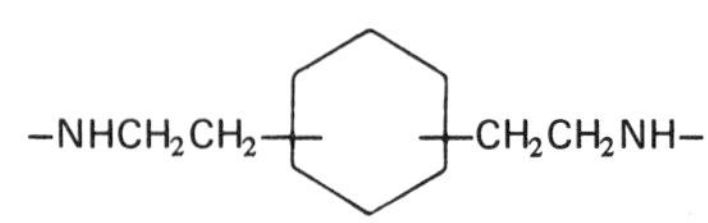

and preferably repeating units of the structure

$$-NHCH_2CH_2 \!-\!\!\!\bigcirc\!\!\!-\! CH_2CH_2NH-$$

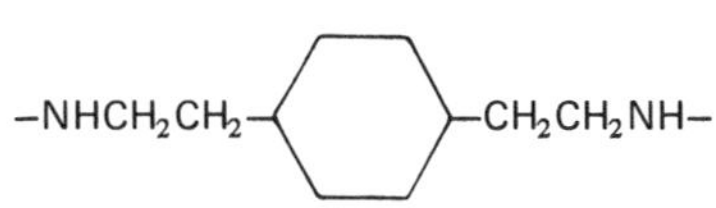

wherein the polyamide is formed from polyamide-forming components having terminal amide-forming groups joined by a divalent hydrocarbon radical having at least 4 carbon atoms.

Example: Equal molar amounts of trans-1,4-cyclohexanebis(ethylamine) and dodecanedioic acid were reacted in water at about 50°C to form aqueous solutions of the salt of trans-1,4-cyclohexanebis(ethylamine) and the specified acid. Each solution was concentrated by evaporation under a pressure of 13 psig until the temperature reached 136°C. The concentrated salt solution was then charged to a closed vessel and heated to remove water and effect polymerization. The pressure and temperature were then raised to about 250 psig and 220°C, respectively. The temperature was increased to about 243°C, while the pressure was held at 250 psig by removal of steam pressure.

The pressure was then reduced to atmospheric over a 25-minute period. The temperature was adjusted to 15°C above the melting point of the polyamide and the polymer allowed to equilibrate for 30 minutes. The polymer was melt spun directly from the bottom of the autoclave to yield a monofilament yarn having good textile properties. The melting point and glass transition temperature of the polyamide fiber were 254° and 120°C (at 30% relative humidity), respectively. The percent retention of the sonic modulus and glass transition temperature values of a fiber are excellent indications of its dimensional stability and are commonly used in the art for this purpose.

Sonic modulus retention values at 30% relative humidity were 37.4% at 30°C, 91% at 45°C, 87% at 60°C, 82% at 75°C, and 71% at 90°C.

Antistatic Fibers

It has been suggested that antistatic properties of fibers of polyamide could be improved by uniformly dispersing in the polyamide between about 1 and 12% by weight of a predominantly branched, chain-extended propylene oxide-ethylene oxide copolymer based on a diamine. However, with incorporation of this antistatic additive in the polyamide, serious problems have been encountered in melt-spinning due to the frequent occurrence of "nubs" or enlarged places in the extruded polyamide filament.

It has been found by *R.L. Wells and L. Crescentini; U.S. Patent 4,051,196; September 27, 1977; assigned to Allied Chemical Corporation* that the occurrence of the nubs in the antistatic polyamide fiber can be greatly reduced by dispersing in the polymer about 0.5 to 12% by weight, based on the weight of the antistatic agent, of a phenol compound represented by the formula:

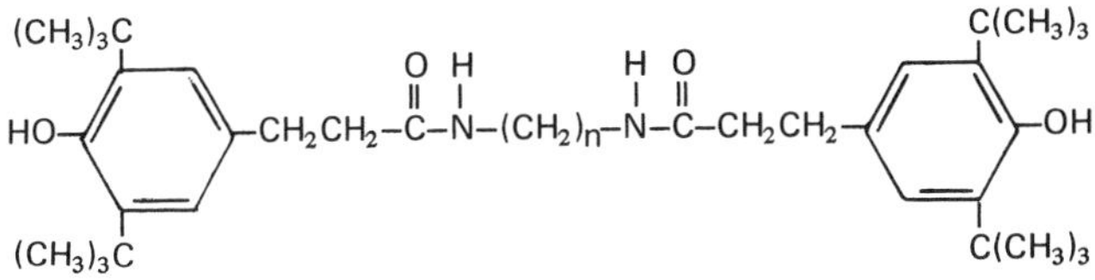

$$(CH_3)_3C \underset{(CH_3)_3C}{\overset{}{\diagdown}} HO-C_6H_2-CH_2CH_2-\underset{O}{\overset{}{C}}\underset{H}{\overset{}{N}}-(CH_2)_n-\underset{H}{\overset{}{N}}-\underset{O}{\overset{}{C}}-CH_2CH_2-C_6H_2-OH \underset{C(CH_3)_3}{\overset{C(CH_3)_3}{\diagup}}$$

where n is 2 to 6.

In comparative tests with antistatic yarns containing various commercially available hindered phenol compounds, the yarns showed significantly greater breaking strength retention after exposure to light in standard tests.

Example 1: This example shows a method of preparing a preferred antistatic additive which is a chain-extended reaction product of a tetrol compound; the molecular weight of the latter is preferably between 4,000 and 50,000. The tetrol compound used in the example is covered by U.S. Patent 2,979,528 and sold as Tetronic 1504.

300 g of Tetronic 1504 (MW 12,500) was placed in a three-neck flask fitted with a thermometer, stirrer, and addition funnel. The Tetronic 1504 was stirred and heated to 105°C, and 4.2 g of dimethyl terephthalate (MW 194.2) was added to the material in the flask. Agitation was continued for about 3 hours at 200°C after the addition was completed. Then the product was cooled to room temperature. It was a soft solid having a melt viscosity of about 10,000 cp at 100°C, measured with the Brookfield viscometer. The viscosity of the original Tetronic 1504 was 200 cp at 100°C.

Example 2: A glass reactor equipped with a heater and stirrer was charged with a mixture of 1,520 g of ε-caprolactam and 80 g of aminocaproic acid. The mixture was then flushed with nitrogen and was stirred and heated to 255°C over a 1-hour period at atmospheric pressure to produce a polymerization reaction. The heating and stirring was continued at atmospheric pressure under a nitrogen sweep for an additional 4 hours to complete the polymerization. During the final

30 minutes of the polymerization, 1.7 g of N,N'-hexamethylenebis(3,5-di-tert-butyl-4-hydroxy-hydrocinnamamide) and 48 g of the antistatic compound of Example 1, were added to the polycaproamide and stirring was continued to thoroughly mix the additives throughout the polymer. Nitrogen was then admitted to the glass reactor and a small pressure was maintained while the polymer was extruded from the glass reactor in the form of a polymer ribbon. The polymer ribbon was subsequently cooled, pelletized, washed and then dried. The polymer was a white solid having a relative viscosity of about 55 to 60 as determined by a concentration of 11 g of polymer in 100 ml of 90% formic acid at 25°C (ASTMD-789-62T).

The polycaproamide pellets containing the antistatic agent and phenol additive were melted at about 285°C and the melt-extruded under a pressure of about 1,500 psig through a 16-orifice spinnerette, each of the orifices having a diameter of 0.014 inch, to produce a 250-denier fiber. The fiber was then collected at about 1,000 fpm and was drawn about 3.5 times its extruded length to produce a 70-denier yarn. For convenience, this yarn hereinafter will be called Yarn A. A control yarn containing the antistatic agent but no additional additive was produced in the same manner as described above. For convenience, this yarn will be called Yarn B. A second control yarn containing no antistatic compound and no phenol compound was produced in the same manner as above; this yarn will be called Yarn C.

Yarn A, Yarn B and Yarn C were woven into conventional plain weave fabrics. The fabrics were cut into fabric test samples having a width of 3 inches and a length of 9 inches. The fabric samples were tested for their antistatic properties in accordance with the general procedure described in the *Technical Manual of the American Association of Textile Chemists and Colorists,* 45, 206-207 (1969). This test procedure is entitled "Electrostatic Clinging of Fabrics: Fabric-to-Metal Test" and is numbered AATCC 115-1969.

In accordance with this test, Yarn C showed poor antistatic properties, i.e., the average time for fabric samples to decling from metal completely on their own was over 325 seconds after 5 to 25 wash cycles. In contrast, Yarn A and Yarn B both showed excellent antistatic properties, for example, average time for fabric samples to decling from metal completely on their own was about 130 seconds after 25 wash cycles. Yarn A, Yarn B and Yarn C were also tested for the number of nubs per pound as shown in Example 3.

Example 3: This example outlines the method used for locating, identifying and calculating the nubs per pound in Yarn A, Yarn B and Yarn C as prepared in Example 2. In this method a nub is defined as an enlarged place in a filament which is not more than several filament diameters in length. This method may be used for either monofilament or multifilament yarns; however, it is not applicable to most types of crimped yarn.

In accordance with the test, the 70-denier yarn is drawn directly from the package by means of an air aspirator and is passed through an opening of known width, specifically, 0.0030 inch in width. Such an opening is conveniently provided by use of a ceramic cleaner gap, which is well-known in the art. The presence of a nub is detected when it stops the yarn passage through the opening. The filaments are separated and the cause of the yarn stopping identified as a nub or as the twisted end of a broken filament. For representative results, about 75 g of yarn is passed through the gap and the number of nubs counted.

Yarn Sample	Nub Count per Pound of Yarn
Yarn A	3,300
Yarn B	16,700
Yarn C	2,250

The polyamide yarn made without additives had a relatively low nub count of 2,250 nubs per pound of yarn. Addition of the antistatic compound to the polyamide caused the nub count to increase to 16.700 per pound of yarn. However, the addition of the antistatic compound plus the phenol compound of the process reduced the nub count to 3,300 nubs per pound of yarn.

Retention of Properties During Laundering Due to Low Water Absorption

R.N. MacDonald; U.S. Patent 4,018,748; April 19, 1977; assigned to E.I. DuPont de Nemours and Company has developed polyamides having the formula

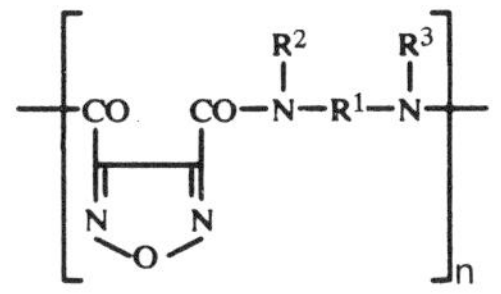

where R^1 is alkylene of 2 to 12 carbon atoms or an arylene or a bisarylene group of 6 to 18 carbon atoms and R^2 and R^3 are hydrogen, alkyl of up to 6 carbon atoms or when R^1 is an alkylene group, R^2 and R^3 jointly can be alkylene of 2 to 12 carbon atoms; and n is the degree of polymerization, and having an inherent viscosity of from 0.1 to 10 measured in 0.1% solution in chloroform at 25°C.

Preferred embodiments can be represented by the formula

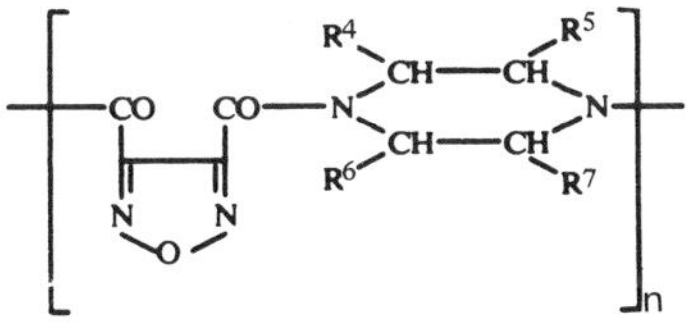

wherein R^4, R^5, R^6 and R^7 are each hydrogen or an alkyl group of up to 4 carbon atoms. Particularly preferred compounds are derived from piperazine and trans-2,5-dimethylpiperazine.

Example 1: *Preparation of 1,2,5-Oxadiazole-3,4-Dicarbonyl Chloride* — In a glass reactor under a blanket of nitrogen was placed 120 ml of thionyl chloride and 15.8 g of 1,2,5-oxadiazole-3,4-dicarboxylic acid [Grundman, *Ber.* 97 (2), 575-8, 1964]. The mixture was heated to reflux and 0.5 ml of dimethylformamide was added. In the ensuing vigorous reaction HCl and SO_2 were given off. Refluxing was continued for about 1 hour until all the free acid was dissolved. Excess thionyl chloride was removed at 50°C under vacuum and the residue distilled at 35°C/0.5 mm. Two redistillations gave 13.5 g of 1,2,5-oxadiazole-3,4-dicarbonyl chloride in the form of a colorless liquid boiling at 62° to 64°C/9 mm. Infrared absorption at 1,760 cm^{-1} (C=O) and at 1,525 cm^{-1} and 1,440 cm^{-1} (−C=N) was consistent with the proposed structure.

Analysis—Calculated for $C_4O_3N_2Cl_2$: C, 24.7%; N, 14.4%; Cl, 36.40%; H, 0.0%; Found: C, 25.5%; N, 13.21%; Cl, 36.75%; H, 0.9%.

Example 2: *Polyamide from trans-2,5-Dimethylpiperazine and 1,2,5-Oxadiazole-3,4-Dicarbonyl Chloride* — In a glass-jacketed blender a mixture of 6.46 g (0.0567 mol) of trans-2,5-dimethylpiperazine, 4.00 g (0.1 mol) of sodium hydroxide in 50 ml of aqueous solution, 120 ml of water, and 30 ml of alcohol-free chloroform was stirred rapidly at 9°C as a solution of 9.75 g (0.05 mol) 1,2,5-oxadiazole-3,4-dicarbonyl chloride in 35 ml of anhydrous, alcohol-free chloroform was added all at once. The temperature rose immediately to 27°C and dropped to 13°C as stirring was continued for 5 minutes. The system was poured into 2 liters of acetone to precipitate sodium chloride which was filtered off.

Evaporation of some of the filtrate gave 0.8 g of water-insoluble polyamide. It was washed with water and dried at 80°C/5 mm overnight. Its inherent viscosity was found to be 0.54 (0.1% in chloroform at 25°C) and its stick temperature on a gradient bar was 235°C. No glass transition could be observed. Its thermogravimetric air weight losses were 5% at 253°C and 50% at 376°C.

Analysis—Calculated for $C_{10}H_{12}N_4H_3$: C, 50.83%; H, 5.08%; N, 23.72%; Found: C, 50.37%; H, 5.36%; N, 22.62%.

Infrared absorption was consistent with the proposed polyamide: 3.35 and 3.40 μ for saturated C−H; 6.08 μ for carbonyl; 6.43 and 6.65 μ for ring C=N; and 7.22 μ for C−CH$_3$.

Water absorption of a tough film cast from chloroform was found to be 3%, a property favorable to retention of fiber properties in laundry cycles.

Noncrystalline, Spinnable Polyamides

C.A. Drake; U.S. Patent 4,025,493; May 24, 1977; assigned to Phillips Petroleum Company has developed diacids having the formulas

```
                          R"
                          |
                          CH₂
   (1)                    |
               HOOC−R−C−R−COOH
                          |
                          XQ

                          R"
                          |
                        H−C−XQ
   (2)                    |
               HOOC−R−C−R−COOH
                          |
                          H

                              R"
                              |
                        H     CH₂
   (3)                   |     |
              HOOC−R'−C——C−R−COOH
                        |     |
                        XQ    H
```

wherein the radicals R, which can be the same or different, are alkylene radicals having 1 to 16 carbon atoms per radical, R' is an alkylene radical $-C_nH_{2n}-$, n being an integer from 0 to 15, with the further provision that there are at least 6 carbon atoms in the chain separating the two carboxyl groups $-COOH$, R'' is hydrogen or an alkyl radical having 1 to 6 carbon atoms per molecule; X is selected from the group of phenylene, naphthylene, cyclohexylene and decahydronaphthylene, Q is selected from the group of radicals consisting of hydrogen, halogen, alkyl, dialkylamino, alkoxide, hydroxy, alkylthio, and mercapto, which are useful precursors for diesters, polyamides and polyesters.

An embodiment of the process consists in polyamides and polyesters, consisting essentially of repeating units of the formula

(4)

$$-\!\!\left[Z'\!-\!Z\!-\!Z'\!-\!\underset{\underset{O}{\|}}{C}\!-\!Y\!-\!\underset{\underset{O}{\|}}{C}\right]\!\!-$$

in which Y is individually selected from the group consisting of

$$-R-\underset{\underset{XQ}{|}}{\overset{\overset{R''}{|}}{\underset{|}{C}}}-R-, \qquad -R-\underset{\underset{H}{|}}{\overset{\overset{H-C-XQ}{|}}{\underset{|}{C}}}-R-, \qquad \text{and} \qquad -R'-CH-\underset{\underset{XQ\ H}{|}}{\overset{\overset{R''}{|}}{\underset{|}{C}}}-R-;$$

and Z' is $-O-$ or $-NH-$. Z is a divalent radical individually selected from the group of substituted or unsubstituted alkylene, cycloalkylene, arylene, alkarylene, alkylenearylene, alkylcycloalkylene, alkylenecycloalkylene and cycloalkylenearylene radicals having 2 to 20 carbon atoms per radical Z; and wherein R, R', R'', Q, X and n have the same meaning as defined above in connection with the diacids of this process. These polymers contain several of such groupings of formula (4), so that their molecular weight preferably is in the range of about 10,000 to 500,000.

The diacids as previously described, alone or in admixture with other diacids, can be reacted with primary diamines to produce polyamides in accordance with formula (4) in which Z' is $-NH-$.

Example 1: *Preparation of the Unsaturated Dinitrile* — A mixture of unsaturated dinitriles was produced essentially as described in U.S. Patent 3,840,583, Example 4. The product mixture was distilled to give a purified product mixture containing approximately 54 wt % 5-methylene-1,9-nonanedinitrile, 36 wt % 5-methyl-4-nonene-1,9-dinitrile, 6 wt % 2-methyl-4-methylene-1,8-octanedinitrile, 2 wt % 2,4-dimethyl-3-octene-1,8-dinitrile, 2 wt % 2,4-dimethyl-4-octene-1,8-dinitrile and minor amounts of other isomers.

Preparation of the Phenyl-Substituted Dinitrile — In a 2-liter reactor, 105 g of the abovedescribed purified mixture of unsaturated dinitriles and 1,200 ml of benzene were mixed. Aluminum chloride (240 g) was added slowly over a 30-minute period to the vigorously stirred solution. Following aluminum chloride addition, the reaction mixture was heated at 80°C for 2 hours. The reaction mixture was poured into an ice-water mixture which was then extracted with benzene. The benzene extract was then washed with water, dried over anhydrous magnesium sulfate and distilled. A mixture containing approximately 90 wt % 5-methyl-5-phenyl-1,9-nonanedinitrile, 10 wt % 2,4-dimethyl-4-phenyl-1,8-

octanedinitrile and minor amounts of other isomers was obtained in 83% of theoretical yield.

Preparation of the Dicarboxylic Acid — 120 g of the product prepared in accordance with the previous procedure and being a dinitrile mixture consisting essentially of 90 wt % 5-methyl-5-phenylnonanedinitrile and 10 wt % 2,4-dimethyl-4-phenyloctanedinitrile, as well as 160 g potassium hydroxide were charged into a 1-liter glass reactor containing 670 g of ethylene glycol. The mixture was stirred and heated to 160°C until the ammonia evolution ceased, after which the heating was continued for 2 additional hours. After cooling, the resulting mixture was acidified with hydrochloric acid while cooling by addition of ice. The reaction mixture was extracted with ether. The ether extract was washed with water. The ether was evaporated. 136 g of solid material were produced thereby.

This solid material was recrystallized twice from hot benzene. 126 g (91% of theoretical maximum yield) of the solid material was thereby produced. The solid material produced consisted of 90 wt % 5-methyl-5-phenylnonanedioic acid and 10 wt % 2,4-dimethyl-4-phenyloctanedioic acid. The melting point of the mixture produced was 136° to 138°C. The elemental analysis for the mixture was—Calculated for $C_{16}H_{22}O_4$: C, 69.1%; H, 7.9%. Found: C, 69.0%; H, 7.9%.

The mixture prepared in accordance with this example for the following example is referred to as PMHD.

Example 2: *Polyamide from PMHD, Adipic Acid and 2,2-Bis(4-Aminocyclohexyl)Propane* — Into a 1-liter stainless steel reactor, 58.38 g PMHD, 83.44 g 2,2-bis(4-aminocyclohexyl)propane, 20.45 g adipic acid and 40 g distilled water were charged. The charge thus contained the two acids PMHD and adipic acid in a molar ratio of 60:40. The reactor was flushed with nitrogen, sealed and pressured to 60 psig with nitrogen. The system was heated to 210°C over 50 minutes while the pressure reached 140 psig. The temperature of 210°C was maintained in the system for an additional hour while the pressure rose to 170 psig.

Thereafter, the system was heated to 210°C over 45 minutes and vented sufficiently to maintain 500 psig pressure. The temperature of 310°C was maintained for an additional hour. Then the system was vented over a 30-minute period until it reached atmospheric pressure. The system then was maintained at the atmospheric pressure and at 310°C for 30 minutes. Vacuum thereafter was applied to the reactor reducing the pressure therein to 20 torrs over 30 minutes. The conditions of 20 torrs pressure and 310°C temperature were maintained for 1½ hours. Thereafter, the product was blown down under 200 psig pressure into a tiny extrudate.

The resulting polymer exhibited an inherent viscosity of 0.73 with a polymer melting temperature of 245°C. No crystalline melting point could be observed. The following properties were measured on the polyamide:

Property	ASTM Method	Result
Density, g/cc	–	1.0764
Flexural modulus, psi x 10^{-3}	D790-66	351

(continued)

Property	ASTM Method	Result
Tensile yield, psi	D638-68	11,150
Elongation, %	D638-68	10
Izod impact, ft-lb/in notch	D256-56	0.35
Heat distortion, °F	D648-56	320
Hardness, Shore D	D2240-68	87

Nylon Terpolymer with Improved Resistance to Zinc Chloride

R.U. Pagilagan; U.S. Patent 4,076,664; February 28, 1978; assigned to E.I. Du Pont de Nemours and Company disclosed a terpolyamide resin having components of nylon 612 or 610, nylon 6I or 6T (wherein I is an isophthalic acid moiety and T is a terephthalic acid moiety) and nylon 69, 610, 636, or 612; the terpolyamide is especially resistant to zinc chloride and is useful for making molded articles and extruded shapes such as flexible tubing.

The resin has the following repeating segments

(A)

$$-NH(CH_2)_6-NH-\overset{\overset{\displaystyle O}{\|}}{C}-(CH_2)_x-\overset{\overset{\displaystyle O}{\|}}{C}-$$

wherein x is 8 or 10;

(B)

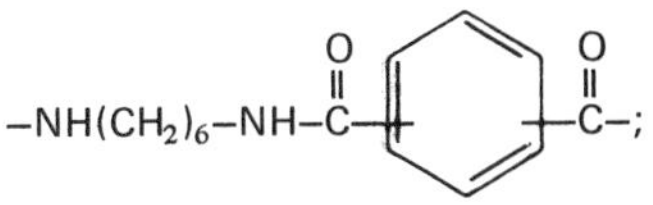

$$-NH(CH_2)_6-NH-\overset{\overset{\displaystyle O}{\|}}{C}-\underset{}{\bigcirc}-\overset{\overset{\displaystyle O}{\|}}{C}-;$$

and

(C)

$$-NH(CH_2)_6-NH-\overset{\overset{\displaystyle O}{\|}}{C}-R-\overset{\overset{\displaystyle O}{\|}}{C}-$$

where R is a hydrocarbon radical having 7, 8, 10 or 34 carbon atoms distributed therein, wherein the resin has an inherent viscosity of at least 1.0 and segment (A) is present in an amount up to 80% by weight, based upon the resin weight, segment (B) is present in an amount between about 15 and 25% by weight, based upon the resin weight, and segment (C) is present in an amount between about 3 and 10% by weight, based upon the resin weight.

In one embodiment, the terpolyamide resin has up to 75% by weight based upon the resin weight, of segment (A), between about 15 and 20% by weight upon the resin weight of segment (B), and between about 5 and 10% by weight, based upon the resin weight of segment (C).

In the following examples inherent viscosity data are obtained by measuring the viscosity of the terpolyamide solution relative to that of the solvent alone, and inherent viscosity is calculated from the following equation

$$\text{Inherent Viscosity} = \frac{\ln \left(\text{Viscosity of Terpolyamide Solution} / \text{Viscosity of Solvent} \right)}{C}$$

wherein C is the concentration expressed in grams of terpolyamide resin per 100 milliliters of solution and ln is the natural logarithm.

The zinc chloride test was carried out as follows. Molded specimens of terpolyamide resin having a rectangular form and a length of 5 inches, a width of one-half inch and a thickness of one-eighth inch were bent around the surface of a mandrel having a diameter of one inch and secured thereto. Each bent specimen was immersed in a 50% aqueous zinc chloride solution and inspected periodically to determine its resistance to zinc chloride.

Example 1: A terpolyamide resin having an inherent viscosity of 1.45 and containing 80 wt % of nylon 612 segments, 15 wt % of nylon 6I segments, and 5 wt % of nylon 69 segments were prepared as described below.

53.6 parts by weight of nylon 612 salt having a concentration of 40.0 wt % was charged into an agitated tank. Thereafter, 72 parts by weight of water and 5.6 parts by weight of hexamethylenediamine having a concentration of 80 wt % were added to the tank. To the mixture of nylon salt was added 2.1 parts by weight of azelaic acid and 6.1 parts by weight of isophthalic acid. The resulting mixture of nylon salts was then heated to about 70°C. The pH of the mixture of nylon salts was measured and adjusted to 8.75 by adding additional amounts of hexamethylenediamine. 6 ml of Antifoam FG-10 emulsion was then added and the mixture of nylon salts was charged into an autoclave after which it was heated to 200°C and a pressure of 180 psig in about 10 minutes. For 75 minutes the water in the form of steam was allowed to escape while maintaining pressure in the autoclave at 180 psig.

When the temperature of the polymer reached 246°C, the pressure in the autoclave was then slowly reduced within 58 minutes to atmospheric pressure. 13.1 parts by weight of Santicizer 8 (a mixture of N-ethyl and p-toluenesulfonamide, Monsanto Co.) was added and the terpolyamide was held for 35 minutes at atmospheric pressure and the temperature was allowed to rise to 268°C. Vacuum was then applied to the terpolyamide for 24 minutes; during this step the temperature of the batch rose to 270°C.

The terpolyamide resin was then extruded through a die, quenched in cold water, and cut into small pellets. The pellets were then surface coated with 0.2 wt % octylene glycol, followed by 0.35 wt % potassium iodide, 0.05 wt % cuprous iodide, and 0.05 wt % aluminum distearate lubricant. The octylene glycol was used to facilitate adhesion of the additives onto the pellets. The terpolyamide resin pellets were then injection molded into rectangular test specimens having planar dimensions of 5 inches by ½ inch and a thickness of ⅛ inch.

Ten specimens of the terpolyamide resin moldings were tested for zinc chloride resistance as described above. All the specimens withstood 32 days immersion in 50 wt % zinc chloride solution without exhibiting any cracks.

Analysis of the polymer showed that it contained 17.5 wt % of the Santicizer 8 plasticizer.

Example 2: *Comparative* — A homopolymer of nylon 612 plasticized with 17.1 wt % Santicizer 8 and an inherent viscosity of 1.56 was prepared in accordance with the procedure described in Example 1. Five specimens were tested for zinc chloride resistance. After 3 days in 50 wt % zinc chloride solution, all the specimens exhibited cracks. This illustrates the vulnerability of plasticized nylon 612 to zinc chloride solutions.

Thermoplastic Hydrogels

Thermoplastic hydrogels are prepared by *D.R. Cowsar and A.C. Tanquary; U.S. Patent 4,064,086; December 20, 1977; assigned to National Patent Development Corporation* by making condensation polymers of spirolactones of the formula:

(1)

$$\begin{array}{cc}
\overset{O}{\overset{\|}{C}}\text{---}(CR_2)_{n'} & (CR_2)_{\overline{n}}\overset{O}{\overset{\|}{C}} \\
& C \\
O\text{---}(CR_2)_{m'} & (CR_2)_{\overline{m}}O
\end{array}$$

with a difunctional compound capable of opening the lactone rings. The thermoplastic hydrogels can be crosslinked. In the formula the total of n and m and also of n' and m' is 2 to 5 and the R groups are H or hydrocarbyl with not over three R groups being hydrocarbyl. Most preferably all R groups are H and n, m, n' and m', are all 1.

The difunctional compound is characterized by at least two active hydrogen-containing groups which can be primary amino, secondary amino, alcoholic hydroxyl, phenolic hydroxyl, or mixtures thereof.

The resulting thermoplastic polymeric hydrogel is characterized by the units designated by (2) and (3) below:

(2)

$$\begin{array}{c}
\overset{}{OH} \\
\overset{O}{\overset{\|}{}}\quad (CR_2)_{m'} \quad \overset{O}{\overset{\|}{}} \\
\text{---}(\overset{O}{\overset{\|}{C}}\text{---}(CR_2)_{\overline{n'}}\overset{|}{C}\text{---}(CR_2)_{\overline{n}}\overset{\|}{C}\text{---}) \\
(CR_2)_m \\
OH
\end{array}$$

(3)

$$\text{---}(A\text{--}R'\text{--}A)\text{---}$$

wherein R' is a divalent radical, e.g., alkylene, alkylidene, cycloalkylene, arylene including mono- and polyaromatic ring (e.g., phenyl and naphthyl rings) structures which can be bridged rings or fused rings, divalent heterocyclic radicals (aliphatic and cyclic) and the like; and wherein each A variable represents the oxy group, the primary amino group, or the secondary amino group.

Example: The multistep synthesis of 4,4'-spirobibutyrolactone was carried out in two reaction series. The intermediate, 3,3-bis(cyanomethyl)oxetane, was from the first series and the 4,4'-spirobibutyrolactone was isolated from the second series.

Reaction Series One — Tribromoneopentyl alcohol, 975 g (3.0 mols), was added to a solution of 217 g (3.3 mols) of potassium hydroxide pellets in 100 ml of water and 1,000 ml of methanol. The mixture was heated to refluxing with stirring overnight (18 hours) after which it was cooled to room temperature and 1,000 ml of cold water was added. Stirring was stopped and the mixture was allowed to separate into two phases. The aqueous alcohol phase (top layer) was decanted and discarded. The remaining heavy oil, 740 g, was crude 3,3-bis(bromomethyl)oxetane.

To the oil was added 1,000 ml of 95% ethanol and 455 g (6.6 mols) of sodium cyanide. The mixture was heated to reflux with stirring for 36 hours and then filtered while hot to remove the sodium bromide by-product. The ethanol solution was cooled in an ice water bath to effect crystallization of the product. The first crop (110.7 g) of crude 3,3-bis(cyanomethyl)oxetane was collected by filtering. The filtrate was evaporated in vacuo and the soluble part of the solid residue was taken up in 1,000 ml of boiling benzene. The hot benzene solution was treated with 5 g of decolorizing carbon, filtered, and cooled to effect crystallization of the second crop of product.

The second crop (79.5 g) of 3,3-bis(cyanomethyl)oxetane was collected by filtering and combined with the first crop. The total yield of buff-colored platelets was 190.2 g, 46.6% of theory.

Reaction Series Two — Bis(cyanomethyl)oxetane, 343.5 g (2.5 mols), was added to a solution of 220 g (5.5 mols) of sodium hydroxide in 1,200 ml of water and the mixture was heated at 80°C with stirring for 48 hours. Ammonia gas was evolved during the reaction and when the gas could no longer be detected with wet test paper, hydrolysis was assumed to be complete.

The solution was acidified with about 210 ml of concentrated sulfuric acid to adjust the value of pH to 2. The acidified mixture was heated at 80° to 90°C with stirring overnight (15 hours). The reaction mixture was then cooled slowly (with stirring) to about 15°C to effect crystallization of the product. The crude product, 268 g, was collected by filtering, washed with ice water and air dried. The crude product was recrystallized once from 500 ml of p-dioxane to give 234.8 g of white crystals, MP 208° to 209°C by capillary melting point technique and 216°C by DSC. The yield of pure 4,4'-spirobibutyrolactone was 62.5% of theory based on crude 3,3-bis(cyanomethyl)oxetane and 29% overall based on tribromoneopentyl alcohol.

Preparation and Characterization of a Series of Hydrophilic Polyamides — When 4,4'-spirobibutyrolactone was slowly heated in a test tube with equimolar portions of primary diamines, exothermic addition reactions occurred yielding viscous polymer melts.

The polymers prepared by this simple technique were generally of low molecular weight, with inherent viscosities in the range of 0.03 to 0.11 dl/g (at 0.5 g/dl in DMF).

Preparation of Crosslinked Resins — The hydrophilic polyamides prepared from 4,4'-spirobibutyrolactone could be covalently crosslinked via the hydroxyl groups by treating them in DMF solution with a diisocyanate to form urethane linkages. The crosslinking of poly[dodecamethylene-3,3-bis(hydroxymethyl)glutaramide] with hexamethylene diisocyanate and 4,4'-methylenebis(phenylisocyanate) is illustrated below.

Crosslinked resins were made by adding 5 and 10 mol % of hexamethylene diisocyanate to solutions of the polymer in DMF. The mixtures were stirred well and placed in an oven at 95°C for approximately 2 hours. The resulting gels were precipitated in water in a high speed blender and then washed repeatedly with water. The resulting pea-sized pieces of the crosslinked resins were dried at 90°C in a vacuum oven and stored in a desiccator. One batch of crosslinked resin was similarly prepared with 5 mol % of 4,4'-methylenebis(phenylisocyanate).

Most of the characterizations of the crosslinked polymers were carried out on film specimens that were prepared by melt pressing the materials.

The average water uptakes of film specimens (5 each) made from resins cross-linked with 5 and 10 mol % of hexamethylenediisocyanate were 31.5 and 22.5% respectively.

Tensile Properties and Hydrolytic Stability — Tensile strength, elongation at break and modulus were determined on various microtensile film specimens of the cross-linked resins that had various compositions and pretreatments prior to evaluation. These pretreatments were designed to test the effects of crosslink density, im-mersion in water, heat cycling (i.e., repetitive melt processing), and prolonged exposure at 40°C to water and buffered solutions of pH 4 and pH 8.

The first series of samples was crosslinked with 5 mol % hexamethylenediiso-cyanate, HMDI; the second series was crosslinked with 5% of 4,4'-methylene-bis(phenylisocyanate), MBPI, and the third group was crosslinked with 10% HMDI. The values of tensile strength, modulus, and elongation for the untreated samples in each series quantitatively reflect the effect of increased crosslinking. With increased crosslinking, the polymers become stronger, stiffer, and harder and have lower elongations. Samples crosslinked with 15 and 20% of HMDI were prepared and while not brittle, they were quite hard.

The samples crosslinked with 5% HMDI have lower values of modulus than those crosslinked with 5% MBPI. The 5% MBPI sample would be expected to have a higher moisture uptake and a higher modulus than a sample prepared with 10% HMDI.

As was expected, the wet tensile strength of the polymer, as shown for the 10% HMDI crosslinked sample, is considerably lower than the dry tensile strength. A similar reduction in modulus is also observed but not in elongation. The water uptake of the sample that was tested wet was about 22.5 wt %.

Given a certain level of crosslinking density, as in the case of the 10% HMDI crosslinked sample, water soaking has little effect on tensile strength and elonga-tion and no effect on modulus. The slightly higher tensile strength and lower elongation may reflect extraction of residual monomers which were acting as plasticizers. On the other hand, the 5% HMDI crosslinked sample showed a marked decrease in tensile strength, possibly due to extraction of polymer or plasticization.

The samples exposed to pH 4 showed marked reductions in tensile strength ac-companied by a lesser change in modulus and elongation. This indicates sus-ceptibility to acid hydrolysis. From the fact that the reduction was much higher in the case of the 5% crosslinked sample, the amide linkage, which is beta to a hydroxyl, may be especially susceptible to acid hydrolysis. If the urethane link-age were the weak point of the network, a more substantial decrease in the ten-sile strength of the 10% crosslinked sample would probably have been observed.

Exposure to pH 8 is not as detrimental. The 10% crosslinked sample showed a general hardening as indicated by the slightly increased tensile strength, the marked increase in tensile modulus, and a considerable reduction in elongation.

The thermal stability of the 10% HMDI crosslinked sample was determined by subjecting the samples to three 15-minute press cycles at 150°C. Although the shear was not severe, the ability of the material to withstand a lengthy thermal exposure is demonstrated by the small decrease in tensile strength and elongation and the slight increase in modulus of dry samples. The wet samples showed no change in modulus, a small increase in tensile strength, and a small reduction in elongation.

POLYIMIDES AND ADDITIONAL
NITROGEN-CONTAINING POLYMERS

POLYIMIDES

Oxindole Diamines

Oxindole diamines are prepared by a process developed by *F.P. Darmory and M. DiBenedetto; U.S. Patent 4,016,173; April 5, 1977; Ciba-Geigy Corporation* comprising reacting an isatin derivative with an anilide and a catalyst (Friedel-Crafts) in an inert solvent at a temperature of from 50° to 150°C with subsequent hydrolysis of the reaction product to yield the oxindole diamine. These compounds are useful in preparing polyimide-type polymers. The oxindole diamine derivatives are represented by the formula:

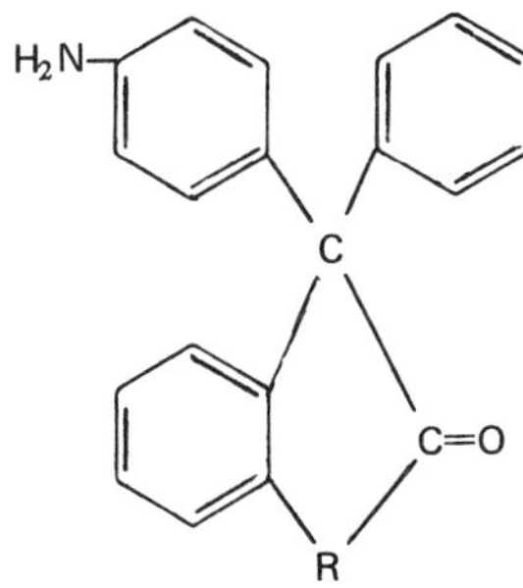

wherein R is N-(lower)alkyl or N-aryl.

In preferred oxindole diamines of this process, R is N-aryl. Especially preferred is 3,3-di(p-aminophenyl)-1-phenyloxindole.

Example 1: *3,3-Di(p-Aminophenyl)-1-Phenyloxindole* — (a) 1-Phenylisatin is prepared as follows. To a solution of 1,590 g (12.5 mols) of oxalyl chloride in

20 liters of dry methylene chloride at 0°C was added, with stirring, a solution of 2,115 g (12.5 mols) of diphenylamine in 10 liters of dry methylene chloride.

After the addition was completed, the mixture was brought to room temperature and stirred for 18 hours. The solution was then concentrated in vacuo. The residual oil was taken up to 30 liters of nitrobenzene, and this solution was treated with 3,340 g (25 mols) of aluminum chloride. This solution was then heated, with stirring, at 120°C for 4 hours. Vigorous hydrogen chloride evolution occurs throughout this operation.

(b) The preparation of 3,3-di(p-aminophenyl)-1-phenyloxindole is as follows. The cooled solution from (a) was treated with 3,000 g (14.1 mols) of s-diphenylurea and an additional 3,340 g (25 mols) of aluminum chloride. The mixture was maintained, with stirring, at 100°C for 18 hours.

The cooled solution was poured, with vigorous stirring, into a mixture of 100 liters of hexane and 100 liters of water. The precipitated solid was collected and washed with 10 additional liters of hexane to remove all residual nitrobenzene. The collected solid was suspended in 50 liters of acetic acid and treated with 0.5 liter of water and 9 liters concentrated sulfuric acid. The mixture, becoming homogeneous on heating, was then refluxed for 18 hours. The cooled solution was filtered, and the filter cake was washed with 2 liters of acetic acid. The combined filtrates were then treated carefully with 16 kg of sodium carbonate and concentrated, in vacuo, to dryness. The remaining solids were taken up in 100 liters of boiling 1 N hydrochloric acid. The solution was cooled and filtered. The filtrate was made basic with concentrated ammonium hydroxide and the precipitated solid was collected.

The crude diamine was recrystallized from 50 liters of methyl cellosolve with Darco decolorization (1 kg) and water (70 liters) precipitation. The product had a melting point of 276° to 278°C. Analysis – Calculated for $C_{26}H_{21}N_3O$: C, 79.77%; H, 5.41%; N, 10.74%. Found: C, 78.87%; H, 5.65%; N, 10.45%.

Example 2: To a solution of 0.01 mol of 3,3-di(p-aminophenyl)-1-phenyloxindole in 45 ml distilled N-methylpyrrolidone, under nitrogen, was added 3.222 g (0.01 mol) of 3,4,3',4'-tetracarboxylic benzophenone dianhydride (BTDA) in portions over a 15-minute period. The solution was then stirred for about 15 hours at room temperature and under nitrogen.

The reaction vessel was then immersed in a 200°C oil bath. Thermal equilibrium was rapidly established at 185°C, and the reaction mixture was maintained at that temperature for 3 hours. The reaction vessel was swept out by a strong nitrogen flow during the first few minutes of the imidization so as to remove all traces of water formed in the reaction. The vessel was again swept out after 10 minutes, 30 minutes and 1 hour. A soluble polyimide was obtained which had an intrinsic viscosity of 0.60 at 25°C and a glass transition temperature of 355°C as determined by torsional braid analysis.

Films were cast from the polyimide solution onto glass and aluminum and heated in a forced air oven at 200°C for several hours to drive off the solvent. The coatings obtained were clear, tough, and flexible and all coatings were able to be dissolved in the solvent from which they were prepared.

When the same coatings were heated to 300°C for ½ hour, they were still tough, clear and flexible; however, they were no longer soluble.

The polyimide was aged isothermally in a forced air oven at 300°C. The percent weight less was minor after 600 hours. The polyimide powder which is obtained by precipitation from solution with acetone and dried in a vacuum oven at 80°C may be molded by heating the powder in a cavity mold at 410°C and a pressure of about 5,000 psi.

Chain-Extending Phenolic End-Capped Polyimides

According to *G.F. D'Alelio; U.S. Patent 4,026,871; May 31, 1977; assigned to University of Notre Dame* aromatic polyimides with phenolic end groups are chain-extended (molecular weight increased) by reacting them with formaldehyde, compounds capable of generating formaldehyde under the reaction conditions or hexamethylene tetramine in the presence of a catalytic amount of lime. These polyimides can be shaped and formed prior to the chain-extending.

The aromatic polyimides appropriate for chain-extending are formed by the reaction of aromatic dianhydrides, aromatic diamines and phenolic compounds such as aminophenol.

Example 1: Preparation of phenolic-terminated oligomeric polyimide (BTAP-1) and reaction of 3,3',4,4'-benzophenonetetracarboxylic acid dianhydride (BTCA), 4,4'-oxydianiline (ODA) and p-aminophenol (2:1:2) is as follows. In the m-cresol:benzene azeotropic apparatus consisting of a 100 ml flask equipped with a Dean-Stark trap, magnetic stirrer, condenser, thermometer, heating mantle and dropping funnel, was placed a solution of BTCA (3.222 g, 0.01 mol) in 25 ml of m-cresol and 15 ml of benzene. After warming to 50°C, a solution of ODA (1.001 g, 0.005 mol) in 15 ml of m-cresol was added, forming an immediate yellow precipitate. Further heating did not dissolve the precipitate, and, after 10 minutes, a slurry of freshly purified p-aminophenol (190° to 193°C MP, 1.019 g, 0.01 mol) in 10 ml of m-cresol was added.

At reflux the solid dissolved forming an orange solution. During 2 hours of reflux, 0.20 ml of water was collected and a precipitate formed. Then, the solvents were removed on a rotary flash evaporator and the residue was vacuum-dried at 170°C, yielding a yellow solid, 4.907 g (99%). Its infrared spectrum was consistent with the structure expected for the compound. It softened slightly at 50°C, was almost completely melted at 265° to 280°C, and rehardened at 283°C. It was soluble in hot m-cresol, hot N,N-dimethyl acetamide (DMAC) and hot sulfolane.

Analysis — Calculated for $C_{58}H_{30}N_4O_{13}$: C, 70.30%; H, 3.05%; N, 5.65%. Found: C, 70.04%; H, 3.01% N, 5.73%.

The following describes the polymerization of BTAP-1. A mixture of BTAP-1 with 10 to 12% paraformaldehyde and 1 to 5% lime cures into an insoluble intractable polymer when heated at the melting point of BTAP-1 on a hot plate.

Example 2: Preparation of phenol-terminated oligomeric polyimide (BTAP-4) and reaction of BTCA, 3,3'-sulfonyldianiline (SDA-3,3) and p-aminophenol (5:4:2) is as follows. According to the m-cresol:benzene technique there was allowed to react BTCA (6.4446 g, 0.02 mol), SDA-3,3 (3.9730 g, 0.016 mol) and p-aminophenol (0.8730 g, 0.008 mol). There was obtained 10.0418 g (95%) of a pale yellow powder (BTAP-4) which was soluble in m-cresol, DMAC and sulfolane. In hot dioxane, BTAP-4 formed a separate oily layer. On a Fisher-Johns melting point apparatus BTAP-4 softened at 210°C, melted at 240° to 260° and did not harden on being heated at 300°C for 40 minutes. The lowest temperature at which a sample would melt completely when dropped onto the pre-heated stage was 250°C. Analysis — Calculated for $C_{145}H_{72}N_{10}O_{35}S_4$: C, 65.90%, H, 2.75%; N, 5.30. Found: C, 65.45%; H, 2.88%; N, 4.99%.

The following describes the polymerization of BTAP-4. A mixture of BTAP-4 with 10 to 12% paraformaldehyde and 1 to 5% lime cures into an insoluble, intractable polymer when heated at the melting point of BTAP-4 on a hot plate.

Chain-Extending Amine End-Capped Polyimides

In a related process by *G.F. D'Alelio; U.S. Patent 4,058,505; November 15, 1977; assigned to University of Notre Dame du Lac* aromatic polyimides with amine end groups are chain-extended by reacting them with aromatic di- or tri-anhydrides. The reaction can either be at a temperature above the melting point of the reactants or in solvents for the reactants.

Example: The preparation of amine-terminated oligomeric polyimide (BTAT-3) and the reaction of BTCA and ODA (7:8) are as follows. In the m-cresol:benzene azeotropic apparatus was placed 4,4'-oxydianiline (ODA), 3.204 g (0.016 mol), in 20 ml of m-cresol and 10 ml of benzene. After warming to 60°C, a solution of 3,3',4,4'-benzophenonetetracarboxylic acid dianhydride (BTCA), 4.512 g (0.014 mol), in 30 ml m-cresol was added. A copious yellow precipitate formed immediately. The reaction mixture was heated to reflux and the solid material dissolved, forming an orange solution. After 2 hours of reflux, 0.30 ml of water had been collected and a precipitate had formed in the reaction flask.

After cooling, the reaction mixture was concentrated on a rotary flash evaporator and then vacuum-dried at 170°C to give 7.0960 g (~100%) of a yellow solid, which was slightly soluble in m-cresol and insoluble in sulfolane and DMAC. On a Fisher-Johns melting point apparatus it softened slightly at about 210°C, partially melted by 235°C and rehardened to a granular solid at 240°C.

Analysis — Calculated for $C_{215}H_{110}N_{16}O_{43}$: C, 72.18%; H, 3.10%; N, 5.48%. Found: C, 70.26%; H, 3.26%; N, 6.11%.

The preparation of oligomeric anhydride (BTOD-1) and the reaction of BTCA and ODA (2:1) are as follows. Into a 100 ml three-necked, round-bottom flask equipped with a magnetic stirrer, thermometer, condenser, gas inlet tube, dropping funnel, etc., there was placed, under nitrogen atmosphere, a solution of BTCA, 6.444 g (0.02 mol), in 25 ml of DMAC. Then, a solution of ODA, (2.00 g (0.01 mol), in 15 ml of DMAC was added over a period of 15 minutes. The reaction, which was exothermic, was maintained at 40°C during the addition. Following the addition, it was heated at 85° to 90°C for 15 minutes.

To this clear amber-colored solution, acetic anhydride, 3.06 g (0.03 mol) was added and the mixture was heated to 125°C. Within 15 minutes, a yellow precipitate formed.

After heating the reaction mixture for 1 hour, the solvents were removed in a rotary flash evaporator. The residual light yellow solid was washed with anhydrous ether and dried in a vacuum oven at 140°C to afford a quantitative yield. It softened slightly on a Fisher-Johns melting point apparatus at 120°C and did not melt when heated to 300°C. The product was soluble in m-cresol and N-methyl-2-pyrrolidone and only slightly soluble in boiling benzonitrile, acetophenone or DMAC.

Its infrared spectrum shows the peaks for $-C=O$ of the anhydride group at 4.50 and 5.63 μ, and for the imide $C=O$, at 5.82 μ.

Analysis — Calculated for $C_{46}H_{20}N_2O_{13}$: C, 68.32%; H, 2.49%; N, 3.47%. Found: C, 68.27%; H, 2.82%; N, 3.79%.

The melt reaction of BTAT-3 and BTOD-1 is as follows. An intimate mixture of equimolar amounts of BTOD-1 and BTAT-3 was placed between glass plates and placed on a block preheated to 250°C. The sample melted and rehardened within 3 minutes. Then, as the temperature was raised at 5°C/min, the sample resoftened at 290°C and rehardened at about 325°C.

Chain-Extending Nitrile End-Capped Polyimides

In another process by *G.F. D'Alelio; U.S. Patent 4,060,515; November 29, 1977; assigned to University of Notre Dame du Lac* aromatic polyimides with nitrile end groups are chain-extended by polymerization using Lewis acid salts as catalyst. Molecular weights are increased with little or no by-product formation. These polyimides can be shaped and formed prior to the polymerization.

The aromatic polyimides appropriate for chain-extending are formed by the reaction of aromatic dianhydrides, aromatic diamines and nitrile compounds such as aminobenzyl cyanide and aminobenzonitrile.

Apparatus and abbreviations are the same as in the preceding patents.

Example 1: The preparation of nitrile-terminated oligomeric polyimide BTAN-6 and the reaction of BTCA, MDA-4,4 and AN (9:8:2) are as follows. In a m-cresol:benzene azeotropic apparatus was placed 2.1751 g (0.0135 mol) and 4-aminobenzonitrile (AN), 0.1772 g (0.0030 mol), in 25 ml of m-cresol and 10 ml of benzene.

The mixture was brought to reflux and maintained at reflux for 30 minutes. A solution of 4,4'-methylene dianiline (MDA-4,4'), 1.1896 g (0.0120 mol), in 15 ml of m-cresol was added and the mixture was refluxed for 30 minutes. At the end of the reflux period, the theoretical amount of water had been collected and the reaction mixture was a clear yellow solution. The reaction mixture was then added dropwise to approximately 100 ml of methanol. The precipitated oligomer was digested three times in 100 ml of hot methanol, filtered and vacuum-dried at 70°C for 24 hours. A pale yellow powder, BTAN-6, 2.1102 g (94%) was obtained. On a Fisher-Johns melting point apparatus, BTAN-6 melted over the range 230° to 290°C. The drop melt was 270°C. It was soluble in m-cresol and sulfolane and became swollen in DMAC and dioxane.

Example 2: The curing of nitrile-terminated oligomeric polyimides BTAN-6 is as follows. An intimate mixture ($\sim$0.25 g) of the nitrile-terminated oligomeric polyimide BTAN-6 with Cu_2Cl_2 (5% by weight) was prepared in a Wig-L-jig apparatus. The sample was placed into a test tube, the tube flushed with nitrogen and capped with a nitrogen-filled balloon. The tube was then placed into a metal block preheated to and electronically maintained at 300°C. After 2 hours the tube was removed and cooled. The Thermal Gravimetric Analysis in air of the sample was then performed at 10°C/min on a Du Pont 900. The break was at 365°C; percent residue was 100% at 300°C, 95% at 400°C; and 0% at 500°C.

Chain-Extending Schiff Base-Terminated Polyimides

G.F. D'Alelio; U.S. Patent 4,055,543; October 25, 1977 describes aromatic polyimides with Schiff base end groups which are chain-extended by heating the polyimides, preferably in the presence of Lewis acid salts as catalyst.

The aromatic polyimides appropriate for chain-extending are formed by the reaction of aromatic dianhydrides, aromatic diamines and a Schiff base monoamine such as p-amino-benzylidene-aniline, p-amino-benzylidene-4-vinylaniline and p-aminobenzylidine-4-ethynylaniline.

Example: The preparation of Schiff base-terminated oligomeric polyimide is as follows. According to the m-cresol:benzene technique there was allowed to react 3,3',4,4'-benzophenonetetracarboxylic acid dianhydride (BTCA), 6.4446 g (0.02 mol), 3,3'-sulfonyldianiline (SDA-3,3), 3.9730 g (0.016 mol), and an equivalent molar amount of p-aminobenzylidene-aniline, $NH_2C_6H_4CH=NC_6H_5$, 1.568 g. The corresponding Schiff base-terminated oligomer is obtained.

Similarly, when an equivalent molar amount of p-aminobenzylidene-4-vinylaniline, $H_2NC_6H_4CH=NC_6H_4CH=CH_2$, 1.765 g, or an equivalent molar amount of p-aminobenzylidene-4-ethynylaniline, $H_2NC_6H_4CH=NC_6H_4C\equiv CH$, 1.75 g, is used, the corresponding Schiff base-terminated oligomers are obtained. The aminoaryl Schiff bases are readily prepared by the procedure given by Rossi, *Gazz. chim., Ital.,* 44, 263 (1966).

The Schiff base-terminated oligomers couple similarly to nitrile-terminated oligomers. They couple readily when heated in the range of 200° to 300°C for 30 minutes to 2 hours, depending on the nature of the oligomer and of the Schiff base termini. The coupling reaction is accelerated markedly by the addition of catalytic quantities of Lewis acid salts, from about 0.15 to 3 wt %, such as $AlCl_3$, $SbCl_3$, $SbCl_5$ or any of the numerous Lewis acid salts well-known as alkylation, isomerization or polymerization catalysts. For reasons of economy and relative ease of handling, zinc chloride, zinc sulfate, and the copper salts are preferred as coupling catalysts for the Schiff base-terminated oligomeric polyimides.

Thermally Stable, Highly Fused Imide Compositions

Thermally stable, highly fused imide polymers are prepared by *F.E. Arnold and F.L. Hedberg; U.S. Patent 4,045,409; August 30, 1977; assigned to The Secretary of the Air Force* by reacting 2,2'-bis(phenylethynyl)-4,4'-diaminobiphenyl or 2,2'bis(phenylethynyl)-5,5'-diaminobiphenyl with an aromatic dianhydride. The polyimides are particularly suitable for use in high-temperature applications, such as in the fabrication of fiber-reinforced structural composites, fibrous materials and protective coatings.

Example: A mixture of 2,2'-bis(phenylethynyl)-5,5'-diaminobiphenyl, 0.292 g (0.76 mmol) and bis[4-(3,4-dicarboxyphenoxy)phenyl] sulfone dianhydride, 0.412 g (0.76 mmol) was dissolved in 250 ml of dried (molecular sieves) N,N-dimethylacetamide. The reaction mixture was stirred, under an atmosphere of dry nitrogen, at room temperature for 24 hours. Acetic anhydride (10 ml) was added to the reaction mixture and it was heated at 130°C for 1.5 hours. The reaction mixture was allowed to cool to room temperature and the polymer isolated by precipitation of the reaction mixture into 2 liters of methanol. The polymer, after being washed with methanol, and dried at 80°C for 24 hours, exhibited an inherent viscosity (0.5% solution in N,N-dimethylacetamide at 30°C) of 0.11.

Analysis — Calculated for $(C_{56}H_{30}N_2O_8)_n$: C, 75.50%; H, 3.39%; N, 3.14%; S, 3.60%. Found: C, 75.55%; H, 3.25%; N, 3.05%; S, 3.40%.

Analysis of the polymer by differential scanning calorimetry showed an exothermic reaction maximizing at 237°C. Prior softening of the polymer at 200° to 225°C was indicated by both thermomechanical analysis and softening-under-load measurements. After curing the polymer at 240° to 250°C for 24 hours, a glass transition temperature (Tg) of 350° to 360°C was measured by differential scanning calorimetry and softening-under-load.

The foregoing example demonstrates that such polyimides are cured at moderate temperatures by intramolecular reactions. During the curing process no volatile by-products are evolved, thereby eliminating the possibility of void formation. Thus, the polyimides are not subject to the disadvantages of conventional aromatic-heterocyclic systems which must be cured by interchain chemical reactions. Also, the data show that the polyimides on curing provide modified polymers with no softening point below their decomposition temperature.

Because of their outstanding properties, the polyimides are particularly suitable for use in fabricating molded articles, e.g., by vacuum molding, such as fiber-reinforced structural composites. The imide polymers can also be used in forming fibers and protective coatings by conventional methods.

Heat-Stable Polyimide Resin

M. Bargain; U.S. Patent 4,064,192; December 20, 1977; assigned to Rhone-Poulenc, SA, France describes heat-stable resins having good mechanical and electrical properties combined with chemical inertness at temperatures of 200° to 300°C, which resins are resins of a three-dimensional polyimide which is obtained by reacting, at between 50° and 350°C, a bisimide of the general formula:

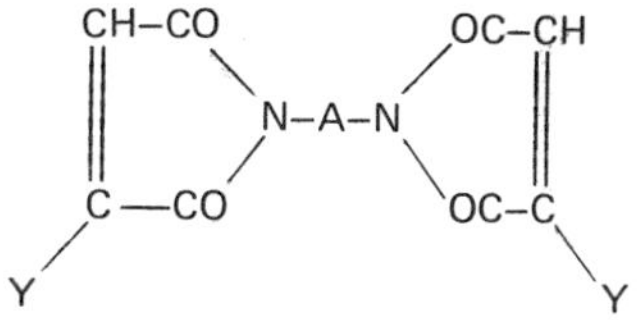

in which Y denotes H, CH_3 or Cl, and A represents a divalent organic radical possessing at least two carbon atoms, a polyamine of the general formula $R(NH_2)_x$ in which x represents an integer at least equal to 2 and R denotes an organic radical of valency x, and an alazine of the general formula:

$$G-CH=N-N=CH-G$$

in which G represents a monovalent aromatic radical, and a polymerizable mono-mer other than a bisimide, containing at least one polymerizable vinyl, maleic, allyl or acrylic $-CH=C<$ group in amounts such that if N_1 represents the number of mols of bisimide used, N_2 represents the number of mols of polyamine used and N_3 represents the number of mols of alazine used, the ratio:

$$\frac{N_1}{2N_2/x \ + \ N_3}$$

being at least 1.3, x being defined as above.

Example: 71.6 g of N,N'-diphenylmethane-bismaleimide, 3.34 g of benzalazine and 12.7 g of 4,4'-diaminodiphenylmethane are intimately mixed. The mixture is thereafter spread on a metal plate and kept for 45 minutes in a chamber heated to 145°C and then for 14 hours at 130°C.

After cooling, the prepolymer is finely ground (particle diameter less than 100 μ); a powder of softening point 140°C is obtained. 25 g of this powder are in-troduced into a cylindrical mold (76 mm diameter) which is placed between the platens of a press which have beforehand been heated to 250°C. The whole is kept at this temperature for 1 hour under a pressure of 200 bars.

After release from the mold while hot, the article is subjected to a supplementary heat treatment at 250°C for 24 hours. After cooling, it has a flexural breaking strength of 15.6 kg/mm^2 (25.4 mm span) at 25°C. After a heat test lasting 1,000 hours at 250°C, this strength is still 14.8 kg/mm^2.

Polyimides Reacted with Polycarboxylic Acids

According to *I. Forgo, A. Renner and A. Schmitter; U.S. Patent 4,076,697; Feb-ruary 28, 1978; assigned to Ciba-Geigy Corporation*, polyaddition products with imide groups are made by reacting certain polyimides (preferably maleimides) with polycarboxylic acids in the presence of basic compounds. The preferred embodiment relates to the use of catalysts as basic compounds. In particular, tertiary amines, secondary amines or mixed tertiary-secondary amines can be used.

A further embodiment relates to the use of primary polyamines as basic com-pounds. In the latter case, the polyamine participates in the polyaddition mech-anism, that is to say, the molecules are incorporated into the polyaddition prod-ucts. In this latter case, catalysts can also be used additionally.

The products can be utilized particularly in the fields of surface protection, the electrical industry, laminating processes, and the manufacture of foamed plastics, and in the building industry.

Example 1: The manufacture of a trisimide is as follows. 294 g (3.0 mols) of maleic anhydride, dissolved in 800 ml of dioxane, are initially introduced into a reaction vessel equipped with a stirrer and thermometer. A solution of 371 g

(1 mol) of tris-(4-aminophenyl)phosphate, dissolved in 2.5 liters of dioxane, is added dropwise to the above solution at 10° to 20°C over the course of 4 to 5 hours. After completion of the addition, the mixture is stirred for a further 1½ hours and the reaction product is then filtered off, washed with chloroform and dried. 669 g of a yellowish substance of melting point 127° to 130°C are obtained. According to the analytical data, this substance is a tris-maleamic acid and has the following structure:

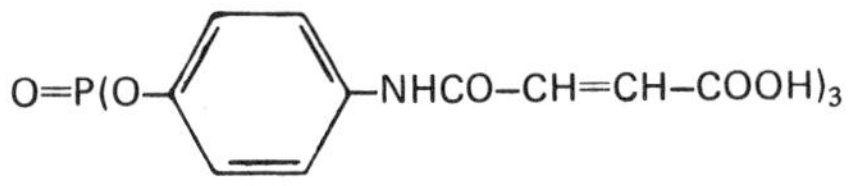

$$O=P(O-\langle\ \rangle-NHCO-CH=CH-COOH)_3$$

85 g of sodium acetate and 1.1 liters of acetic anhydride are initially introduced into a reaction vessel equipped with a stirrer and thermometer and are warmed to 60°C by means of an oil bath. 945 g of the trismaleamic acid are added in portions to this solution over the course of 30 minutes in such a way that the reaction temperature does not exceed 90°C.

After completion of the addition, the mixture is allowed to cool to room temperature and thereafter a mixture of 2 liters of isopropanol and 0.7 liter of water is added dropwise to the reaction product which has partially crystallized out. The substance which has precipitated is filtered off, washed with isopropanol and water until free from acid and dried. 532 g of a substance of melting point 173.5° to 177°C are obtained; according to analytical data, this substance is the trismaleimide of tris(4-aminophenyl)phosphate and has the following structural formula:

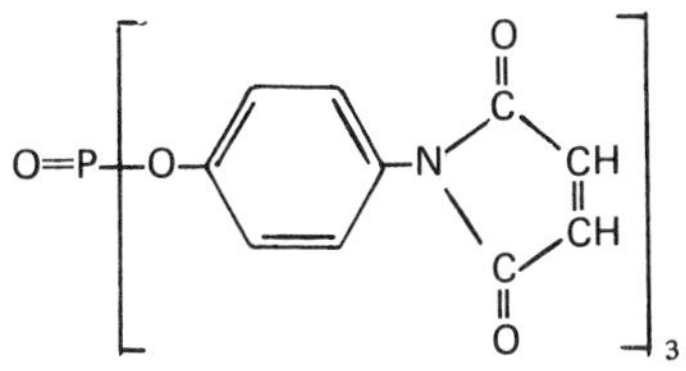

Example 2: 86.0 g (0.24 mol) of N,N'-4,4'-diphenylmethane-bismaleimide, 3.51 g (0.024 mol) of adipic acid and 4.48 g (~5% by weight, relative to the total amount of the reactants) of tetramethyldiaminodiphenylmethane are mixed well, fused in a round-bottomed flask at 180°C and ~1 to 3 mm Hg, poured into a mold of 150 x 150 x 4 mm dimensions and heated to 190°C for 14 hours. The mixture is then post-cured for 4 hours at 220°C. The resulting product is mechanically stable and has a heat distortion point (ISO/R 75) above 300°C.

The polyadducts manufactured according to the above process in all cases have a higher heat distortion point than the polyadducts which have been manufactured according to French Patent 1,555,564. In most cases, the products also have a superior flexural strength and a superior impact strength. The superiority with regard to the heat distortion is particularly marked in some cases.

Example 3: 252 g (1 mol) of N,N'-4,4'-diphenylmethane-bismaleimide, 30.8 g (0.3 mol) of adipic acid and 2.8 g of a silicone stabilizer, Si 3,193, are ground in a ball mill and then advanced in a kneader at 160°C for 40 minutes.

The mixture thus obtained is ground with 2.8 g of triethylenediamine in a ball mill. 0.5 g of azodicarboxylic acid amide (blowing agent) are admixed to 15 g of the resulting powder and this mixture is again ground. 10 g of this powder are introduced into a metal mold prewarmed to 160°C. After curing for 1 hour at 160°C, followed by 1 hour at 180°C, the mold is opened. The resulting foamed molding is found to have filled the mold well. It has a density of 0.3 g/cm³. A uniform pore structure is obtained.

Elastomeric Flame-Resistant Polymers

In a process developed by *J.J. Waldmann; U.S. Patent 4,122,046; October 24, 1978* elastomeric, chemically stable, flame-resistant polymers are obtained by polymerizing acrylamide or mixtures of acrylamide and at least one ethylenically unsaturated monomer copolymerizable therewith in the presence of a catalyst having the structure:

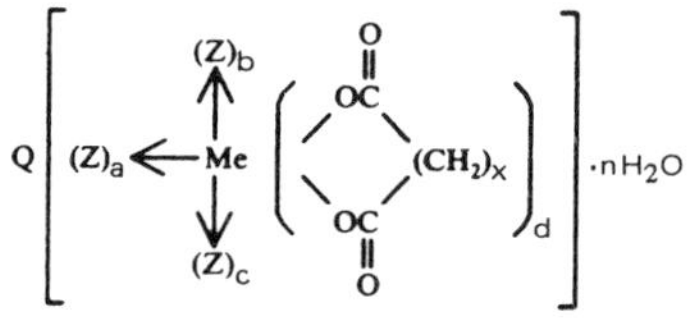

$$Q \left[(Z)_a \leftarrow \underset{(Z)_c}{\overset{(Z)_b}{Me}} \left(\overset{OC}{\underset{OC}{\diagdown}} (CH_2)_x \right)_d \right] \cdot nH_2O$$

wherein Q is Na, K, NH_4, Ca or Ba; n is 2 to 5; Me is Ti, Mn, Mo, Zr or W; x is 0 or 1; a, b, c each is 0 or 1; a + b + c + d is valence of Me; and Z is O, H_2O, $-OH$; and $-OOC(CH_2)_x COO-$.

The catalyst may be used alone as well as with UV, electron beam, ionizing radiation and combinations of these and with general redox systems known in the literature.

These catalysts, in contrast to free-radical initiators, polymerize acrylamide and mixtures of acrylamide and ethylenically unsaturated monomers via the imine-linkages.

While the exact mechanism of the polymerization reaction is not clearly under-stood, it is believed that it proceeds via amide anions which are probably formed by an H^+ shift from the amino group to the ceto carbonyl. While there is no desire to be held to any particular theory of operation, it has, nevertheless, been demonstrated that polymer compositions prepared by the method are polyimides as opposed to C–C chained polymers containing amide appendages and possess unique and in many respects, superior properties to those prepared by the here-tofore known methods.

Among the suitable catalysts contemplated for use in the method can be included the following:

$$K[Mn(C_2O_4)_2(H_2O)_2] \cdot 2H_2O$$
$$K[Mn(C_3H_2O_4)_2(H_2O)_2] \cdot 2H_2O$$
$$(NH_4)_2[TiO(C_2O_4)_2] \cdot 3H_2O$$
$$K_2[TiO(C_2O_4)_2] \cdot 2H_2O$$

(continued)

$$Na_2[TiO(C_2O_4)_2]\cdot 3H_2O$$
$$Ba[TiO(C_2O_4)_2]\cdot 4H_2O$$
$$Ca[TiO(C_2O_4)_2]\cdot 5H_2O$$
$$Na_2[WO_2(OH)_2(C_2O_4)]\cdot nH_2O \ (n = 1 \text{ to } 3)$$
$$(NH_4)_2[WO_3(C_2O_4)]\cdot H_2O$$
$$(NH_4)_2[MoO_3)(C_2O_4)]\cdot nH_2O \ (n = 1 \text{ to } 3)$$
$$Na_2[(MoO_3)(C_2O_4)]\cdot 3H_2O$$
$$K_4[Zr(C_2O_4)_4]\cdot 5H_2O$$

Of these oxalate complexes, the first two compounds are preferred for they produce the most efficient reaction. Moreover, they readily decompose to CO_2 at room temperature, which makes them most desirable in the production of foamed polyimides.

Advantageously, flame retardants are included in the polymerization reaction. The preferred types of flame retardants are those which bond chemically to the polymer so as to provide prolonged flame resistance to the polymeric material. Illustrative of such polymers are halogenated dienes, such as hexachlorobutadiene, poly(phosphoramidopentachlorodienylurea), poly(phosphamidopentachlorodienylurea), poly(phosphorylnitride) and halogenated, unsaturated organic salts and esters such as 1,1-dichlorovinyldiethylphosphate, bis(β-chloroethyl)-vinylphosphate, triethanolaminebisallylchlorophosphoric acid and Phosgard C22R. In general, flame retardants are added in amounts ranging from 0.5 to 25% by weight of the monomers.

Example 1: The following compounds, by weight ratios, are homogenized at room temperature:

Sulfuric/phosphoric acid, 8/2 ratio	25
Urea oxalate or derivatives	3
Acrylamide or derivatives	12.5
Hexachlorobutadiene	5.5
Dimethylformamide/H_2O	1.5/0.5
Zinc methacrylate	6.0
Complex oxalic, $K[Mn(C_2O_4)_2(H_2O)_2]\cdot 2H_2O$	0.08
2-(N,N-diethanolamino)ethyl acrylate	3.5
Potassium persulfate	0.04
Poly(phosphoramidopentachlorodienylurea)	5.5

These materials are stirred in an open container until a homogeneous solution is obtained and is continued to be stirred for an additional 120 minutes. When the solution begins to gel, the solution is poured into a form and polymerization allowed to go to completion forming an elastomeric solid.

Heating this solid to approximately 200°C in an oven or a tunnel, not requiring any special atmospheres, forms a fibrous foamed material in a few minutes.

Example 2: The process of Example 1 is followed with the following compositions:

H_2SO_4/H_3PO_4, 4/1 ratio	3
Urea, derivatives	5
Acrylamide or derivatives	10

(continued)

Dimethylformamide	1
Epoxy-5,5-dimethylhydantoin acrylate	7.7
Acrylonitrile	1.5
Barium methacrylate	8.2
Poly(phosphorylnitride)	5.5
Hexachlorobutadiene	5
Complex oxalic, $K[Mn(C_3H_2O_4)_2(H_2O)_2]\cdot 2H_2O$	0.01
Potassium persulfate	0.01

Example 3: Example 2 is repeated, except that persulfate is not used, but ultra-violet light in the range of 3000 to 7000 A or ionizing radiation of about 0.5 Mrad per hour is used. The time of reaction is 5 to 12 minutes to form the polymer material.

Example 4: To 43 units of phosphoric acid gradually is added 45 units of urea, while the mixture is stirred constantly. To decrease the viscosity of the solution the temperature is increased to about 30°C. When the urea is completely dissolved, there is added another dimethylsulfoxide solution containing, in parts by weight:

Acrylamide	180
Zinc methacrylate	25
Acrylonitrile	53
Poly(phosphorylnitride)	10
Methyl methacrylate	40

Continuous stirring is maintained. Using another 9 to 14 units of methyl methacrylate, 0.5 to 1.5 units of $K[Mn(C_2O_4)_2(H_2O)_2]\cdot 2H_2O$ is brought to a solution. This solution is then added to the previous mixture. The temperature is raised to 40°C; meanwhile, the solution is stirred constantly. The solution will gel after some time, at which the stirring is stopped. After the material has fully polymerized, it is foamed at about 200°C.

All the crosslinked polymeric products produced by Examples 1 through 4 are fibrous materials thermostable to 150°C and flame-resistant. When the fibrous material is held in a Bunsen burner flame (over 95% oxygen), the fibrous material carbonizes directly without flaming or developing much smoke, and when removed from the flame, the material does not flame and does not burn.

The polymeric products, when subjected to chemical solubility tests for 216 hours at room temperature, were found not soluble in the following chemicals: water, aqueous alkaline solution (N/10 NaOH), sulfuric acid (conc. 98%), formic acid (conc. 90.3%), phosphoric acid (conc.), acetic acid (conc.), oxidant mixture $K_2Cr_2O_7$-H_2SO_4 (conc.), acetic anhydride, benzene, aromatic hydrocarbons (gasoline), cyclohexanone, chlorobenzene, tetrahydrofuran (THF), methanol, hexane, dichloromethane (methylene chloride), ethylene chlorobromide, pyridine, dimethylformamide, dimethylsulfoxide and ammonium hydroxide (conc. 28.9%).

The specific gravity of the products falls in the range of approximately 0.018 to 0.5 and the products exhibit no liquid absorption. Further, the products are characterized by a moisture content of 2.5 to 3.5% (no static electricity build-up) and high thermal expansion and shock absorption properties.

AMINO RESINS AND RELATED POLYMERS

Flame-Retardant, Color-Fast Formaldehyde-Amino Fiber

*S. Tohyama, Y. Masuda and T. Nogi; U.S. Patent 4,145,371; March 20, 1979;
assigned to Toray Industries, Inc., Japan* disclose a highly flame-retardant fiber
of excellent color-fastness, which is made from the mixture composition of 30
to 60 parts of an amino resin composed of a condensation product of formaldehyde with melamine and other amino compounds selected from urea, dicyandiamide and benzoguanamine; and 70 to 40 parts of polyvinyl alcohol.

The incorporation of 0.1 to 4.0% by weight of phosphorus by the addition of
a reactive organic phosphorus compound to the spinning dope or to the wet filaments before drawing causes noticeable enhancement of the flame-retardance.
Curing and crosslinking may be carried out during the yarn-making process of
spinning, drawing and heat-setting or after the yarn-making process by treatment
in an acidic formaldehyde environment.

The method which was used for evaluation of flammability is as follows:

Class of Flammability: About 50 mg of the crumbled fiber staple was put into an
upper part of a lighter flame about 2 cm long for about 2 seconds, and the state
of burning of the sample was observed and graded into the following five classes:

(1) The sample flares up at once when it is put into the lighter flame,
    and the flame does not extinguish even after the lighter is removed.
    General cellulose fiber belongs to (1).

(2) The sample keeps burning while it is in the flame of the lighter,
    but extinguishes within 5 to 10 seconds after the lighter flame
    is removed.

(3) The sample extinguishes within 5 seconds.

(4) The sample extinguishes within 2 seconds.

(5) The sample does not seem to flare up even in the flame of the
    lighter, and no flare can be seen when it is removed from the
    flame of the lighter.

Example 1: The preparation of P_1 solution is as follows. In a 10 liter flask, the
following mixture was stirred at 80°C for 2 hours: 1,734.4 parts 80% aqueous
tetrakis(hydroxymethyl)phosphonium chloride (THPC) solution; 771.2 parts
urea; and 6,720 parts water. Each gram of the solution obtained (P_1 solution)
contained 0.0836 g of urea and 0.02448 g of phosphorus (P). The parts mentioned are by weight.

Example 2: The amino resin solution is prepared as follows. In a 1 liter separable flask, 63.6 parts of melamine, 12.2 parts of urea, 5.5 parts of dicyandiamide,
62.3 parts of 37% formaldehyde aqueous solution and 3.5 parts of 1N sodium
hydroxide aqueous solution were mixed and reacted in 370.5 parts of water at
80°C for 1 hour. 44.7 parts of the P_1 solution and 55.7 parts of 35% hydrochloric acid aqueous solution were then added, in that order, and the reaction
was kept at 80°C for 70 minutes. A resin solution was obtained where the
methylene content of the amino condensate was 82%.

Example 3: The preparation of spinning dope is as follows. 97.3 parts of polyvinyl alcohol (the degree of polymerization, 1,800; the degree of saponification, over 98.5 to 99.4%) was dissolved in 366 parts of water at 90° to 95°C. This polyvinyl alcohol solution was cooled to 70°C and then 618 parts of the P_1 solution (80°C) was added and mixed, while vigorously stirring the mixture for 1 hour, the temperature being gradually reduced to 30°C. The dope was then further stirred slowly for 16 hours at 30°C. The characteristic properties of the dope obtained above are as follows: total polymer content, 18.0%; mol ratio of the amino compounds (melamine/urea/dicyandiamide), 74.8/18.7/6.5; weight ratio of the amino resin to polyvinyl alcohol, 50/50; phosphorus content (P atom-based), 0.6%; and viscosity at 30°C, 22.5 poises.

Example 4: The spinning and fiber characteristics are as follows; the dope obtained above was filtered and defoamed before spinning under the following conditions: spinneret, diameter 0.08 mm, 200 holes; extrusion rate, 3.5 g of dope per minute; coagulating system, 3 baths of the same composition in weight, Na_2SO_4/NaOH/water, 25/2/100, bath 1, 30°C; bath 2, 30°C; bath 3, 50°C; drawing temperature, 210° to 230°C (set temperature of a heat plate); total draw ratio, 8.8; and heat-set temperature, 235°C (surface temperature of a heat roller).

The fibers obtained by the above process (drawn fibers) were next dipped in an aqueous solution of 10% sulfuric acid, 3.7% formaldehyde aqueous solution and 20% sodium sulfate for 30 minutes at 50°C. The fibers were then washed with cold water and boiling water repeatedly, followed by drying to obtain cured fibers. The characteristics of the drawn (uncured) and cured fibers are as follows:

	Drawn Fibers	Cured Fibers
Denier	2.6	3.2
Tenacity, g/den	5.2	3.3
Breaking elongation, %	10.9	23.0
Initial Young's modulus, g/den	129	67
Flammability	–	5
Colorfastness*	–	3

*According to JIS-L-0844, A-2, indicating slight running of color

Amine-Terminated Polymers and Formation of Block Copolymers

According to the process developed by *W.L. Hergenrother, R.A. Schwarz, R.J. Ambrose and R.A. Hayes; U.S. Patent 4,157,430; June 5, 1979; assigned to The Firestone Tire & Rubber Company*, polymers of anionically polymerized monomers such as conjugated dienes, vinyl-substituted aromatics, olefinic-type compounds and heterocyclic nitrogen-containing compounds, are produced and end-capped with a polyisocyanate or polyisothiocyanate. Such end-capped polymers are then reacted with compounds containing an amide such as lactam to give an imide-type end group.

The imide-type-terminated polymer is hydrolyzed to form a stable amine-terminated polymer, which may be utilized as a composition of matter or stored for a short period of time to an extended period of time and reacted with other various polymers and monomers, or various combinations of monomers to form various block or graft polymers. That is, the amine polymer may be subsequently reacted with any amine-reactive compound, such as with a polyisocyanate or polyisothiocyanate and a lactam in the presence of a known anionic lactam polymerization catalyst to give a blocked nylon copolymer.

Similarly, other block or graft copolymers may be obtained by reacting amine-reactive compounds such as various monomers or polymers with the terminated amine polymer and examples of amine-reactive polymers include polyepoxy, polyurea-aldehyde, polyphenol-aldehyde, polyamide, polyurea-urethane, polyurethane, polyimide, polyurea and similar polymer segments.

Example 1: The pilot plant preparation of 1,2-amine-terminated polybutadiene is as follows. A 500 gal stainless steel reactor fitted with a single contour blade turbine conforming to the reactor bottom was used. The reactor was baffled at a level requiring 100 to 200 gal of material to be effective. The reactor was cooled to approximately $0°F$ by circulating coolant through the jacket. The reactor was then charged with 360 lb of dimethyl ether and 9.5 lb of 15% n-butyllithium in hexane. A butadiene monomer, dried to less than 50 ppm water by circulation through a column packed with 3 A molecular sieves, was metered into the reactor at a rate such that the temperature was maintained at $40°F$.

After an initial exotherm to $48°F$, the desired temperature was maintained and a total of 200 lb of butadiene was charged over 8 hours. A sample taken after completion of butadiene addition showed 1,2-polybutadiene of greater than 97% 1,2-microstructure and a dilute solution viscosity of 0.20 dl/g on an 0.5% solution in toluene at $25°C$. To the active lithium polymer cement was added 8 lb of 2,4-toluene diisocyanate [Hylene T (Du Pont)] over 2.3 minutes, with maximum stirring.

After sampling for analytical study, 40 lb of 25 wt % caprolactam in benzene was charged over 5 minutes and dried to less than 50 ppm water by circulation through 3 A molecular sieves. The reaction mixture was diluted by addition of about 200 lb of dry hexane as the dimethyl ether was removed by distillation. Distillation was accomplished by heating the mixture at $220°F$. When essentially all of the dimethyl ether end portion of the hexane had been removed, 1.45 kg of distilled water was added and the reaction mixture was cooled, stabilized with antioxidants and dropped into drums as a 54.5% solids cement.

Titration of a polymeric base indicated that 49.5% of the material was amine-terminated, in good agreement with the value of 51% noncoupled material, measured by gel permeation chromatography.

Example 2: The preparation of the 34.5/65.5 1,2-polybutadiene-nylon 6 block copolymer is as follows. To a stainless steel resin kettle of 500 ml capacity, having a mechanical stirrer and a vacuum takeoff, was added 102 g of a cement of a 1,2-polybutadiene amine-terminated polymer in hexane approximately 73.5% solids at a molecular weight of about 14,700 g/mol, along with 150 g of flake caprolactam. The mixture was warmed in a warm water bath from about $30°$ to $50°C$ and a vacuum of 30 to about 80 mm of mercury applied to remove the bulk of the hexane.

The resulting mixture was then heated from about $90°$ to $105°C$ in a Wood's metal bath and a vacuum of less than 2 mm of mercury was applied and maintained for a 16-hour period with stirring. Nitrogen was then admitted to the resin kettle and a slight nitrogen purge maintained during the remainder of the preparation. To the stirred dry mixture was added 0.45 g of sodium hydride (58.1% dispersion in mineral oil, 11 mmol).

After 5 minutes of mixing, 0.75 ml of toluene diisocyanate (Hylene TM), 5.27 mmol, was added. The temperature was slowly raised to 165° to 175°C. Agitation was continued until the temperature reached 120° to 130°C and the mixture became too viscous to stir. The polymerization was carried out for 3 hours at 165° to 175°C. Upon cooling, the copolymer was removed from the resin kettle and compounded by milling. The milling recipe contained 100 parts of the copolymer, 150 parts of silica 325 mesh, 1.5 parts of Z6075 silane, and 2.0 parts of Dicup R.

After curing of 20 minutes at 350°F, the compounded block copolymer gave the following physical properties: 16,210 psi flexural strength; 9.88×10^5 flexural modulus; Izod, notched of 0.38 ft-lb/in; Izod, unnotched of 4.2 ft-lb/in; Gardner impact strength of 3.7 in-lb; heat distortion temperature at 264 psi of 222°C; and a Rockwell hardness E scale of 79.

Flame-Retardant Melamine-Aldehyde Resins

Y. Nihongi and N. Yasuhira; U.S. Patent 4,088,620; May 9, 1978; assigned to Kuraray Co., Ltd., Japan describe flame-retardant and infusible fibers having a water-swelling degree lower than 2.0, obtained by spinning precondensates of melamine-aldehyde resins and thereafter curing the resulting fibers.

The degree of swelling is defined as W/W_0, wherein W is the weight of the fiber determined after immersing the dry fiber in a solvent for a fixed period of time and W_0 is the weight of the same totally dry fiber. The degree of swelling decreases with the progress of the crosslinking reaction, with the minimum, of course, being 1. A polymer which is not crosslinked has an infinite degree of swelling, since the polymer would be dissolved in the solvent.

Water is a solvent for the melamine-aldehyde precondensate. The technique used to determine this degree of swelling is as follows. A dry fiber having a weight of W_0 is immersed in water at 20°C for 16 hours and then dehydrated in a centrifugal separator operating at 3,000 rpm and a centrifugal acceleration of 1,000 g for 5 minutes. The resulting weight of this fiber is termed "W" and the ratio of W/W_0 provides the degree of swelling, which must be less than 2.0, preferably less than 1.5. When the degree of swelling is higher than 2.0, the fiber has been insufficiently cured with the result that its tensile strength and flexibility may be too small to provide a fiber having the desired properties.

In addition to having a degree of water-swelling of less than 2.0, the birefringence (hereinafter abbreviated as Δn) thereof is less than 0.02. Generally, the value of Δn of a polymeric fiber corresponds to the degree or extent of orientation of the polymer constituting the fiber and it may generally be observed that as the drawing or stretching ratio, during the manufacture of the fiber, increases, the larger the values of Δn and tenacity of the fiber will become. This is because the higher the drawing ratio, the higher the degree of orientation, generally. Contrary to these facts, with the fiber of the process, although the value of Δn is small, below 0.02, the fiber has a tenacity of from 1.6 to 5.0 g/den and sufficient flexibility.

Example: An aqueous solution of N-methylol melamine having a 70% resin concentration was prepared by dissolving N-methylol melamine resin [Sumirez Resin 607 (Sumitomo Chemical Industrial Corporation)], the resin having a degree of methylolation of 3.0, in water at 60°C with gentle agitation for 1 hour. Solution was aged at 25°C for 4 days until the viscosity was 450 poises measured at 35°C using a B-type rotation viscometer. Reaction spinning this solution was performed by spinning the solution through a nozzle into an atmosphere of air maintained at a temperature of 210°C. The holes in the nozzle had a diameter of 0.25 mm.

A fiber was obtained which was 100% melamine-formaldehyde resin. Its whiteness value was 0.85; it had a denier of 16.5, a tenacity of 2.3 g/den, an elongation of 11% and a degree of swelling (in water as discussed above) of 1.22. The value of Δn was 0.002 and it was observed by x-ray analysis that the fiber was completely amorphous. After testing, it was determined that this fiber was not combustible and upon being contacted with a flame, a gentle glowing was observed and only an extremely small amount of smoke was generated.

Upon removal of the flame, the flow ceased instantly with no smolder being observed. No tendency to fuse was observed and the fiber showed a tendency to decompose gradually only under strong heating to a temperature of 500°C. The thermal shrinkage of the fiber, at 300°C in air, was 6% and the dye-ability of the fiber was satisfactory. Specifically, the absorption of an acid dye in a dyeing bath at 98°C was 100%. The light fastness of the fiber dyed with the dye was observed as grade 6, measured according to JIS-L0843-71.

The moisture absorption of the fiber at 20°C and 65% relative humidity was 6.2%, a value which is near to that of cotton. On the other hand, the fiber exhibited no shrinkage when placed in hot water at 120°C.

Polyheterocyclic Polymers from Tetraamino Pyridines

A.H. Gerber; U.S. Patent 4,079,039; March 14, 1978; assigned to Horizons Research Incorporated describes soluble precyclized precursors to polybenzimidazoles and similar polyheterocyclic polymers which are prepared by reacting (a) an acid derivative such as bis(acid halide) or mono(acid halide)anhydride or dianhydride of a di-, tri-, or tetrabasic acid with (b) an acid salt of a tetraaminopyridine in which the four position is substituted with an alkyl group, or with an acid salt of an α,α-N,N'-disubstituted tetraaminopyridine, or with an acid salt of an N,N'-disubstituted bipyridyl compound.

The polybenzimidazoles are derived from the tetraamines represented by formulas 1 and 2 and can be represented by formulas 3 and 4:

(1)

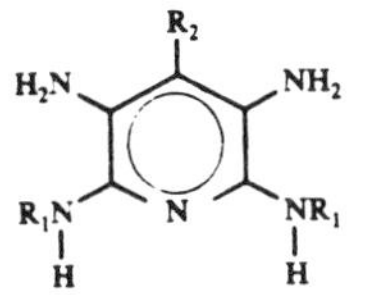

(2)

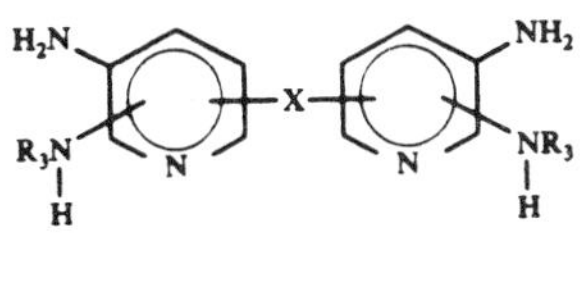

(3)

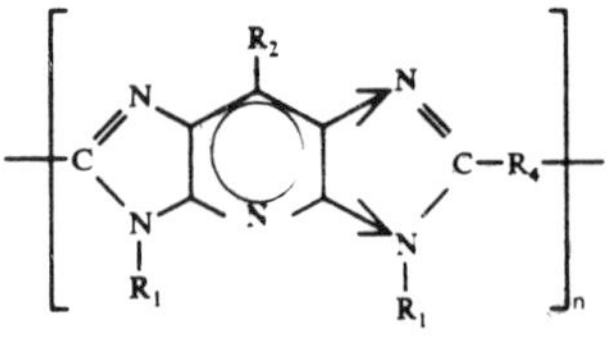

(4)

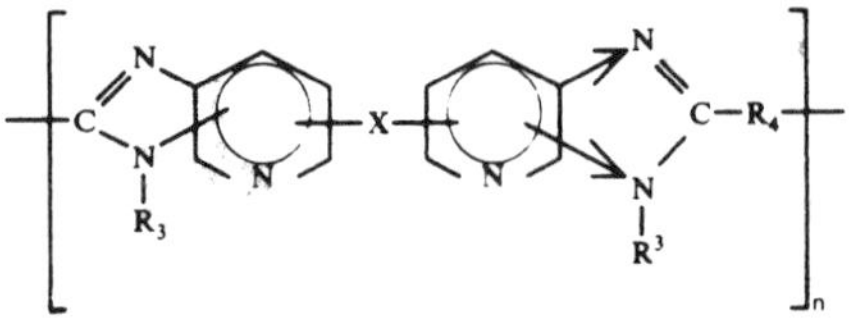

wherein n is an integer greater than 3; R_1 is a monovalent member selected from the group consisting of hydrogen, alkyl, substituted alkyl, alkenyl, substituted alkenyl, arylalkyl, aryl, heteroaryl, heteroarylalkyl, substituted arylalkyl, substituted aryl and substituted heteroaryl, with aryl and heteroaryl including monocyclic, linear bicyclic and fused ring structures; typical substituents within the scope of this process include: methyl, phenyl, pyridyl, F(aromatic), Cl(aromatic), $-CN$, $-COOH$ and its salts, $-COOC_6H_5$, $-SO_3H$ and its salts, $-SH$, thioaryl, thioalkyl, $-CH=CHC_6H_5$, and N,N-(dialkylamino); and all of the R_1's of formula 1 need not be identical.

R_2 is a monovalent member selected from the group consisting of hydrogen, methyl, ethyl, propyl, butyl and pentyl; R_3 is a monovalent member selected from the group consisting of H and alkyl of 1 to 5 carbon atoms; and X is zero or a divalent radical selected from alkylene of 1 to 3 carbon atoms, $-S-$, $-O-$, and

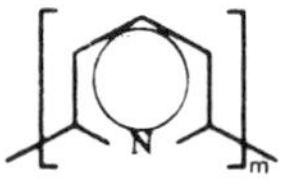

where m is equal to 1 or 2; each R_3NH- group is located ortho to an amino group and the pyridyl rings containing the amino groups are joined via the 2,2', 3,3' or 2,3' positions with the proviso that X can equal

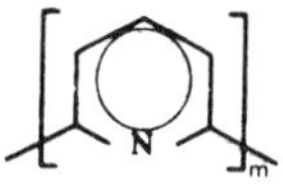

where the pyridyl groups are joined 2,2', and where X equals zero both R_3's cannot be hydrogen when the pyridyl groups are joined 2,2'; and all of the R_3's of formula 2 need not be identical; and R_4 is a divalent paraffinic, perfluoroalkyl, perfluoropolyalkylene oxide, alkenyl, aromatic or inorganic/organic radical including acylic paraffinic, cycloparaffinic, carbocyclic radicals and heterocyclic radicals having a single, multiple or fused ring structure, the multiple ring structures including polyarylenes with 2 to 9 aryl rings in which the aryl groups are bonded directly to each other or bridged by a divalent member selected from the group consisting of alkylene with up to 3 carbon atoms, perfluoroalkylene of 2 to 10 carbon atoms, $-O-$, $-C=O-$, $-S-$,

$-CH=CH-$, 5- and 6-membered heteroaromatics containing at least 1 nitrogen atom, and mixtures thereof, and substituted aromatic radicals where the substituents are selected from lower alkyl, F, Cl, $-CN$, $-SO_3H$, and

the inorganic/organic radicals consisting of ferrocenyl, carboranyl, and biaryls separated by at least 1 phosphorus atom or by at least 1 silanyl or siloxanyl group, and mixtures thereof; R_7 represents H, lower alkyl, or phenyl; R_1, R_2, R_3 and X are as defined in formulas 1 and 2; and the symbol $\rightarrow$ indicates possible isomerism.

The polymers represented by formulas 3 and 4 are preferably prepared from the linear precyclized polyamides 5 and 6, respectively, and acid salts thereof, by a cyclodehydration reaction:

(5)

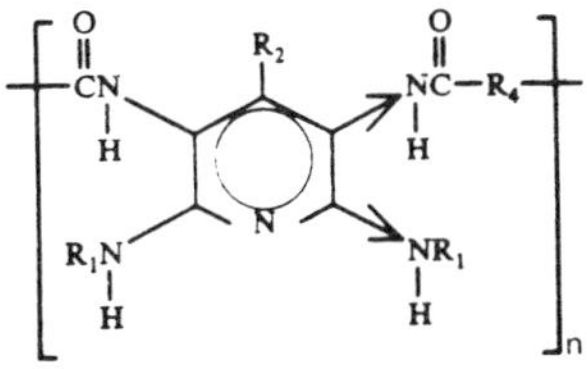

(6)

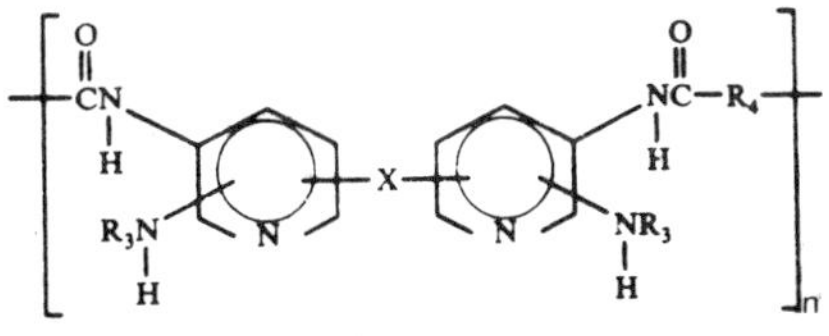

wherein n is an integer greater than 3; each of R_1, R_2, R_3 and R_4 has the meanings previously given to it in formulas 3 and 4.

Example: Precyclized polymer and polybenzimidazole from 3,5-diamino-2,6-di-(anilino)pyridine dihydrochloride and isophthaloyl chloride are prepared as follows. Isophthaloyl chloride, 5.16 g (0.0254 mol) was added over 5 minutes under a nitrogen atmosphere to a stirred cold mixture of 9.25 g (0.0254 mol) 3,5-diamino-2,6-di(anilino)pyridine dihydrochloride and 45 g of N-methylpyrrolidinone. The reaction was maintained at 5° to 10°C for several hours and then kept at room temperature for 21 hours. The polymer solution was poured into 200 ml methanol with good stirring.

The precipitate of precyclized polymer was filtered, washed well with methanol twice and vacuum dried overnight at 50° to 55°C. The polymer, 9.8 g, was obtained as a yellow-green powder which was soluble in N,N-dimethylformamide (DMF), dimethylsulfoxide (DMSO), and formic acid. Analysis — Calculated for $C_{25}H_{19}N_5O_2 \cdot HCl$: C, 65.5%; H, 4.4%; N, 15.3%; Cl, 7.8%. Found: C, 67.5%; H, 4.6%; N, 15.4%; Cl, 2.9%.

This polymer was isolated as its neutral salt as follows. The hydrochloride polymer, 3.0 g, was dissolved in DMF, 25 ml, treated with triethylamine, 0.7 g, and then precipitated into methanol and purified by further washing with methanol and vacuum drying as described above. The neutral precyclized polymer was soluble in DMF, DMSO and formic acid.

Analysis — Calculated for $C_{25}H_{19}N_5O_2$: C, 71.3%; H, 4.5%; N, 16.6%. Found: C, 71.1%; H, 4.7%; N, 16.6%.

Similarly, other acid salts of the above precyclized polymer may be formed by replacing the dihydrochloride salt by the hydrobromide or methanesulfonate, or trifluoroacetate salts.

The above precyclized hydrochloride polymer, 3.0 g, was converted to the cyclized polybenzimidazole, 2.55 g, by heating for 2 hours under vacuum at each of the following temperatures: 150°, 200°, 300° and 350°C. The resulting polybenzimidazole was soluble in formic acid, trifluoroacetic acid and methanesulfonic acid. Analysis — Calculated for $C_{25}H_{15}N_5$: C, 77.9%; H, 3.9%; N, 18.2%. Found: C, 77.8%; H, 4.0%; N, 18.1%.

This polymer possessed outstanding thermo-oxidative stability retaining 99% of its weight after 500 hours isothermal aging in air at 600°F (316°C). In comparison, the polybenzimidazole derived from 2,3,5,6-tetraaminopyridine·3HCl and isophthaloyl chloride retained only 87% of its weight after 300 hours and poly-[2,2'-(m-phenylene)-5,5'-bibenzimidazole] retained only 57% of its weight after 200 hours in air at 600°F.

An amine-terminated precyclized hydrochloride polymer was obtained by reaction of isophthaloyl chloride (0.0229 mol) with 3,5-diamino-2,6-di(anilino)pyridine dihydrochloride (0.0254 mol) as described above. This polymer could be further extended in molecular weight by reaction in N-methylpyrrolidinone with one of the following: 2,6-naphthalene diacid dichloride, dodecanedioic acid dichloride, or 3,3',4,4'-benzophenonetetracarboxylic dianhydride.

Poly(Bisbenzimidazobenzophenanthroline)

An overall process is provided by *E.C. Chenevey and H.T. Hanson; U.S. Patent 4,005,058; January 25, 1977; assigned to Celanese Corporation* wherein poly(bisbenzimidazobenzophenanthroline) is formed on an expeditious basis and is recovered from the reaction mass in substantially pure form. Appropriate quantities of 1,4,5,8-tetracarboxynaphthalene dianhydride and 3,3',4,4'-tetraaminobiphenyl are condensed in a reaction zone, while dissolved in molten antimony trichloride, and the resulting reaction mass containing the dissolved product is discharged into a precipitation zone containing hydrochloric acid.

The resulting solid product is separated and dried. The liquid portion, which is separated from the solid product, contains antimony trichloride dissolved in hydrochloric acid and is subjected to distillation which substantially separates these components. The antimony trichloride is recycled to the reaction zone, and the hydrochloric acid is recycled to the precipitation zone.

Example: Poly(bisbenzimidazobenzophenanthroline) is formed by the condensation of 1,4,5,8-tetracarboxynaphthalene dianhydride and 3,3',4,4'-tetraaminobiphenyl to yield a fully cyclicized polymer, one isomer of which is illustrated in the following equation:

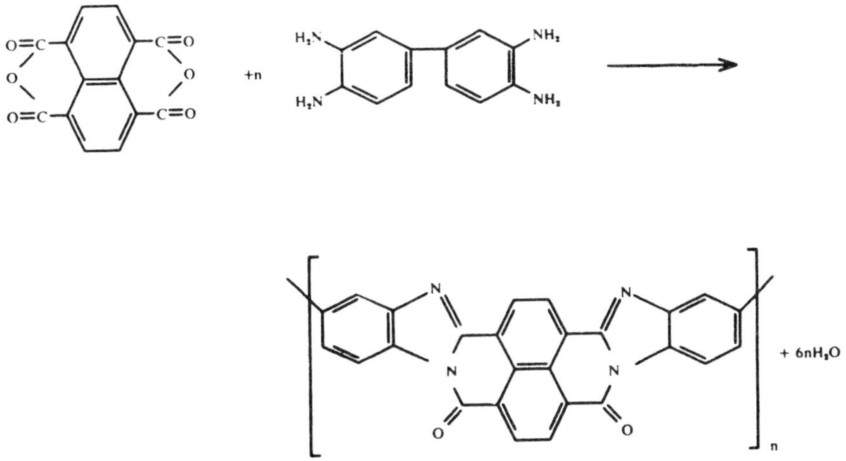

The specific isomer illustrated may be identified as poly[6,9-dihydro-6,9-dioxo-bisbenzimidazo(2,1-b:1',2'-j)benzo(1mn)(3,8)phenanthroline-2,13-diyl]. It will be apparent to those skilled in the art that various additional isomers will also be produced during the condensation reaction. During the description of the example, reference is made to Figure 4.1.

Equimolar quantities of 1,4,5,8-tetracarboxynaphthalene dianhydride and 3,3',-4,4'-tetraaminobiphenyl are added to glass-lined reaction zone **1** and are dissolved in molten antimony trichloride. The reaction zone is provided with a nitrogen purge and an anchor-type agitator. The reactants are dissolved in the molten antimony trichloride in a total concentration of 4% by weight based upon the total weight of the reactants and the antimony trichloride, and the contents of the reaction zone are maintained at 180°C for 5 hours, while undergoing agitation. The pressure in the reaction zone is maintained at atmospheric pressure with a slight nitrogen purge.

The resulting reaction mass, while at a temperature of 180°C, gradually is discharged through line **2** to precipitation zone **4**. The gradual discharge of the reaction mass requires about ½ hour and is accomplished by partially opening the reactor discharge valve.

Line **2** is heated to prevent premature chilling of the reaction mass. The reaction mass which is introduced into the precipitation zone **4** contains the resulting poly(bisbenzimidazobenzophenanthroline) product and antimony by-products dissolved in molten antimony trichloride.

Figure 4.1: Process for Formation of Poly(Bisbenzimidazobenzophenanthroline)

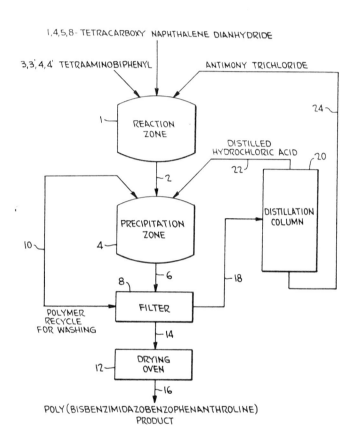

Source: U.S. Patent 4,005,058

The precipitation zone **4** is provided with a Cowles high-speed cutting agitator. Within the precipitation zone is provided hydrochloric acid having a hydrogen chloride content of about 20% by weight, i.e., constant boiling hydrochloric acid, which is maintained at a temperature of about 80°C. The relative quantity of hydrochloric acid to reaction mass is about three times by volume.

Upon contact with the hydrochloric acid, the poly(bisbenzimidazobenzophen-anthroline) is precipitated as a string-like material which is chopped by the high-speed agitator. The antimony trichloride and antimony by-products are dissolved in the hydrochloric acid. The precipitated polymer assumes a configuration re-sembling rice grains and is agitated in the precipitation zone for about 20 min-utes, subsequent to the addition of the reaction mass.

The contents of the reaction zone next are passed via line **6** to screen filter **8** where the poly(bisbenzimidazobenzophenanthroline) particles are separated from the liquid portion. The polymeric particles are recycled from the filter **8** to precipitation zone **4** through line **10**, where they are sequentially washed several times with constant boiling hydrochloric acid, then with deionized water, and finally are passed to drying oven **12** via line **14**. The drying oven is of the vacuum type and is maintained at about 180°C.

After 90 minutes, the final product is discharged via line **16**. It is found that the final product has an inherent viscosity of 2.5 as determined by capillary viscometry of an 0.2 wt % solution of polymer in 97% sulfuric acid at 25°C. The particulate product also contains less than 100 ppm antimony.

The liquid resulting from the initial filtration separation at filter **8**, as previously described, is passed through line **18** to distillation column **20**. The distillation column **20** is glass-lined with a center feed. From the top a stream of substantially pure hydrochloric acid containing less than 100 ppm antimony trichloride is distilled and is recycled via line **22** to precipitation zone **4**. From the bottom a stream of molten antimony trichloride is recycled via line **24** to reaction zone **1**.

Polyion Complex

According to *E. Tsuchida and Y. Osada; U.S. Patent 4,137,217; January 30, 1979*, a fibrous polyion complex or a polyion complex shapable to a mass or a film is prepared by reacting a polycation having polymerization degree of not less than 5 or polyamine having a polymerization degree of not less than 6 with a polycarboxylic acid having a polymerization degree of not less than 5. At least one of the reactants may preferably be of a long chain polymer if a fibrous polyion complex is desired.

Example 1: 2 g of N,N,N',N'-tetramethylethylene-p-xylyldiammonium dichloride polymer having a molecular weight of about 1×10^4 was dissolved into 1,000 ml of water. The solution was agitated by a magnetic stirrer at the speed of several hundred rotations per minute. When a solution obtained by dissolving into 50 ml of water 1.3 g of polymethacrylic acid having a molecular weight of about 7×10^4 was gradually introduced dropwise into the solution above, no precipitation of complex was observed.

After polymethacrylic acid was dropwise mixed in a predetermined amount into the resultant solution, agitation was continued for 30 minutes. When the polyion complex solution was allowed to stand for about 2 days at room temperature and in the dark, a powdered deposit was observed. After the lapse of an additional 3 days, the powdered deposit was coagulated to finally obtain a fibrous network. The network showed an optical anisotropy when observed under a polarizing microscope, and the width of the fiber was about 10 to 150 μ.

Example 2: 3 g of methacrylic acid and 5.1 g of N,N,N',N'-tetramethylethylene-p-xylydiammonium dichloride polymer having a molecular weight of about 10,000 were dissolved in water in a reaction flask, provided with stirrer, nitrogen gas-introducing tube and thermometer, having a volume of 1,000 ml to obtain a solution of 700 ml. The reaction was adjusted to pH 7 by the use of NaOH.

The resultant solution was heated to 50°C and nitrogen gas was passed over the solution for about 30 minutes to remove oxygen in the system. 0.1 g of an initiator potassium persulfate was directly dissolved into 5 ml of water and the resultant solution was charged into the solution to inititate polymerization. After the lapse of 2.5 hours, a transparent, glutinous or jelly-like polyion complex was deposited at the bottom of the flask. The degree of polymerization was 20 to 30%.

The deposited product was allowed to stand at room temperature for a suitable time period to cause contained water to be evaporated to permit viscosity to be increased. When it was taken or drawn out with a suitable method, fibers were continuously spun, and polyion complex fibers were manufactured. It was also possible to continuously spin the deposited product directly without evaporating the water content. The thickness of the fiber was about 10 to 80 μ. The thickness of the fiber is controllable according to the spinning requirement.

UREA AND URETHANE RESINS

Resorcinol-Terminated Urea-Formaldehyde Resins

According to *E.A. Blommers and R.H. Moult; U.S. Patent 4,032,515; June 28, 1977; assigned to Koppers Company, Inc.,* a urea-formaldehyde, resin-based adhesive that is resistant to heat and hydrolysis and, hence, is useful for exterior use in plywood and laminated timbers, consists of a resorcinol-terminated urea-formaldehyde resin having the general formula:

where n is 0 to 10, that is cured by conventional hardeners under neutral or alkaline conditions.

Example 1: The preparation of methylene diurea is as follows. A 4 liter resin flask equipped with a stirrer and condenser was charged with 1,540 g urea; 1,540 ml H_2O; 22.5 ml HCl (conc); and 250 g of 37% formaldehyde. After being stirred well, the mixture was allowed to sit at room temperature for 2 days.

The mixture was neutralized to pH 7 with 50% NaOH and the heavy white precipitate collected by filtration. The impure product was dried (925 g yield) and then extracted with 1,500 ml of ethanol. The alcohol-insoluble portion was again dried. The yield of white powder was 471 g (MP >270°C).

Example 2: The preparation of dimethylol methylene diurea is as follows. A 300 ml resin flask equipped with stirrer, thermometer and condenser was charged with 32.4 g (0.4 mol) 37% formaldehyde (<1% MeOH) and the pH adjusted to 8.5 with 50% NaOH. To this solution was added 26.4 g (0.2 mol) of purified methylene diurea over a period of 1½ hours. The temperature remained at 22° to 23°C, but the solution became milky. 3 ml of Solox and 5 ml of H_2O were added. The pH was adjusted to 8 with 50% NaOH and the milky solution allowed to sit at room temperature overnight. The product was a paste.

Example 3: The preparation of resorcinol-terminated dimethylolmethylene diurea (resorcinol/DMMDU–1/0.6) is as follows. A 300 ml resin flask, equipped with stirrer, thermometer and condenser, was charged with 36.7 g resorcinol (0.33 mol), 36.7 g H_2O and 0.5 g p-toluene sulfonic acid. The pH was around 1. To this solution was slowly added the pasty methylene diurea thinned with 37 ml Solox. The addition took 2 hours and the temperature was not permitted to go above 32°C. After the addition was complete, the milky solution was stirred at room temperature for ¾ of an hour. The pH was then adjusted to 7 with 50% NaOH and water was removed overnight in a forced draft oven at 45°C. The highly viscous product had a solids content of 90.5%. The percent free resorcinol was 7.27 (by IR).

Polyurethane-Amides Based on Isocyanate-Terminated Prepolymers

A.M. Reader; U.S. Patent 4,031,035; June 21, 1977; assigned to Celanese Corporation produces polyurethane-amide resins based upon isocyanate-terminated prepolymers of bis(2-hydroxyethyl) terephthalate by the reaction of bis(2-hydroxyethyl) terephthalate with a molar excess of an isocyanato-derivative having the general formula: $R(NCX)_{m+1}$, wherein R is a polyvalent organic radical containing between 2 and about 22 carbon atoms; X is an oxygen or sulfur atom; and m is a positive integer having a value of 1 to about 5, preferably 1 to 2.

Example 1: The preparation of catalyst for bis(2-hydroxyethyl) terephthalate synthesis is as follows. A catalyst is prepared from Montrek 600E by mixing 24 g of the material with 19.5 g of terephthalic acid, as well as 30 ml of water so that good mixing can be obtained.

After stirring the mixture for about 1 hour, it is placed on a rotary film evaporator for the removal of the water and a thick solid recovered, which is the phthalic acid salt of the hydroxyethylated polyethyleneimine. 40% aqueous solution of hydroxyethylated polyethyleneimine [Montrek 600E (Dow Chemical Company)] is prepared by reacting polyethyleneimine having a molecular weight of about 40,000 to 60,000 with ethylene oxide.

Example 2: The preparation of bis(2-hydroxyethyl) terephthalate is as follows. A 3 liter stirred autoclave is charged with 600 g (3.6 mols) of fiber-grade terephthalic acid; 1,600 ml of chlorobenzene. 12.4 g of the terephthalic acid salt of hydroxyethylated polyethyleneimine as prepared in Example 1, and then purged with nitrogen. Liquid ethylene oxide (8.6 mols) is then pumped in and the reactor heated to 175°C by passing steam through internal coils. The temperature is maintained at 175°C for about 30 minutes with the pressure varying during the period from about 215 psig at the beginning of the period to 80 psig at the end of the period.

After the 30 minute period, the reaction is terminated by pressuring the contents of the auotclave into a vessel where ethylene oxide is flashed and then the unreacted terephthalic acid and other solids removed by filtration. Cooling of the filtrate to about 30°C yields about 760 g of bis(2-hydroxyethyl) terephthalate (dry basis). Conversion of the terephthalic acid changed to the diester product is about 91 mol %.

Example 3: The preparation of polyisocyanate is as follows. In a 500 ml resin kettle equipped with stirrer, condenser with drying tube, thermometer and nitrogen inlet, is placed 168 g (1.0 mol) of hexamethylenediisocyanate dissolved in 300 ml of 1,1,2,2-tetrachloroethane. To this is added 127 g (0.5 mol) of bis-(2-hydroxyethyl) terephthalate. The system is flushed with nitrogen, and the reaction is conducted with stirring under a nitrogen atmosphere. The temperature is slowly raised from room temperature to about 60° to 80°C to insure complete reaction by bringing the reactants into solution. The mixture is maintained at temperature for 30 minutes.

The reaction product, which is a diisocyanate formed by capping 1 mol of bis-(2-hydroxyethyl) terephthalate with 2 mols of hexamethylenediisocyanate, is isolated by evaporating the 1,1,2,2-tetrachloroethane solvent under reduced pressure in a rotary evaporator. In some cases, it is convenient to use the reaction product in solution without separating it from the solvent.

Example 4: The preparation of polyurethane resin is as follows. The reaction product of Example 3, without isolating it from the S-tetrachloroethane solvent, is used as the starting material for a poly(ester/amide/urethane) of high molecular weight in the following manner: the reaction product is transferred to a 1,000 ml resin kettle, equipped similarly to the one of Example 3. 1 mol of adipic acid (146 g) is added and the mixture is diluted with an additional 300 ml of sym-tetrachloroethane to reduce the viscosity during reaction to permit evolution of carbon dioxide.

The system is flushed with nitrogen and heated slowly with stirring under nitrogen using a heating mantle. The temperature is raised from room temperature at 2° to 3°C per minute to 80°C and maintained at reaction temperature for 1 hour. Evaporation of the solvent produces a clear, tough polymer. Films cast from methylene chloride solution have excellent properties, and useful fibers are dry-spun from this solvent.

Radiation-Curable Acrylated Polyurethane

According to *L.E. Hodakowski and C.H. Carder; U.S. Patent 4,131,602; December 26, 1978; assigned to Union Carbide Corporation*, radiation-curable acrylated polyurethane is produced by (a) producing an isocyanate-terminated intermediate by coreacting an organic diisocyanate with a combination of organic tri/tetraol and organic diol, the combination being chosen from polyester tri/tetraol-polyether diol and polyether tri/tetraol-polyester diol combinations; and (b) reacting the isocyanate-terminated intermediate with an hydroxyacrylate such as 2-hydroxyethyl acrylate. Unexpectedly, the oligomer has desirably low viscosity, yet cures upon exposure to radiation to a coating having good physical properties.

Example 1: The preparation of acrylated polyurethane based on polyether diol/-polyester triol is as follows. To a 3-liter, four-neck flask fitted with a stirrer, thermometer, condenser and dropping funnel, there were charged 555 g of isophorone diisocyanate, 0.3 g of dibutyltin dilaurate, and 940 g of 2-(N-methyl-carbamoyloxy)ethyl acrylate. The mixture in the flask was heated to 40°C and there was added a mixture of 450 g of poly-epsilon-caprolactone triol [MW (av) 900; Hydroxyl No. (av) 187 mg KOH/g; Acid No. 0.25] and 1,000 g of poly-oxypropylene glycol [MW (av) 1,000; Hydroxyl No. (av) 111.4 mg KOH/g; Acid No. 0.1].

The mixture of poly-epsilon-caprolactone triol and polyoxypropylene glycol was added at a rate which maintained the reaction temperature at between 45° and 55°C. There were then fed to the flask 187.5 g of 2-hydroxyethyl acrylate. The reaction was continued with stirring until the free isocyanate level was 0.18 wt %.

Example 2: The preparation of acrylated polyurethane based on polyether triol/polyester diol is as follows. To a 3-liter, four-neck flask, fitted with stirrer, thermometer, condenser and dropping funnel, there were charged 555 g of isophorone diisocyanate, 697 g of 2-(N-methylcarbamoyloxy)ethyl acrylate and 0.25 g of dibutyltin dilaurate. The contents of the flask were heated to 41°C. There was then added to the flask a polyol mixture consisting of 354 g of polyoxypropylene triol [MW (av) 708; Hydroxyl No. (av) 237.5 mg KOH/g; Acid No. 0.05] and 530 g of poly-epsilon-caprolactone diol [MW (av) 530; Hydroxyl No. (av) 212 mg KOH/g; Acid No. 0.4].

The polyol mixture was fed at a rate which maintained the reaction temperature below 55°C. After the addition of the polyol mixture was complete, 187.5 g of 2-hydroxyethyl acrylate were added and the reaction was continued until the free isocyanate content reached a level of 0.6 wt %.

Example 3: Radiation-curable coating compositions were produced using the acrylated polyurethanes of Examples 1 and 2 plus 1% by weight of a di-sec-butoxyacetophenone. The viscosity of each coating composition was determined using a Brookfield model LVT viscometer with a No. 3 spindle.

Each coating composition was drawn down on release paper to a thickness of about 4 mils using a stainless steel rod. The coating composition on the substrate was cured by 1.9 seconds of exposure to medium-pressure mercury arc lamps delivering 500 W/ft^2. Tensile properties are given in the table below. The formulation containing the acrylated polyurethanes of Example 1 which contained no diluent other than residual 2-(N-methylcarbamoyloxy)ethyl acrylate which was used as a reaction medium in the production of the acrylated polyurethane, displayed a surprising combination of application viscosity and tensile properties.

While the tensile strength of the formulation containing the acrylated polyurethanes of Example 2 was considerably lower, it was nonetheless within the useful range for many commercial end uses. Generally, the coatings of the examples showed an unexpectedly good combination of tensile properties and desirably low application of viscosity in view of the low concentrations of monoacrylate diluents used.

	Example 1	Example 2
Viscosity, cp at 25°C	25,800	6,640
Tensile strength, psi	1,800	400
Elongation, %	105	80

POLYHYDRAZIDES AND POLYAZOMETHINES

Anisotropic Dopes of Polyamide Hydrazides

P.W. Morgan; U.S. Patent 4,070,326; January 24, 1978; assigned to E.I. Du Pont de Nemours and Company describes film- and fiber-forming optically anisotropic polyamide-hydrazide solutions (dopes) which are prepared in a solvent system comprising concentrated sulfuric acid or a mixture thereof with fluorosulfonic acid.

Example: This example illustrates (a) the preparation of the polyamide-hydrazide from p-aminobenzoyl hydrazide and terephthaloyl chloride; (b) an anisotropic solution thereof; and (c) fibers thereof.

To a stirred solution of 27.21 g (0.18 mol) of p-aminobenzoyl hydrazide in an ice-cooled mixture of 300 ml of hexamethylphosphoramide and 300 ml of N-methylpyrrolidone are added 36.54 g (0.18 mol) of terephthaloyl chloride.

After 15 minutes, a clear, viscous solution exists. The cooling bath is then removed and the stirred solution is allowed to warm up during the next 1¾ hours. The solution is then combined with water to precipitate the polymer which is collected, washed separately with water and alcohol, and dried in vacuo at 90°C. There is obtained 56 g of product, η_{inh} = 5.3 (dimethyl sulfoxide/5% LiCl).

An anisotropic spinning solution is prepared by first combining, at room temperature and over an 0.5 hour period, a 50 g sample of the polyamide-hydrazide described above with 200 g of concentrated sulfuric acid (100.65% by weight H_2SO_4) to form a composition containing 20% polymer. This is transferred to a twin-cell spinning unit, described in U.S. Patent 3,767,756, mixed (9 passes), and spun according to the procedure as shown below.

The anisotropic solution prepared above, maintained at room temperature, is extruded through a spinneret (10-hole, each hole of 0.002" diameter, maintained at room temperature) into an aqueous coagulating bath maintained at 1°C and positioned ¼" vertically below the face of the spinneret. The filaments emerging from the coagulating bath are washed with water and wound up at the rate of 606 ft/min.

The washed and dried filaments exhibit the following tensile properties, measured as described in U.S. Patent 3,836,498: T/E/Mi/den: 4.5/3.8/244/2.4. The polymer has the following repeating units in equimolar amounts.

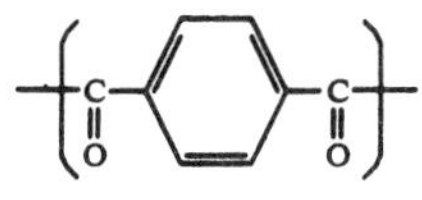

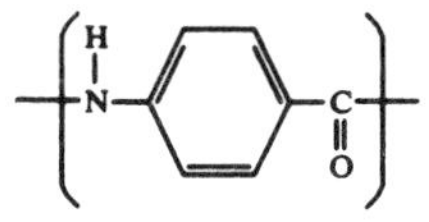

Anisotropic Dopes of Copolyhydrazides

In a related patent, *J.D. Hartzler and P.W. Morgan; U.S. Patent 4,039,502; August 2, 1977; assigned to E.I. Du Pont de Nemours & Company* discloses spinning solutions of film- and fiber-forming polymers comprising at least 5% by weight of certain copolyhydrazides in concentrated sulfuric acid, at least 99.5%, but not greater than 102% concentration, or in a mixture thereof with fluorosulfonic acid. The copolyhydrazides consist essentially of repeating units of the formula:

The divalent radical R_1 in each repeating structural unit may be the same or different and is selected from the group of aromatic, carbocyclic and aliphatic radicals of from 1 to 12 carbon atoms; 2,5-pyridinediyl radicals or a chemical bond. Some of these solutions are optically anisotropic.

Example 1: This example illustrates the preparation of poly(chloroterephthaloyl/terephthaloyl hydrazide)(1/1) and an anisotropic dope thereof in 100% sulfuric acid.

An ice-cooled solution of chloroterephthalic dihydrazide, 4.56 g (0.02 mol), in 50 ml of hexamethylphosphoramide (HMPA) is prepared in a tubular flask. To this stirred solution (paddle stirrer) is added, with cooling, terephthaloyl chloride 4.06 g (0.02 mol), in portions. Within 30 minutes a viscous paste forms. Lithium carbonate, 1.48 g, is added to the stirred reaction mixture ½ hour later.

After another 15 hours, during which the mixture is allowed to warm to room temperature, a clear, very viscous solution forms. The latter is combined with water to precipitate the polymer which is collected, washed separately with water and with methanol, and dried in a vacuum oven at 80°C. There is obtained 7.2 g of product, η_{inh} = 1.07, having the repeating structural units:

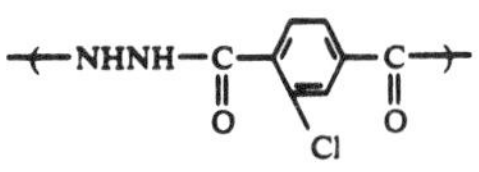

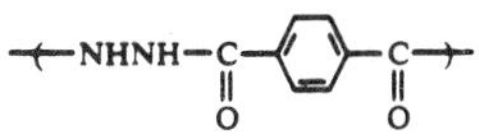

in a 50/50 mol ratio. A 20% solids dope of this polymer in 100% sulfuric acid at 25°C is optically anisotropic.

Example 2: Illustrated in the example is the copolyhydrazide prepared from terephthaloyl dihydrazide and terephthaloyl chloride/2,5-pyridinedicarbonyl chloride (50/50), and an anisotropic dope thereof in 100% sulfuric acid.

An ice-cooled solution of terephthaloyl dihydrazide, 1.94 g (0.01 mol), in a mixture of HMPA (9 ml) and N-methylpyrrolidone-2 (9 ml) is prepared and stirred (magnetic stirrer) in a 50 ml Erlenmeyer flask. To this solution are added, with cooling, 2,5-pyridinedicarbonyl chloride, 1.02 g (0.005 mol), and terephthaloyl chloride, 1.015 g (0.005 mol). Lithium carbonate, 0.74 g, is added 1 hour later.

The stirred reaction mixture is allowed to warm to room temperature in the next hour. After 14 hours more, the cloudy, viscous reaction mixture is worked up as in Example 1 to yield 2.99 g of copolymeric product, η_{inh} = 0.43, comprising the structural units:

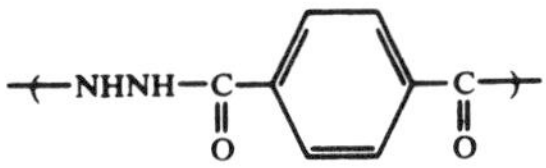

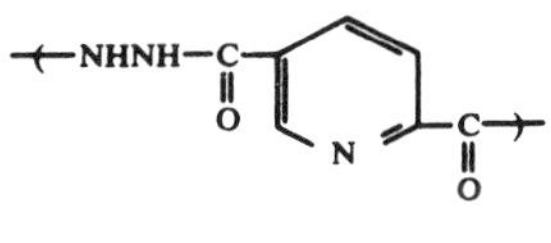

with A/B = 75/25 mol ratio. A 20% solids dope of this copolymer in 99.7% sulfuric acid is optically anisotropic.

Fibers and Anisotropic Melts of Polyazomethines

P.W. Morgan; U.S. Patents 4,122,070; October 24, 1978; and 4,048,148; September 13, 1977; both assigned to E.I. Du Pont de Nemours and Company provides melt-spinnable aromatic and cycloaliphatic polyazomethines and copolyazomethines having polymer melt temperatures below 375°C, inherent viscosities of at least 0.2, and which display optical anisotropy in the molten state.

Also provided are useful oriented fibers spun from these polymers without need for afterdrawing. Many of the as-spun fibers show increased orientation and tenacity of at least 4 g/den and an initial modulus in excess of 400 g/den (and often exceeding 500 g/den). Other shaped articles such as films and bars may be prepared from the polymers.

Example 1: This example illustrates the thermal preparation of poly(nitrilo-2-methyl-1,4-phenylenenitriloethylidyne-1,4-phenyleneethylidyne), which forms an optically anisotropic melt.

In a 12" (30.5 cm) polymer tube are combined 2-methyl-1,4-phenylenediamine, 2.44 g (0.02 mol) and 1,4-diacetylbenzene, 3.24 g (0.02 mol). The reactants are heated at 156°C for 1 hour, then at 205°C for 2 hours, all under a slow nitrogen bleed. The product is collected, broken up, washed separately with water and with methanol in a blender, and dried in a vacuum at 80°C to yield 4.32 g of polymer, η_{inh} = 0.5. The polymer exhibits a polymer melt temperature (PMT) of 370°C.

The PMT is determined on the hot bar, method A, described in *Preparative Methods of Polymer Chemistry*, Sorenson and Campbell, 2nd Ed., Interscience Pub., pp. 57-59 (1968). The polymer may be in the form of particles, chips, film or fiber for this measurement.

Example 2: This example illustrates preparation of copoly(nitrilo-2-methyl-1,4-phenylenenitrilomethylidyne-1,4-phenylenemethylidyne/nitrilo-1,4-phenylenenitrilomethylidyne-1,4-phenylenemethylidyne) (95/5). The copolymer is shaped into a molded bar.

To a stirred solution of 2-methyl-1,4-phenylenediamine, 4.64 g (0.038 mol) and 1,4-phenylenediamine, 0.22 g (0.002 mol), in a mixture of hexamethylphosphoramide, 20 ml; N-methylpyrrolidone-2, 20 ml; and lithium chloride, 2 g, under nitrogen, is added terephthalaldehyde, 5.36 g (0.04 mol). The reaction mixture is stirred for 16 hours at room temperature and then combined with water. The precipitated polymer is collected and washed as in Example 1 and dried in a vacuum at 110°C to yield 8.6 g of copolymer, η_{inh} = 4.7; PMT = 260°C. The copolymer melt is optically anisotropic.

A sample of this product is placed in a bar mold and held at 300°C for 15 minutes. The bar exhibits a flexural strength of 6.3×10^3 lb/in², a flexural modulus of 4.3×10^5 lb/in², and a yield strength of 4.3×10^3 lb/in² (4.43, 302, and 3.02 kg/mm², respectively).

Example 3: This example illustrates the preparation of poly(nitrilo-2-methyl-1,-4-phenylenenitrilomethylidyne-1,4-phenylenemethylidyne). The product forms an anisotropic melt and is spun into strong fibers whose tensile properties are enhanced by a relaxed heat treatment.

A solution of 2-methyl-1,4-phenylenediamine, 77.9 g (0.64 mol) in 200 ml of ethanol is prepared at room temperature. A second solution of terephthalaldehyde, 81.3 g (0.61 mol), is prepared in 200 ml of refluxing ethanol. These solutions are simultaneously poured into a 2 liter beaker; polymer precipitation begins in 1 to 3 minutes. This reaction mixture is permitted to stand overnight at room temperature, under nitrogen. After the ethanol is evaporated, the polymeric residue is washed with 1 liter of water and dried in vacuo at 110°C for 1.5 hours. The dried residue is polymerized further in a heated screw extruder.

A portion of the extrudate is molded into a plug, η_{inh} = 6.0, and spun into air through a 5-hole spinneret [each hole of 0.007" (0.018 cm) diameter; spinneret temperature = 260°C; melt zone temperature (MZT) = 255° to 262°C] and wound up at 600 yd/min (548 m/min, bobbin A); another bobbin, B, is collected at 900 yd/min (822 m/min, MZT = 260°C). For bobbin B fiber, η_{inh} = 7.9. These properties are observed for these as-spun yarns:

Bobbin Source	T (g/den)	E (%)	Mi (g/den)	Den
A	7.3	1.1	916	20.0
B	6.4	0.92	900	15.1

A sample of the yarn from bobbin A is wound on a bobbin wrapped with Fiber-Frax and is heated in an oven, continuously swept with nitrogen, under these successive conditions: from room temperature to 160°C/2 hours; 180°C/2 hours; 200°C/4 hours; 250°C/12 hours. After this treatment, the fiber exhibits these filament properties (average of 15 samples): T/E/Mi/den: 28:3.2/939/4.3. One filament exhibits T/E/Mi/den: 44/4.2/1,118/4.2.

In another treatment a sample of bobbin A yarn is separated into single filaments (5) which are suspended vertically from a copper wire and heated in an oven, continuously swept with nitrogen under these conditions: room temperature to 165°C/40 min; 165° to 230°C/1 hour; 232°C/1.3 hours; 234°C/6.3 hours. After this treatment, the fiber exhibits the following filament properties: T/E/Mi/den: 38/4.4/1,012/3.7. Filament and yarn properties are measured by the procedure shown in U.S. Patent 3,827,998.

ACRYLICS AND MODACRYLICS

ACRYLONITRILE POLYMERIZATION PROCESSES

Polymerization Above 120°C in the Presence of Water

T. Kobashi, M. Ozaki, K. Ono and N. Abe; U.S. Patent 4,049,605; September 20, 1977; assigned to Japan Exlan Company Limited, Japan describe a process for the polymerization of acrylonitrile or a monomer mixture containing acrylonitrile as a main component and at least one other ethylenically unsaturated compound, characterized in that the polymerization is conducted at a temperature above 120°C under a pressure above the vapor pressure generated in the polymerization system under the polymerization conditions in a system in which water is present in a range of 3 to 5 wt % based on the total weight of the monomer(s) and water to produce an acrylonitrile polymer in a substantially molten state. The polymers produced according to the above process have the advantage of being able to be directly shaped by extrusion without the need of solvents.

The polymerization step is greatly simplified, and the amounts of water and heat energy used are greatly reduced compared to prior-art processes. Furthermore, there is an advantage in that the problem of solvent recovery and purification can be avoided since shaped products can be obtained without the use of a solvent.

Since the polymerization is carried out in a homogeneous system, heat transfer is easily effected. Accordingly, it is possible to suppress the accumulation of heat in the polymerization system and prevent a runaway reaction, and further to make the polymerization reaction product uniform.

Since the process can be carried out at a low pressure below 50 atm, the process is industrially extremely advantageous from the viewpoint of the structure of the reaction apparatus and production efficiency. Furthermore, it is a great advantage that the polymerization and melting are simultaneously performed in one step to simplify the process. Other characteristic features of the process include a very low amount of polymerization by-products, which reduces mono-mer loss, and uniformity of the molecular weight of the resulting polymer.

Example 1: To a monomer mixture consisting of 90% acrylonitrile and 10% methyl acrylate, 1% di-tert-butyl peroxide (of which the decomposing tempera-ture for obtaining a half-life period of 10 hours is 124°C) as the catalyst, and 1% 3,5-di-tert-butyl-4-hydroxytoluene as the polymerization retarding agent both based on the weight of the monomer mixture, were added and dissolved therein.

Then, 0.8 part of this monomer solution and 0.2 part of water were put in a hard glass tube, 5 mm in inner diameter and 150 mm in length, having the lower end closed. After the air in the upper space of the glass tube was replaced with nitrogen, the glass tube was sealed. The glass tube containing the polymeriza-tion reactants was allowed to stand in an oil bath. In all cases where the polymeriza-tion temperature was 130° to 220°C, a transparent and viscous polymer was ob-tained which was in a substantially melted state. When the polymerization tem-perature was 115°C, a white chalky polymer (not melted) was obtained. At a polymerization temperature of 100°C, only a slurrylike polymerization product was obtained.

When the abovementioned monomer solution alone (containing no water) was enclosed in a glass tube and polymerized at 160°C for 60 minutes, only a white chalky unmelted polymer was obtained.

Example 2: Continuous polymerization is carried out using a stainless polymer-ization tube, 10 mm in inner diameter, which contains a built-in Kenics Mixer (Kenics Corp.) as the mixing device. One end of the polymerization tube is joined to a plunger pump through the intermediary of a 3 mm ϕ orifice. The polymerization tube is arranged so as to be heated to 155°C with ethylene glycol as the heating medium.

At the start of the polymerization, the nozzle orifice, which is intended for the outlet of the polymerization tube, is closed up, and thereafter the plunger pump is started to supply a monomer solution and water. During this time, the clos-ing condition of the nozzle orifice is controlled so that the pressure in the poly-merization tube can be always maintained about at 10 to 15 kg/cm^2 (gauge pres-sure). After the polymerization has progressed and just reached a point of time at which the polymer begins to extrude itself, the shutting object is removed, whereby a stable continuous extrusion of the polymer is achieved.

By following this procedure, a monomer mixture composed of 90% acrylonitrile and 10% methyl acrylate was subjected to continuous polymerization under the following polymerization conditions: supply rate of the monomer solution, 1.6 parts/min; supply rate of water, 0.4 part/min; catalyst (di-tert-butyl peroxide), 1.0%; and retarding agent (2,6-di-tert-butyl-4-methylphenol), 1.0%.

The acrylonitrile copolymer melt was extruded smoothly from the nozzle orifice provided at the outlet side of the polymerization tube. The catalyst and the retarding agent were supplied after they were dissolved in the monomer solution.

Polymerization Below 120°C in the Presence of Water

T. Kobashi, M. Ozaki and K. Ono; U.S. Patent 4,062,857; December 13, 1977; assigned to Japan Exlan Company Limited, Japan describe a process for producing an acrylonitrile polymer in a substantially melted state by polymerizing a monomer mixture composed mainly of acrylonitrile in the presence of water under a pressure above the autogenous pressure at a temperature of from 80° to 120°C so as to attain a polymerization rate of at least 45%.

Because the process is performed at a relatively low-temperature region (80° to 120°C), inexpensive, safe heat sources can be used. This makes the process industrially very advantageous in energy costs and polymerization operation and from the viewpoint of reaction apparatus and productivity. Furthermore, a remarkable feature is seen in that the simultaneous practice of the polymerization step and the melting step is very advantageous for the simplification of the process.

Other features of the process are: very little by-products in the polymerization step (consequently the reduction of the loss of monomers), the suppression of discoloration of the polymer and the uniformity of the molecular weight of the resulting polymer.

In the example, parts and percentages are by weight unless otherwise specified. The APHA number (American Public Health Association) mentioned in the examples is the calculated value by the APHA standard curve, of the degree of absorbance for a transmitted light of 430 mμ through a sample solution of 0.4 g polymer in 20 ml dimethylformamide. The greater this number, the greater is the degree of discoloration.

Example: There was mixed with a monomer mixture consisting of 91 mol percent acrylonitrile (AN) and 9 mol percent methyl acrylate (MA), 0.5%, based on the monomer mixture, of di-tert-butyl peroxide, as the catalyst, which was dissolved therein. Thereafter, 8 parts of this monomer solution and 2 parts of water were placed in hard glass tubes, each 5 mm in inner diameter and 150 mm in length, with the lower end closed. After the air in the vacant space of the glass tubes had been replaced with nitrogen gas, the glass tubes were fusion-enclosed. Each of the glass tubes containing the reaction mixture was allowed to stand in an oil bath and the mixture solution was polymerized under the various conditions shown in the table on the following page. In every case, a transparent, viscous polymer in a substantially melted state was obtained. The polymerization results are shown in the table.

As apparent from the results, the degree of discoloration of the resulting polymers can be greatly suppressed by employing polymerization temperatures below 120°C.

When only the monomer solution (without the presence of water) was enclosed in the glass tube and polymerized under the same conditions as shown, a white or yellow chalky polymer showing no fluidity was obtained at every polymerization temperature.

No.	...Polymerization... Temperature (°C)	Time (min)	Conversion (%)	APHA No.	Color of the Melt
1	100	420	71.9	120	Colorless
2	115	60	69.7	170	Colorless
3	120	60	82.9	200	Pale yellow
4	140	60	90.4	450	Yellow
5	160	60	90.8	620	Yellow

IMPROVED COLOR AND GLOSS

Modacrylics with Permanent Brilliance and Transparence

E.-A. Albers, W. Fester, and B. Sassenrath; U.S. Patent 4,056,516; November 1, 1977; assigned to Hoechst AG, Germany have found that modacryl filaments and fibers made from acrylonitrile/vinylidene chloride copolymers insoluble in acetone, containing from 20 to 45 wt % of vinylidene chloride and up to 11.5 wt % of further copolymerizable compounds can be obtained by dissolving these copolymers having K values according to Fikentscher [definition cf *Cellulose-chemie 13*, 58 (1932)] of from about 60 to about 90 in a mixture of a usual solvent for acrylonitrile polymers, e.g., dimethyl formamide, dimethyl acetamide, dimethyl sulfoxide or ethylene carbonate, and a nonsolvent for the polymer; and spinning this polymer solution in a special wet spinning process.

This process comprises the use of several, at least three, coagulation baths having gradually reduced contents of polymer solvent, a drawing of the filaments to a multiple of their length in at least one of these baths, and at least one further drawing of the practically solvent-free filaments.

By modacryl filaments and fibers, according to the rules of the Federal Trade Commission of the United States, there are to be understood those filaments and fibers the fiber-forming substance of which is a polymer containing less than 85 wt %, but more than 35 wt % of acrylonitrile.

These modacryl filaments and fibers are especially distinguished by their permanent brilliance and permanent transparence; these properties are maintained even on contact with water having a temperature above 80°C.

Example 1: A 23% spinning solution is prepared using a copolymer of 60% of acrylonitrile and 40% of vinylidene chloride having a K value of 86, prepared by suspension-precipitation polymerization, and a solvent mixture of 92% of dimethyl formamide and 8% of water. The degassed spinning solution is forced through a nozzle having 100 holes of a diameter of 70 μm at a rate of 3.3 cm³/min into a first coagulation bath having a length of 60 cm, containing a mixture of 65% of dimethyl formamide and 35% of water having a temperature of 25°C.

Without an intermediate conveying device, the spinning tow obtained is passed on to a second coagulation bath having also a length of 60 cm, containing 40% of dimethyl formamide and 60% of water at a temperature of also 25°C. The spinning tow is then drawn off the second coagulation bath by means of a draw-off roller at a speed of 4 m/min, passed on to a third bath containing 20% of dimethyl formamide and 80% of water of a temperature of 60°C, drawn three times its length by means of a further draw-off roller, washed with water of 90°C, dried and finally drawn two times its length at 120°C on a pair of rollers. The filament so obtained is steamed for 15 minutes in steam of 110°C with free shrinkage, and it then has the following properties: titer, 363 dtex; tensile strength, 3.5 g/dtex; and elongation at break, 25%.

The filaments obtained have a very good whiteness degree, high brilliancy and complete transparency. At a 120-fold magnification under the microscope, a material boiled in water for 1 hour shows no difference as compared to an unboiled material. Both samples are transparent and free from cavities. A product dyed with Remacryl Blue RL at boiling temperature and dried at 60°C has a remission power at 4,200 of 33.3%, and a product dried at 125°C has a remission power of 33.1% under these conditions. The filaments obtained are therefore stable to boiling. Fibers and filaments are considered stable to boiling if the relative difference in their remission power does not exceed 3%.

Example 2: *(Comparative)* — A spinning solution according to Example 1 is spun under the same conditions into a coagulation bath containing 65% of dimethyl formamide and 35% of water, but after 20 cm of immersion in the bath it is drawn off this coagulation bath and drawn in a drawing bath containing also 65% of dimethyl formamide and 35% of water. All further conditions are as in Example 1. The filament obtained has the following properties: titer, 367 dtex; tensile strength, 2.5 g/dtex; and elongation at break, 19%.

The filaments obtained are transparent and brilliant; at a 120-fold magnification under the microscope, no cavities are observed. On boiling in water, these filaments, however, become dull and opaque. Under the microscope, a filament so treated looks black in transmitted light. Samples dyed with Remacryl Blue RL and dried at 60°C and 125°C give the following remission values: dried at 60°C, 44.0% and dried at 125°C, 36.3%, indicating instability to boiling.

In a related patent, *E.-A. Albers and B. Sassenrath; U.S. Patent 4,056,517; November 1, 1977; assigned to Hoechst AG, Germany* describe modacryl filaments and fibers made from copolymers, insoluble in acetone, of acrylonitrile and from 20 to 45 wt % of vinylidene chloride, containing from 0 to 1.5 wt % of unsaturated organic sulfonic acids or the salts thereof, and from 0 to 10 wt % of further copolymerizable compounds, which copolymers have a K value according to Fikentscher of more than about 95. These modacryl filaments and fibers are also distinguished by their permanent brilliance and transparence; these properties are maintained even on contact with water having a temperature above 80°C.

Example 1: 15 parts of a copolymer of 60% of acrylonitrile and 40% of vinylidene chloride having a K value of 102, prepared by suspension-precipitation polymerization, are dissolved in 85 parts of dimethyl formamide.

This spinning solution is spun through a nozzle having 100 holes of a diameter of 70 μm at a rate of 3.3 cm³/min into a first coagulation bath having a length of 60 cm, containing a mixture of 65% dimethyl formamide and 35% of water having a temperature of 25°C. Without an intermediate conveying device, the spinning tow obtained is passed on to a second coagulation bath having also a length of 60 cm, containing 40% of dimethyl formamide and 60% of water at a temperature of also 25°C. The spinning tow is then drawn off the second coagulation bath by means of a draw-off roller at a speed of 4 m/min, passed on to a third bath containing 20% of dimethyl formamide and 80% of water of a temperature of 60°C, drawn 3 times its length washed with water of 90°C, dried and finally drawn 2 times its length at 120°C on a pair of rollers. The filament so obtained is steamed for 15 minutes in steam of 110°C with free shrinkage, and it then has the following properties: titer, 240 dtex; tensile strength, 3.5 g/dtex; and elongation at break, 20%.

The filaments obtained have a very good degree of whiteness, high brilliancy and complete transparence. At a 120-fold magnification under the microscope, a material boiled in water for 1 hour shows no difference as compared to an unboiled material. Both samples are transparent and free from cavities. A product dyed with Remacryl Blue R L at boiling temperature and dried at 60°C has a remission power at 4,200 of 37.2%, a product dried at 125°C of 37.4% under these conditions. The filaments obtained are therefore stable to boiling.

Example 2: *(Comparative)* — 23 parts of a copolymer of 40% of vinylidene chloride and 60% of acrylonitrile having a K value of 86, prepared by suspension-precipitation polymerization, are dissolved in 77 parts of dimethyl formamide, and this solution is spun as indicated in Example 1. Filaments having the following properties are obtained: titer, 365 dtex; tensile strength, 1.3 g/dtex; and elongation at break, 17%. The filaments are transparent and brilliant, but become opaque and loose their brilliancy on boiling in water.

At 120-fold magnification under the microscope, filaments untouched by hot water have no cavities and pores. Filaments treated in boiling water, however, look black in transmitted light under the microscope. As to the remission power of samples dyed with Remacryl Blue R L and dried at different temperature the following data are obtained: 60°C dried, 51.4% and 125°C dried, 35.2%.

Modacrylics with Improved Dyeability

The process prepared by *R. Miessen, G. Blankenstein, S. Korte and C. Süling; U.S. Patent 4,014,958; March 29, 1977; assigned to Bayer AG, Germany* provides dry-spun modacrylic filaments with improved coloristic properties based on a polymer mixture, which comprises at least one acrylonitrile-vinyl chloride copolymer and a chlorine-containing copolymer containing sulfonic acid ester groups or sulfuric acid ester groups.

Example 1: Example of the production of a polymeric dye-receptive additive of acrylonitrile, vinylidene chloride and N-acryloyl dimethyl taurine. The following solutions and mixtures are used: (1) 4,325 g of dimethyl formamide (DMF) and 175 g of deionized water; (2) 1,000 g of dimethyl formamide, 375 g of N-acryloyl dimethyl taurine (ADT), and 132.5 g of 2-methyl amino ethanol; (3) 1,250 g of acrylonitrile and 875 g of vinylidene chloride; (4) 300 g of dimethyl formamide and 6 g of ammonium peroxy disulfate; and (5) 300 g of dimethyl formamide and 7 g of oxalic acid.

The DMF/water mixture (1) is initially introduced into a 10-liter glass reaction vessel, equipped with a reflux condenser cooled with iced water, a stirrer and a gas inlet pipe, followed by rinsing with nitrogen at 50°C. The ADT-ammonium salt solution (2) obtained by neutralization at 0°C and the monomer mixture (3) are then combined with the DMF/water mixture (1). After the reaction temperature of 50°C has been adjusted, polymerization is initiated by the addition of the initiator solutions (4) and (5).

After 15 hours, a solids content of 24.0 wt % is reached in the solution, corresponding to a conversion of 75%. Following dilution with water, a polymer of the following composition is recovered from the polymer solution with a conversion of 75% by the addition of electrolyte: 47.0 wt % of acrylonitrile; 32.0 wt % of vinylidene chloride; 21.0 wt % of N-acryloyl dimethyl taurine (in the form of the ammonium salt of 2-methyl amino ethanol) (ADT salt); K value, 72; and yield, 1,970 g.

Example 2: *(Comparative)* — A copolymer of acrylonitrile and vinyl chloride, which contains the comonomers in a ratio of 42:58 and which has an intrinsic viscosity in DMF of 0.99 at 25°C, is made into a paste with dimethyl formamide at room temperature, and dissolved over a period of 6.5 hours at 40°C in a stirrer-equipped vessel.

In order to remove gas bubbles, the vessel is evacuated for 30 minutes, resulting in the formation of a spinning solution with a viscosity of 2,000 poises at 40°C. This solution is delivered by a gear pump acting as a metering unit through a spinneret in the form of an annular die which is arranged in a vertical spinning duct and comprises 120 bores with a diameter of 0.15 mm. The solution is pumped at a rate of 38.4 cc/min, and the filaments are taken off at a rate of 100 m/min. The spinning process is carried out at a duct temperature of 145° to 150°C, and at an air temperature of 180°C, the draft applied in the spinning duct amounting to 1:5.5. The resulting filaments are stretched to five times their original length in boiling water, after which the material is treated at 140°C, resulting in 25% shrinkage. Modacrylic filaments characterized by the following values are obtained: tensile strength, 2.02 p/dtex and elongation at break, 40.3%. A fiber saturation value S_F of 1.2 and an absorption rate V of 0.76 are obtained during dyeing with basic dyes.

Example 3: The procedure is as in Example 2 except that the vinyl chloride-acrylonitrile copolymer used is replaced by a mixture of that copolymer and the acrylonitrile-vinylidene chloride-acryloyl dimethyl taurine terpolymer according to Example 1 in a mixing ratio of 92 wt % of the copolymer and 8 wt % of the terpolymer. Dry spinning was carried out under the following conditions: spinning duct temperature, 151°C; air temperature, 182°C; spinneret temperature, 100°C; delivery rate, 39.4 cc of solution per minute; take-off rate, 100 m/min; and in-duct draft, 1:5.2.

The further aftertreatment was carried out in the same way as in Example 2. The filaments obtained have the following values: tensile strength, 1.94 p/dtex and elongation at break, 38%.

A fiber saturation value number S_F of 3.3 and an absorption rate V of 2.07 are obtained during dyeing with basic dyes.

The values characteristic of coloristic behavior are significantly improved in relation to the Comparative Example 2.

Similar results are obtained by wet spinning as disclosed by *R. Miessen, G. Blankenstein, S. Korte and C. Süling; U.S. Patent 4,017,561; April 12, 1977; assigned to Bayer AG, Germany.*

Comparative Example: A copolymer which contains 43 wt % of acrylonitrile and 57 wt % of vinyl chloride and which has an intrinsic viscosity of 0.94 (as measured at 25°C in DMF), is dissolved in DMF at 40°C by means of a screw, so that a solids concentration of 34 wt % and a solution viscosity of 965 poises at 20°C are obtained. The solution is extruded by a gear pump through a spinneret with 150 bores 0.1 mm in diameter into a precipitation bath with a temperature of 5°C consisting of water and dimethyl formamide in a ratio of 1:1. The filaments are run off at a rate of 12 m/min, washed at room temperature and stretched by 400% in water at approximately 95°C, dried at 110°C and, after a permitted shrinkage of 25%, are wound into package form in a steam atmosphere at 130°C. Modacrylic filaments with the following textile values are obtained: tensile strength, 2.31 p/dtex and elongation at break, 29%.

A fiber saturation value S_F of 1.1 and an absorption rate V of 0.77 are obtained during dyeing with basic dyes.

Example: The copolymer described above is replaced by a mixture of a copolymer containing 43 wt % of acrylonitrile and 57 wt % of vinyl chloride, and the terpolymer of the preceeding patent, the mixing ratio of copolymer to terpolymer amounting to 91:9 by weight.

Spinning and aftertreatment were carried out in the same way as in the comparative example. The filament yarn was found to have the following textile values: tensile strength, 2.54 p/dtex; elongation at break, 27%; fiber saturation value S_F, 3.3; and absorption rate V, 3.4.

Delustered Polyacrylonitrile Fibers

S.K. Kostadinov, I.M. Benrey, V.A. Shopov, M.Y. Kostadinova and M.S. Dimitrova; U.S. Patent 4,007,248; February 8, 1977; assigned to DSO "Neftochim," Bulgaria provide a method for obtaining delustered polyacrylonitrile fibers, wherein fibers are formed from a spinning solution, that is, being a mixture of two copolymer solutions, one of them being polyacrylonitrile and the other a ter-grafted copolymer of acrylonitrile and styrene on rubber such as butadiene or butadiene-styrene rubber-ABS-copolymer.

Delustered polyacrylonitrile fibers with an increased degree of whiteness and elasticity are obtained from mixtures containing 90 to 99.8 wt % polyacrylonitrile and 0.2 to 10 wt % ABS copolymers. The copolymers are usually not mixed but only dissolved in DMF or other solvents and the solutions obtained are mixed in definite ratios for obtaining polyacrylonitrile fibers with the content desired. The fibers thus obtained and containing up to 3 wt % (ABS-copolymer) can be additionally delustered with titanium dioxide of a quantity maximum 0.2 wt % (related to both copolymers) for obtaining the dullness. The fibers containing 3 and above 3 wt % ABS-copolymers have the same dullness as those delustered with 0.5 wt % (related to the polyacrylonitrile) titanium dioxide.

Example: To obtain delustered polyacrylonitrile fibers 3.3 dtex containing 99.5% polyacrylonitrile and 0.5% ABS-copolymer (all percentages are by weight) the following procedure is carried out. When filtering, the dimethyl formamide solutions of the polyacrylonitrile are mixed with ABS-copolymer in the following ratio: 75 ml/min 25% solution of the copolymer of polyacrylonitrile (acrylonitrile, 93%; methyl methacrylate, 6%; sodium vinylsulfonate, 1%), containing 0.5% (in relation to the polyacrylonitrile) oxalic acid and 0.02% (related to the polyacrylonitrile) optical dissolvent Uvitex MA and 1.88 ml/min 5% solution from ABS-copolymers.

The formation of the fibers from the spinning solution thus prepared is carried out by the well-known moist method of spinning in a sedimentary bath containing 50% dimethyl formamide and 50% water. The fibers obtained have the following properties: degree of whiteness, 77.8%; tenacity, 25.5 cN/dtex; and elongation, 33%.

Acrylics Which Retain Luster in Boiling Water

Acrylic fibers lose transparency and turn milky in their fiber structures through dyeing, steam treatment or the like, thus developing a defect which is called delustering. Various reasons can be given for this fact but the main reason is that acrylic fibers, because they are generally manufactured by wet spinning, tend to have minute spaces in the fiber structures and the fibers, because they are low in softening temperature, tend to swell relatively easily at a temperature of about 100°C.

K. Kozuka, S. Kurioka, T. Yasumoto, S. Kobayashi, A. Kubota and N. Otoshi; U.S. Patent 4,002,809; January 11, 1977; assigned to Kanegafuchi Kagaku Kogyo KK, Japan provide a method of manufacturing acrylic fibers free from delustering in boiling water, which comprises the steps of forming a spinning solution by dissolving in a solvent of acetone, acetonitrile, dimethylformamide and mixtures thereof, (a) a copolymer consisting of (1) 30 to 80% acrylonitrile- and (2) 70 to 20% vinyl chloride or vinylidene chloride, or (b) a tripolymer consisting of (1), (2) and (3) 3.0% or less by weight of ethylenically unsaturated monomer having hydrophilic groups, and adding one or more compounds selected from the group of (A) homopolymer or copolymer of glycidyl methacrylate and (B) specific organic tin compounds. The solution is spun into a first bath of 60% or less of organic solvent in water, followed by a second bath of 61-85% organic solvent in water and a third bath of less than 60% organic solvent in water. This is followed by drying, drawing out and annealing the yarn by ordinary methods.

Example: An acrylic resin (specific viscosity 0.161 at 30°C of a 2.0 g/l cyclohexanone solution) consisting of 32.0% acrylonitrile, 67.0% vinyl chloride, and 1.0% sodium p-styrene sulfonate was dissolved in acetone so as to be 24.0% in resin concentration, and homopolymer of glycidyl methacrylate (specific viscosity 0.091 at 30°C of a 4.0 g/l acetone solution) was used as an additive to a spinning solution in a ratio of 3.0% to the acrylic resin.

The spinning solution thus prepared was spun by use of a nozzle having 300 holes each 0.10 mm in diameter in a spinning bath consisting of the first bath of an aqueous solution of 10% acetone concentration, the second bath of an aqueous solution of 75% acetone concentration, the third bath of an aqueous solution of 50% acetone concentration, and the fourth bath of water, then dried at 120°C, further thermally drawn out 300% and heat treated at 140°C for five minutes to obtain fibers. The acrylic fibers thus obtained underwent little or no delustering even when treated in boiling water and were excellent also in heat and light resistance.

In contrast thereto, ordinary acrylic fibers produced from a spinning solution prepared from the above acrylic resin by adding 3.0% epoxy-based Epon 834 [Shell Petroleum Company, comprising glycidyl ether of bisphenol A or 2,2-bis(4-hydroxyphenyl)propane glycidyl ether, of molecular weight 624] thereto under the spinning conditions of an ordinary spinning method consisting of the first bath of an aqueous solution of 40% acetone concentration, the second bath of an aqueous solution of 20% acetone concentration, and the third bath of water, or ordinary acrylic fibers spun by the same spinning method without an additive were readily delustered by boiling water treatment and inferior in quality.

Gloss-Stable Modacrylics

It is basically not possible to apply the methods for producing compact, gloss-stable, shrinkage-free acrylic fibers to modacrylic fibers containing less than 85% of acrylonitrile units, because modacrylic fibers have a much greater tendency to develop porous structures, but also they show increased thermal sensitivity. As a result, it is not possible to apply high temperatures for consolidating the fiber structure. Furthermore, gloss cannot be stabilized by the action of dry heat alone on a drawn, porous modacrylic fiber.

G. Lorenz, A. Nogaj, H. Miller and H. Wilsing; U.S. Patent 4,126,603; Nov. 21, 1978; assigned to Bayer AG, Germany describe a process for obtaining gloss-stable modacrylic fibers and filaments comprising 50 to 84% of acrylonitrile, 15 to 48% of vinylidene chloride and 2 to 5% of an olefinically unsaturated sulfonic acid or salt thereof, all percentages being by weight, having a shrinkage in boiling water of at most 0.3%, a reduction in density on treatment with boiling water of no more than 0.015 g/cc and a scattered light component after boiling of at most 35 units compared with a $BaSO_4$ standard of 1,000 units, the increase brought about by treatment with boiling water amounting to no more than 30 units.

Example: A modacrylic polymer with a K-value of 75.4 according to Fikentscher, consisting of acrylonitrile, vinylidene chloride and methallyl sulfonic acid, with a chlorine content of 28.3% and a conductivity of 33.5 μ Siemens in the form of a 1% solution in DMF after treatment with a mixed bed ion exchanger was dissolved in dimethyl formamide to form a 37% solution.

After preheating to 115°C, the spinning solution was dry spun through a 240 bore spinneret, the filaments being run off from the spinning duct at a rate of 250 m/min. The spun material still contained 16% of DMF.

The spun filaments were wetted with water and collected in cans. For aftertreatment, a number of slivers were combined so that a weight per meter of 37 g was obtained after drawing and drying.

The aftertreatment was carried out as follows. The tow was initially drawn in hot water at 94°C in a ratio of 1:1.63 of its original length, washed at 70°C and then redrawn at 96°C in a ratio of 1:2.45. The overall drawing ratio thus amounted to 1:4.0. The drawn tow was treated with an antistatic agent and dried by means of hot cylinders at 160°C in a continuous drying unit, a shrinkage of 15% being allowed. After drying, the tow had a residual shrinkage of 10%. Take-off rate: 42.5 m/min.

The dried tow was crimped in a stuffer box in the usual way and cut into staple fibers approximately 60 mm long. The staple fibers were introduced into a continuous steaming apparatus into which 400 kg/h of steam were introduced. A temperature of 107°C prevailed in the steamer for an excess pressure of 8 mm water column. The steam saturation inside the steamer amounted to 96%. After a residence time of 4.5 minutes, the fibers were removed from the steamer and cooled.

A high gloss modacrylic fiber with a supple, pleasant feel was obtained, its properties being as follows: denier, 6.1 dtex; tensile strength, 2.5 cN/dtex; elongation at break, 49%; loop tensile strength, 0.82 cN/dtex; loop elongation at break, 16%; tack point, 215°C; residual fiber shrinkage, 0%; conductivity of the fibers (dissolved in DMF after treatment with ion exchanger), 34.0 μS; moisture, 1.8%; and yarn shrinkage, 2.7%.

The filaments had a uniform round cross section and did not show any vacuoles under a microscope. Gloss stability was determined by two different methods:

(1) Scattered light measurement—The scattered light of a fiber sample was measured by comparison with an arbitrary barium sulfate standard of 1,000 units. The results obtained were as follows:
    (a) 3 units in the case of an untreated sample; and
    (b) 12 units in the case of a sample boiled in water for 30 minutes.
(2) Visual vacuole assessment by immersing the fibers in a mixture of 70% of o-nitrotoluene and 30% of chlorobenzene, in which a non-dulled pore-free fiber of the abovementioned composition is invisible because of equal light refraction. If a fiber forms vacuoles, the following stages become visible, according to the vacuole population: 0 = no vacuoles, fibers invisible; 1 = individual fibers weakly visible; 2 = a few vacuoles; 3 = distinct vacuole formation, fibers visible; and 4 = heavy vacuole formation, limey appearance. The fiber produced in the example was assessed as follows: (a) untreated, 0; and (b) boiled, 0 to 1. Accordingly, the fiber is gloss-stable.

The fact that dry heat treatment alone is not sufficient for producing gloss-stability is shown by the following test. A sliver sample was removed from the dryer during production of the fibers as described above. This sample was glossy and vacuole-free. Scattered light measurement produced the following results: (a) untreated, 3 units; and (b) boiled, 110 units.

After boiling, the sliver appeared milky and opaque. The aftertreatment installation was then stopped and, 10 minutes later, another sample was removed from the dryer. As a result of the temperature of 160°C prevailing throughout this entire period, the sliver sample was distinctly brown in color, but still glossy.

Scattered light testing of this sample produced the following results: (a) untreated, 6 units; and (b) boiled, 140 units. Accordingly, gloss was not stabilized by the heat treatment.

IMPROVED HOT-WET PROPERTIES

Polycarbonate Resin Incorporated into Acrylic Fibers

The acrylic fiber produced by conventional spinning processes is deficient in hot-wet properties and in elastic recovery. These deficiencies cause various problems in fiber processing and in service use of knit or woven fabrics made from such fiber. Specifically, such fibers, yarns, or knit and woven fabrics are deformed in dyeing to such an extent that they do not survive subsequent processing or have poor commodity value and develop wrinkles that are difficult to remove.

K. Takeya, H. Suzuki and N. Yamawaki; U.S. Patent 4,012,459; March 15, 1977; assigned to American Cyanamid Company disclose a process for preparing acrylic fibers of improved hot-wet properties and improved elastic recovery by finely dispersing in the spin dope a solution of a polycarbonate resin in a solvent immiscible with the spin dope. The acrylic fibers so produced contain elongated fibrils of the polycarbonate resin of 0.05 to 5.0 microns diameter and up to several centimeters long.

The useful polycarbonate resin is one that has the structure

$$H-[ORO-\overset{\overset{\displaystyle O}{\|}}{C}]_n-OROH$$

wherein R is a residue of an aliphatic or aromatic-aliphatic dihydroxy organic compound after esterification to form a polycarbonate and n is an integer sufficient to provide a molecular weight of at least about 18,000, preferably 25,000, and still more preferably about 100,000 or more. These polycarbonates are well-known commercial products that are widely available.

Example: A fiber-forming spinning solution was prepared by dissolving 11 parts of a copolymer of 90% acrylonitrile and 10% methyl acrylate in 89 parts of an aqueous solution of 60% sodium thiocyanate. A solution of a polycarbonate obtained by interfacial polymerization of bisphenol A with phosgene having a number average molecular weight of 120,000 was prepared by dissolving 18 parts of the polycarbonate in 82 parts of tetrachloroethane. The polycarbonate solution was dispersed in the spinning solution using a propeller type agitator. The polycarbonate solution was added in sufficient amount to provide a ratio of 85:15 of acrylic polymer to polycarbonate and mixing was continued until a uniform dispersion was obtained.

The resulting composition was deaerated and spun into an aqueous solution of sodium thiocyanate maintained at –30°C using a spinnerette having 50 orifices each of a diameter of 0.1 mm. The coagulated filaments were subjected to stretching at a stretch ratio of 2 in conjunction with cold water washing and then subjected to stretching at a stretch ratio of 5 in boiling water. After drying to collapse the wet gel structure, the fiber was heat relaxed at 125°C to obtain an acrylic fiber free from devitrification and fibrillation. The fiber, upon microscopic examination, was found to contain numerous fibrils of polycarbonate discontinuously distributed therethrough, the fibrils having diameters ranging from 0.5 to 2 microns and a length to diameter ratio of 10 and greater. The following properties relative to fiber spun from the same acrylic polymer without the polycarbonate additive were obtained.

	With Polycarbonate	Without Polycarbonate
Young's Modulus (grams/denier) in water at 95°C	2.37	0.88
3% elongation elasticity in water at 95°C (%)	80.4	78.6

Two-Stage Stretching Process

Highly shrinkable dry-spun acrylic fibers having shrinkage levels of around 35% are already known. Unfortunately, fibers of this kind have low strength values of the order of 1.5 p/dtex because their high shrinkage values can be obtained only by stretching in water to a maximum of 250% at stretching temperatures below 90°C. In order to retain their high shrinkability, the fibers also have to be dried and crimped under mild conditions during their production, with the result that, in many cases, they show only a minimal power of adhesion. This often has an extremely adverse effect during further spinning of a yarn, in particular, when these fibers are spun without other fibers being admixed.

U. Reinehr, A. Nogaj and G. Lorenz; U.S. Patent 4,108,845; August 22, 1978; assigned to Bayer AG, Germany have found that, by dividing the stretching process into a prestretching and an afterstretching stage with the high shrinkage level of the acrylic fibers of 35% and more kept intact, it is possible to increase the total stretching ratio to approximately 1:4.5 and, hence, to obtain the required fiber strengths of 2 p/dtex and more.

Another important property of the high-shrinkage fibers obtained by the process is their vacuole-free, compact structure. By virtue of this property, finished articles produced from fibers of this kind do not undergo any undesirable changes in color and gloss.

It has been found that all the acrylic fibers produced by the process have a density of about 1.18 g/cc, which indicates the presence of vacuole-free, compact fiber structures.

Example 1: An acrylonitrile copolymer of 93.6% of acrylonitrile, 5.7% of methyl acrylate and 0.7% of sodium methallyl sulfonate was dry-spun by standard methods known in the art. The tow, which had an overall denier of 1,200,000, was stretched in a ratio of 1:1.5 in boiling water and was subsequently washed under tension in 3 successive washing baths at 80°C (washing baths 1 and 2) and 50°C (washing bath 3). The tow was then afterstretched in a ratio of 1:2.0 at a stretching-bath temperature of 75°C, so that the total stretching ratio amounted to 200%, i.e., to three times the original length of the tow. The rate of travel of the tow after the second stretching stage amounted to 50 m/min.

Individual filaments removed from the tow showed a shrinkage of 45.0% in boiling water. The tow was then treated with an antistatic preparation and crimped in a stuffer box into which steam was sprayed. The shrinkage of a number of individual filaments removed from the crimp tow was determined and gave an average shrinkage value of 44.6% in boiling water. The tow was then cut into staple fibers, dried in a dryer at 30° to 40°C, baled and packaged.

The final denier of each individual fiber amounted to 2.4 dtex. The fiber shrinkage of a number of individual filaments amounted to 43.7% in boiling water. Fiber strength was 2.3 p/dtex. Elongation at break was 23%. Fiber strength and elongation at break were measured with a Statigraph IV (Textechno Company). The high-shrinkage fibers were then spun into yarn with yarn counts of 40/1. Yarn constants: tensile strength, 11.5 RKm; elongation at break, 12.5%; satisfactory travel over cards and intersecting gill boxes; and density, 1.174 g/cc.

Example 2: *(Comparison)* — An acrylonitrile copolymer of 91.4% acrylonitrile, 5.2% methyl acrylate, and 3.4% sodium methallyl sulfonate was dry-spun and was stretched in a ratio of 1:2.5 in water at 80°C, washed at 50°C and then further aftertreated in the same way as described in Example 1. The fiber shrinkage in boiling water amounted to 41.8%. Fiber strength 1.5 p/dtex. Although the high shrinkage level required was obtained for a stretching level of around 250%, the required strength of at least 2 p/dtex was not obtained.

If, by contrast, the stretching ratio is increased to 1:3.6 at a stretching bath temperature of 75°C, a fiber strength of 2.1 p/dtex is obtained, whereas the fiber shrinkage amounts to only 28%.

N-Methylol Urethane Compounds as Copolymerizable Additives

The poor dimensional stability of fibers of acrylonitrile polymers and copolymers under hot, wet conditions manifests itself in poor elasticity of stitches of knitted fabrics, poor stability of pressed pleats under washing conditions and creasing or sagging of the finished articles in hot washes.

G.D. Wolf, W. Schnoor, J.-C. Voegele, U. Reinehr and G. Nischk; U.S. Patent 4,100,143; July 11, 1978; assigned to Bayer AG, Germany have found that acrylic fibers which have excellent characteristics under hot, wet conditions may be obtained if one uses as starting materials a copolymer of acrylonitrile and of a copolymerizable N-methylol compound of an unsaturated monourethane or bisurethane or of a copolymerizable alkyl ether of such N-methylol compounds of unsaturated monourethanes or bisurethanes and, optionally, other comonomers.

The N-methylol compounds which appear to be particularly suitable are the N-methylol compounds of unsaturated monourethanes or bisurethanes and the corresponding alkyl ethers corresponding to the following general formula

$$CR^1R^2=CR^3-CR^4R^5-OCON-R$$
$$\underset{CH_2R^6}{|}$$

wherein R represents hydrogen, a straight- or branched-chain C_{1-6} alkyl group, a cycloalkyl group or an optionally substituted phenyl group; R^1 represents hydrogen or a methyl group; R^2 represents hydrogen, a methyl group or the group

$$-CH_2-OCON-R$$
$$| \quad CH_2OR^6$$

R^3 represents hydrogen or the group

$$-CH_2-OCON-R$$
$$| \quad CH_2OR^4$$

R^4 represents hydrogen or a methyl group; R^5 represents hydrogen or the group

$$-CH_2-OCON-R$$
$$| \quad CH_2OR^4$$

and R^6 represents hydrogen, a straight- or branched-chain C_{1-6} alkyl group or a cycloalkyl group.

Example: 390 liters of water were boiled for 30 minutes under an atmosphere of nitrogen in a 450 liter enamel boiler and then cooled to 50°C. 28 kg of acrylonitrile followed by 1.2 kg of N-methyl-N-methylol allylurethane are added at this temperature and the pH is adjusted to 4 using 20% sulfuric acid. Polymerization is initiated by the addition of 0.33 kg of potassium peroxydisulfate and 1.32 kg of sodium disulfite. The suspension is cooled to room temperature after 6 hours and suction filtered through a 250 liter filter. The residue is washed neutral with water and dried in a vacuum at from 50° to 60°C. Yield: 27.8 kg (92.6% of the theoretical yield). K-value: 86.5.

A foil is cast from a 25% solution of the polymer in dimethyl formamide (DMF). The foil is briefly dried and precipitated in a 1% aqueous hydrochloric acid solution. The cast foil is washed in water until neutral and dried in a vacuum at from 90° to 100°C. At the end of these treatments, only small portions of the foil dissolve in DMF and the foil is crosslinked.

The acrylonitrile copolymer described above is dissolved in dimethyl formamide for preparation of a 22 wt % spinning solution. After filtration, the solution is dry spun by the conventional process. The filaments are collected on spools at the outlet of the spinning shaft. Every four spools of filament yarn having a total titer of 2,880 dtex are then combined into one band which is stretched by 1:3.6 in boiling water, washed and treated with antistatic dressing. It is then dried tension-free at 170°C for 10 minutes, at which stage crosslinking of the fiber is initiated. The titer of the individual filament is 3.3 dtex.

Resistance to the Formation of Vacuoles — The density of the fibers is 1.185 g/cc before and after a 10-minute treatment in boiling water. Both the samples which have been treated in boiling water and dried in a vacuum at 40°C for 12 hours and the untreated samples were embedded in anisole. The samples showed no differences in the transparency of the solutions.

Tack Point — Under the melting point microscope, the fibers show no signs of deformation at temperatures below 350°C.

Solubility — The fibers are still completely undissolved after one hour in cold dimethyl formamide and in dimethyl formamide heated to 130°C.

Dyeing — The whole cross section of the crosslinked fibers is completely dyed.

Dimensional Stability — (1) The hydrothermal elongation produced under a tension of 0.30 p/dtex in boiling water was measured on crosslinked fibers and on uncrosslinked comparison fibers. It amounts to 18.2% in the crosslinked fibers and the fibers do not tear whereas the uncrosslinked acrylic fibers tear after an elongation of more than 200%. (2) Ultimate tensile strength and elongation at tearing at 20°C demonstrate the reduced "flow properties" of the fibers according to the process.

	Crosslinked Fibers According to Example	Comparison
Ultimate tensile strength	2.65 cN/dtex	2.52 cN/dtex
Elongation on tearing	20%	54%

Crosslinking with an N-Methylol Alkyl Ether

T. Neukam, U. Reinehr, F. Bentz and G. Nischk; U.S. Patent 4,059,556; Nov. 22, 1977; assigned to Bayer AG, Germany describe a process for the production of crosslinkable polymers of acrylonitrile by polymerizing acrylonitrile with a crosslinking component and optionally other comonomers in strongly polar organic solvents and in the presence of a peroxodisulfate and a 1,3-diketone as initiator system.

The preferred crosslinking component is an N-methylol alkyl ether corresponding to the general formula

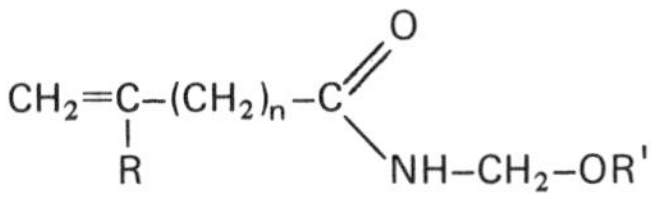

$$CH_2{=}C{-}(CH_2)_n{-}C{\Big\langle}_{NH-CH_2-OR'}^{O}$$
$$\underset{R}{|}$$

in which R represents hydrogen or a methyl group; n = 0 or an integer from 1 to 5; and R' represents a linear or branched-chain alkyl radical with 1 to 10 carbon atoms, preferably with 1 to 4 carbon atoms.

Example 1: 250 parts by weight of dimethyl formamide, 118 parts by volume of acrylonitrile and 5 parts by weight of N-methoxy methyl acrylamide were heated to 40°C. The polymerization reaction was started by the addition of 12 parts by weight of ammonium persulfate and 0.8 part by volume of acetyl acetone, followed by stirring under nitrogen for 21 hours. The viscous solution was then freed from the residual monomers in a thin layer evaporator and diluted to such an extent that it had a viscosity of 480 p (80°C).

Example 2: The polyacrylonitrile solution of Example 1 was filtered and dry spun by a standard method. 70 packages of filament yarns with a total denier of 1,600 dtex were combined to form a sliver and four such slivers combined to form a tow with an overall denier of dtex 448,000. The tow was drawn in a ratio of 1:3.6 in boiling water, washed, treated with antistatic preparation, dried in the absence of tension for 10 minutes at 175°C (crosslinking taking place under the effect of the heat applied) and crimped. The fiber tow had an individual fiber denier of 3.3 dtex.

Under a melting point microscope, the fibers did not show any deformation up to a temperature of 350°C.

Comparisons were made with a dry-spun acrylic fiber with a denier of 3.3 dtex produced from a copolymer of 94 wt % of acrylonitrile, 5% of methyl acrylate and 1% of methallyl sulfonate and drawn in a ratio of 1:3.6 in boiling water.

Pieces of knitting produced from fiber yarns of the process and from the comparison fibers were washed with a gentle detergent, spin-dried and dried at room temperature 5 times at 60°C (according to DIN 54 010) and 5 times at 95°C water temperature (DIN 54 011). The dimensional changes were then determined.

The dimensional changes for the process fabric at 60°C were: length, -4%; width, +11%. For the comparison, the values were -14% and +40%. At 95°C, the process fabric showed changes of -6% and +18%, compared to -20% and +60% for the comparison fabric. As can be seen from the dimensional changes, a distinct improvement in dimensional stability is obtained with the crosslinked fibers.

Spinning with a Polyoxazoline

E. Radlmann, U. Reinehr, P. Hoffmann, D. Arlt and G. Nischk; U.S. Patent 4,131,724; December 26, 1978; assigned to Bayer AG, Germany provide crosslinked synthetic fibers and filaments of acrylonitrile copolymers with improved dimensional stability which may be produced by a simple heat treatment at a certain stage of the aftertreatment process during the processing cycle by which the fibers are formed.

The process comprises dissolving an acrylonitrile copolymer containing at least 20 mval of carboxyl groups per kg of polymer together with at least one polyoxazoline, in a polar organic solvent at a temperature of from 20° to 120°C, the molar ratio of carboxyl groups to oxazoline groups amounting to at least 1:1, spinning the resulting solution into filaments by a wet or dry spinning process and effecting crosslinking during drying of the filaments at a temperature of from 120° to 190°C.

Example: 15 parts of an acrylonitrile/acrylic acid copolymer containing 555 mval of carboxyl groups per kg of polymer (4.0 wt % of acrylic acid), together with 0.899 part of 1,4-bis(Δ-2-oxazolin-2-yl)-benzene, are dissolved in dimethyl formamide at 80°C to form a 28 wt % solution. The solution is filtered and dry spun by a conventional method at a duct temperature of 160°C.

The spun filaments are collected on bobbins and doubled to form a tow. The tow is then drawn in a ratio of 1:3.6 in boiling water and washed in boiling water for 3 minutes under light tension. An antistatic preparation is then applied, followed by drying for 10 minutes at 165°C in a screen drum dryer with 20% permitted shrinkage, the heat effect initiating the crosslinking reaction. The tow is then cut into fibers with a staple length of 60 mm. The individual fibers have an individual denier of 3.3 dtex and are insoluble in dimethyl formamide, even after 1 hour at 130°C. Under a melting point microscope the fibers do not show any deformation up to 350°C. The fibers have a density of 1.187 g/cc which they retain after treatment for 15 minutes in boiling water. The fibers can be deeply dyed with a blue dye and do not show any significant difference when visually compared with a standard commercial-grade acrylic fiber containing a dye-receptive additive (composition: 94 wt % of acrylonitrile, 5 wt % methyl acrylate, and 1 wt % of sodium methallyl sulfonate).

Important fiber properties are compared below with those of a standard commercial-grade acrylic fiber.

Fiber Properties	Crosslinked Sample	Comparison
Strength, p/dtex	2.30	2.61
Elongation, %	40	48
Tack point, °C	>280	215–220
Degree of swelling, %*	1,550	soluble
Flexlife test, min	1.39	5–7
Loop tensile strength, p/dtex	1.11	1.50
Loop elongation at break, %	14	16

*Degree of swelling = increase in weight after 2 hours at 50°C in DMF.

Removable Compound Providing High Moisture Absorption and Water Retention

E. Radlmann, U. Reinehr and G. Nischk; U.S. Patent 4,143,200; March 6, 1979; assigned to Bayer AG, Germany provide a process for the production of filaments and fibers having a moisture absorption of at least 7% and a water retention capacity of at least 25% by dry-spinning an acrylonitrile copolymer containing more than 50 mval of carboxyl groups, from a solvent which contains 5 to 50 wt % of a compound with properties defined herein, washing the compound added to the solvent out of the freshly spun filaments and fibers and partly or completely converting the carboxyl groups into the salt form.

The substances or mixtures of substances added to the solvent should have a boiling point higher than that of the solvent by preferably about 50°C, they should be miscible with water and the solvent, preferably in any proportions, and they should be nonsolvents for the polymer, i.e., the copolymer should at the most undergo only slight dissolution in the liquid. The good solubility in water is important to ensure complete removal of the substance during the aqueous after-treatment of the fibers. Furthermore, it is advantageous to select compounds which do not form an azeotropic mixture with the spinning solvent used so that they can be recovered as far as possible quantitatively.

Suitable compounds include, e.g., monosubstituted or polysubstituted alkyl ethers and esters of polyhydric alcohols, such as diethylene glycol monomethyl or dimethyl ether, diethylene glycol monoethyl or diethyl ether, diethylene glycol, triethylene glycol, tripropylene glycol, triethylene glycol diacetate, tetraethylene glycol, tetraethylene glycol dimethyl ether, glycol ether acetate, e.g., butyl glycol acetate, high boiling alcohols, e.g., 2-ethylcyclohexanol, esters or ketones, trimethylolpropane, mannitol, sorbitol, glucose or, preferably, glycerol, or mixtures thereof.

In addition to good filament properties such as high tensile strength, elongation on tearing and dye absorption capacity, the filaments show a hitherto unknown combination of high water retention capacity with high moisture absorption.

It is possible, by the method described, to obtain types of filaments having combinations of properties far superior to those of cotton. This is of great practical importance because these two factors are important physical properties for textiles used in clothing. One advantage of the filaments according to this process compared with cotton filaments is that cotton which has absorbed a large quantity of water has a wet feel, whereas these filaments, by virtue of their porous core and sheath structure and their hydrophilic character, allow the water to diffuse into the core so that textiles worn next to the skin feel comparatively dry even under conditions of heavy perspiration and are comfortable to wear.

The moisture absorption based on the dry weight of the filaments is determined gravimetrically. The samples are exposed to an atmosphere of 21°C and 65% relative humidity for 24 hours. To determine the dry weight, the samples are then dried to constant weight at 105°C. The water retention capacity is determined in accordance with DIN specification 53 814.

Example: 2.85 kg of an acrylonitrile/acrylic acid copolymer composed of 90% of acrylonitrile and 10% of acrylic acid (139 mval of carboxyl groups per kg) are dissolved in a mixture of 10.00 kg of dimethyl formamide and 2.15 kg of glycerol at 80°C for 1 hour, filtered and dry spun by known methods at a shaft temperature of 160°C. The spun goods are collected on spools and doubled to form a cable still containing 13.9% of glycerol. The cable is then stretched in a ratio of 1:3.6 in boiling water, washed in boiling water under a slight tension for 3 minutes, thereupon passed under a light tension through an aqueous bath containing about 10 wt % of sodium carbonate at 25°C for 5 minutes and finally again washed in boiling water for 3 minutes. An antistatic dressing is then applied and the cable is then dried in a sieve drum dryer at a maximum temperature of 130°C and under conditions permitting 20% shrinkage, and it is then cut up into staple fibers 60 mm in length.

The individual filaments having a titer of 3.3 dtex have a moisture absorption capacity of 9.2% and a water retention capacity of 92%, an ultimate tensile strength of 1.8 p/dtex and an elongation on tearing of 25.9%. Under an optical microscope, the fibers show a clear core and sheath structure of irregular cross section. The proportion of residual solvent in the filaments is less than 0.2% and the proportion of glycerol still left in the filaments is less than 0.6%. The filaments can be dyed to a deep color.

FLAME-RETARDANT POLYMERS

Stable to Heat and Light

T. Yamazaki, S. Kurioka, T. Hatano, Y. Higashiyama and S. Asada; U.S. Patent 4,007,232; February 8, 1977; assigned to Kanegafuchi Kagaku Kogyo KK, Japan provide an acrylic synthetic fiber comprising a copolymer of a polycomponent system consisting essentially of 40 to 65 wt % acrylonitrile, vinyl chloride and vinylidene chloride in the ranges defined by the formula

$$102.8 < A + B + 1.29C < 107.4$$

wherein A is the weight percent of a mixture of olefinic monomers excluding vinyl chloride, vinylidene chloride and monomers having dyeable chemical sites in their molecular structures, and which olefinic monomers mainly comprise acrylonitrile, B is the weight percent of vinyl chloride and C is the weight percent of vinylidene chloride.

Further, there may be added to the polymer 0.3 to 10 wt % of antimony oxide and/or stannic acid, based on the weight of the copolymer.

Examples 1 and 2; Comparative Examples 1 through 4: In each example, emulsion polymerization was carried out using a pressure proof polymerization vessel of 15 liter inside capacity. The polymerization conditions were as follows. For 100 parts of a mixture of monomers, 700 parts of water were used. A polymerization initiator comprising a combination of ammonium persulfate and sodium hydrogen sulfite was used. Sodium alkylbenzene sulfonate was used as the emulsifier. The temperature of polymerization was 43°C. The polymerization time was 5 hours. In Table 1, there are listed six specimens of acrylic copolymers obtained by emulsion polymerization using the above conditions and having the specified components.

Table 1

Ex. No.	..Copolymer Composition (wt %)...				A + B + 1.29C	Content of Chlorine (Cl %)
	AN	VCl	VdCl$_2$	NaMAS		
1	58.0	28.5	13.1	0.4	103.4	25.8
2	58.2	20.3	21.0	0.5	105.6	26.9
1*	57.8	36.3	5.4	0.5	101.1	24.6
2*	57.6	0	42.1	0.3	111.9	30.8
3*	58.1	41.3	0	0.6	99.4	23.5
4*	58.5	10.9	30.2	0.4	108.4	28.3

Note: AN is acrylonitrile, VCl is vinyl chloride, VdCl$_2$ is vinylidene chloride, and NaMAS is sodium methallyl sulfonate.

*Comparative.

A spinning dope was prepared by dissolving each copolymer shown in Table 1 into dimethyl formamide as a solvent and adding 3 wt % of stannic acid and 3 wt % of a straight polymer of glycidyl methacrylate. Then wet spinning was carried out in each example, extruding the dope through a nozzle into a coagulating bath consisting essentially of 60% aqueous solution of dimethyl formamide to coagulate. The coagulated filament was continuously washed with water, dried, stretched and finally thermally treated.

The nonflammability, antidevitrifying, lightproof, antirusting and antiyellowing properties were tested for on the obtained fibers and the results thereof are shown in Table 2.

Table 2

Ex. No.	A + B + 1.29°C	..Property.. Nonflammability	Anti-Devitrifying	Light-Proof	Anti-Rusting	Anti-Yellowing
1	103.4	Superior	x	x	x	x
2	105.6	Superior	x	x	x	x
1*	101.1	Not superior	x	x	x	x
2*	111.9	Not superior	x	y	z	y
3*	99.4	Not superior	x	x	x	x
4*	108.4	Not superior	x	y	y	y

Note: x is superior; y is inferior; and z is not as superior.

*Comparative.

Modacryl Copolymer with Improved Thermal Stability and Whiteness

B. Huber, H.-J. Kleiner and H. Neumaier; U.S. Patent 4,052,551; October 4, 1977; assigned to Hoechst AG, Germany describe filaments and fibers, the fila-ment-forming substance of which is a copolymer of from 35 to 85 wt % of acrylonitrile, from 5 to 45 wt % of vinyl chloride, vinyl bromide and/or vinyl-idene chloride, from 5 to 30 wt % of carboxyphosphinic acid derivatives of the formula

(1)

$$\begin{array}{c} R_1 \\ \diagdown \\ \diagup \\ R_2 \end{array} \overset{O}{\underset{\parallel}{P}}\text{-}O\text{-}CH_2\text{-}CH_2\text{-}O\text{-}\overset{O}{\underset{\parallel}{C}}\text{-}\underset{\underset{R_3}{|}}{C}\text{=}CH_2$$

where R_1 is lower alkyl having up to 8 carbon atoms, or $CH_2Cl\text{-}$, R_2 is lower alkyl having up to 8 carbon atoms in which a hydrogen atom may be substituted by a halogen atom in case of R_1 being $CH_2Cl\text{-}$, and R_3 is hydrogen or methyl, and from 0 to 15 wt % of other unsaturated compounds having an activated double bond; the weight percentages being relative to the total amount of mono-mers, and a process for their manufacture. The filaments obtained are self-ex-tinguishing and excel furthermore by their thermostability and their high degree of whiteness.

Example 1: The following amounts of the following substances were introduced into a steel vessel having a capacity of 1 liter: 250 ml/h of a monomer mixture consisting of 72 parts by weight of acrylonitrile, 15 parts by weight of vinylidene chloride and 13 parts by weight of a compound of the formula

$$\begin{array}{c} CH_3 \\ \diagdown \\ \diagup \\ C_2H_5 \end{array} \overset{O}{\underset{\parallel}{P}}\text{-}O\text{-}CH_2\text{-}CH_2\text{-}O\text{-}\overset{O}{\underset{\parallel}{C}}\text{-}CH\text{=}CH_2$$

250 ml/h of a solution of 10 g of sodium acetate, 15 g of sodium-methallyl sul-fonate and Mohr's salt in 1,750 ml of water, which solution was adjusted by means of sulfuric acid to a pH of 2.5, varying amounts of a solution of potas-sium peroxodisulfate in water and a solution of sodium disulfite in water.

The weight ratio of potassium peroxodisulfate to sodium disulfite was 1:4. The amounts required depended on the intended polymerization degree.

The polymerization was carried out at 55°C with pressure. The polymer suspen-sion formed was discharged continuously after a residence time of 1 hour, thus obtaining a conversion rate of about 80%. The polymer was carefully washed and dried.

The relative viscosities were measured at 25°C on a 0.5% solution in dimethyl formamide.

The polymer was introduced into the dimethyl formamide at 0° to -20°C with agitation, and agitation was continued for one-half hour at 60°C. Subsequently, the solution was filtered and degassed. The polymer content of the spinning solutions was 15 to 30 wt %, relative to the total solution. The spinning solu-tion was forced through a nozzle having 100 holes of a diameter of 80 μ, into a coagulation bath consisting of 65% of dimethyl formamide and 35% of water. The temperature of the spinning bath was 30°C. The filaments so obtained were drawn to about 4 times their length in two further hot baths having a lower dimethyl formamide content, washed with water in further baths and dried on hot godets. After drying, a further drawing by 25% of their length was carried out.

The degree of whiteness of these filaments was determined; the quotient of the reflectance at 426 nm (R_{426}), divided by the brightness value, serving as measure for the degree of whiteness.

The thermostability was tested on the polymer powder, and in part on the filaments spun therefrom. The polymer powder was screened and exposed to thermal strain at 150°C in a drying cabinet. The reflectance was measured by means of the reflectance photometer Elrepho. The brightness was determined by means of the colorimetric filter FMY/C adjusted to a calibrated MgO working standard, and the value R_{426} was determined with the aid of the filter R 42 (main wavelength 426 nm).

Example 2: In accordance with Example 1, a further copolymer was prepared for a comparison and filaments were spun from this material. The phosphorus compound used was

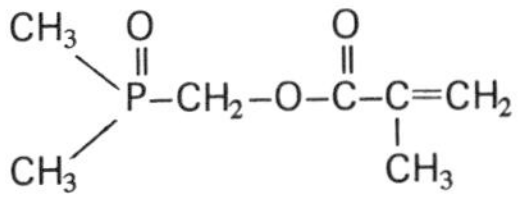

$$\begin{array}{c} CH_3 \\ {}^{\diagdown} \\ {}^{\diagup} \\ CH_3 \end{array} \overset{O}{\overset{\|}{P}} - CH_2 - O - \overset{O}{\overset{\|}{C}} - \underset{\underset{CH_3}{|}}{C} = CH_2$$

Example 3: Example 1 was repeated, but 14 parts by weight of a compound of the following formula

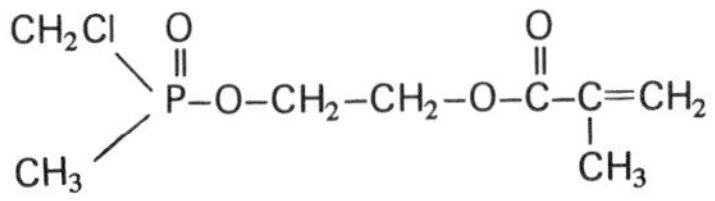

$$\begin{array}{c} CH_2Cl \\ {}^{\diagdown} \\ {}^{\diagup} \\ CH_3 \end{array} \overset{O}{\overset{\|}{P}} - O - CH_2 - CH_2 - O - \overset{O}{\overset{\|}{C}} - \underset{\underset{CH_3}{|}}{C} = CH_2$$

containing 71 parts by weight of acrylonitrile and 15 parts by weight of vinylidene chloride were copolymerized.

The results show that the thermostability of modacryl filaments and fibers and the copolymers used as basic material thereof depends considerably on the kind of the phosphorus derivative employed. At about the same phosphorus content in the copolymer, the flame protection is practically equal when phosphinic acid derivatives and phosphine oxide derivatives are used; however, when the copolymer is subjected to thermal strain, the material modified with phosphine oxide derivatives shows heavy discoloration which becomes evident by the indicated brightness values.

Furthermore, the test results indicate that in the case of fibers of copolymers containing compounds of formula (1) where R_1 is CH_2Cl-, no discoloration can be measured. Fibers according to Example 3 are completely transparent upon visible examination, while the filaments according to Example 1 are slightly turbid. The filaments of comparative Example 2 are heavily discolored.

Nonfoaming Emulsion

T. Kobashi and K. Masuhara; U.S. Patent 4,036,803; July 19, 1977; assigned to Japan Exlan Company Limited, Japan have developed an acrylic synthetic fiber having excellent flame retardancy and transparency, without causing aggregation and foaming, by introducing into a spinning solution of an acrylonitrile polymer a halogen-containing polymer emulsion obtained by a special emulsion polymerization process.

The flame-retardant acrylic synthetic fiber is produced by mixing a spinning solution composed of an inorganic solvent solution of an acrylonitrile polymer with an emulsion of a polymer of a vinyl halide and/or vinylidene halide and wet-spinning the resulting mixed solution. The emulsion is obtained by polymerizing a vinyl halide and/or vinylidene halide, if necessary plus other unsaturated monomers copolymerizable therewith, in an aqueous medium of a pH below 4 using a water-soluble catalyst, in the presence of a water-soluble polymer containing monomer units consisting essentially of an ethylenically unsaturated carboxylic acid or a salt thereof and monomer units consisting of an ethylenically unsaturated sulfonic acid or a salt thereof.

Example: Methacrylic acid and sodium p-styrenesulfonate were mixed in the ratio of 70:30. To 100 parts of this monomer mixture, 2 parts of ammonium persulfate, 1.8 parts of sodium metabisulfite, 0.0015 part of ferrous chloride and 230 parts deionized water were added, and the resulting mixture was subjected to polymerization at 70°C for 1 hour under stirring. The thus-obtained solution of the water-soluble polymer was directly dissolved in water and used in the subsequent emulsion polymerization.

The formulation used in the emulsion polymerization was as follows, with all parts by weight: vinylidene chloride, 38; water-soluble polymer, 2; ammonium persulfate, 0.2; sodium metabisulfite, 0.18; ferrous chloride, 0.0006; and deionized water, 60.

The prescribed amounts of the water-soluble polymer and ferrous chloride were first dissolved in 55 parts of deionized water. This solution was fed to a polymerization vessel, and after the prescribed amount of vinylidene chloride was further fed, stirring was started. Solutions obtained by dissolving the ammonium persulfate and sodium metabisulfite in 2.5 parts of deionized water respectively were added dropwise respectively to the polymerization vessel, and then the polymerization was started. The speed of addition of these catalyst solutions was so controlled that the addition was completed in one hour. Since the pH of the polymerization system was below 3, no particular adjustment was made. The polymerization reaction was carried out at atmospheric pressure at 30°C for 2 hours.

Eleven parts of the thus-obtained polyvinylidene chloride emulsion of an average particle diameter of about 50 mμ and 16 parts of an acrylonitrile copolymer consisting of 88% acrylonitrile and 12% vinyl acetate were mixed, and further 8 parts of water and 65 parts of an aqueous 58% solution of sodium thiocyanate were added. After stirring at 70°C for 1 hour, an acrylic spinning solution in which polyvinylidene chloride particles were dispersed finely and uniformly was obtained, without causing foaming. In the spinning solution thus obtained, no formation of masses due to aggregation of polyvinylidene chloride particles occurred, and the solution was very useful as a spinning solution for producing filaments.

The spinning solution was extruded into a coagulating bath of an aqueous 10% sodium thiocyanate solution at 0°C to form filaments. The filaments, after being washed with water, were stretched 10 times the length, dried and then subjected to wet heat relaxing treatment at 115°C, with the result that an acrylic synthetic fiber having very good flame retardancy and excellent transparency was obtained.

Troubles such as filter clogging, spinnerette clogging or foaming in the coagulating bath did not occur.

High Luster and Low Light Transmission

A. Maranci; U.S. Patent 4,081,498; March 28, 1978; assigned to American Cyanamid Company describes acrylic fibers of a fiber-forming first acrylonitrile polymer containing at least 50% acrylonitrile, a flame-retardant amount of a halogen-containing vinyl monomer and any balance of a halogen-free vinyl monomer having heterogeneously dispersed therein a small amount of an incompatible, halogen-free, second acrylonitrile polymer containing at least 70% acrylonitrile and one or more halogen-free vinyl monomers. The fibers are prepared by wet-spinning an intimate mixture of the polymers separately dissolved in aqueous inorganic solutions of the same salt following conventional procedures but including a hot-wet relaxation of the stretched wet-gel filaments prior to drying.

The provision for small amounts of a halogen-free polymer within the fiber-forming halogen-containing polymer with which it is incompatible coupled with the provision for relaxation of the stretched wet-gel filaments prior to drying results in a flame-retardant acrylic fiber having high luster and low light transmission. The latter property provides a desirable low level of apparent soiling tendencies in the resulting fiber.

Comparative Example A: A spinning solution was prepared containing 10% of a fiber-forming polymer of composition 81.1% acrylonitrile, 9.2% methyl methacrylate, and 9.7% vinylidene chloride in 90% of an aqueous solution of 46% sodium thiocyanate. The solution had a viscosity of 34 poises at 28°C and was extruded through a spinnerette having 10 orifices, each of 200 microns diameter, into an aqueous 12% sodium thiocyanate solution maintained at -2°C, to form filaments. The filaments were continuously withdrawn from the bath, stretched at a stretch ratio of 2, washed with water, and drawn a second time in water at 99°C so as to provide a cumulative stretch ratio of 12 and a denier of 9.6.

The stretched filaments were dried in a free-to-relax state at 127°C dry bulb and 60°C wet bulb and then further relaxed in saturated steam at 130°C. The filaments obtained had a high degree of luster and a high degree of light transmission, indicating that the filaments had an undesirably high level of apparent soiling tendencies.

Comparative Example B: A portion of the wet-stretched filaments of Comparative Example A, prior to drying were first exposed to saturated steam at 110°C in a free-to-relax state for 10 minutes and thereafter dried at 100°C dry bulb and 36°C wet bulb. The filaments obtained were highly delustered but had a very low light transmission, thus indicating that the apparent soiling tendencies were low but the luster value of 9.2 was unsatisfactory.

Example 1: A solution of 11.2% of a polymer having a composition of 89.3% acrylonitrile and 10.7% methyl methacrylate, was prepared in 88.8% of an aqueous solution of 40% sodium thiocyanate. Five parts of this solution were mixed with 95 parts of the solution used in Comparative Example A. The mixed solution was turbid, thus indicating incompatibility between the two solutions. The mixed solution was spinnable, however, and stretched wet gel filaments were made following the procedure of Comparative Example A.

The wet-stretched filaments were then exposed to saturated steam and dried as in Comparative Example B. The filaments obtained had a high degree of luster, 19.8%, and a low degree of light transmission, indicating low apparent soiling tendencies.

Example 2: The procedure of Example 1 was followed in every material detail except that 10 parts of the solution of Example 1 were mixed with 90 parts of the solution of Comparative Example A. The resulting fiber had a low degree of light transmission, thus indicating a desirably low apparent soiling tendency, and a high degree of luster, 17.5%.

In addition to the light transmission and luster properties reported in the examples above, the fiber obtained in each of the examples had a desirable level of flame retardancy when tested according to standard procedures.

Reduced Corrosive Effects

L.G. Robinson; U.S. Patent 4,077,929; March 7, 1978; assigned to E.I. Du Pont de Nemours and Company produces flame-resistant filaments by preparing a hot solution of a halogen-containing acrylonitrile polymer composition in an inert organic solvent for the composition, extruding the solution to form filaments and removing the solvent from the filaments. The improvement provides for modification of the solution to greatly reduce corrosive effects of the solution on metallic equipment used in the process.

The acrylonitrile polymer compositions to which the process is applicable are compositions containing at least 40 wt % polymerized acrylonitrile monomer units and at least 3 (preferably at least 10) wt % halogen, particularly chlorine or bromine. The composition may be a halogen-containing acrylonitrile polymer in which the halogen is bonded directly to the polymer chain, or it may be a mixture of an acrylonitrile polymer with a halogen-containing organic compound.

Water and a phosphorus compound are included in the hot solution of halogen-containing acrylonitrile polymer composition in inert organic solvent. The phosphorus compound is phosphoric acid, or phosphorous acid, or a triorganophosphite compound consisting of carbon, hydrogen, oxygen and phosphorus. The amount of water should be 1 to 10 wt % (preferably 2 to 5) based on the weight of inert organic solvent plus water. The amount of phosphorus compound should be 0.005 to 0.1 (preferably about 0.02 to 0.03) mol per kilogram of polymer in the composition.

Surprisingly, the rate of corrosion of stainless steel by solutions prepared in accordance with the process is as much as a hundredfold lower than the rate of corrosion experienced with solutions containing no water or phosphorus compound. The water and the phosphorus compound have a synergistic effect upon one another in reducing corrosion. In the absence of water, corrosion is at best reduced slightly by adding a phosphorus compound to the solution of the flame-resistant polymer. In the absence of the phosphorus compound, corrosion is reduced moderately by adding water to the solution; but up to a 20-fold additional reduction is achieved by adding the phosphorus compound along with the water.

High Sticking Temperature and Low Shrinkage

D.S. Gibbs; U.S. Patent 4,164,522; August 14, 1979; assigned to The Dow Chemical Company provides a flame-retardant acrylic fiber by utilizing an additive whose presence only moderately affects the desirable inherent physical properties, such as high sticking temperature and low shrinkage, of an unmodified acrylic fiber.

A crosslinked vinylidene chloride polymer microgel powder is incorporated in acrylic fibers as a flame-retardant additive. The microgel powder is recovered from a latex obtained by emulsion polymerizing in sequence:

(1) a first monomer mixture comprising about 85 to about 95 parts by weight of vinylidene chloride, about 5 to about 15 parts by weight of a copolymerizable ethylenically unsaturated comonomer, and a minor amount of a copolymerizable crosslinking polyfunctional comonomer, wherein the polymer resulting from the polymerization of the first monomer mixture has a gel content in the range of about 1 to about 50%;

(2) a minor amount of a polyfunctional comonomer for providing graft sites on the product of (1); and

(3) about 10 to about 25 wt %, based on the weight of the first monomer mixture, of a second monomer mixture comprising about 85 to about 95 wt % acrylonitrile and about 5 to about 15 wt % of an ethylenically unsaturated comonomer copolymerizable with acrylonitrile.

The microgels in the resulting latex have a diameter less than about one micron.

A spinning solution for the preparation of flame-retardant acrylic fibers comprises (a) a fiber-forming polymer containing at least 85 wt % acrylonitrile; (b) a solvent for the fiber-forming polymer; and (c) the crosslinked vinylidene chloride polymer microgel powder.

Example 1: *Preparation of Crosslinked Vinylidene Chloride Polymer Microgel Powder —*

Initial Water Phase: 1,800 g distilled water, 15 g Aerosol MA 80 emulsifier (80% active), pH adjusted to 3.5 acetic acid;

Reducing Agent: 9.75 g Hydrosulfite AWC in 1,000 g aqueous solution, feed rate, 10 g/hr;

Initiator: 13.5 g of 70% tert-butyl hydroperoxide (TBHP) in 1,000 g aque-our solution, feed rate, 10 g/hr;

Aqueous Emulsifier Stream: 236 g Dowfax 2Al emulsifier (45% active) in 1,600 g aqueous solution, used 800 g in 20 hours, feed rate, 40 g/hr;

Monomers for Seed Latex: 50 g acrylonitrile (VCN), 450 g vinylidene chloride (VDC), 10 g 1,3-butylene glycol dimethacrylate (BGDM), used 150 g in seed latex reaction;

First Monomer Mixture: 200 g VCN, 80 g BGDM, 3,800 g VDC, used 2,500 g in 20 hours (feed rate = 125 g/hr);

Polyfunctional Monomer: 85 g 1,3-butadiene added at end of monomer feeding;

Second Monomer Mixture: 315 g methyl acrylate, 1,785 g VCN, used 375 g in 3 hours (feed rate, 125 g/hr).

The initial water phase was introduced into a two-gallon reactor and placed un-der vacuum of about 25 mm Hg for 10 minutes while being heated to 40°C with agitation at 100 rpm. When the water phase had reached a temperature of about 40°C, the vacuum was shut off and 150 g of seed latex monomer was in-troduced. Immediately, the reducing agent and initiator streams were introduced at 10 g/hr. When the seed latex reaction had proceeded to approximately 4 psi pressure drop from the maximum pressure achieved, introduction of the first monomer mixture at 125 g/hr and the emulsifier stream at 40 g/hr was begun and continued for 20 hours while concurrently pumping reducing agent and in-itiator each at 10 g/hr.

After 20 hours of feeding the first monomer mixture, this stream was shut off and 85 g of butadiene was shot into the reactor. Within 15 minutes, pumping of the second monomer mixture was begun at 125 g/hr while continuing to feed emulsifier at 40 g/hr and the reducing agent and initiator each at 10 g/hr. The second monomer mixture feed and the emulsifier stream were shut off after 3 hours and the initiator and reducing agent were pumped for 3 additional hours at 10 g/hr each to complete the reaction. The latex was cooled to room tem-perature to prepare for polymer recovery.

The microgel powder was recovered from the latex by a common coagulation method: 7,000 ml water and 35 g alum were mixed and heated to 50°C. Then, with vigorous agitation, the 2,700 ml latex was slowly added and the tempera-ture was raised to 70°C and held at that temperature for 5 minutes. The mixture of water, alum, and polymer crumb was then rapidly cooled in ice and the wet polymer powder was collected in a centrifuge and washed with water for 10 min-utes. Thereafter, the powder was air dried at room temperature.

Example 2: *Evaluation of the Microgel Powder as a Flame-Retardant Additive* — A spinning solution was prepared by first adding 148 g of the powder obtained in Example 1 to 650 g of dimethyl formamide and thoroughly mixing the in-gredients at room temperature in a Tekmar Dispax high shear agitator. The mix-ture was then heated to 50°C to ensure complete dispersion. The so-formed dis-persion was slightly turbid, but uniform and relatively low in viscosity.

Next, 202 g of a fiber-forming polymer comprising a polymer used to prepare the commercial acrylic fiber Creslan T61 was added and the mixture was stirred for 3 hours at a rate sufficient to maintain the temperature below 85°C (no external heating was required).

The resulting spinning solution was spun into fibers using a spinning unit having a reservoir, pump, filter, and spinnerette. The spinnerette was immersed in a coagulation bath comprising a mixture of 16 parts water and 84 parts dimethyl formamide at about 25°C. The fibers were then passed over a wash roll, oriented under steam, and then relaxed under steam and gathered on a hot chrome roll at 105°C and then wound on spools.

The resulting fibers, having been subjected to a net draw of 5.0, were tested for physical properties. For comparison, fibers prepared in a similar manner, but without adding the vinylidene chloride polymer microgel powder, were tested. The results of these tests are shown below.

	Process Fiber	Comparison Fiber
Percent shrink	17.8	16.5
Overall stretch ratio	5.0	5.0
Denier per filament	3.17	2.94
Tenacity, g/den	2.04	3.94
Percent elongation	19.4	14.9
Percent Cl	25	0
Limiting oxygen index	24	17

CELLULOSIC POLYMERS

SUPERABSORBENT FIBERS

Coating of Water-Insoluble, Water-Absorbent Polymers

A technique has been found by *A.R. Reid; U.S. Patent 4,128,692; December 5, 1978; assigned to Hercules Incorporated* which greatly facilitates the incorporation of superabsorbent polymeric materials into structures in which their superabsorbent properties can be utilized. It has been found that if the superabsorbent material is precipitated from a gel onto the surface of a fibrous cellulose material and dehydrated with a water-miscible nonsolvent, discrete coated cellulose fibers are recovered which exhibit extremely good absorbency properties as to both rate of absorption and to volume of fluid which can be absorbed.

The process comprises adding a superabsorbent material of the type described above to an aqueous suspension of a long fiber cellulose furnish while agitating, continuing agitation for a time sufficient to allow substantially all of the superabsorbent material to form an aqueous gel slurry having the long fiber cellulose furnish suspended therein, precipitating the superabsorbent material onto the surface of the long fiber furnish by adding a water-miscible nonsolvent material to the slurry, and dehydrating the coated fibers with a water-miscible nonsolvent.

Applicable superabsorbent materials include any water-insoluble, water-swellable polymers, including synthetic polymers such as crosslinked acrylamide-sodium acrylate copolymers. The superabsorbent materials of choice in this process are based on polysaccharides, either natural or synthetic. Materials of this class include, e.g., crosslinked, normally water-soluble cellulose derivatives which are crosslinked to water-insoluble, water-swellable compounds such as crosslinked sodium carboxymethylcellulose (CMC), and crosslinked hydroxyethylcellulose, crosslinked partial free acid CMC, and cellulose, starch, and guar gum grafted with acrylamide and acrylic acid salts in combination with divinyl compounds, e.g., methylenebisacrylamide.

The most preferred materials are the CMC derivatives, either crosslinked sodium

CMC or partial free acid CMC. Both of these materials are known to be highly absorbent.

In evaluating the absorbent performance of the products, two tests are used principally. These are referred to as the CAP test which measures absorbent capacity and initial rate of absorption, and the Syringe test which measures absorption rate and wicking ability.

The apparatus employed for the CAP test consists of a Buchner fritted glass funnel, with a rubber tube attached to its neck; the tube is attached at the other end to a 50 ml burette. The burette is filled with the test solution, and the level of liquid is allowed to rise until it just makes contact with the bottom of the frit in the funnel. The level of liquid in the burette can be anywhere from 0 to 60 cm below the bottom of this frit.

The test sample is placed on top of the frit and a weight exerting a pressure of from 0.1 to 0.4 psi is applied to the sample. The test is then begun, and the loss of fluid in the burette is monitored as a function of time to give the rate of absorption. When equilibrium is reached, the capacity is calculated by dividing the total fluid absorbed at equilibrium, or at the end of 45 min, by the weight of the polymer sample. The conditions used with the CAP test for this work were: (1) pressure exerted on the sample was 0.11 psi; (2) all of the tests were done with the liquid in the burette 2 cm below the fritted glass initially. This level was allowed to continually change as absorption occurred; (3) pore size of the frit was about 4 to 5.5 μ.

In carrying out the Syringe test, a 10 cc calibrated syringe is filled with 1.0 g of test sample and compressed with the syringe plunger to give a uniform column of material. The volume to which the material was compressed varied with the bulk of the sample. For most fibrous samples, the compressed volume was 5 cc, but a few very bulky samples could be compressed only to about 8 cc. Granular materials occupied a volume between 1 and 3 cc.

The syringe, without the plunger or a needle, is immersed to the 1 cc mark in a beaker of dyed blue test solution. The rate of uptake of the test solution is observed, and either the time required for a 5 ml rise or the volume attained at 30 min is recorded.

In the example absorbent properties are demonstrated with a 1% NaCl solution to simulate human body fluids.

Example: In a Waring Blendor jar containing 400 ml of water was dispersed 1 g of Grade 85 Chemical Cotton (Hercules Incorporated). To this was added 9 g of partial free acid CMC (made from Grade 85 Chemical Cotton) and stirring was continued for 5 min. The slurry was then allowed to stand at room temperature for 10 min. After another minute of stirring at low speed, the aqueous slurry was transferred to a 2-liter beaker.

To this was added, with agitation, 600 ml of acetone. The blender jar was rinsed with 200 ml of additional acetone which was also added to the 2-liter beaker. After 10 min of low speed stirring in the beaker, excess liquid was removed by alternately pressing and decanting supernatant fluid. The sample was then steeped three times in 600 ml aliquots of acetone for about 5 min each time. Excess acetone was then removed via pressing and decanting and the sample was dried in vacuum at 60°C for 1.5 hr. The procedure was repeated using varying amounts of chemical cotton and partial free acid CMC.

CAP test data showed that the equilibrium absorption capacity of the coated chemical cotton reaches a maximum at about 90% add-on of partial free acid CMC. Also, coated samples containing between about 60 and 90% partial free acid CMC have absorption capacities equal to or greater than the partial free acid itself. The coated products containing 40 to 90% partial free acid CMC have faster initial rates of absorption than the partial free acid CMC by itself. Lower Syringe Test values for the coated samples indicated better wicking ability and faster rate of absorption than the partial free acid CMC alone.

Alloy Fibers Containing Polyvinylpyrrolidone

Known in the art are alloy fibers, consisting of sodium carboxymethylcellulose and regenerated cellulose, which can be employed in various articles which are intended to absorb body liquids.

According to *F.R. Smith; U.S. Patents 4,041,121; August 9, 1977; 4,136,697; January 30, 1979; both assigned to Avtex Fibers Inc.* high fluid-holding alloy fibers are prepared by mixing an aqueous solution of polyvinylpyrrolidone with a filament-forming viscose, shaping the mixture into fibers, coagulating and re-generating the shaped fibers and thereafter drying the same. Viscose constitutes the major portion of the mixture and the shaped alloy fibers are coagulated and regenerated by known means, and preferably in an acid bath containing sulfuric acid and sodium sulfate.

Zinc sulfate is often incorporated in the bath as well as other coagulation mod-ifiers, as desired. No special finishes and/or drying procedures are required to render the alloy fibers in a form which can be carded without difficulty.

The viscose which is employed in making the alloy fibers is, desirably, of a com-position as is used in making conventional regenerated cellulose fibers. The com-position of such viscose is well documented in the prior art and, in general, is produced by reacting alkali cellulose with carbon disulfide, with the resulting sodium cellulose xanthate being diluted with aqueous caustic to provide the resulting viscose with a desired cellulose and alkali content.

The polyvinylpyrrolidone may be the sole high polymeric additive in the viscose or it may be used together with other water-soluble (including aqueous alkali-soluble) high polymers. Preferably these are anionic polymers such as polymeric acids or salts (e.g., alkali metal salts) thereof, e.g., salts of carboxyalkylcelluloses (such as sodium carboxymethylcellulose), salts of polyacrylic acids, (including polyacrylic acid or polymethacrylic acid homopolymer, or copolymers of acrylic and/or methacrylic acid with one or more other monomers such as acrylamide or alkyl acrylates), salts of copolymers of maleic or itaconic acid with other monomers such as methyl vinyl ether, or naturally occurring polycarboxylic polymers, such as algin.

Example 1: Using conventional rayon spinning equipment, aqueous solutions of polyvinylpyrrolidone, designated as K-60 (GAF Corporation) and having an average molecular weight of about 160,000 and K-value of 50 to 60, were separately injected by a metering pump into a viscose stream during its passage through a blender and the blend thereafter extruded. During this the blend was subjected to high mechanical shearing. The viscose composition was 9.0% cellulose, 6.0% sodium hydroxide and 32% (based upon the weight of the cellulose) carbon disulfide. The viscose ball fall was 56 and its common salt test was 7.

The mixtures of viscose and polyvinylpyrrolidone were extruded through a 720 hole spinneret into an aqueous spinning bath consisting of 7.5% by weight of sulfuric acid, 18% by weight of sodium sulfate, and 3.5% by weight of zinc sulfate. After passage through the spinning bath, the resulting continuous tow was washed with water, desulfurized with an aqueous solution of sodium hydrosulfide, washed with water, acidified with an aqueous HCl solution, and again washed with water. The still wet multifilament tow was cut into staple fibers and, without any further treatment, dried.

The fluid-holding capacity of sample fibers, made with various approximate proportions (tabulated below) of cellulose and polyvinylpyrrolidone in the spinning solution, was determined using the following test procedure.

Sample staple fibers were carded or otherwise well opened and then conditioned at 75°F and 58% relative humidity. 2 g of such alloy fibers were placed in a one-inch diameter die, pressed to a thickness of 0.127 inch, and maintained in this condition for 1 min. This compressed pellet of fibers was removed from the die and placed on a porous plate of a Buchner funnel. The upper surface of the pellet was then engaged with a plunger which was mounted for free vertical movement, the plunger having a diameter of one inch and a weight of 2.4 pounds.

The funnel stem was connected by a flexible hose to a dropping bottle from which water was introduced into the funnel to wet the pellet of fibers. Control over the water flow was exercised by the position of the dropping bottle. After an immersion period of 2 min, the water was permitted to drain from the fiber pellet for 3 min, after which the still wet pellet was removed from the funnel and weighed. One-half of the weight of water in the sample pellet is a measure of the fluid-holding capacity of the fibers, expressed in cc/g.

The test results of sample fibers, as described above, are set forth in the table below.

Sample	Cellulose	Polyvinyl-pyrrolidone	Fluid-Holding Capacity (cc/g)	Percent Water Retention*
A	100	0	3.06	105
B	95	5	3.16	112
C	90	10	3.52	121
D	80	20	4.15	145
E	70	30	4.69	186
F	65	35	4.68	178
G	60	40	4.65	190

*Percent water retained by the loose mass of fibers after centrifuging at 1 g for 3.5 minutes.

Example 2: Example 1 was repeated, but a 1:1 blend of a 9% aqueous solution of the polyvinylpyrrolidone with a 9% solution of sodium CMC (Hercules grade 7 MF in 6% NaOH, DS 0.7) was injected. Various amounts of this blend were used; the proportions of cellulose; polyvinylpyrrolidone; and CMC were varied as follows: 100:0:0; 95:2.5:2.5; 90:5:5; 85:7.5:7.5; 80:10:10. A portion of the resulting fibers was finished with a 0.5% water solution of Span 20 (sorbitan monolaurate); and then dried; a second portion was made somewhat alkaline by washing in 1% aqueous solution of sodium bicarbonate, then rinsed in water before finishing with the 0.5% Span 20 solution and drying. The presence of the additive gave improved fluid-holding capacity measured by the Syngyna test (described by G.W. Rapp in a June 1958 publication of the Department of Research, Loyola University, Chicago, Illinois), e.g., the 80:10:10 blend treated with sodium bicarbonate gave a fluid-holding capacity well above 6 cc/g.

Alloy Fibers Containing Alkali Metal Salts of Alginic Acid

In a related patent, *F.R. Smith; U.S. Patent 4,063,558; December 20, 1977; assigned to Avtex Fibers Inc.* describes a method for making absorbent alloy fibers of regenerated cellulose containing a uniform dispersion of alkali metal salts of alginic acid.

Example: An aqueous solution of sodium alginate (prepared from a granular form (Keltex) was made by dissolving Keltex in water to give 3% solution. The solution was injected into viscose, whereby the spinning solution contained 11.1% sodium alginate, based on cellulose. The fibers produced were subsequently processed in different ways and evaluated for fluid-holding capacity.

A portion of the resulting alloy rayon tow was treated with 1% aqueous Na_2CO_3 and 1% Span 20 in one solution and dried. Sample staple fibers were carded, or otherwise well opened, and then conditioned at 75°F and 58% relative humidity. 2 g of such alloy fibers were placed in a one-inch diameter die, pressed to a thickness of 0.127 inch, and maintained in this condition for 1 min. This compressed pellet of fibers was removed from the die and placed on a porous plate of a Buchner funnel. The upper surface of the pellet was then engaged with a plunger which was mounted for free verticle movement, the plunger having a diameter of 1 inch and a weight of 2.4 pounds.

The funnel stem was connected by a flexible hose to a dropping bottle from which water was introduced into the funnel to wet the pellet of fibers. Control over the water flow was exercised by the position of the dropping bottle. After an immersion period of 2 min, the water was permitted to drain from the fiber pellet for 3 min, after which the still wet pellet was removed from the funnel and weighed. One-half of the weight of water in the sample pellet is a measure of the fluid-holding capacity of the fibers, expressed in cc/g. This measurement is defined as the potential ratio.

A blend of equal parts of this sample fiber with a three denier crimped rayon (U.S. Patent 3,046,983) had a potential ratio of 4.28 cc/g. The rayon control sample (no alginate loading) had a potential ratio of 2.7 cc/g.

Alloy Fibers Containing Polyacrylic Acid Salts

According to *F.R. Smith; U.S. Patent Reissue 30,029; June 12, 1979; assigned to Avtex Fibers Inc.* a mass of alloy fibers of polyacrylic acid salt of alkali metals or ammonium and regenerated cellulose, useful for absorbing fluids, are prepared by mixing a caustic solution of polyacrylic acid with viscose, spinning the mixture into fibers and obtaining dry fibers in the alkaline state. The fibers are advantageously dried with an alkaline lubricating finish thereon and then processed into dressings, sanitary napkins, tampons and diapers.

Example: A sodium polyacrylate solution (12.5% solids having 10,000 to 20,000 cp viscosity) was injected through a metering pump into the viscose stream of a spinning machine. The viscose composition was 9.0% cellulose, 6.0% sodium hydroxide and 32% carbon disulfide, based on the weight of the cellulose. The viscose ball fall was 70 and its common salt test was 8.

The mixture was spun through a 720 hole spinneret into an aqueous spinning bath consisting of 7.5% by weight of sulfuric acid, 18% by weight of sodium sulfate and 3.5% by weight of zinc sulfate. The alloy fibers passed through the bath and were washed with water, desulfurized and washed again with water. The wet gel fibers were then passed through an alkaline finish bath consisting of 1% by weight of sodium carbonate and 1% by weight of Span 20. The fibers were cut, dried and carded. The fluid holding capacity was tested for fibers having different amounts of sodium polyacrylate in the alloy fibers using the test described in the preceding patent.

The results are set forth in the following table along with results for other fibers prepared as above except that these alloy fibers were prepared with polyacrylic acid without the formation of the sodium salt in the dried fiber product.

Sodium polyacrylate, % BOC*	0	10	20
Fluid held, cc/g	2.75	4.55	6.05
Polyacrylic acid, % BOC*	0	10	20
Fluid held, cc/g	2.50	3.25	3.85

*BOC means based on the weight of the cellulose in the alloy fiber.

From the above data, it is seen that the absorbent mass of alloy fibers, as disclosed herein, has good fluid holding capacity and the sodium salt is necessary to provide distinctly better fluid holding results.

Alloy Fibers Containing Salts of Acrylic Acid/Methacrylic Acid Copolymers

T.C. Allen and D.B. Denning; U.S. Patent 4,066,584; January 3, 1978; assigned to Akzona Incorporated have found that alkali metal and ammonium salts of copolymers of acrylic and methacrylic acid incorporated into the viscose solution improves the absorbency and fluid retention properties of the resulting fiber and, most unexpectedly, cause the fiber spun from the viscose to retain its cohesiveness to such an extent that the fibers can be easily processed on carding equipment, whereas fibers incorporating equal amounts of the sodium salt of polyacrylic acid did not card satisfactorily on conventional carding equipment.

Example: A solution of a 90/10 copolymer of acrylic acid and methacrylic acid was prepared and injected into a viscose solution at a concentration of 5% cellulose in viscose, thoroughly mixed with the viscose and spun into a conventional acid spinbath containing 8.5% sulfuric acid, 5.0% of $MgSO_4$, 3.0% of $ZnSO_4$, 18.2% of Na_2SO_4 and 30 to 35 ppm laurylpyridinium chloride (LPC) at 49° to 51°C to coagulate and regenerate the cellulose to give an 1,100 denier yarn containing 480 filaments. The resulting yarn was then run through a fresh hot water bath at 93° to 95°C and stretched 37% in the bath. The yarn was then collected in a pot in cake form, washed at 30°C for 40 min, at 52°C in 0.50% aqueous sodium sulfide containing 0.05 to 0.10 sodium hydroxide, for 80 min at 30°C in water, for 40 min at 30°C in 0.01% acetic acid, for 40 min at 40°C in 0.2% solution of an emulsified mineral oil controlled to a pH of 7 to 8, hydroextracted for 4.5 min, and dried at 70° to 80°C overnight.

Additional samples of fiber were prepared in the same manner, but using 80:20 and 50:50 (weight ratio) copolymers of acrylic acid and methacrylic acid. In the table below, the water retention values of the fibers are compared to the same fiber without added copolymer.

Polymer Added	Water Retention Value*
None	0.97
AA/MAA 90:10	1.36
AA/MAA 80:20	1.30
AA/MAA 50:50	1.27

*g of water retained per g fiber.

Alloy Fibers Containing Salts and Esters of Acrylic and Methacrylic Acids

According to *A.W. Meierhoefer; U.S. Patent 4,104,214; August 1, 1978; assigned to Akzona Incorporated*, an alloyed cellulosic fiber containing an alkali metal salt or ammonium salt of a copolymer or terpolymer of acrylic acid and/or methacrylic acid and an aliphatic ester of at least one of the acids is prepared by a process wherein the copolymer or terpolymer is mixed with a viscose solution and the mixture is extruded through a spinneret into a conventional spin bath and processed into staple fibers which are adapted to be used in absorbent articles.

An unsaturated aliphatic dicarboxylic acid such as maleic acid or anhydride may be incorporated in the copolymer or terpolymer. Also, a copolymer or terpolymer of acrylic acid or methacrylic acid and an aliphatic ester of acrylic acid or methacrylic acid may contain radicals obtained from acrylamidoalkane sulfonic acid or an unsaturated phosphorus acid.

The percent water retention as indicated by the secondary swelling of a rayon fiber may be determined by soaking 2 to 3 g of previously washed and dried rayon fiber in water, and removing excess water by centrifuging at a force of 2,500 to 3,500 times gravity for 15 min in stainless steel sample holders. These holders are 22 mm i.d. x 25 mm deep, with screw caps to cover both ends.

Space is provided in the centrifuge cup below the sample holder to contain the excess water which is removed from the yarn during centrifuging. The extracted fiber is placed in a preweighed weighing bottle; the weight of the swollen fiber is obtained and, after drying overnight at 105°C, the weight of the dry fiber is determined. The percent swelling is then determined by use of the following equation:

$$Q = \frac{(\text{Swollen weight} - \text{dry weight})}{\text{dry weight}} \times 100$$

The saline retention value (SRV) is determined by the same procedure as the water retention value except that a 1% aqueous solution of sodium chloride is substituted for water.

Example 1: In this example a rayon fiber which does not contain any alloying polymer or copolymer was prepared for comparison.

A ripened viscose solution containing 8.4% cellulose, 4.8% sodium hydroxide, and 2.3% sulfur was extruded through a spinneret having 480 holes to produce filaments of 1,100 denier. The spinbath contained 5.5% sulfuric acid, 24% sodium sulfate, and 0.95% zinc sulfate in water. The temperature of the spinbath was 50°C. After passing through the spinbath, the resulting fibers were further processed, stretched 37% and cut into staple fibers. The pH of the fiber was adjusted with sodium bicarbonate to provide an alkaline fiber which was treated with a 0.3% aqueous solution of Tween 20 to surface finish the fiber.

After conditioning, water retention (WRV), as secondary swelling, and saline retention value determinations were made on the staple fiber. The WRV of the sample was 71% and the SRV was 67%.

Example 2: Example 1 was repeated except that 10% by weight of an acrylic acid polymer as a 19% aqueous solution having a Brookfield viscosity of 9,900 cp determined with a No. 2 spindle at 3 rpm was mixed with the viscose solution by injection just prior to extrusion. The WRV was 112, and SRV was 95.

Example 3: The sodium salt of 2-acrylamido-2-methylpropane sulfonic acid (AMPS) was prepared by adding 225 parts of AMPS to a cold (10°C) solution of 43.5 parts sodium hydroxide in 450 parts water. To the solution of sodium AMPS was added 25 parts n-butyl acrylate (nBA), 50 parts isopropanol, and 200 parts water. The solution was heated to 65°C, while purging with nitrogen. Potassium persulfate, 0.5 parts, was added to the mixture and the temperature maintained at 65°C for 1½ hr. The resultant polymer solution had a Brookfield viscosity of 3,000 cp (no. 2 spindle, 6 rpm at room temperature of about 20°C).

Sufficient of the resulting aqueous copolymer was mixed with viscose as in Example 1 to give an equivalent of 10% by weight of AMPS. A staple fiber was then prepared by the procedure used in Example 1. The WRV was 125, and SRV was 111.

Polyacrylonitrile-Modified Cellulose Subjected to Aging and Soaking

J.W. Adams; U.S. Patent 4,151,130; April 24, 1979; assigned to American Can Company found that the fluid absorptive and retentive properties of polyacrylonitrile-modified (PAN-modified) cellulose fibers prepared by previously known methods and particularly by the procedure which involves an alkaline hydrolysis of PAN-modified fibers at high solids concentrations (30 to 60% total solids) may be greatly enhanced by certain soaking, washing, swelling and aging procedures carried out subsequent to the hydrolysis.

Example 1: 1,500 g of bleached southern softwood kraft pulp were added together with 20 liters of water and 1.8 g of ferrous ammonium sulfate hexahydrate to a steam jacketed ribbon blender reactor equipped with a reflux condenser. The pH was adjusted to 3.9 with 10% sulfuric acid. The reactor contents were blanketed with a nitrogen atmosphere and heated to 90°C to purge the system. The pulp slurry was then cooled to 60°C and 3.75 liters of inhibitor-free acrylonitrile and 38 ml of 10% hydrogen peroxide were added under agitation. After 1 hr at moderate reflux, the unreacted acrylonitrile monomer was allowed to distill off, leaving a product comprising PAN-modified cellulose fibers in a polymer-to-fiber ratio of about 2 to 1 from which the aqueous reaction medium was removed by filtration to leave the fibrous product in a damp state.

The PAN-modified cellulose was subjected to an alkaline hydrolysis carried out at high solids content as follows: 300 g of the polymer-grafted cellulose fibers (dry basis) dampened with 200 g of water were agitated in a ribbon blender while in a crumb-like state and a solution of 150 g of sodium hydroxide in 350 g of water were added to the moist graft polymerized cellulose fibers. The resulting mass of 45% solids concentration was heated at 90°C for 90 min under agitation in the ribbon blender. At the end of the hydrolysis period, the reaction product was a crumb-like, damp, porous mass containing cellulose fibers having grafted thereon primarily the sodium salt of polyacrylic acid.

In the above hydrolysis reaction, the solids concentration is at all times so high that the fluid content is insufficient to disperse the solid polymer-grafted fibers, which are merely dampened by the aqueous medium and remain throughout the reaction in a crumb-like state so that the hydrolysis product is a crumbly, porous mass which is easily dried or may be readily washed with water which quickly penetrates the porous structure of the mass and is capable of rapid removal therefrom by filtration.

The above reaction product from the high solids hydrolysis of polymer-grafted cellulose fibers exhibits varying properties in absorbency for aqueous fluids, as measured by WRV, SRV and water absorption rate (WAR), and also varies in gel strength dependent on the factors previously mentioned, as will be hereinafter described.

In this example, for instance, the hydrolysis was carried out at 45% solids concentration. The resultant product was found to be very receptive to the aging, swelling and conditioning process by means of which the absorbency of the material, as measured by WRV and SRV, was improved by 250% over the absorbency of the material as taken from the reaction vessel at the completion of the hydrolysis. The physical consistency of the product, as measured by the gel strength of resistance to deformation or displacement under stress, was completely satisfactory for use in absorptive pads such as disposable diapers, sanitary pads, tampons and incontinence pads.

It has been found, however, that carrying out the hydrolysis at 65% solids concentration yields a product which, although having a very high gel strength, has a substantially lower initial absorptivity and is very much less responsive to the process, the total degree of improvement in absorptivity achieved by the process when applied to this product being in the range of 50% or less.

Furthermore, if the hydrolysis is carried out at about 30% solids concentration, e.g., the product, although responsive in its absorptivity aspect to the conditioning process is deficient in gel strength. Because of this factor, as the product becomes saturated with aqueous fluids during use as a diaper, for example, the particulate structure tends to break down under pressure and the absorbent is subject to displacement within the confines of the diaper outer layers.

Example 2: *Aging The Hydrolyzed Product* — The fluid absorptive and gel strength characteristics of polyacrylate modified cellulose fibers thus derived are altered by storing the hydrolysis product at room temperature for varying periods of time. Increasing the length of the period of storage results in an increase in water absorption rate as well as both the water and salt retention values. These changes, however, are accompanied by corresponding decreases in the gel strength of the polymer-modified material, so that it has not been found possible to obtain, by merely employing simple aging procedures, an absorptive product exhibiting both desirably high absorbency properties and acceptably high structural integrity, as measured by gel strength.

In accordance with the process, the crumb hydrolyzed products are allowed to age at room temperature or moderately elevated temperatures for a period of time sufficient to increase the absorptive properties thereof as compared to the fresh hydrolyzed crumb product. In general, the products are aged for a period of at least one, and preferably, three to six weeks before being subjected to the washing and soaking procedure.

A crumb-hydrolyzed product produced as in Example 1 at about 50.5% solids concentration was aged at room temperature for 70 days. Samples removed at various intervals were soaked twice in water, dried and analyzed for PAM and polyacrylic acid (PAA) contents.

Test results indicate that hydrolysis is substantially about 83% complete before any appreciable aging and slowly increases to about 87% after storage for 70 days. The disappearance of carboxyl groups during this period (as indicated by the PAA%) suggests that a crosslinking via esterification with cellulose hydroxyl groups is occurring. It is also believed that during aging, the firm gel developed during hydrolysis gradually weakens resulting in severance of some bridging chains and reformation of macromolecules.

Example 3: *Steeping and Swelling of Aged Product* — The aged hydrolyzed product is subjected to a steeping and swelling process and is carried out by placing a quantity of the hydrolysis product in at least 20 times its weight of water and, after gentle stirring to ensure complete wetting of each of the particulate masses of fibers, the aqueous slurry is allowed to remain quiescent for a period of time ranging from 10 min to as much as an hour or more. The water temperature may vary from cold to moderately warm, ambient temperature being quite suitable. Boiling and strong agitation should be avoided because they tend to fracture and fragment the particulate entities. As the steeping period progresses, the expansion and swelling of the granular masses of fibers may be visually observed, the volume occupied by the solid component of the slurry increasing by from about 15% to as much as 40% within a period of about 15 min.

The receptiveness of the polymer-modified fibers to the swelling process is variable, depending on the prior history of the particular sample being treated, and this variable receptiveness is responsible for the variation in the volume change observed.

When the steeping period is complete and no further swelling is evident, the swelled, water saturated gel-like granules or particles may be separated from the fluid medium by filtration or centrifugation and subsequently dried or, if desired, the steeping step may be repeated before drying in order to insure a more complete removal of impurities such as salts, alkali and the like from the gel particles. During the second steeping in an aqueous medium, little or no further swelling of the solid granules will be observed.

Example 4: *Drying Hydrolyzed, Aged, Swollen Product* — Drying may be carried out at moderately elevated temperatures in a circulating hot air oven, by passing hot air through the damp solids while suspended on a traveling foraminous carrier, or by similar conventional techniques. Moderate care should be exercised to avoid undue fragmentation of the fibrous gel granules before and during drying and excessively high air temperatures should also be avoided because they have been found to deleteriously affect the absorptivity of the granules.

For example, a sample of polyacrylate-modified cellulose fibers treated in accordance with the steeping procedure outlined above and dried in an air stream having an air temperature of 250°F was found to have a WRV of 55, whereas, the same material dried in 325°F air had a WRV of only 14. In general, temperatures below 300°F will be satisfactory herein.

After completion of the steeping and swelling process, the polyacrylate-modified cellulosic fiber material, which is completely saturated with water, is comprised of clumps or agglomerates of fibers in particulate masses which may be described in physical appearance as wet granules of a firm, white, swollen, gel-like substance. Drying of the gel under conditions which do not crush or pulverize the individual grains gives an amber, coarsely granular material which is easily rewet by water and which absorbs aqueous fluids very rapidly and in very large amounts.

By the process a significant improvement in the absorptive properties of the polymer-modified cellulosic fiber material may be obtained over and above that resulting from merely aging the product and the improvement obtained is accomplished without a significant loss of gel strength. It is therefore possible to stop the aging of the hydrolysis product mass at a point at which the gel strength is still acceptably high and to subject the mass at that time to the soaking procedure in water which greatly enhances the absorptivity of the material.

The final product, then, exhibits the hitherto unattainable combined properties of a very high absorptivity and a strong gel structure so that the product is an excellent absorbent but will not lose its structural integrity and become pasty or glutinous when saturated with fluid.

IMPROVEMENT IN OTHER PROPERTIES

Flameproofness

K. Mimura, A. Kawai, Y. Kametani and T. Nakahama; U.S. Patent 4,066,730; January 3, 1978; assigned to Mitsubishi Rayon Co., Ltd., Japan prepare regenerated cellulose fibers having very good flameproofness by adding to a viscose, before spinning, a polyphosphonate having the following formula:

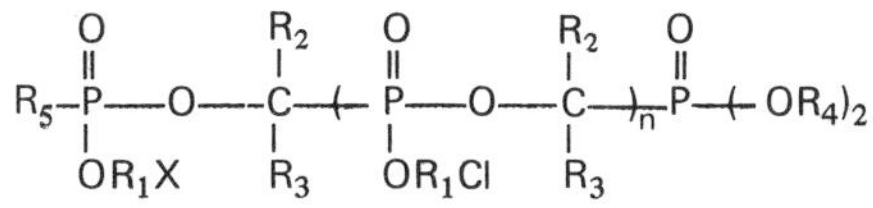

$$R_5-\overset{\overset{O}{\|}}{\underset{\underset{OR_1X}{|}}{P}}-O-\overset{\overset{R_2}{|}}{\underset{\underset{R_3}{|}}{C}}\left(\begin{array}{c}\overset{\overset{O}{\|}}{P}\end{array}\right.-O-\overset{\overset{R_2}{|}}{\underset{\underset{R_3}{|}}{C}}\left.\begin{array}{c}\end{array}\right)_n\overset{\overset{O}{\|}}{P}-\left(OR_4\right)_2$$

wherein R_1 is ethylene or propylene; R_2 and R_3 are methyl or ethyl which may be the same or different; R_4 is an alkyl group having 1 to 4 carbon atoms, or an aralkyl group in which the alkyl substituent has 1 to 4 carbon atoms and the hydrogen atoms in the alkyl and aralkyl groups may be substituted with chlorine or bromine; R_5 is an aromatic group having 1 to 20 carbon atoms or an unsaturated or saturated aliphatic or alicyclic group having 1 to 20 carbon atoms or a combination thereof and R_5 may contain carbonyl group, sulfonyl group or nitrogen; X is halogen, and n is an integer from 1 to 1,000.

Example 1: A 1-liter separatory flask provided with a stirrer, a thermometer, a reflux condenser, a dropping funnel, and a nitrogen flow inlet device was charged with 253 parts of purified ethylene chlorophosphite, 386 parts of purified methyl-chloroform, and 15.5 parts of purified tris(2-chloropropyl) phosphite. Then, a dropping funnel was charged with 116 parts of purified acetone and 1.9 parts of purified tert-butyl chloride.

While under a sufficient flow of dried nitrogen gas, the outside of the flask was cooled in an ice-water bath and the mixed solution of acetone and tert-butyl chloride was dropped with stirring into the flask during which addition the temperature within the flask was kept at 5° to 10°C. After addition of the reagent was completed, the reactants were allowed to stand at room temperature for about 48 hr. The polymer obtained existed as 384.9 parts of a solid portion and had the following structure with mean molecular weight 6,000:

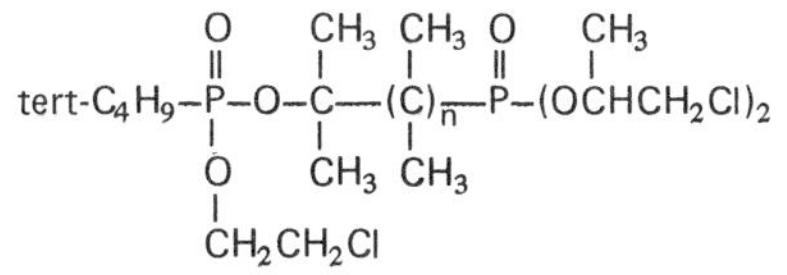

$$tert\text{-}C_4H_9-\overset{\overset{O}{\|}}{\underset{\underset{CH_2CH_2Cl}{\underset{|}{O}}}{P}}-O-\overset{\overset{CH_3}{|}}{\underset{\underset{CH_3}{|}}{C}}-(\overset{\overset{CH_3}{|}}{\underset{\underset{CH_3}{|}}{C}})_n-\overset{\overset{O}{\|}}{P}-\overset{\overset{CH_3}{|}}{(OCHCH_2Cl)_2}$$

Example 2: A viscose which was prepared by adding 48% based on the weight of cellulose of carbon disulfide to alkali cellulose and which had a cellulose concentration of 7.5% and an alkali concentration of 4.2% was ripened until the viscosity reached 200 poises and the salt point reached 16. To 100 parts of the prepared viscose was added 3.5 parts of a solution of the polyphosphonate prepared in Example 1 in methylchloroform. The solution was mixed to homogeneity. Thereafter, the viscose was spun into a coagulation bath containing 18 g/l of sulfuric acid, 70 g/l of sodium sulfate, and 0.4 g/l of zinc sulfate at 30°C.

The filaments withdrawn from the coagulation bath were stretched by 100% in a second bath containing 2 g/l of sulfuric acid at 80°C and were successively treated in a third bath containing 5 g/l of sulfuric acid at 60°C to complete regeneration. Thereafter, the filaments were subjected to the conventional after-treatment such as desulfurization, bleaching, acid treatment, lubrication, and drying.

In the example, essentially no sticky materials were recognized as forming during the spinning step which is a problem when conventional polyphosphonates are used.

Moreover, in the example, the retention of the polyphosphonate (flameproofing agent) in the fibers was 95.5%, which is extremely high. Furthermore, there were no troubles in the yarn spinning operation of the obtained fibers and the limit of the oxygen index (LOI) of the fibers was 27.5 which indicates that the fibers had excellent flameproofness.

Comparative Example: The same apparatus described in Example 1 was used. The flask was charged with 281 parts of purified propylene chlorophosphite and 410 parts of purified dichlorethane. The dropping funnel was charged with 116 parts of purified acetone and 0.4 part of water. The acetone solution was added to the flask under the same conditions as described in Example 1 to effect polymerization. The polyphosphonate obtained had a mean molecular weight of 15,000 and the molecular weight distribution was very broad. A 50% solution of the polyphosphonate in dichloroethane was added to a viscose under the same conditions described in Example 2 and the viscose was spun under the same conditions described in Example 2. In this comparative example, the retention of the polyphosphonate in the fibers was 80%.

Thermal Stability

C.-M. Kuo, R.H.S. Wang and R.T. Bogan; U.S. Patent 4,137,201; January 30, 1979; assigned to Eastman Kodak Company disclose a new thermal stabilizer system for cellulose esters and their commercially useful formulated compositions which can be molded, shaped and otherwise processed in a thermoplastic condition at elevated temperatures to provide products having (1) better molecular weight retention and (2) less discoloration after thermal processing when compared to similar cellulose ester compositions containing prior art stabilizers.

The stabilizing system comprises: (A) at least one cyclic phosphonite compound having the formula:

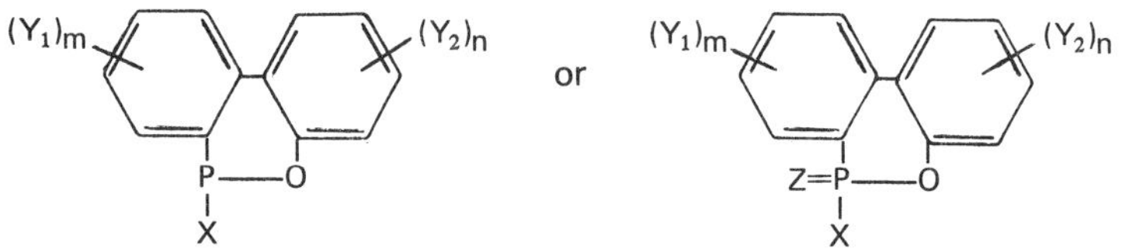

wherein X is hydrogen, hydroxyl, amino, alkyl having 1 to 22 carbon atoms, alkoxy having 1 to 22 carbon atoms, alkylthio, aryloxy or arylthio having 1 to 22 carbon atoms; Y_1 is alkyl having 1 to 18 carbon atoms; Y_2 is halogen, alkyl having 1 to 18 carbon atoms, alkoxy having 1 to 18 carbon atoms, nitro, cyano or sulfonic acid radical; and Y_1 and Y_2 combined with a biphenyl ring form a phenanthrene ring; Z is an oxygen or sulfur atom; and m and n are whole numbers ranging from 0 to 4; in combination with (B) a conventional antioxidant and (C) an acid accepting epoxy compound.

The stabilizing combination which is particularly useful for stabilizing cellulose esters is the combination of 9,10-dihydro-9-oxa-10-phosphaphenanthrene-10-oxide (HCA), tetrakis[methylene(3,5-di-tert-butyl-4-hydroxyhydrocinnamate)methane] (Irganox 1010), and either an epichlorohydrin/bisphenol A type expoxy resin (Epon 815) having a viscosity (25°C) of 5 to 7 poises, an epoxide equivalent of 175 to 195, and an average molecular weight of ~330 or neopentyl glycol di-glycidyl ether. The cyclic phosphonite compounds can be obtained commercially or prepared as in U.S. Patent 3,702,878.

Example 1: 100 parts of cellulose acetate (CA-400-25) having an acetyl content of about 40 wt % and 25 seconds viscosity (ASTM D1343) was thoroughly mixed with 36.7 parts of diethyl phthalate (plasticizer) and the amounts of stabilizers listed below. Compression molded plates were made from these mixtures and properties of the cellulose acetate plastic such as roll color, heat test color, and heat test inherent viscosity (I.V.) were determined.

The roll color was determined using a Gardner Color Difference Meter on the pressed plates. The heat test color was determined by heating the acetate plastic to 205°C for 1 hr and comparing the heated samples with known standards. The higher numbers indicate more discoloration. The heat test I.V. of the formulations was determined after the cellulose acetate plastic had been heated to 250°C for 30 min. These test results show that a cellulose ester formulation having relatively good thermal stability is improved further by use of the stabilizer system as shown below.

. . . . Stabilizer*			. . . . Properties.		
HCA	Irganox 1010	Epon 815	Roll Color	Heat Test Color	Heat Test I.V.
0	0	0	14.2	40	0.90
0.15	0.1	0.5	12.1	25	0.80
0.15	0.1	1.0	12.3	25	1.14
0.15	0.1	1.5	12.4	25	1.22

*Parts per 100 parts by weight of cellulose acetate.

Example 2: Cellulose acetate (CA-400-25) plastic compositions were prepared in a manner similar to that described in Example 1, except that the tests were run by heating the cellulose ester plastic compositions at 205° to 208° in a heated block for 1 hr, and the increase in plastic color was determined using Gardner Color Standards. The test results indicate that prior art thermal stabilizers such as neopentyl phenyl phosphite (NPPP) are not nearly as effective as the HCA-based stabilizers.

Cellulose Acetate Solutions with Reduced Gelling Tendency

According to *A.F. Turbak, J.P. Thelman and A.B. Auerbach; U.S. Patent 4,118,350; October 3, 1978; assigned to International Telephone and Telegraph Corporation* the tendency of cellulose acetate solutions containing methylolated polymers of melamine or guanamine to prepolymerize or gel is reduced by the addition to the solutions of an amine oxide. The amine oxide also acts to reduce the viscosity of the cellulose acetate solutions.

In addition to 4-methyl-morpholine oxide, other useful amine oxides are N-coco-morpholine oxide; the pyridine oxides such as 4-methoxypyridine-N-oxide, pyridine-N-oxide, 3-hydroxypyridine-N-oxide; and aliphatic amine oxides such as bis(2-hydroxyethyl)octadecylamine oxide and dimethylhexadecylamine oxide. The amount of the amine oxide should be at least about 0.5%, based on the weight of the methylolated melamine or guanamine polymer in the solution.

Example: An acetone solution containing 18% by weight of cellulose acetate (39.4% acetyl), 1.8% of a capped methylol melamine polymer and 0.018% (1.0% based on melamine) polymer of 4-methylmorpholine oxide was heated at 100°C for 16 hr in a sealed tube. The capped methylol melamine polymer was a commercially available melamine-formaldehyde polymer prepared from 3.2 mols of formaldehyde for each mol of melamine with approximately 2 mols of methanol used in capping. The polymer had a degree of polymerization of 2, an approximate molecular weight of 380, was 72% alkylated with methyl groups and contained 28% free methylol groups.

Ball fall viscosities were determined for the above solution before and after heating. The ratio of final viscosity/initial viscosity was 1.6, as compared to 13.4 for a control with no additives. The 4-methylmorpholine oxide had no effect on the ultimate cure of the fibers.

FURTHER APPLICATIONS

Admix for Elastomeric Materials

A process is provided by *R.W. Posiviata and J.A. Johnston; U.S. Patent 4,125,493; November 14, 1978; assigned to The Gates Rubber Company* for preparing a generally dry fibrated admix for reinforcing or filling viscoelastomers such as the natural or synthetic rubbers or blends thereof. The admix is dispersed in the elastomer to yield a cured fiber-elastomer composite having improved physical characteristics for products such as hose, tires, or power transmission belts.

The fibrated admix is prepared by blending a mass of sized synthetic fibers, nonregenerated cellulosic fibers, or combinations thereof with conditioners which aid in strengthening, adhering, and dispersing the fibers when mixed with a viscoelastomer.

A suitable mass of synthetic fibers may be prepared by fibrilizing or classifying the fibers with a mechanical action device to a desired length. Limp high tenacity fibers like aramid and polyester may be prepared by stiffening them with a stiffening agent, such as dried and heat-cured solution of blocked phenyl isocyanate. Stiffer fibers such as nylon, fiber glass, softwood cellulose, or hardwood cellulose may be handled without a stiffening treatment.

Some fiber sources such as shredded newsprint do not require a high degree of fiberizing and screening because the size distribution of newsprint fiber is generally satisfactory for most fiber-elastomer composites. The inherently stiffened (with lignin) newsprint material may be easily defiberized by impacting with blades. If desired, however, the newsprint material may be mechanically fiberized and fibrilized with a hammer mill or similar mechanical action device.

Conditioners are added to the fibrous material in a high intensity blender such as one with spinning blades. In the case of cellulose, a polar liquid such as water or ethylene glycol may be added in small quantities. The liquid is adsorbed by the fibers and acts as a vehicle for coating the fibers with a water-soluble adhesive, (if desired). A partitioning agent like carbon black or clay may be added to the fiber mixture to help separate or partition the individual fibers from each other. When shredded newspaper is used as the fiber material, the carbon black also aids in fiberization and fibrilization during blending. Oil may then be added to the mixture to concentrate the fibrated admix into a smaller volume and minimize free carbon black dust. The oil also aids in dispersing the fiber in a viscoelastic mass.

The fibrated admix may then be used with formulated elastomeric powders or bulk elastomers to establish either elastomeric master batches or fully compounded stocks. In either case, the fibrated admix is dispersed in a viscoelastomer with mechanical means such as a cold-feed extruder, Banbury mixer, mill or the like. The fibers are substantially evenly dispersed in the elastomer and generally oriented in the direction of elastomer flow during mixing.

A fiber-elastomer composition like rubber exhibits increased physical characteristics such as higher secant modulus than was attainable in the prior art at the same volume of fiber loading. The improved characteristics are believed to result from improved dispersion and packing of the fiber in the viscoelastomer rather than being directly dependent on fiber length or aspect ratio. While fiber aspect ratio is an inherent characteristic of all fibers and while some physical properties are always linkable thereto, it is believed that the morphological properties of the fibers are more important for imparting reinforcement to elastomers like rubber.

An advantage of the process is that expensive processing steps such as liquid slurrying and elastomer coating of fibers (e.g., with latex coagulating or rubber friction coating of fabric with a calender) prior to incorporation into a viscoelastomer are eliminated.

Asbestos Substitute

G. Cederqvist and U. Aberg; U.S. Patent 4,159,224; June 26, 1979; assigned to Rockwool Aktiebolaget, Sweden provide a method for producing shaped bodies of fibrous materials using mineral fibers, cellulosic fibers, and zero fibers. The zero fibers are extremely thin and short cellulosic fibers produced either synthetically or as natural by-products from the cellulose industry. These bodies can be used as substitutes for asbestos fiber products, and display physical properties, such as resistance and rigidity, comparable to the asbestos products. The fibrous materials are produced using an aqueous dispersion of the mineral fibers, cellulose fibers, and zero fibers, whereafter the dispersion is dewatered, shaped and dried. The mineral fibers include stone wool fibers, slag fibers, glass wool fibers and so on.

At too high a content of cellulose material, however, a dimensional instability will be introduced. For these reasons, the proportions of mineral wool and of cellulose material must be found by a compromise method. The result of extensive investigations for finding this compromise has been that the participation of mineral fibers should not be more than 90% by weight and preferably not more than 70% by weight of the total amount of fibers, and also not be less than 30% by weight of the total amount of fibers.

Example 1: A fiber suspension was prepared from 50% mineral fibers and 50% Kraft cellulose without any addition of zero fibers. On a continuous vira a sheet of the fiber composition was shaped and dewatering simultaneously took place on the vira. The sheet thereafter was dried as much as possible. The rigidity in all directions was unsatisfactory. To increase the rigidity, a binding means in the form of a latex solution in water was added in an amount of 25% of the initial weight of material and the product was again carefully dried to about 95% dry weight of material. It then displayed a surface weight of 300 g/m², and it showed a good evenness of surface, a good pulling rigidity and a good tearing rigidity. The splitting rigidity against strains in a direction perpendicular to the plane of the product, on the contrary, was completely unsatisfactory, and was only 3.2 kg/cm². The product produced therefore was classified as less satisfactory which was to be expected.

Example 2: A fibrous composition was prepared in the same way as in Example 1 above, however with the difference, that all of the quantity of cellulose fibers comprised zero fibers. It was expected that a poor rigidity would be obtained, especially poor wet rigidity because there was no long fibrous Kraft cellulose present. It also proved difficult to dewater the fiber composition on a continuous vira because the product did not withstand, in its wet state, the strains which are unavoidable during such a treatment.

Example 3: A fibrous suspension was prepared comprising 60% of mineral fibers, 20% of zero fibers and 20% of Kraft fibers of cellulose. The product was treated on a vira and by subsequent drying as described in Example 1 above, but without any addition of binding means. A plane shaped body was obtained having a surface weight of 300 kg/m² and with excellent rigidity properties against both pulling and tearing. The rigidity under a strain in a direction perpendicular to the plane of the shaped body, however, now had increased to 11 kg/cm², which completely corresponded to the high demands for such a rigidity under normal treatment, and the product therefore was classified as completely satisfactory.

Rayon Fibers Containing Starch

F.R. Smith; U.S. Patent 4,144,079; March 13, 1979; assigned to Avtex Fibers Inc. discloses rayon fibers made by spinning a viscose containing dissolved starch.

The starch-containing rayon fibers are suitable for a great many uses. Fabrics made entirely therefrom have been found to be capable of being washed repeatedly (e.g., 50 washes with household detergent in an automatic washing machine, using standard laundering conditions). The effects of such washing have been found not to differ significantly from those with ordinary rayon. Under the light microscope, the starch-containing fibers appear to be of homogeneous chemical nature; e.g., on iodine staining (indicating the presence of starch), the staining is found to be uniform throughout the cross-section of the fiber.

With ordinary rayon dyes (e.g., vat dyes, such as mayvat blue BFC; reactive dyes, such as procion yellow MX4G; and direct dyes, such as Solantine Red 8BLN), the starch-containing fibers dye well, usually more intensely than ordinary rayon and with more substantivity, thus requiring less dyestuff to attain a given desired change. Moisture regain (measured at 75°F and 58% RH) is, for fibers containing about 10% starch based on cellulose (boc) in the range of about 11 to 12% (ordinary rayon is usually within the same range).

The fibers are resistant to removal of the starch; e.g., when a mass of the fibers (of 10% starch content boc) is soaked for about ½ hr at room temperature in about thirty times its weight of a 1 N aqueous solution of NaOH, the fibers swell to a considerably greater extent than ordinary rayon fibers; but when the soak liquid is then poured off, neutralized with HCl or H_2SO_4 and tested for the presence of starch by the conventional iodine test, it shows only a very faint color indicating that the starch content of the soak liquid is less than 50 ppm.

The fibers behave well in processing, such as in high-speed carding, to form a card web suitable for bonding into a nonwoven fabric, (e.g., by impregnation with a latex of polymeric bonding agent). The fibers may be used to form structures in which they are the sole fibers or they may be blended with other fibers.

Fabrics made from the starch-containing fibers may be used for such purposes as cover stock for diapers and pads; tampons; industrial wipes; food filters; felts; surgical sponges; prep balls and swabs (medical); and spun lace-like nonwovens.

The textile fabrics may be woven or knitted and may be used in home furnishings (such as draperies and upholstery); in apparel (such as adult and children's wear, e.g., shirting, blouses, underwear, and interlinings, like neckwear, lapels, etc); for domestic uses (such as sheeting, linens, and towels), or for industrial uses (e.g., in hose reinforcements) or other purposes (e.g., tarpaulins, tentage materials, and wall coverings).

Example: Alkaline starch solution was prepared by mixing a slurry of corn starch grains in water with an 18% NaOH aqueous solution at about 20° to 25°C, to give a translucent viscous solution comprising 13% starch and 4% NaOH.

In a conventional viscose mixer, viscose containing 9.2% cellulose, 6.2% NaOH, 32% CS_2 boc, and about 0.5% TiO_2 boc, was prepared by dissolving xanthated alkali cellulose in aqueous NaOH and mixing for about 2 hr. A quantity of the alkaline starch solution described above was then added to the viscose in amount such that the resulting viscose-starch blend contained 10% starch based on the weight of the cellulose.

Mixing was continued for 1 hr and the solution was aged for about 24 hr at about 19°C, (including a period of about 12 hr for vacuum deaeration). The solution was filtered both before deaeration and after, and directly pumped (e.g., within a half hour) through the spinnerette. At the spinnerette, the ball fall viscosity of the viscose-starch blend was about 90 and its salt test value was about 8. The solution was spun (through 12,000 circular spinnerette holes 0.0025 inch in diameter) into an aqueous spin bath containing 7 to 8% H_2SO_4, about 1.5% $ZnSO_4$, and about 21% Na_2SO_4 at 55°C.

The tow formed in the spin bath was passed around a driven roll and then pulled (by a second driven roll) through a stretch bath containing 3% H_2SO_4 aqueous solution at about 90°C. The stretch bath is continuously replenished by spin bath carried into it by the tow, and by additions of water from time to time. The exit speed (i.e., the speed at the surface of the second driven roll) was 60 m/min, and the speed ratio of the first and second driven rolls was such that the tow was stretched about 60 to 75% in the stretch bath.

The length of travel of the tow in the spin bath was about 0.5 m and in the stretch bath about 2 m. After leaving the driven roll, the tow dropped into a cutter and the resulting cut fibers dropped into flowing hot water (about 85° to 90°C) where relaxation (and crimping) occurred. The fibers were taken up as a blanket, washed with hot water and desulfurized (with a conventional solution of sodium polysulfide), rewashed, a conventional staple finish solution, made from Red Oil, was applied, and the fibers were then dried in hot air (e.g., at about 90°C).

41 samples, each comprising 10 single fibers, were tested for tensile properties. The results (averaged) were set forth in the table below.

Additional tests showed crimps ranging from 9.4 to 12.6 per inch, for an average of 10.95. The denier per filament of the fibers was about 1.5.

Properties	Conditioned*	Wet
Tenacity, g/den	2.84	1.55
Elongation, %	19.15	23.83
Breaking energy, g-cm/cm/den	0.33	0.19

*ASTM 75°F, 57% relative humidity.

OLEFINIC AND VINYL POLYMERS

POLYMERIZATION PROCESSES

Increasing Reaction Rates in Bulk Polymerization

A method of increasing polymerization reaction rates during the production of a fiber-thermoplastic matrix by in situ bulk polymerization is described by *R.R. Casper and M.P. Marander; U.S. Patent 4,148,949; April 10, 1979; assigned to Weyerhaeuser Company* allowing polymerization to at least 25% polymer content by wt of the matrix in less than 8 min.

In a fibrous web of less than 0.25" thickness, saturated with a liquid phase polymerizable composition containing a vinyl monomer and a thermally activated free radical initiator at ½ to 10% by wt of the composition, polymerized by the application of controlled temperature and pressure conditions, reaction rates of less than 8 min are attained by: (a) initiating polymerization with the matrix containing a monomer concentration in excess of that which if completely polymerized is desired in the final product matrix; and (b) polymerizing the monomer contained in the matrix until the desired level of polymer is attained while, simultaneously with the polymerizing step, removing excess monomer at a controlled rate such that the reaction is sustained until the desired level of polymer in the finished matrix is attained.

The resulting matrix is then treated further to remove any excess monomer and is suitable for molding by the application of pressure and heat into useful products.

To carry out the process of saturating and polymerizing the fiber-monomer composition matrix on a continuous basis, a process and apparatus such as disclosed in Figure 7.1 can be used. A roll of fiber web is unwound from roll **10** and fed through and around preheaters **12** and **14**. The fiber web is then passed over idler rolls into a saturation tank **16** holding the liquid polymerizable composition and a free radical initiator. The web may have a polymer incorporated therein prior to saturation with the liquid polymerizable composition.

Figure 7.1: In Situ Bulk Polymerization Process

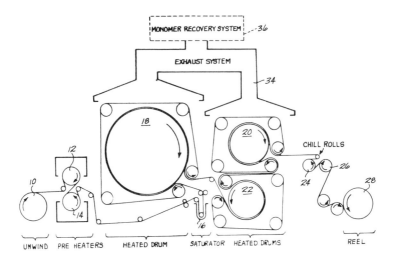

Source: U.S. Patent 4,148,949

From tank **16** the fiber web is brought into contact with a hot, temperature controlled drum **18** which raises the temperature of the web and the polymerizable composition incorporated therein to initiate polymerization. Drum **18** is filled with hot oil which is continuously circulated through a temperature control system. The large quantity of hot oil in the drum acts as a heat sink which can add or remove heat from the polymerization zone.

The web is maintained in close contact with the drum by a continuous belt of stainless steel or other suitable material. A series of oil filled heat radiating coils flank portions of the belt to control its temperature. The web then passes over idler rolls to subsequent hot drums **22** and **20**, chill rolls **24** and **26**, and is wound on roll **28**. Drum **22** is also filled with hot oil and is temperature controlled in the same manner as is drum **18**. Drum **20** is filled with saturated steam which condenses to add heat to the polymerization zone, or vaporizes to absorb heat should the reaction temperature get too high.

Chill rolls **24** and **26** cool the web to end polymerization and to permit satisfactory winding of the web on roll **28**. Any convenient number of heated and chilled rolls may be used. Thus the temperatures of hot drums **18**, **20** and **22** are continuously maintained at levels sufficient to initiate and maintain polymerization of the polymerizable composition in the web matrix without incurring substantial monomer losses or causing damage or charring of the product. Exhaust system **34** collects any escaping monomer which is recovered by monomer recovery system **36**.

Example: A cellulosic fiber web comprising a paperboard 0.040" thick with a weight of approximately 265 lb/3,000 ft^2 is fed through the saturation tank.

The polymerizable composition contained in the saturation tank is styrene that has been partially polymerized to 39% by wt polymer solids and 61% by wt styrene monomer. Benzyl peroxide initiator is added at a rate of 3 lb/100 lb of liquid resin. The liquid resin composition has a viscosity of 550 cp at 75°F. The impregnated sheet contains liquid polymerizable composition to the extent of approximately 290 lb/3,000 ft^2 and is ready for polymerization. The sheet-matrix is continuously fed into the temperature-controlled section of the process. The first drum **18** is held at a temperature of 285°F. On the first drum the matrix is increased in temperature to initiate the polymerization reaction but its temperature is always held below the boiling point of the containing composition monomer.

The web travels to the second heated drum **22** which is also at 285°F, whereupon polymerization is continued until essentially complete. The matrix then travels to the third drum which is heated to 320°F, which temperature drives the polymerization reaction essentially to completion and evaporates substantially all of the remaining monomer. The resulting final sheet contains resin to the extent of approximately 205 lb/3,000 ft^2.

Dried Latex Polymer from Aqueous Emulsion

D.B. Korzenski, B. Vallino, Jr. and W.E. Zarnecki; U.S. Patent 4,035,347; July 12, 1977; assigned to Nalco Chemical Company provide a method for preparing substantially dry homopolymers and copolymers which comprises forming a polymeric latex or water-in-oil emulsion and partially inverting the latex by regulated contact time with water in the time span 0.5 to 10.0 seconds and preferably less than 1 second. A preferred operation is carried out in a static mixer where the contact time for inversion is regulated by the diameter or length of a cylindrical tube containing static baffles.

The process may be briefly summarized by the following series of steps: (a) forming a water-in-oil emulsion from water which contains dissolved therein a water-soluble ethylenic-unsaturated monomer, thereby producing an aqueous monomer phase which has a concentration of from 75 to 95% by wt of the emulsion; an inert hydrophobic liquid of 4 to 24% by wt of the emulsion; a water-in-oil emulsifying agent in a concentration of from 0.1 to 15% by wt of the emulsion; and a free radical initiator; (b) heating the emulsion under free radical forming conditions to polymerize the water-soluble ethylenic-unsaturated monomer contained in the emulsion; (c) polymerizing the monomer in the water-in-oil emulsion to produce a polymeric latex.

The water-in-oil emulsion containing polymer is then contacted with water for a period of 0.1 to 10.0 seconds and preferably less than 1 second. The water contains dispersed therein 0.1 to 10.0% by wt based on polymer of a water-insoluble surfactant to provide a partially inverted emulsion. Preferably the water contains 0.1 to 5.0% surfactant, based on polymer. It is desired to add enough water to reduce the solids level of the emulsion to 15 to 25%, preferably 20% by wt. This involves the addition of 30 to 40% by wt of water, based on total emulsion.

Example: A series of experiments were performed in which polymeric latexes were prepared and then drum dried upon a pilot plant sized unit. The formulations consisted of 35 parts polymer, 30 parts oil, and 35 parts water. 4% octyl phenol reacted with 3 mols EtO is added as activator. The drum speed was 6 rpm. Drying temperature was 325°F. Samples were 0.001" thick. Results were as follows:

Polymer	Oil	Dissolution Time, min
Acrylamide	Kerosene	10
Acrylamide/ acrylic acid	Isopar M	7
Acrylamide/ dimethylamino- ethyl methacrylate	Toluene	5.5

Comparisons, dry polymers produced according to U.S. Patent 3,284,393, dissolved in water in from 1 to 1½ hours.

Formation of Block Copolymers from Amine-Terminated Polymers

According to *W.L. Hergenrother, R.A. Schwarz, R.J. Ambrose and R.A. Hayes; U.S. Patent 4,157,429; June 5, 1979; assigned to The Firestone Tire & Rubber Company* polymers of anionically polymerized monomers such as conjugated dienes, vinyl substituted aromatics, olefinic type compounds, and heterocyclic nitrogen-containing compounds, are produced and end capped with a polyisocyanate or polyisothiocyanate. Such end capped polymers are then reacted with compounds containing an amide such as lactam to give an imide type end group.

The imide type terminated polymer is hydrolyzed to form a stable amine terminated polymer which may be stored for a short period of time to an extended period of time and reacted with other various polymers and monomers, or various combinations of monomers to form various block or graft polymers. The amine polymer may be reacted with any amine-reactive compound such as with a polyisocyanate or polyisothiocyanate and a lactam in the presence of known anionic lactam polymerization catalysts to give a blocked nylon copolymer.

Similarly, other block or graft copolymers may be obtained by reacting amine-reactive compounds such as various monomers or polymers with the terminated amine polymer.

The preparation of the end capped anionically produced polymers will be more fully understood by referring to the following examples.

Example 1: *1,2-Polybutadiene* — To a clean, dry, nitrogen purged 28 oz beverage bottle was added 600 cc of purified tetrahydrofuran, and 77.6 g of butadiene. After cooling to –20°C, 1.52 cc of 1.64 M (2.49×10^{-3} mol) n-butyllithium in a hexane solution was added and held at –20°C for 4 hours until the butadiene had polymerized. To this lithium polybutadiene, which was 85% 1,2-microstructure, was added rapidly 5.92 cc of a 1.22 M (7.21×10^{-3} mol) toluene diisocyanate solution in toluene. The resulting polymer had an average MW of 31,000.

Example 2: *1,4-Polybutadiene* — To a clean, dry 28 oz beverage bottle was added 500 to 600 cc of purified toluene, 98.1 g of butadiene and 1.43 cc of 1.75 M

(2.5 x 10^{-3} mol) n-butyllithium in a hexane solution. The bottle was stirred magnetically overnight at 25°C before rapidly adding 2.52 cc of 0.995 M toluene diisocyanate (2.5 x 10^{-3} mol). This solution was stirred for 1 hour and the product was precipitated with methanol. The resulting 1,4-polybutadiene had a viscosity average MW of 56,000 g/mol and vinyl content of about 10%.

The abovedescribed polyisocyanate or polyisothiocyanate end capped polymers are further reacted with an amide compound to give an imide type terminated polymer and then hydrolyzed to produce a stable amine-terminated anionically prepared polymer.

Example 3: *34.5/65.5 1,2-Polybutadiene-Nylon 6 Block Copolymer* – The preparation of the block copolymer from an amine terminated polymer can be carried out in any reaction vessel capable of withstanding temperatures up to about 200°C, having provisions for mechanical stirring, and capable of maintaining an inert atmosphere and small pressures. The starting materials of an amine terminated polymer and a preferred nylon monomer or prepolymer may be charged in any order as liquids, solids, melts or combinations thereof. Of course, the starting materials must be dried, either separately or in combination to remove water and other low molecular weight impurities.

Solutions or liquids may be dried by contacting them with activated molecular sieves or activated alumina whereas solids and melts are preferably dried by subjecting them to a vacuum of less than 2 mm of Hg at temperatures up to 120°C and preferably at a vacuum of less than 1 mm of Hg at 100°C for 5 to 16 hours.

To a stainless steel resin kettle of 500 mm capacity having a mechanical stirrer and a vacuum take-off was added 102 g of a cement of a 1,2-polybutadiene amine terminated polymer in hexane (approximately 73.5% solids at a MW of about 14,700) along with 150 g of flake caprolactam. The mixture was warmed in a warm water bath from 30° to 50°C and a vacuum of 30 to 80 mm Hg applied to remove the bulk of the hexane. The resulting mixture was then heated from 90° to 105°C in a Wood's Metal Bath and a vacuum of less than 2 mm Hg was applied and maintained for a 16 hour period with stirring.

Nitrogen was then admitted to the resin kettle and a slight nitrogen purge maintained during the remainder of the preparation. To the stirred dry mixture was added 0.45 g of sodium hydride (58.1% dispersion in mineral oil–11 mmol). After 5 minutes of mixing, 0.75 mm of toluene diisocyanate (Hylene TM, 5.27 mmol) was added. The temperature was slowly raised to between 165° and 175°C. Agitation was continued until the temperature reached 120° to 130°C and the mixture became too viscous to stir. The polymerization was carried out for 3 hours at 165° to 175°C. Upon cooling, the copolymer was removed from the resin kettle and compounded by milling.

The milling recipe contained 100 parts of the copolymer, 150 parts of silica 325 mesh, 1.5 parts of Z6075 silane, and 2.0 parts of Dicup R. After curing for 20 minutes at 350°F, the compounded block copolymer gave the following physical properties: flexural strength of 16,210 psi, flexural modulus of 9.88 x 10^5 psi, Izod, notched of 0.38 foot pounds per inch, Izod, unnotched of 4.2 foot pounds per inch, Gardner impact strength of 3.7 inch pounds, heat distortion temperature at 264 psi of 222°C, and a Rockwell hardness E scale of 79.

IMPROVEMENT IN PROPERTIES

Flame-Retardant Fiber

A. Ohmori, M. Ando and A. Akiyama; U.S. Patent 4,079,036; March 14, 1978; assigned to Kuraray Co., Ltd., Japan have discovered that the addition of calcined stannic acid having a specific crystal size to polyvinyl alcohol (PVA) and polyvinyl chloride (PVC) mixed polymer fiber renders the fiber more flame-retardant than with conventional stannic acid and at the same time does not cause discoloration of the fiber in the course of heat treatment, unlike the discoloration characteristic with prior art stannic acid treated fibers. The fiber also evidences excellent water-proofness, strength and weather resistance.

The following tests were used to evaluate the products.

Flame-Retardancy (Limiting Oxygen Index—LOI): As test samples, knitted cloths (wt: 250 g/m^2) made of filament yarn (1,000 filaments, 2,000 den) spun from a spinning solution of PVA, PVC and SnO$_2$ compounds with a weight ratio PVA:PVC:SnO$_2$ equal to 67:33:2, were employed.

The LOI, defined as the minimum oxygen concentration in percent required to maintain the combustion in a gaseous mixture of oxygen and nitrogen, was measured for each of the test samples by using a ON-1 type device (Toyo Rika Company Limited). A high LOI value corresponds to a high degree of flame-retardant property for the test samples.

Discoloration by Heat Treatment: The PVC/PVA mixed fiber containing stannic compounds is subjected to dry heat drawing for 18 sec at 227°C so that the fiber is drawn to 3 times its original length and then to dry heat setting until the fiber length decreases by 10%. The degree of discoloration of the fiber caused by such heat treatment was determined by visual inspection and ranked in the following 5 grades.

Grade 5	None
Grade 4	Very little
Grade 3	Slight
Grade 2	Rather considerable
Grade 1	Considerable

Example 1: Metastannic acid (Showa Kako Co., Ltd.) was calcined for 3 hours at 500°C to obtain a calcined product with a crystal size of 70 A. The resulting calcined product was added to an aqueous solution of PVA with a DP of 1,700 and a saponification degree of 99.9%. The compound was incorporated in the PVA by heating the mixture. The mixture was then cooled to 70°C and added to a PVA-grafted PVC emulsion with a mean particle size of 250 A obtained by emulsion polymerization, so as to obtain a spinning solution with a wt ratio of PVA:PVC:calcined stannic acid equal to 60:40:1.5 and a total polymer concentration of 18%.

The spinning solution thus obtained was defoamed and extruded into a coagulating bath containing 420 g/l of Glauber's salt and 0.03 g/l of sulfuric acid. The resulting fiber was subjected to roller drawing, wet heat drawing, water washing, drying, dry heat drawing and shrinking so as to produce a fiber consisting of

1,000 (2 den) filaments with an LOI of 43%. The same fiber which was subsequently formalized so that the formalization degree was equal to 33 mol % based on PVA, had good flame-retardancy equal to 35% LOI, which was somewhat lower than that of the same fiber prior to formalization.

Example 2: Metastannic acid (Shin Nippon Kinzoku Kagaku KK) was calcined for 1.5 hours at 450°C to obtain a calcined product having a crystal size of 55 A. The calcined product was then pulverized to a mean particle diameter of 1 μ dispersed into water. To this aqueous dispersion were added boric acid and PVA with a polymerization degree of 1,700 and a saponification degree of 99.4 mol %, and the compounds were dissolved by heating and then cooled to 65°C. To this mixture was added a PVC emulsion with a pH of 7 and a mean particle size of 350 A, obtained by emulsion polymerization using sodium laurylsulfate as an emulsifying agent.

The resulting spinning solution with a weight ratio of PVA:PVC:H_3BO_3:calcined stannic acid equal to 67:33:1.2:2 and a total polymer concentration of 17% was defoamed and extruded into a coagulating bath containing 50 g/l of caustic soda and 250 g/l of Glauber's salt to obtain a fiber consisting of 1,000 (2 den) filaments through the consecutive steps of roller drawing, neutralization, wet heat drawing, water washing, drying, dry heat drawing and shrinking. The fire-retardancy and resistance to discoloration of the resultant fiber were measured and were found to be excellent (LOI–39% and grade 4, respectively). The fiber also had a strength of 6.5 g/den and a water-proofness which was completely acceptable even without acetalization.

Stability to Heat and Light

A stabilized polyolefin is disclosed by *G. Cantatore; U.S. Patent 4,104,248; August 1, 1978; assigned to Montefibre SpA, Italy* characterized in that it contains as stabilizer an organic polyamine having the general formula:

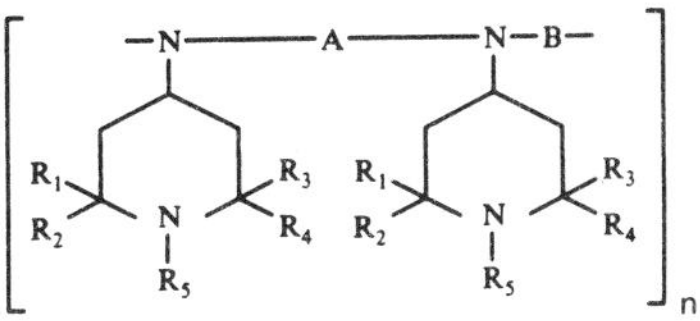

wherein R_1, R_2, R_3 and R_4 are equal to or different from each other and are an alkyl group having from 1 to 4 carbon atoms; R_5 is hydrogen or an alkyl group having from 1 to 4 carbon atoms; A is an alkylene group having from 2 to 10 carbon atoms; B is a divalent aliphatic, cycloaliphatic, aromatic or alkylaromatic radical, which can contain hetero atoms such as O, S, N and P either in the chain or as side substituents; and n is a whole number between 2 and 1,000. The organic polyamine has an inherent viscosity of between 0.01 and 1 dl/g, and is employed in an amount equal to or less than 5% by wt.

Example: *Preparation of N,N'-Bis(2,2,6,6-Tetramethyl-4-Piperidyl)Ethylenediamine* – 162.75 g (1.05 mols) of 2,2,6,6-tetramethyl-4-piperidone dissolved in 200 cc of methanol, 30 g (0.5 mol) of ethylenediamine dissolved in 40 cc of methanol, and 0.5 g of Pt (at 10% on carbon) were introduced into a 1 liter autoclave, and the mixture hydrogenated at 80°C and 50 atm pressure for 2 hours.

After removal of both catalyst and solvent the residue was distilled, thus obtaining 155.5 g (92%) of product, having a boiling point of 150° to 151°C/0.1 mm Hg; a melting point of 80° to 81°C, and a found N content of 16.51% (calculated for $C_{20}H_{42}N_4$ = 16.56%).

Preparation of the Polyamine — To a solution of 33.8 g (0.1 mol) of N,N'-bis-(2,2,6,6-tetramethyl-4-piperidyl)ethylenediamine in 50 cc of methanol were added 9.25 g (0.1 mol) of epichlorohydrin. This mixture was heated under reflux for 10 hours, adding during the last 8 hours of heating, in small regular portions, 4 g (0.1 mol) of sodium hydroxide in the form of tablets.

At the completion of the addition of NaOH, the mixture was heated under reflux for a further 2 hours. It was then filtered to separate the sodium chloride formed in the reaction. The filtrate was dried by removing the methanol, first at atmospheric pressure at 70° to 100°C and then by heating for 4 hours at 120°C and 1 mm Hg.

In this way there was obtained 39 g of a brittle resinous product, showing a light yellow color and an inherent viscosity of 0.14 dl/g and a content of nitrogen of 13.6%. The inherent viscosity was determined at 25°±0.1°C with a solution of 0.5% in chloroform, using a Deseux-Bischoff viscosimeter.

Stabilization Test: 25 g of the abovedescribed polyamine were dissolved in 100 cc of methanol. The solution thus obtained was mixed together with 5 kg of polypropylene having an intrinsic viscosity of 1.65 dl/g, a residue on heptane extraction of 96.5%, and an ash content of 80 ppm, and 5 g of calcium stearate. This mixture was granulated in an extruder in an oxygen-free atmosphere at 180°C, and was then spun under the following conditions: temperature of the screw, 250°C; extruding head temperature, 230°C; temperature of the spinneret, 230°C; and maximum pressure, 35 kg/cm².

The spinneret used has 40 holes of 0.8 mm diameter and a length of 4 mm. The filaments coming out of the spinneret were gathered at a rate of 500 m/min and were stretched at 130°C in a steam atmosphere with a stretch ratio of 3.3. The filaments thus obtained showed the following characteristics: titer, 17 dtex; tenacity, 2.8 g/dtex; elongation at break, 85%; and tenacity after 1,400 hours of exposure to a Weatherometer, 1.8 g/dtex. The Weatherometer has a xenon arc lamp of 6,000 W, a RH of 30±5% and a black panel temperature of 63±3°C.

For comparison purposes it should be noted that with the prior art stabilizers known so far, the tenacity is halved after about 500 hours of exposure.

Heat-Resistant, Infusible Fibers

Infusible and insoluble styrene copolymeric fibers are prepared by *I. Kimura, K. Ohtomo and K. Shirane; U.S. Patent 4,007,250; February 8, 1977; assigned to Kanebo, Ltd., Japan* by copolymerizing styrene with an olefinic compound having at least one functional group selected from haloalkyl group, amino group, carboxyl group, carbonic acid ester group, carbonic acid halide group, hydroxyl group, amido group, nitrile group and halogen atom; melt spinning the resulting styrene copolymer into fibers and treating the formed fibers with a crosslinking agent to form three dimensional crosslinkages in the fibers.

As compared with the very brittle untreated fibers not subjected to the three dimensional crosslinking treatment, the infusible and insoluble crosslinked fibers have a much higher acid resistance and are considerably more flexible and have much higher strength and elongation. Furthermore, the fibers exhibit excellent whiteness and have a silklike gloss.

When an amino compound is used as a crosslinking agent in the three dimensional crosslinking treatment, the dyeability with an acidic dyestuff is very good. The fibers are insoluble in a solvent and when these fibers are heated, e.g., by a burner at a high temperature, the fibers are only carbonized and are not fused. Accordingly, the fibers can be used for electric insulating materials, reinforcing material for plastics by utilizing the excellent heat resistance as well as for general purpose cloths.

Example: A mixture of 70 parts of styrene and 30 parts of chloromethylstyrene (a mixture of ortho and para isomers) was added with 2.5 parts of benzoyl peroxide, and subjected to a copolymerization reaction at 80°C for 12 hours. The resulting copolymer was melted at 150°C under a nitrogen atmosphere, and the pressure was reduced to 5 mm Hg to distill off unreacted monomer. The thus obtained copolymer had a viscosity η_{rel} of 0.32 in benzene.

The copolymer was made into chips, and the chips were extruded through an extruder of 20 mm diameter. The extruded filaments were taken up on a bobbin at a rate of 800 m/min to obtain an undrawn filament of 75 den/25 ft. The undrawn filament was taken out from the bobbin and immersed in a 30% solution of pyridine in methanol in a bath ratio of 1:200 at a temperature of 20°C, then heated up to 60°C in 2 hours and further kept at 60°C for 5 hours to effect a crosslinking treatment. The filament was taken out from the bath, washed with methanol and water repeatedly, and dried at 60°C under a reduced pressure to obtain an insoluble and infusible filament.

The resulting filament having a silklike gloss was subjected to an extraction with acetone to measure the amount of uncrosslinked portion. 97% was acetone-insoluble. When the filament was treated in a hot air circulating type drier kept at 200°C for 3 hours, the filament did not substantially color.

The strength and elongation of the filament were determined to be about 10 times of those of the untreated filament. When the filament is heat treated at 200°C for 3 hours, the elongation is somewhat decreased, but the strength and bending strength are maintained sufficiently high, showing that the filament is excellent in heat resistance. When the filament was heated to redness on a spatula by a burner, the filament was merely carbonized.

When the filament was dyed with 2% by wt of an acid dye (Rocceline, Sumitomo Kagaku Co.) based on the weight of the filament in a bath ratio of 1:100 at a temperature of 95°C for 150 minutes under acetic acid acidity, the dye receptivity was 52%.

Good Dyeing Affinity

P. Couchoud; U.S. Patent 4,039,634; August 2, 1977; assigned to Rhone-Poulenc-Textile, France discloses shaped structures, e.g., filaments, fibers or films based on poly(vinylidene fluoride) possessing good dyeing affinity for acid and metal-containing dyestuffs and comprising a fiber and filament forming mixture of poly(vinylidene fluoride) and a copolymer containing 80 to 90% by wt of methyl methacrylate and 5 to 20% by wt of a compound which contains a quaternary ammonium group and is copolymerizable with methyl methacrylate.

In the example, the specific viscosity is measured using a solution of polymer of concentration 2 g/l at 25°C in dimethylformamide, and the dyestuffs are identified by their C.I. reference taken from *Colour Index,* 1971 edition.

The LOI index of textile samples is measured by means of a method based on ASTM Standard Specification B-2863-70, relating to the measurement of the LOI index of plastic test pieces, but using textile samples requiring a rectangular frame of internal dimensions 5 x 16 cm.

The light-fastness test using the xenotest is the subject of ratified French Standard Specification NF G 07,067 (ISO Recommendation R 105/V—1969). The test relating to fastness to washing at 60°C is the subject of French Standard Specification NF G 07,015 (ISO Recommendation 105/IV—1968).

Example: 186 parts of methyl p-toluenesulfonate and 50 parts of acetone are introduced dropwise, by means of a dropping funnel, with stirring, at a temperature below 25°C, into a reactor containing a solution of 157 parts of N-dimethylaminoethyl methacrylate diluted with 50 parts of acetone. The quaternized product precipitates; it is recovered, rinsed, dried and then recrystallized from ether.

26 parts of the product obtained above, 154 parts of methyl methacrylate, 420 parts of dimethylformamide and 0.1%, relative to the weight of monomers, of azo-bisisobutyronitrile are then introduced into a reactor. Polymerization is carried out with stirring, under nitrogen, for 20 hours at 60°C.

The polymer obtained is precipitated in water; 120 parts of polymer, of specific viscosity 0.96, containing 275 base meq derived from the quaternary moiety per kg, are obtained. A spinning solution of concentration 27.6%, in dimethylformamide is prepared which contains 20 parts of the copolymer prepared above, containing 91% of methyl methacrylate and 9% of N-trimethylethylammonium methacrylate p-toluenesulfonate, and 80 parts of poly(vinylidene fluoride), of melt flow index 190.

This solution is heated to 60°C and is extruded through a spinneret with 64 orifices of diameter 0.07 mm into a coagulating bath, kept at 20°C, containing 57 parts of dimethylformamide and 43 parts of water. As they issue from the coagulating bath, the filaments are first stretched in air at ambient temperature at a ratio of 3:4 and then in boiling water at a ratio of 1:7, washed with water countercurrently at ordinary temperature, and finally dried on rollers kept at 120°C.

The filaments obtained possess the following properties: dry tenacity, 16 g/tex; elongation, 20%; base meq/kg of polymer, 55; and LOI index, 28.

The filaments were dyed with Acid Blue C.I. 40, Acid Black C.I. 60, Orange C.I. 19, and Orange C.I. 92. The dyeing conditions were 1 hour at 105°C in a bath ratio of 1:50 (1 g of material/50 cm^3 of bath), in the presence of 1 cm^3/l of 80% strength formic acid, after desizing and vaporizing for 15 minutes at 110°C, the dye bath containing 2% of dyestuff relative to the weight of material. After dyeing, the filaments are rinsed for 15 minutes at 70°C using a bath containing 1 cm^3/l of oxyethylated cetyl alcohol and 1 g/l of Na$_2$CO$_3$.

Orange C.I. 19 resulted in a light-fastness of 3; the other dyes gave values of 6 to 7. All washing-fastness values were 5/4–5 or 5/5.

Increased Absorbency

J.R. Gross; U.S. Patent 4,041,020; August 9, 1977; assigned to The Dow Chemical Company discloses a composition which is useful to form water swellable articles of a carboxylic type synthetic polyelectrolyte which consists of a solvent such as lower alcohols, water, or mixtures thereof, and 5 to 6%, preferably 15 to 40% by wt based on the solvent of a carboxylic copolymer which contains in the copolymer 25 to 98% by wt based on the total weight of the copolymer of an alkali metal salt of an olefinically unsaturated monosulfonic or monocarboxylic acid, 2 to 50% by wt of an olefinically unsaturated monocarboxylic acid, and 0.1 to 5.0% by wt of an N-substituted acrylamide or methacrylamide wherein the substituent group is a hydroxymethylene or an alkoxymethylene group having 1 to 8 carbons in the alkyl group.

The articles are crosslinked by heating and/or removing substantially all of the water from the precursor composition. The absorbent articles are useful as surgical sponges, diapers, tampons, meat trays, bath mats and the like.

Example 1: Acrylic acid (19.48 g), deionized water (61 g), 50% sodium hydroxide (19.15 g), N-methylolacrylamide (0.41 g of 60% solution) and sodium persulfate (0.5 g) are charged to a screw-cap 4 oz bottle. The bottle is sealed, shaken and placed in a 50°C water bath for 16 hours. The extremely viscous solution was then diluted with an equal volume of water and spread on a mirror finish chrome plate with a 30 mil draw bar.

After air drying to a clear film, the film was further dried in a 150°C oven for 2 hours. 1 g of this cured film absorbed 49 g of 0.27 N NaCl solution to give a firm, readily filtered gel.

Example 2: Acrylic acid (5.0 g), sodium styrene sulfonate (19.87 g), deionized water (75 g) and N-methylolacrylamide (0.21 g of 60% solution) were charged to a 4 oz screw-cap bottle and treated as in Example 1. The dried and cured film of this heteropolymer absorbed 31 g of salt solution per gram of film.

Following the procedure of Example 1, 9.95 g of sodium styrene sulfonate, 0.05 g of N-methylolacrylamide, 30 g of water and 0.5 (5% by wt) g of acetic acid were reacted for 16.5 hours at 50°C. A film was cast from the resulting solution and cured at 150°C for 5 hours.

The film was found to be completely water-soluble and not water swellable. This is believed to illustrate the fact that a copolymerized acid is needed to achieve a crosslinked and water swellable product.

Spontaneous Crimp

Conjugate filaments which exhibit a high degree of spontaneous crimp and good modulus are described by *R.K. Gupta, G. Jurkiewitsch and L.J. Logan; U.S. Patent 4,115,620; September 19, 1978; assigned to Hercules Incorporated.* The filaments comprise at least 2 components, one of which is a blend of polypropylene with 5 to 50% of certain low MW hard resins derived from hydrocarbons having at least 4 carbon atoms or rosin derivatives and the other component is polypropylene or a blend of polypropylene and a different amount of the hard resin or rosin derivative.

The rosin derivatives which are preferred can be grouped into six classes. The first of these classes comprises rosins which have been modified by hydrogenation, disproportionation, polymerization, condensation with unsaturated carbocyclic compounds to form resinous condensation adducts or combinations of such modifying treatments. Some typical representative members of this class include hydrogenated rosin, disproportionated rosin, polymerized rosin, specifically dimerized rosin, hydrogenated disproportionated rosin, hydrogenated dimerized rosin, condensation adducts of rosin with styrene, divinyl benzene, diisopropenyl benzene, α-methyl-p-methyl-styrene or cyclopentadiene, as well as the hydrogenated condensation adducts thereof.

The second of these classes comprises the individual resin acids which are the resin acid components of the hydrogenated rosin and disproportionated rosin of the first class. The principal members of this class include dihydroabietic acid, tetrahydroabietic acid, dehydroabietic acid, dihydrodextropimaric acid, tetrahydrodextropimaric acid, dihydroisodextropimaric acid, and tetrahydroisodextropimaric acid.

The third of these classes comprises the hydroabietyl alcohol esters of the modified rosins of class (1) above and the resin acids of class (2) above. Some typical representative members of this class include the hydroabietyl alcohol ester of hydrogenated rosin, disproportionated rosin, dihydroabietic acid, tetrahydroabietic acid, dehydroabietic acid, and the like.

Dirosin amine which constitutes the fourth class of rosin derivatives may be prepared by the hydrogenation of rosin nitrile over a nickel catalyst at temperatures above 200°C with removal of ammonia.

The monoamides which constitute the fifth class of rosin derivatives include N-dehydroabietyl hydrogenated rosin amide, N-dihydroabietyl hydrogenated rosin amide, N-tetrahydroabietyl hydrogenated rosin amide, N-dehydroabietyl disproportionated rosin amide, N-dihydrodextropimaryl dimerized rosin amide, N-dehydroabietyl dihydroabietic acid amide, N-dihydroabietyl dehydroabietic acid amide, N-tetrahydroabietyl tetrahydroabietic acid amide, and the like.

The diamides which constitute the sixth class of rosin derivatives include the diamide of hydrogenated rosin and ethylene diamine, the diamide of disproportionated rosin and ethylene diamine, the diamide of dehydroabietic acid and ethylene diamine, the diamide of tetrahydroabietic acid and ethylene diamine, and the like.

The filaments prepared according to this process have a crimp potential of 10 to 120 or more depending upon composition, fiber processing temperatures and filament geometry. Crimp potential, as used herein, refers to the inherent crimping tendency of the conjugate fiber, normalized for the fiber denier. Thus, the crimp potential (CP) is described by the equation CP = (CPI) $(\sqrt{\text{denier/filament}})$ where CPI is the crimp frequency measured as the number of crimps per inch of straight fiber. Crimp frequency is determined by counting the number of crimps between 2 arbitrary marks on a crimped fiber and then dividing the number by the stretched out (straight) length in inches between the marks. Usually a satisfactory bicomponent filament will have a bicomponent crimp potential (BCP) of 20 to 100.

Example: A first component was prepared by blending 25 parts of pulverized hydrogenated polyterpene resin with 75 parts of polypropylene resin and extruding the blend into pellets. The polypropylene resin had an intrinsic viscosity of 2.4, a density of 0.905, and contained 0.5%, based on the total sample weight, of a stabilizer combination comprising the acid-catalyzed reaction product of 2 mols of nonylphenol and 1 mol of acetone, the reaction product comprising a mixture of isopropylidene-bis(nonylphenol), and 2-(2'-hydroxyphenyl)-2,4,4-trimethyl-5-6'-dinonylchroman and 0.1% of calcium stearate. The hydrogenated polyterpene resin had a softening point of 118°C, an average MW (Rast) of 790 and an iodine value of 17. A second component containing only the stabilized polypropylene resin was also prepared.

The above first and second components were then melted separately and fed by separate extruders and metering pumps to a bicomponent fiber spinning head of round cross section operated at 250°C and spun as conjugate filaments, the components being disposed side-by-side in a 50:50 relationship. The spun filaments were drawn down and packaged as 9 denier per filament yarns (each yarn containing 35 filaments) at 940 ft/min. The spun yarn was drawn 3.0X using differentially driven feed and draw rolls., the feed roll temperature being about 110°C, and that of the draw roll being about 140°C, and then was packaged under sufficient tension to prevent contraction of the yarn. The drawn denier was 3.0.

The yarn was then heat treated under zero tension in a hot air oven for 10 minutes at about 141°C. The heat treated yarn had a crimp frequency of 55, a bicomponent crimp potential of 95, a Young's modulus of 30.8 g/den and could be self-bonded at temperatures of about 147°C. Visual inspection of the yarn indicated that no void formation had taken place as a result of the drawing.

Improved Sizing

K. Kajitani, T. Moritani, K. Moritani and M. Shiraishi; U.S. Patent 4,172,930; October 30, 1979; assigned to Kuraray Co., Ltd., Japan disclose a size for textile fiber which comprises a modified polyvinyl alcohol type resin obtainable by the alkali saponification of a copolymer comprising a vinyl ester and an ethylenically unsaturated dicarboxylic acid substantially free from the monoester, diester and anhydride as essential and predominant copolymeric units.

Example 1: Vinyl acetate and itaconic acid were radical-polymerized in methanol by the delayed feeding method and after the addition of a polymerization inhibitor, the unreacted monomers were removed azeotropically with methanol to obtain a methanolic solution of the corresponding modified polyvinyl acetate. This solution was saponified by the addition of a methanolic solution containing a precalculated amount of sodium hydroxide at 40°C for a predetermined time. With the progress of saponification, the reaction system gained in viscosity until a polyvinyl alcohol gel precipitated out to give a nonhomogeneous system. The gel was roughly fragmented, washed well with methanol, methyl acetate, acetone or the like, dried in the air at room temperature and further dried under reduced pressure at 50°C.

The dried gel was mechanically pulverized and further dried. By the above procedure was obtained a sample powder of the modified polyvinyl alcohol resin. The sample powder was dissolved in a sufficient amount of water to give a concentration of 10 to 20 wt %.

Then, this size solution was cast on a polyester film (Diafoil 100) intimately wrapped around a hot drum (70°C) having a circumference of 3 m and dried in situ to produce a size film as thick as 50 μ. This laminated film was slit to suitable dimensions and heat-treated in a hot-current drier at 130°C for 10 min, after which the size film was peeled off and tested for various physical properties. To measure the adhesive affinity of the size for polyester, the laminated film was cut into a rectangular piece and with a pressure-sensitive reinforcing tape stuck to the sized face of the laminated film, the size film was peeled off the polyester substratum. The peel strength thus measured was taken as the adhesive strength of the resin.

The modified polyvinyl alcohol containing the ethylenically unsaturated dicarboxylic acid, which is exemplified by itaconic acid, has both excellent solubility after the heat-treatment and the necessary adhesive affinity for polyester. This adhesive affinity, in particular, is remarkably high, that is more than 10 times that of the conventional polyvinyl alcohol. It is easy to see that the high adhesive affinity and excellent solubility are responsible for the high performance of the modified polyvinyl alcohol as a size for textile fiber.

Example 2: 100 kg of the same itaconic acid-modified polyvinyl alcohol as used in Example 1 was dispersed and dissolved in 1,500 liters of water and 5 kg (as pure) of Makonol TS-253 (Matsumoto Yushi Seiyaku Co., Ltd.) was added. Using the resultant size solution, a polyester/cotton (65/35) yarn was sized and woven into cloth.

The sizing workability was excellent, without the formation of size scums, skinning of the size solution or other troubles. The divisibility of the sized yarn was also satisfactory and yarn could be woven with a high weaving efficiency, being free from yarn breaks, fluffing, wale streaks, etc. The fabric was desized, scoured and dyed to obtain a pink-colored fabric. This finished fabric had no dyeing specks and was judged to have been well desized. Weaving results: yarn breaks per hour, 0.28; weaving efficiency, 91.6%.

ADDITIONAL FIBER-FORMING POLYMERS

POLYLACTONES AND RELATED POLYMERS

Copolymers of D,L-Lactide and Epsilon-Caprolactone

Copolymers which can be fabricated into films, fibers and structural shapes are prepared by *R.G. Sinclair; U.S. Patents 4,045,418; August 30, 1977; and 4,057,537; November 8, 1977; both assigned to Gulf Oil Corporation* by copolymerizing an optically inactive lactide, i.e., D,L-lactide, and epsilon-caprolactone in the presence of a tin ester of a carboxylic acid. A copolymer prepared from 85 parts by weight of D,L-lactide and 15 parts by weight of epsilon-caprolactone is a thermoplastic elastomer. A copolymer prepared from 90 parts by weight of D,L-lactide and 10 parts by weight of epsilon-caprolactone is a rigid, clear, thermoplastic solid.

Example 1: *60/40, D,L-Lactide/Epsilon-Caprolactone* — 60 g of purified, dry D,L-lactide melting at 115° to 128°C is placed in a glass ampoule. To the D,L-lactide in the glass ampoule is then added 1 ml of benzene (calcium hydride-dried) containing 0.2 g of pure stannous caprylate. The ampoule is then evacuated to remove solvent and back flushed with nitrogen. 40 g of pure epsilon-caprolactone distilling at 56° to 57°C/0.35 torr is then added to the D,L-lactide and catalyst in the ampoule. The contents of the ampoule are thoroughly mixed and degassed. The ampoule is then sealed in vacuo at its constriction by means of a torch.

The ampoule and its contents are immersed in an oil heating bath. The contents of the ampoule are vigorously agitated to provide a homogeneous melt while the melt viscosity is still low. The contents of the ampoule are heated for 1 hour at 135° to 145°C and then 5 days at 115° to 126°C. The ampoule is allowed to cool to allow the copolymer to solidify. The glass ampoule is then shattered and the glass is removed from the solidified copolymer. The resulting copolymer is a transparent colorless, weak, tacky material. The product is definitely a copolymer since its properties are distinctly different from a mere physical blend of the two homopolymers of D,L-lactide and epsilon-caprolactone. The weight average molecular weight ($\overline{M}_w$) as determined by Gel Permeation Chromatography (GPC) is 154,300. The number average molecular weight ($\overline{M}_n$) is 38,100. The copolymer is melt-formed by conventional thermoplastic techniques into transparent, colorless, tough, sparkling sheets.

Example 2: *75/25, D,L-Lactide/Epsilon-Caprolactone* – The procedure of Example 1 is repeated except that 270 g of D,L-lactide, 3.6 ml of benzene containing 0.72 g of stannous caprylate and 90 g of epsilon-caprolactone are placed in the ampoule, evacuated, sealed and heated for 89 hours at 118° to 122°C. The resulting copolymer is a light yellow, transparent, void-free elastomer. The yield of product is 352 g after removing the glass. The copolymer has GPC molecular weights of $\overline{M}_w$ = 174,400 and $\overline{M}_n$ = 61,500. The copolymer is easily melt-formed by conventional thermoplastic techniques into clear, sparkling sheets, films and other useful items of commerce. Objects formed from the copolymer are slowly biodegradable.

To illustrate the moldability of this copolymer, the copolymer is cut into small slices on a bale cutter. 50 g of the sliced copolymer is placed in a 6" x 6" x 0.75" mold, preheated to 130°C. The heated copolymer is then compression-molded in a Preco Press for 5 minutes at 130°C. Excessive heating and the presence of moisture is to be avoided during molding in order to avoid an adverse affect on the molecular weight of the copolymer. The molded copolymer has GPC molecular weights of $\overline{M}_w$ = 131,500 and $\overline{M}_n$ = 45,100.

Example 3: *Comparative* – 75/25 D,L-Lactide/Delta-Valerolactone: In a manner similar to Example 2, 75 g of D,L-lactide is charged into an ampoule with 1 ml (0.2 g/ml) of anhydrous stannous caprylate solution in benzene. The benzene is removed by evacuating the system. After removal of the benzene, 25 g of delta-valerolactone is added. Following mixing, degassing and sealing of the ampoule, the contents of the ampoule are heated by placing the ampoule in an oil bath at 126° to 138°C for 24 hours. Examination of the product shows a 97% conversion of monomer to polymer. The polymer is a colorless, transparent, weak, pliable copolymer having GPC molecular weights of $\overline{M}_w$ = 79,200 and $\overline{M}_n$ = 30,500. The maximum tensile strength of the copolymer is 270 psi.

In comparing the copolymer of Example 3 with the copolymer of Example 2, it will be noted that the molecular weights and tensile strength of the D,L-lactide/epsilon-caprolactone copolymer of Example 2 are surprisingly superior to the molecular weights and tensile strength of the D,L-lactide/delta-valerolactone copolymer of Example 3.

Throwaway objects made from copolymers of the process are environmentally attractive in that they slowly degrade to harmless substances. All of the copolymers degrade in a matter of several hours in boiling water. When placed in a Weather-O-Meter (alternating 30 minutes of light and water at 22° to 49°C), the copolymers become tacky after overnight exposure. When subjected to prolonged exposure at a constant temperature of 72°F (22.2°C) and humidity (50%), the copolymers show a marked change in physical characteristics.

The copolymers are sensitive to ambient moisture. When copolymers were prepared from 75 to 90 parts by weight of D,L-lactide and 10 to 25 parts by weight of epsilon-caprolactone, they exhibited drastic changes in their physical properties after 4-months exposure at a constant temperature of 72°F (22.2°C) and 50% relative humidity.

Dicarboxydi(Hydroxymethyl)Diphenylmethane Dilactones

A. Onopchenko, J.G. Schulz, and E.T. Sabourin; U.S. Patent 4,104,281; Aug. 1, 1978; assigned to Gulf Research & Development Company disclose a process for preparing isomeric dicarboxydi(hydroxymethyl)diphenylmethane dilactones wherein benzophenone-3,4,3'4'-tetracarboxylic dianhydride (BTDA) is subjected to hydrogenation in an ether or ester carrier in the presence of a hydrogenation catalyst pretreated in an ether carrier.

Example 1: BTDA was subjected to hydrogenation with a nickel catalyst (NiO-104P, manufactured by Harshaw Chemical Company). The catalyst was pretreated by heating, while stirring, a slurry containing the catalyst and a carrier in a hydrogen atmosphere in a 1 liter autoclave. The reaction mixture was cooled to room temperature and depressured to atmospheric pressure and BTDA was added thereto. The slurry containing the carrier, BTDA and catalyst were stirred and heated in a hydrogen atmosphere.

The autoclave at the end of the hydrogenation period was cooled to room temperature and depressured to atmospheric pressure and the contents thereof were filtered to separate catalyst and any unreacted BTDA therefrom. The filtrate was then concentrated from about one-third to about one-quarter of its volume by heating in a rotary evaporator and the resultant filtrate was filtered to recover crystalline product. The dilactone product that was recovered was twice crystallized from ethyl acetate.

Example 2: A total of 28 g of a mixture of dilactones prepared in accordance with the procedure of Example 1, and 6 g of ethylenediamine were charged into a flask containing 100 ml of N-methylpyrrolidine and 5 g of benzene. The flask was fitted with a Dean-Stark trap filled with 50 ml of benzene. The reaction mixture was heated under reflux (around 140°C) for 4.5 days. A total of 4.3 g of water was collected. After stripping off N-methylpyrrolidine under pressure, the residue amounted to 30.2 g. The crude polymer was ground into powder washed several times with about 100 ml portions of boiling isopropanol to remove last traces of solvent and air dried for six hours, followed by drying in a vacuum at 100°C for 10 hours. Analysis of the product indicated the reaction proceeded as follows:

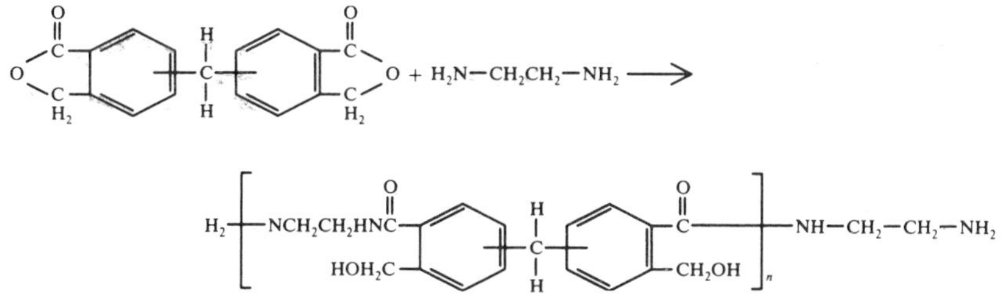

The resin was found to have a molecular weight in excess of 10,000 and was distinctive in its virtual insolubility in conventional solvents, such as methanol, acetone, ethyl acetate, tetrahydrofuran and benzene. The molten resin was drawn into flexible fibers capable of being cold drawn to impart additional mechanical strength thereto suitable for the preparation of fabrics therefrom.

Bioresorbable Unsymmetrically Substituted 4-Dioxane-2,5-Diones

According to *T.A. Augurt, M.N. Rosensaft, and V.A. Perciaccante; U.S. Patent 4,033,938; July 5, 1977; assigned to American Cyanamid Company* 3,6-substituted 1,4-dioxane-2,5-diones may be polymerized to give living tissue absorbable, hydrolytically degradable, surgically useful polymers. These polymers have predominantly regular rather than random spacings of side chains, may be stereoregular and tend toward higher crystallinity than randomly sequenced polymers. A polymer of 3-methyl-1,4-dioxane-2,5-dione has the same empirical formula as an equimolecular copolymer of lactic and glycolic acid but has unique physical properties resulting from its more regular steric configuration.

Example 1: *Synthesis of 3-Methyl-1,4-Dioxane-2-5-Dione* — 1 mol of chloroacetic acid (94.5 g), 1 mol of D,L-lactic acid (107.0 g of 85% water solution), and 8 g of Dowex 50W-X ion exchange resin (equivalent to 1 ml concentrated H_2SO_4), and 200 ml benzene were refluxed and the theoretical amount of water collected in a Dean-Stark trap. The solution was allowed to cool to room temperature and the ion exchange resin was filtered off. The benzene was removed on a rotary evaporator with vacuum. The unreacted chloroacetic acid was sublimed out at 0.2 to 0.4 torr. The O-chloroacetyl-D,L-lactic acid was distilled at 108° to 118°C at 0.2 to 0.3 torr. After recrystallization from toluene, the O-chloroacetyl-D,L-lactic acid has a MP of 72° to 74°C.

3.34 g (0.02 mol) O-chloroacetyl-D,L-lactic acid and 2.02 g (0.02 mol) triethylamine were dissolved in 670 ml dimethylformamide. The solution was heated to 100°±5°C for 6 hours and allowed to cool to room temperature. The solvent was distilled off under vacuum yielding a reddish colored semisolid residue. The product was removed by extraction with acetone leaving solid triethylamine hydrochloride.

The acetone extract was evaporated yielding a reddish colored oil which solidified on standing to a reddish yellow solid. It was recrystallized by dissolving in warm isopropanol and cooling to –25°C. The D,L-3-methyl-1,4-dioxane-2,5-dione had a melting point of 64° to 65°C (0.7 g). It was further purified by sublimation at 0.01 torr at 50° to 60°C. The yield was 0.3 g, melting point 63.8° to 64.2°C; % carbon found was 46.57, 46.10 calculated; % hydrogen found was 4.73, 4.60 calculated. The NMR spectrum showed that the product was essentially racemic as would be expected.

Example 2: *Polymerization of D,L-3-Methyl-1,4-Dioxane-2,5-Dione at 180° C —* To a glass tube was added 6.0 g of D,L-3-methyl-1,4-dioxane-2,5-dione and 1.2 ml of an ether solution containing 0.1 mg of $SnCl_2 \cdot 2H_2O$ per ml. The ether was vaporized and removed, and the tube sealed. The sealed tube was placed in an oil bath at 180°±2°C for 4 hours, cooled and broken. The cooled tube contents were dissolved in 120 ml of boiling acetone and the solution added dropwise to 1,200 ml of methanol. The resulting precipitated poly(D,L-3-methyl-1,4-dioxane-2,5-dione) was dried in vacuo for 2 days at 25°C. The resulting polymer weighed 1.4 g (23% conversion) and had an inherent viscosity in hexafluoroacetone sesquihydrate of 1.19 dl/g (0.5 g/100 ml) at 30°C.

Example 3: *Implantation of Poly(D,L-3-Methyl-1,4-Dioxane-2,5-Dione* — Strips of the poly(D,L-3-methyl-1,4-dioxane-2,5-dione) of Example 2 were prepared and implanted in rabbits. Tissue reaction was unremarkable in each case. The estimated extent of absorption of duplicate samples of these strips and of poly-glycolic acid (PGA) control strips is shown below.

		. . . . Percent Absorbtion. . . . .
		Poly(D,L-3-Methyl-1,4-
Days	PGA	Dioxane-2,5-Dione
15	50,50	0,0
30	85,85	50,50
45	100,100	85,100

The use of the homopolymers and copolymers permits an increase in the range of absorption characteristics available for surgical devices.

Bioresorbable Polydioxanone

Synthetic absorbable sutures and other surgical devices are prepared by *N. Doddi, C.C. Versfelt, and D. Wasserman; U.S. Patent 4,052,988; October 11, 1977; assigned to Ethicon, Inc.* from polymers of p-dioxanone and 1,4-dioxepan-2-one, and alkyl substituted derivatives thereof. Monofilament sutures of oriented fibers are characterized by good tensile and knot strength and a high level of flexibility and softness. The sutures have good in vivo strength retention and are slowly absorbed without significant tissue reaction.

Example 1: *(A) Preparation of p-Dioxanone* — Metallic sodium is dissolved in a large excess of ethylene glycol to obtain a glycolate which is further reacted with about 0.5 mol of chloroacetic acid per mol of sodium to yield the sodium salt of the hydroxy acid. Excess ethylene glycol and by-products of the reaction are removed by distillation and by washing with acetone. The sodium salt is converted to the free hydroxy acid by the addition of hydrochloric acid, and the resulting sodium chloride is removed by precipitation with ethanol followed by filtration.

The hydroxy acid filtrate is slowly heated up to about 200°C, preferably in the presence of $MgCO_3$, to remove alcohol and water by distillation. Upon further heating at atmospheric pressure the p-dioxanone is formed and distills over at a head temperature of between about 200° to 220°C. The purity of the crude di-oxanone product is generally about 60 to 70% as determined by gas chromatography and yields are in the order of 50 to 70%.

The crude p-dioxanone is further purified to about 98% by redistillation, and finally purified to 99+% by multiple crystallizations and/or distillation.

(B) Polymerization of p-Dioxanone — Highly purified p-dioxanone is polymerized in the presence of an organometallic catalyst such as diethyl zinc or zirconium acetylacetonate to obtain high molecular weight, fiber-forming polymers according to the following typical procedure.

0.1 mol (10.2 g) of dry, 99+% pure p-dioxanone monomer is weighed into a dry flask under an inert atmosphere of dry nitrogen and 0.36 ml of 0.138 M di-ethyl zinc in heptane are added. The monomer to catalyst ratio is calculated as 2,000:1. After completely mixing the catalyst and monomer, the flask is swirled at intervals over a period of about one hour or less at room temperature until initiation and polymerization is evident by the occurrence of gelation. The flask is then connected to a vacuum of about 14" of Hg. The sealed flask is maintained at 80°C in a constant temperature bath for about 72 hours to complete the polymerization. The resulting polymer is characterized by an inherent viscosity of 0.70 measured on a 0.1% solution of polymer in tetrachloroethane at 25°C, a glass transition temperature T_g of –16°C, a melting temperature T_m of 110°C, and a crystallinity of 37%.

(C) Polymer Extrusion — The polymer obtained in the preceding step is thoroughly dried and melt extruded through a spinnerette using conventional textile fiber spinning procedures to obtain one or more continuous monofilament fibers suitable for use as synthetic absorbable sutures. The spun filaments are drawn about 5 times at a temperature of about 43°C to increase molecular orientation and enhance physical properties, particularly tensile strength. The drawn monofilaments having a diameter of about 11 mils corresponding to a size 2-0 suture are characterized by an inherent viscosity of 0.64, a crystallinity of 30%, a straight tensile strength of 36,600 psi, an elongation of 99.4%, and a knot strength of 31,900 psi.

Example 2: *In Vivo Absorption* — Two 2 cm segments of monofilament fiber from Example 1, having a diameter corresponding to size 2-0 suture were implanted aseptically into the left gluteal muscles of 24 female Long Evans rats. The implant sites were recovered after periods of 60, 90, 120 and 180 days and examined microscopically to determine the extent of absorption.

After 60 days the suture cross sections were still transparent and intact. The tissue reactions were slight and most sutures were encapsulated with fibrous tissue. The sutures at this period remained birefringent under polarized light.

At 90 days the sutures were becoming translucent and had lost some of their birefringent properties. A few of the suture cross sections stained pink (eosinophilic) around the periphery and the edges were indistinct, indicating the onset of absorption. The tissue reactions generally consisted of a fibrous capsule and a layer of macrophages interposed between it and the suture surface.

At 120 days the sutures were translucent, most cross sections had taken on an eosinophilic stain, and the sutures appeared to be in the process of active absorption. The tissue reactions consisted of an outer layer of fibroblasts with an interface of macrophages several cell layers thick. Absorption at 120 days was estimated to be approximately 70% complete.

At 180 days, absorption of the suture was substantially complete. The incision healed with minimal adverse tissue reaction.

Example 3: *In Vivo Strength Retention* — Segments of the sutures were implanted in the posterior dorsal subcutis of female Long Evans rats for periods of 14, 21, and 28 days. The sutures were recovered at the designated periods and tested for straight tensile strength.

Tensile strengths ranging from 3.08 to 6.47 lb at implantation, decreased by 17.8 to 66.6% after 28 days.

PHENOLICS

Cured Novoloid Fiber

H.D. Batha and G.J. Hazelet; U.S. Patent 4,076,692; February 28, 1978; assigned to American Kynol, Inc. present a cured novoloid fiber and the process for its manufacture which comprises blending a crosslinking agent with a novolac resin at a temperature below about 40°C, rapidly melting the blend, fiberizing the melted blend before it can cure and curing the resulting fibers by exposing them to an acidic gas.

Novolac as used herein is a thermoplastic resin manufactured from a phenol and an aldehyde by the use of acid catalyst and excess phenol.

Example 1: A novolac resin having an average particle size smaller than 200 mesh is blended with about 9% of hexamethylenetetramine. The blend is then pressed to form a solid block of resin. A stream of air at a temperature of about 350°F is blown over the edge of the block. The impinging air melts the surface of the block and blows fibers directly off the solid resin.

Example 2: Blown fibers from Example 1 are charged into a two liter kettle and are heated from 27° to 180°C over a period of 10 minutes in the presence of 100% BF_3 gas at a pressure of 1 atmosphere. The resulting fibers are completely cured after the expiration of the 10 minute time period.

Example 3: The fibers formed in accordance with Example 1 are heated in a kettle at a temperature ranging from 30° to 77°C over a period of 3 minutes. The fibers soften and melt in the absence of the acid gas.

Examination of the foregoing examples clearly shows that novoloid fibers can be rapidly prepared from a blend of a novolac resin with a crosslinking agent provided that melting the blend occurs simultaneously with or immediately prior to fiberization. In accordance with this process, novoloid fibers can be obtained in less than 10 minutes whereas prior art methods required times measured in hours to obtain the final cured novoloid fiber. In addition, the fibers manufactured in accordance with the process are evenly cured through the fiber thus obtaining increased heat resistance, strength and uniformity.

Flame-Resistant Epoxy-Modified Novolak Filaments

H. Koyama and I. Kimura; U.S. Patent 4,021,410; May 3, 1977; assigned to Nippon Kynol Inc., Japan disclose a process for producing a flame-resistant and antifusing cured phenolic continuous filament, characterized by melt-spinning a molten resin of an uncured epoxy-modified novolak resin, and then curing the melt-spun filament obtained.

Example: A 1.0 liter separable flask equipped with a reflux condenser and a stirrer was charged with 570 g (1.0 mol) of novolak having an average molecular weight of 570, which was obtained by polycondensing p-cresol with formalin, 48 g (1.2 mols) of sodium hydroxide and 432 g of water. Then 97.1 g (1.05 mols) of epichlorohydrin was added thereto while heating and stirring at 70°C. The temperature was raised to 97° to 99°C in the course of 15 minutes and the heating and stirring were continued at this temperature for 90 minutes. After the reaction, the reaction mixture was thoroughly washed with hot water at 60°C and heated under reduced pressure, finally 10 mm Hg at 160°C to remove water and low boiling substances.

The novolak-type epoxy resin obtained had an average molecular weight of 650 and a melting point of 135° to 140°C.

The novolak-type epoxy resin was then melted at 160°C, defoamed, and then extruded at a melting temperature of 145°C through a spinneret provided with four orifices, each having a diameter of 2.0 mm. The resulting filaments were then wound on a bobbin at a take-up velocity of 800 m/min.

The filaments thus obtained were immersed in a mixed solution of 500 ml of ethanol and 500 ml of dimethylformamide for 16 hours, and thereafter, the temperature was gradually raised to 90°C over a period of 3 hours and further maintained at 90° to 95°C for 60 minutes. Then, after washing with water, the filaments were dried at 80°C for 30 minutes.

For comparison, a novolak resin having an average molecular weight of 620, which was obtained by polycondensing phenol with formalin in the presence of a sulfuric acid catalyst, was melted at 160°C, defoamed and then extruded at a melting temperature of 135°C through a spinneret provided with four orifices, each having a diameter of 2.0 mm. The resulting filaments were then wound on a bobbin at a take-up velocity of 800 m/mm.

The phenol filaments thus obtained were immersed in a mixed solution comprising 500 ml of 35% by weight of hydrochloric acid and 500 ml of 37% by weight of formalin, and thereafter, the temperature was gradually raised to 90°C over a period of 6 hours and further maintained at 90° to 95°C for 60 minutes. Then, the filaments were immersed in an aqueous solution of 1.5% by weight of ammonia at 70°C for 5 hours, washed with water and dried at 80°C for 30 minutes.

The diameter, strength, elongation and Young's modulus were measured with respect to the epoxy phenol filaments and to phenol filaments as a control. The results obtained are shown below.

	Epoxy-Modified	Control
Filament diameter, μ	17.2	16.9
Strength, kg/cm^2	1,530	1,620
Elongation, %	36.5	7.6
Young's modulus, kg/cm^2	41,000	96,000
Knot strength, kg/cm^2	1,480	600
Bending strength*	4,500	700

*Number of bending cycles at breakage under a load of 0.5 g/den.

As seen from the above table, the epoxy filaments according to the process have a higher elongation and lower Young's modulus than the phenol control filaments and have a very flexible texture.

Heat-Resistant Novolak Modified with Nitrogen-Containing Compound

I. Kumura and S. Yata; U.S. Patent 4,115,364; September 19, 1978; assigned to Nippon Kynol Incorporated, Japan describe a process for producing phenolic novolak filaments having improved heat resistance which comprises melt-spinning a fiber-forming phenolic resin and curing the resulting filaments, wherein at least one compound containing at least one group selected from the group consisting of active hydrogen-containing amino, amide, thioamide, ureylene and thioureylene groups and derivatives of these groups is applied to the phenolic resin before, during or after the curing treatment.

The modified phenolic filaments have remarkably improved heat resistance and bending property as well as their inherent flame-resistant and antifusing properties. They can be directly used in the form of monofilaments, multifilaments, or tows, but can also be used in the form of fibers cut to the desired lengths. They can be used as spun yarns either alone or in admixture with known filaments or fibers, or in the form of twisted yarns. They can also be made into various filamentary structures such as knitted or woven fabrics or nonwoven fabrics either alone or in admixture with known filaments.

The modified phenolic filaments or fibers find applications in apparel fields, interior decoration fields such as curtains or carpets, in the form of woven, knitted or nonwoven fabrics, and paper, or as electrically insulating sheets.

Example: A novolac resin was prepared in accordance with a customary method by condensing formaldehyde with a slightly excessive amount of phenol in the presence of a catalytic amount of oxalic acid. The resin was purified by removing the impurities and the residual phenol. The purified resin had a number average molecular weight of 960. The purified resin was coarsely pulverized and charged into a vessel adapted to be externally heated. The vessel was connected to a nozzle having 32 holes each with a diameter of 1.5 mm through a gear pump. The vessel was heated externally at 180°C and the molten resin was extruded through the nozzle. The extrudate was taken up in the filament form on a rotating bobbin at a take-up speed of 1,100 m/min. The filaments obtained had an average diameter of 9 microns.

50 g of the filaments were dipped at 20°C in a solution of 100 g of dimethylol urea in 1 liter of 18% hydrochloric acid, and the solution was heated to 95°C in the course of 1.5 hours. The uncured filaments obtained were withdrawn, and without washing, dipped at 20°C in a solution of 100 g of dimethylol urea in a mixture of 200 g of 35% hydrochloric acid and 800 cc of methanol in a vessel equipped with a reflux condenser. In the course of 30 minutes, the external temperature was raised to 85°C and the solution inside the vessel was boiled for an additional 30 minutes. The filaments were washed twice with 500 cc of methanol, and then repeatedly with warm water until the filaments did not show acidity, followed by drying. The curing degree of the filaments was measured, and found to be 21%. Analysis of nitrogen contained in the cured filaments show that 8.5% of the weight increase of the filaments as a result of the curing reaction was ascribed to the introduction of urea.

The resulting cured filaments had a tensile strength of 1.5 g/den and an elongation of 55%, and showed no flame generation even when exposed to flame. The bending strength of the filaments was 2,400.

2 g of the filaments were made into a fiber ball having a packing density of about 0.15 g/cm^3. The fiber ball was allowed to stand in an air circulating dryer held at 200°C, and the temperature of the inside (center) of the filaments was continuously

measured. The external temperature reached 200°C in 23 minutes. The test was continued for 100 hours; however, the temperature of the fiber ball did not go beyond 200°C. When the fiber ball was taken out from the dryer after this procedure, no apparently great change was observed either on the peripheral portion of the fiber ball or in the center of the ball. After the end of the test, the filaments had a tenacity of 1.6 g/den and an elongation of 15%. The weight loss was 1%.

High Strength Novolaks

According to *T. Kimura, M. Watanabe, Y. Yamakawa, S. Mukai, K. Takenaka, and M. Kita; U.S. Patent 4,079,113; March 14, 1978; assigned to Mitsubishi Chemical Industries, Ltd., Japan* fibers or films with high mechanical strengths are produced from novolac which are obtained by reaction of resols with phenols under neutral or acidic conditions.

The term resol used herein is intended to mean a condensation product obtained by interacting an aldehyde and a phenol in the presence of a basic catalyst, i.e., a phenol alcohol having reactive methylol groups. In general, typical of the aldehyde useful in preparing a resol of the type mentioned above is formaldehyde. Apart from this, paraformaldehyde, polyoxymethylene, trioxane, furfural and the like may also be employed. These aldehydes may be used singly or in combination.

The phenols suitable for the purpose are those which contain one or more phenolic hydroxyl groups in one molecule thereof and which have three or more free sites at ortho and para positions to the phenolic hydroxyl group. In order to produce a three dimensional crosslinked structure by curing, it is necessary to use phenols which have three or more free sites at the ortho and para positions. Examples of the phenols include phenol, cresol, chlorophenol, phenylphenol, bisphenol A, phenolphthalein, resorcinol, methylresorcinol, hydroquinone, naphthol, etc. These phenols may be used singly or in combination of two or more.

Example 1: *Preparation of Resol* — To a mixture of 61 parts of phenol and 158 parts of an aqueous 37% formalin solution (formalin/phenol = 3 by molar ratio) was added sodium hydroxide in an amount of 5 mol % to the phenol, followed by reaction under heating conditions of 70°C for 3 hours to obtain resol A-1. The conversion of the formaldehyde was 73%.

Example 2: *Preparation of Novolac* — 190 parts of phenol and 6 parts of oxalic acid were added to each of the resols A-1 in the acidic condition for reaction at 98°C for 2 hours, followed by washing three times with 50 parts of hot water at 80°C for removing unreacted materials, oxalic acid and its salt from the reaction mixture. Then, the reaction system was heated up to 170°C for evaporation and thermally treated for 1 hour under a maximum reduced pressure of 10 mm Hg.

The thus obtained novolac was subjected to a melt spinning treatment using a nozzle with the number of starts (hole) of 30 and an aperture of 0.25 mm diameter, in which treatment the novolac was heated to 135° to 160°C and the resulting melt was drawn into a fiber at a winding rate of 500 m/min. The resulting novolac fibers were immersed at 35°C in an aqueous solution containing

17.5% by weight of hydrochloric acid and 17.5% by weight of formaldehyde. The temperature of the solution was then raised gradually from 35° to 100°C over a period of 2 hours and maintained at a temperature of 98° to 102°C for 10 hours. The resultant cured fibers were treated with an aqueous solution containing 2.0% by weight of ammonia and 50% by weight of methanol at a temperature of 60°C for 60 minutes, followed by washing with water and drying (single curing method).

Example 3: *Comparative* — 150 parts of phenol, 105 parts of aqueous 37% formalin solution and 7.5 parts of an aqueous 10% oxalic acid solution were mixed with each other for reaction at 98°C for 4 hours. The reaction mixture was heated for evaporation until the temperature of the reaction system reached 170°C, and further thermally treated for 1 hour under a maximum reduced pressure of 10 mm Hg to obtain a linear novolac. The thus obtained linear novolac was melt spun and cured to obtain a cured novolac fiber.

The cured phenol fibers of Example 2 and the comparative example were each formed into a spinning yarn with a single yarn number count of 20. The properties of the respective spinning yarns were determined. The test results are shown below.

	Example 2	Example 3
Breaking strength, g	311	185
Breaking elongation, %	11.4	3.2
Limiting oxygen index, %	31.9	31.8

The properties were measured as follows: The breaking strength was measured by a method as prescribed in JIS-L 1074; the breaking elongation also by the method as in JIS-L 1074; and the Limiting Oxygen Index value by a method as prescribed in JIS-K 7201. Though the phenol fibers obtained by the method of the process have excellent mechanical strengths and elongation even in the form of a single yarn, the excellency is more pronounced when the fibers are formed into spinning yarns.

Linear Condensates of Phenol and Formaldehyde

J. Löbering; U.S. Patent 4,119,611; October 10, 1978; assigned to Zimmer AG, Germany discloses a two-stage process for making a linear condensate of phenol and formaldehyde which can be extruded as a film or thread. In the first stage, phenol and formaldehyde are reacted at about 90°C to produce a clear low-viscosity phenol-methylol liquid, the pH of which is adjusted to a value greater than 3, say between 3.5 and 7.

In the second stage, the product of the first stage is added in aliquot portions to molten phenol having a pH of 3 or less, while maintaining the temperature at about 100°C. After all the phenol-methylol has been added, the mixture is slowly heated as the viscosity increases up to about 150°C until there is no further increase in viscosity. The highly viscous end product can be extruded as a film or thread.

Example 1: In a reactor containing 620 g phenol, 250 g of formaldehyde are added under stirring and the whole reaction mixture heated up to about 90°C. The pH of the batch is adjusted to a value of 6 to 7 by adding sodium formate-formic acid. At the same time, 5 g of $CaCl_2$ are added. Phenol-methylol is formed which contains 1.25 mols formaldehyde per mol of phenol and which remains liquid at a temperature of 15°C and is stable at a temperature of 90°C.

In the second reactor, 165 g of phenol are melted and maintained at a temperature of 90°C. To this molten phenol, 4 g of $CaCl_2$ are added and the pH value of this batch is adjusted to a value of 2.

The phenol-methylol produced in the first stage is added at a rate of 8 ml/min to the molten phenol. After about 90 min, the addition of phenol-methylol to the phenol is finished. The mol ratio of phenol to formaldehyde in the reaction mixture then amounts to about 1 to 1.1-1.08. During the entire period of time of the addition of the phenol-methylol, the reaction mixture is maintained at a temperature of 85° to 90°C by cooling the reaction vessel. After the addition is completed, the temperature of the reaction mixture is increased to 120°C in a period of 2 hours to 140°C in 2 more hours and finally up to 160°C.

The condensate thus produced is extruded as a high viscosity material, then ground and the ground material reduced in size to powder. The powder is washed with water and afterwards dried.

The end product has a melting point of 125°C and a degree of polymerization of 11.2.

Example 2: The procedure is analogous to the one of Example 1. The only difference is that the pH of the molten phenol of stage 2 is adjusted between 1 and 2 by means of phosphoric acid.

After the reaction is finished, the condensate melt is transferred by means of compressed air to a vessel containing water, where the melt immediately solidifies. This condensate has a mol ratio phenol to formaldehyde of 1 to 0.909, a PF No. of 1.1, a degree of polymerization of 11, average molecular weight of 1,334 and a specific viscosity of 1.81 measured in glycerin.

This product as a melt can be spun to threads at a temperature of 150°C. The condensate also can be ground and used to produce coating or compression molding compounds. The condensate can also be ground, washed with water, i.e., by means of vacuum cell filters and dried at temperatures below the melting points, preferably under vacuum.

The advantages obtained by this process are that it makes possible obtaining a condensate of phenol and formaldehyde with a high degree of polymerization, a high degree of purity and a good uniformity by means of a process which can be carried out on an industrial scale and which is practically operable to make a product suitable for the production of coatings, films and threads.

SULFUR-CONTAINING POLYMERS

Bis-Amidoalkanesulfonic Acids

L.E. Miller and D.L. Murfin; U.S. Patent 4,034,001; July 5, 1977; assigned to

The Lubrizol Corporation provide bis-amidoalkanesulfonic acids and their salts which find use in applications employing amides and sulfonic compounds. Bis-acrylamides and the like are polymerizable monomers whose polymers are useful as gelling agents for aqueous systems, as well as for the preparation of fibers, plastics, resins, etc.

Example 1: The preparation of the sulfonic acids of this process are illustrated by this example. (Salts of the acids may be prepared by neutralization.) All parts are by weight unless otherwise indicated.

To a solution of sulfur trioxide (160 g, 2 mols) in a solvent mixture consisting of dioxane (352 g) and ethylene dichloride (750 ml) there is added at –20°C iso-butyraldehyde (144 g, 2 mols). After the addition, the reaction mixture is allowed to warm to room temperature (25°C) and is maintained at that temperature for one hour. To the mixture there is then added acrylonitrile (2,190 g, 30 mols) and 96% aqueous sulfuric acid (294 g, 3 mols). An exothermic reaction occurs and the temperature of the reaction mixture is maintained at 40°C until the reaction is complete. The product of such reaction is 1,1-bis(acrylamido)-2-methylpropane-2-sulfonic acid.

The reaction mixture is then cooled to 0°C and treated with anhydrous ammonia in an amount sufficient to neutralize the sulfonic acid. The mixture is further diluted with either ethylene dichloride or acrylonitrile, whereupon the ammonium salt of the sulfonic acid is precipitated. The precipitate is collected on a filter and the product is purified by dissolving it in methanol, filtering the methanol solution, evaporating the methanol from the filtrate so as to recover the ammonium salt, washing the ammonium salt with acetone and ether and then drying the ammonium salt at 45° to 55°C in vacuum.

Example 2: The preparation of a polymer of this process is illustrated by the following example. A copolymer of ammonium 1,1-bis(acrylamido)-2-methyl-propane-2-sulfonate and ethyl acrylate is obtained by adding aqueous ammonium persulfate (0.25 g) and aqueous sodium metabisulfite (0.11 g) to a mixture of water (300 ml), the ammonium salt (0.5 g) and ethyl acrylate (49.5 g) at 40°C. The copolymer is isolated from the mixture as a resinous solid.

The polymers are useful in the preparation of plastics, fibers, resins, and other polymeric compositions. Because of the presence of multifunctional groups in the bisamides, the polymers derived therefrom possess new and useful properties. For example, the sulfonic group imparts dye susceptibility and antistatic properties to the polymer. Consequently, a fibrous composition containing the polymers of the bisamides has improved dye susceptibility properties and antistatic properties. A useful fibrous composition is thus exemplified by a fiber blend consisting of 96% by weight of nylon (polycaprolactam) or polyacrylonitrile and 4% by weight of a fibrous polymer of 1,1-bis(acrylamido)-2-methylpropane-2-sulfonic acid.

Polyamide/PVA/Sulfoxide Hollow Fibers

According to *E.F. Steigelmann, R.D. Hughes and J. Gabor; U.S. Patent 4,039,499; August 2, 1977; assigned to Standard Oil Company* hollow membrane fibers are formed from a mixture of N-alkoxyalkyl polyamide, polyvinyl alcohol, (PVA), di(lower alkyl) sulfoxide and water. The fibers are useful for separating chemicals, for example, aliphatically-unsaturated hydrocarbons, from mixtures containing them.

The compositions are generally comprised of hydrophilic, fiber-forming amounts of the polyamide and polyvinyl alcohol, about 30 to 85, preferably about 40 to 70 weight percent of N-alkoxyalkyl polyamide; and about 15 to 70, preferably about 30 to 60 weight percent polyvinyl alcohol based on the total weight of these components. The composites contain sufficient of the di(lower alkyl) sulfoxide to provide an intimate, compatible admixture of the polyamide and polyvinyl alcohol suitable for forming the membrane. These compositions may often contain about 70 to 400 weight percent of the di(lower alkyl) sulfoxide, preferably about 90 to 250 weight percent, based on the total weight of the polyamide and polyvinyl alcohol.

The water in the mixture is generally a sufficient amount to inhibit the degradation of the mixture of polymers and di(lower alkyl) sulfoxide at elevated temperatures, but the amount is not so large that the mixture is incompatible or no longer of fiber-forming consistency. The amount of water present in the composition is often about 1 to 30 weight percent, preferably about 2 to 20 weight percent, based on the total of the water and di(lower alkyl) sulfoxide.

Example: In this example extrusion of the polymer mixture was conducted under a nitrogen pressure between 200 and 1,000 psi on a feed tank and with the extruder having a heated head. Hollow fibers of the polymer blend were formed by extrusion through a die having an opening in its center. During extrusion, air or nitrogen was blown through the center of the fiber by passage through a hypodermic needle extending into the opening in the middle of the die. After extrusion, the fibers were stretched under their own weight by allowing the fibers to drop below the extruder head. The stretched fibers were crosslinked by immersion in a 3% p-toluene sulfonic acid in 10% aqueous sodium sulfate bath for 60 minutes at 55°C. The fibers were then washed repeatedly with water to remove the salt from them and allowed to dry.

Fibers were made while using two different extruder heads. One head (Head 1) had a hole 0.067" in diameter with a 0.031" o.d. needle in the center of the hole, while the other head (Head 2) had a hole 0.040" in diameter having a 0.020" o.d. needle in the center of the hole. During extrusion the polymer compositions are forced through the annular space between the opening in the extruder head and the needle, and air is forced through the needle to keep the extruded hollow fiber from collapsing.

The composition employed was formed by mixing the named ingredients at ambient temperature and then raising the temperature of the mixture to approximately 260°F to effect melting. The mixture is heated at this temperature for at least 1 hour while undergoing vigorous stirring before extrusion. The composition contained the following: Polyvinyl alcohol (0 to 0.5% acetate), 80 g; Nylon, Belding BCI-819, an N-methoxymethyl 66 nylon, 120 g; DMSO, 200 ml; and H_2O, 20 ml.

The polyvinyl alcohol employed was Borden's high molecular weight (0 to 0.5% acetate) grade, and was determined to have a number average molecular weight of about 12,360 by gel permeation chromatography.

Hollow fibers up to 50 feet long were made and were free of holes in the fiber wall and free of plugs in the fiber bore.

Polysulfones

The products of the process developed by *M.E.B. Jones; U.S. Patent 4,094,867; June 13, 1978; assigned to Imperial Chemical Industries Limited, England* are polymers containing repeating units wherein a sulfone group is tied to two aromatic residues. The uncrosslinked products are thermoplastic materials, generally of high softening point, which may be used in any suitable process known for fabricating plastic material. Those of high molecular weight may be tough solids which are substantially inert to a wide variety of chemicals, both acid and alkaline. They may be melt-spun to give fibers and filaments or cast from solution in suitable solvents to give films. They may be admixed with other suitable ingredients such as pigments, heat and light stabilizers, plastizers, lubricants, mold-release agents and fillers and may be blended with other polymeric materials if desired.

The polymers, which are generally thermally stable at very high temperatures, even above their melting points, consist of repeating units having the structure

where Z is an oxygen atom or a sulfur atom or a direct link, and R_1, R_2, R_3 and R_4 are each selected from the group consisting of halogen atoms, alkyl groups containing from 1 to 4 carbon atoms and alkoxy groups containing 1 to 4 carbon atoms.

Example: 734.84 parts (2 mols) of diphenyl ether-4,4'-disulfonyl chloride were fused with 308.52 parts (2 mols) of diphenyl at 90°C under a slow stream of nitrogen in a heated vessel and after stirring for 30 minutes, 4 parts of freshly sublimed ferric chloride were added to the melt. The catalyst dissolved rapidly on stirring with vigorous evolution of hydrogen chloride. The reaction temperature was raised rapidly but the mixture solidified at a bath temperature of about 180°C. The reaction temperature was raised further to 280°C at which temperature the mixture was still solid. The total reaction time was 40 minutes.

The mixture was allowed to cool and the product was then broken up and stirred with 7,850 parts of boiling isopropanol. The insoluble product was filtered off and the process was repeated twice. On drying, the yield was 880 parts of the polymer having a reduced viscosity of 0.15.

Reduced viscosities are measured on solutions of polymer (2 g) in dimethylformamide (100 cm^3) at 25°C.

The polymer was shown to be amorphous by x-ray examination and could be solvent cast from dimethylformamide to give transparent films.

Aromatic Sulfide/Sulfone Polymers

High molecular weight aromatic sulfide/sulfone polymers are produced by *R.W. Campbell; U.S. Patents 4,016,145; April 5, 1977; and 4,102,875; July 25, 1978; both assigned to Phillips Petroleum Company* by reacting a dihalo aromatic sulfone, an alkali metal sulfide other than lithium sulfide, an organic amide, and an alkali metal carboxylate. Use of the alkali metal carboxylate results in polymers of high molecular weight and satisfactory melt flow properties having utility as coatings, films, molded objects, fibers, and the like.

In the following examples, values for inherent viscosity were determined at 30°C in a 3:2 mixture by weight of phenol and 1,1,2,2-tetrachloroethane at a polymer concentration of 0.5 g/100 ml solution. Values for glass transition temperature (T_g) were determined on premelted and quenched polymer samples by differential thermal analysis. The values for polymer-melt temperature (PMT) were determined by placing portions of the polymer on a heated bar with a temperature gradient. The name poly(p-phenylene sulfide/sulfone) is used to describe an aromatic sulfide/sulfone polymer having the following recurring units in the polymer molecule:

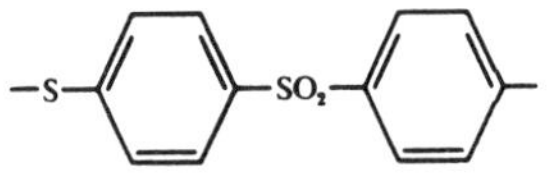

Example 1: In a control run outside the scope of this process, 65.2 g (60% assay, 0.5 mol) sodium sulfide, 0.2 g sodium hydroxide (to react with sodium bisulfide and sodium thiosulfide present in trace amounts in the sodium sulfide), and 158.3 g N-methyl-2-pyrrolidone were charged to a stirred 1 liter autoclave, which was then flushed with nitrogen. Dehydration of the mixture by heating to 205°C yielded 19 ml of distillate containing 18.2 g water. To the residual mixture were charged 143.6 g (0.5 mol) bis(p-chlorophenyl) sulfone (melting point, 146° to 147°C) and 40 g N-methyl-2-pyrrolidone.

The resulting mixture was heated for 5 hours at 200°C at a pressure of 40 to 45 psig. The reaction product was washed repeatedly with hot water and dried at 80°C under nitrogen in a vacuum oven to obtain a yield of 118.5 g of amorphous poly(p-phenylene sulfide/sulfone) having an inherent viscosity of 0.26, a T_g of 205°C and a PMT of 271°C.

Example 2: In a run within the scope of this process, 65.2 g (60% assay, 0.5 mol) sodium sulfide, 0.2 g sodium hydroxide (to react with sodium bisulfide and sodium thiosulfate present in trace amounts in the sodium sulfide), 51.0 g (0.5 mol) lithium acetate dihydrate, and 158.3 g N-methyl-2-pyrrolidone were charged to a stirred 1 liter autoclave, which was then flushed with nitrogen. Dehydration of the mixture by heating to 205°C yielded 41 ml of distillate containing 33.0 g water.

To the residual mixture were charged 143.6 g (0.5 mol) bis(p-chlorophenyl) sulfone (melting point, 146° to 147°C) and 40 g N-methyl-2-pyrrolidone. The resulting mixture was heated for 5 hours at 200°C at a pressure of 25 to 40 psig. The reaction product was washed repeatedly with hot water and dried at 80°C under nitrogen in a vacuum oven to obtain a yield of 118.8 g of amorphous poly(p-phenylene sulfide/sulfone) having an inherent viscosity of 0.40, a T_g of 215°C and a PMT of 271°C.

Thus, based on inherent viscosity, the poly(p-phenylene sulfide/sulfone) produced in this example was of much higher molecular weight than that produced in Example 1, in which lithium acetate dihydrate was not employed.

Samples of the poly(p-phenylene sulfide/sulfone) produced in this example were compression molded at a temperature of 290°C. Properties of the molded specimens are shown below.

Density, g/cc	
(ASTM D 1505-68)	1.4000
Flexural modulus, psi x 10^{-3},	
(ASTM D 790-70)	383
Tensile break, psi	
(ASTM D 638-68)	11,970
Elongation, %	
(ASTM D 638-68)	8
Izod impact strength, ft-lb/in	
notch, (ASTM D 256-72)	0.48
Heat deflection temperature, °C	
at 264 psi (ASTM D 648-56)	184
Hardness, shore D	
(ASTM D 2240-68)	86

As shown in the table, the poly(p-phenylene sulfide/sulfone) exhibited a good balance of properties, the heat deflection temperature being particularly outstanding.

Branched Arylene Sulfide Polymer

According to *J.T. Edmonds, Jr. and L.E. Scoggins; U.S. Patent 4,116,947; Sept. 26, 1978; assigned to Phillips Petroleum Company* branched arylene sulfide polymers of low melt flow are produced by employing (1) a p-dihalobenzene; (2) a polyhalo aromatic compound having more than two halogen substituents per molecule; (3) an alkali metal sulfide; (4) an N-alkyl lactam; and (5) controlled amounts of water in the presence or absence of a sodium carboxylate. The resulting polymers without prior curing can be fabricated into shaped products having desirable properties.

Examples 1 through 4: Branched poly(phenylene sulfide) (PPS) was prepared in the following manner. To a 2 gal autoclave equipped with stirrer were charged 7.765 mols hydrated sodium sulfide (60% assay), sufficient sodium hydroxide to react with sodium bisulfide present as impurity in the sodium sulfide, and 31.05 mols N-methyl-2-pyrrolidone (NMP).

The autoclave was then flushed with nitrogen and all but 1 mol of the water of hydration per mol of sodium sulfide was removed by distillation at atmospheric pressure from the resulting mixture by heating the mixture to a temperature of 204° to 215°C, the distillate thus obtained comprising primarily water, together with a minor amount of NMP. To the residual mixture were added 1.00 mol p-dichlorobenzene (DCB) per mol of sodium sulfide employed, 0.008 mol of

1,2,4-trichlorobenzene (TCB) per mol sodium sulfide employed, an amount of water within the range of 0.00 to 1.25 mols per mol sodium sulfide, and 5.18 mols NMP. The resulting mixture was heated, under autogenous pressure, at 204° to 205°C for 2 hours and then at 265° to 266°C for 3 hours. The reaction product was then cooled, washed with hot water, and dried to obtain the desired branched PPS.

Comparative Examples — Examples 2 and 3 represent control runs outside the scope of this process, for the preparation of branched PPS. These control runs were conducted as described above except that no free water was added prior to the polymerization step. Example 4 represents a control run outside the scope of this process, for the preparation of linear PPS, this control run being carried out as described except that neither TCB nor water was added prior to the polymerization step.

Melt flow values, in g/10 min, were: Example 1, 389; Example 2, 740; Example 3, 1,100; and Example 4, 2,545.

Thus, the polymer product obtained in Example 1 by the process had a melt flow within the desired range of about 1 to 700 g/10 min and was of lower melt flow than those obtained in Comparative Examples 2 through 4.

The values for melt flow were determined by the method of ASTM D 1238-70, modified to a temperature of 316°C using a 5 kg weight, the value being expressed as g/10 min.

SILICONES

Silicone Elastomer

A composition curable to a silicone elastomer is prepared by *C.-L. Lee, M.T. Maxson and L.F. Stebleton; U.S. Patent 4,162,243; July 24, 1979; assigned to Dow Corning Corporation* by mixing a triorganosiloxy-endblocked polydimethyl-siloxane fluid where the triorganosiloxy is dimethylvinylsiloxy or methylphenyl-vinylsiloxy, a reinforcing silica having surface-treated organosiloxane groups which contain 0.05 to 0.32% by weight vinyl, a fluid organohydrogensiloxane, a platinum catalyst and optionally a platinum catalyst inhibitor.

The polydimethylsiloxane fluid has a major peak molecular weight of 68,000 to 135,000, a disperity index greater than 3.8, the lowest molecular weight species between 854 and 3,146 and the highest molecular weight species between 174,000 and 370,000. These compositions are readily extrudable under low pressure and cure to high-strength elastomers with a high durometer.

Example: A silicone elastomeric composition was prepared by mixing by hand-stirring 100 parts of a methylphenylvinylsiloxy-endblocked polydimethylsiloxane fluid, 0.24 part of a chloroplatinic acid complex of symmetrical tetramethyldi-vinyldisiloxane diluted with methylphenylvinylsiloxy-endblocked polydimethyl-siloxane to provide a platinum catalyst having about 0.7 wt % platinum, 0.1 part of 3,5-dimethyl-1-hexyn-3-ol, 1.44 parts of a trimethylsiloxy-endblocked polyorganosiloxane having an average of five methylhydrogensiloxane units and three dimethylsiloxane units and then 40 parts of a treated fume silica. The resulting composition was further mixed by giving it three passes on a three-roll mill to disperse the filler. The mixture was then deaired and yielded an easily extrudable material with a paste-like consistency.

The silica was prepared by placing 543 g of a fumed silica having a surface area of about 400 square meters per gram and which had been dried for 22 hours at 200°C into a flask equipped with a stirrer and addition funnel. To the flask, there was added enough dry toluene to cover the silica. The flask contents were then stirred forming a slurry. To the stirring slurry, 27.1 g of water was added. A mixture of 5.43 g of symmetrical-tetramethyldivinyldisilazane and 103.1 g of hexamethyldisilazane was then slowly added to the stirring slurry. The slurry was then stirred for 48 hours. The resulting slurry was poured into open glass containers which allowed the toluene to evaporate. The resulting residue was then heated for 8 hours at 150°C in an air-circulating oven to remove additional toluene and any other volatiles.

Comparative Example — A silicone elastomeric composition was prepared as described above using the same ingredients and amounts, except a treated silica which contained no vinyl radical was used. This treated silica was prepared in a closed container by mixing 100 parts of a fume silica having a surface area of about 250 square meters per gram, 5 parts of water and 20 parts of hexamethyldisilazane. The resulting mixture was agitated for 4 hours and then heated for 4 hours at 140°C under reduced pressure to remove any volatiles.

The above-defined silicone elastomeric compositions were cured by pressing and curing for 15 minutes at 175°C and then postcured by heating at 150°C for 16 hours. The physical properties of the resulting cured sheets were determined by ASTM D 412 for tensile strength and elongation, ASTM D 625, Die B, for tear strength and ASTM D 2240 for durometer, Shore A scale. The 100% modulus was determined by measuring the tensile stress at 100% strain. The measured physical properties were as shown in the table below. The tensile strength and 100% modulus are recorded in megapascals (MPa) and the tear strength is recorded in kilonewtons per meter.

	Example	Comparative Example
Durometer	65	38
Tensile strength, MPa	6.97	7.53
Elongation, %	454	838
100% modulus, MPa	2.65	0.60
Tear strength, kN/m	35.0	32.9

Silicone-Containing Durable Press Resin

Silicone-containing durable press resin compositions are prepared by *H.E. Griffin; U.S. Patent 4,170,581; October 9, 1979; assigned to Dow Corning Corporation* by polymerizing a polydimethylsiloxane which has been emulsified in an aqueous solution of dimethylolethylene urea or dimethyloldihydroxyethyleneurea. The resulting homogeneous aqueous compositions have improved resistance to separation after freeze-thaw cycling.

Example: A mixture of 1,000 parts of a 40% aqueous solution of dimethyloldihydroxyethyleneurea (DMDHEU) resin (Resin WNM from BI-Chem Division of

Burlington Industries) and 35.6 parts of dodecylbenzenesulfonic acid (BioSoft S-100 from Stephan Chemical Co.) was mixed with 950 parts of cyclopolydimethylsiloxanes using moderate agitation for 20 minutes. The mixture was then homogenized once at 7,000 psig and once at 3,500 psig using a laboratory homogenizer. No particles could be seen in this emulsion using a magnification of 900X. The homogenized mixture was then mixed with an additional 35.6 parts of dodecylbenzenesulfonic acid in 70 parts of DMDHEU resin solution and the resulting mixture comprising 2.2 parts of siloxane per 1.0 part of DMDHEU was heated at 80°C for 4 hours while being agitated.

The mixture was then cooled to 45°C and the sulfonic acid was neutralized with 43.3 parts of triethanolamine dissolved in 83.8 parts of DMDHEU resin solution. The neutralized homogeneous aqueous composition comprising a polydimethylsiloxane and a durable press resin was then further mixed with 116 parts of a 70% solution of octylphenoxypolyethoxy ethanol in water (Igepal CA-897 from GAF) mixed with 30 parts of DMDHEU resin solution. The product was free of visible particles at 900X magnification and showed no oiling or separation after a month at room temperature.

Four textile-treating compositions were prepared by mixing 2.0 parts of the above prepared homogeneous aqueous composition, 8.7 parts of DMDHEU resin solution (Resin WNM), 0.2 part of octylphenoxypolyethoxy ethanol, (Igepal CA-897), 2 parts of a curing catalyst for the durable press resin and 87 parts of water. Compositions 1 and 2 used an aqueous solution of $MgCl_2$ (Curite Mg from Procter Chemical) as the curing catalyst while Compositions 3 and 4 used an aqueous solution of $Zn(NO_3)_2$ (Curite Zn from Procter Chemical) as the curing catalyst. Compositions 2 and 4 additionally contained 0.1 part of methyltrimethoxy silane as a crosslinking agent for the hydroxy-endblocked polydimethylsiloxane emulsion polymer.

For comparison, a textile-treating bath was prepared by the prior-art method and was designated Composition 5. This composition was prepared by mixing 10.0 parts of the 40% resin solution that was used to prepare the above homogeneous mixture and 85.6 parts of water. To this mixture was mixed 2.0 parts of a 40% silicone emulsion that was prepared as above except that water was substituted for Resin WNM in the process, 2.0 parts of Curite Mg, 0.1 part of acetic acid, 0.1 part of methyltrimethoxy silane and 0.2 part of a tin catalyst comprising dibutyltindi(isooctylmercapto acetate) in water.

Compositions 1 to 5 were used to treat 50/50 polyester/cotton fabric samples by padding samples at 40 psi with one of the compositions and curing the padded samples at 350°F (177°C) for 5 minutes. Each fabric sample received approximately 0.5% polydimethylsiloxane and 10% DMDHEU resin based on the weight of the fabrics.

Each sample was then washed five times (AATCC Test Method 124-1975) and tested for flat appearance (AATCC Test Method 124-1975) and shrinkage (AATCC Test Method 135-1973).

The samples treated with textile-treating baths prepared by the method of this process had essentially equivalent shrinkage and slightly inferior flat appearance compared to the sample treated per the prior art as noted in the table below.

Composition Number	Flat Appearance	...Shrinkage (%)...	
		Warp	Fill
1	3.5	0.5	0.8
2	3.4	0.6	0.8
3	3.3	0.7	0.5
4	3.4	0.8	0.6
5*	3.8	0.5	0.6

*Prior art composition for comparison purposes only.

COMPANY INDEX

INVENTOR INDEX

U.S. PATENT NUMBER INDEX

4,065,441 - 225	4,097,546 - 123	4,119,614 - 9
4,066,584 - 309	4,098,768 - 197	4,122,046 - 254
4,066,620 - 54	4,098,774 - 142	4,122,063 - 48
4,066,630 - 62	4,100,142 - 37	4,122,070 - 274
4,066,730 - 314	4,100,143 - 289	4,122,072 - 90
4,067,850 - 77	4,100,145 - 141	4,122,107 - 19
4,067,855 - 105	4,101,145 - 22	4,124,468 - 159
4,067,856 - 10	4,101,399 - 93	4,125,493 - 318
4,067,857 - 12	4,101,447 - 144	4,125,523 - 136
4,070,326 - 271	4,101,517 - 72	4,126,603 - 285
4,070,425 - 179	4,101,525 - 60	4,127,560 - 24
4,071,486 - 155	4,101,526 - 25	4,127,565 - 74
4,072,664 - 221	4,101,531 - 149	4,128,533 - 20
4,072,665 - 223	4,102,875 - 352	4,128,534 - 21
4,073,778 - 147	4,104,214 - 310	4,128,535 - 97
4,075,172 - 186	4,104,248 - 329	4,128,692 - 304
4,075,182 - 141	4,104,260 - 150	4,130,521 - 156
4,075,262 - 56	4,104,263 - 19	4,130,541 - 46
4,075,269 - 209	4,104,281 - 339	4,130,545 - 102
4,076,664 - 239	4,104,324 - 165	4,130,552 - 12
4,076,692 - 343	4,105,645 - 144	4,131,601 - 14
4,076,697 - 252	4,106,098 - 40	4,131,712 - 177
4,076,783 - 84	4,107,149 - 95	4,131,724 - 292
4,077,929 - 300	4,107,150 - 96	4,133,800 - 3
4,077,944 - 24	4,107,154 - 149	4,133,801 - 101
4,077,945 - 30	4,108,828 - 121	4,137,201 - 316
4,077,946 - 161	4,108,845 - 288	4,137,217 - 267
4,079,036 - 328	4,110,316 - 42	4,141,880 - 207
4,079,039 - 261	4,111,869 - 160	4,141,882 - 103
4,079,045 - 88	4,112,016 - 201	4,143,200 - 293
4,079,046 - 32	4,113,708 - 180	4,144,079 - 320
4,079,113 - 346	4,113,794 - 131	4,145,461 - 111
4,080,317 - 4	4,115,231 - 188	4,148,949 - 323
4,081,430 - 229	4,115,350 - 46	4,151,130 - 311
4,081,498 - 299	4,115,357 - 110	4,151,223 - 75
4,082,724 - 98	4,115,364 - 344	4,155,889 - 109
4,083,827 - 156	4,115,370 - 230	4,157,429 - 326
4,083,893 - 123	4,115,371 - 23	4,157,430 - 258
4,083,894 - 212	4,116,942 - 7	4,159,224 - 319
4,085,091 - 51	4,116,943 - 219	4,162,243 - 354
4,086,212 - 95	4,116,947 - 353	4,162,346 - 209
4,087,408 - 70	4,118,350 - 317	4,164,522 - 301
4,088,620 - 260	4,118,374 - 167	4,168,602 - 133
4,092,301 - 157	4,118,470 - 119	4,169,062 - 113
4,094,867 - 351	4,119,611 - 347	4,170,581 - 355
4,096,124 - 35		

NOTICE

Nothing contained in this Review shall be construed to constitute a permission or recommendation to practice any invention covered by any patent without a license from the patent owners. Further, neither the author nor the publisher assumes any liability with respect to the use of, or for damages resulting from the use of, any information, apparatus, method or process described in this Review.

ADHESIVE TECHNOLOGY 1980
Developments Since 1977
Edited by S. Torrey

Chemical Technology Review No. 148

This volume describes new developments in adhesives manufacture and processing since the appearance of our previous title *Adhesives Technology Annual,* published in 1978.

While the information is derived from U.S. patents, the coverage is actually worldwide in scope, as nowadays over 35% of all processes patented in the U.S. are developed by foreign investigators from practically every industrial nation.

As its predecessor, this is a practical, useful manual. It reflects the efforts and skills of many talented inventors. Its continuing purpose is to present the necessary chemistry, as well as changing technology and applications, notably the replacement of organic solvents by less toxic and less polluting carriers, to serve the varied interests of the makers and users of adhesives.

The following is a partial, condensed table of contents, including **chapter headings and examples of some subtitles** and, in parentheses, the number of processes per topic.

ISBN 0-8155-0787-9

500 pages

NONWOVEN MATERIALS 1979
Recent Developments
by M.T. Gillies

Chemical Technology Review No. 141

Historically the concept of nonwovens was strongly associated with that of disposables. Materials that absorb, stretch and breathe, but can be sterilized and are priced for discarding, are still the mainstay of the nonwoven goods industry. Yet more and more nonwoven durables are sold each year for perhaps two reasons: manufacturing techniques employed for the entanglement of fibers in the nonwoven processes yield a less expensive product than other processes; and nonwovens can be engineered with facility for specific end uses.

This book is roughly divided in two sections. The first section covers general methods for making and binding nonwoven webs which may be varied widely according to the type of product desired, while the second section emphasizes the manufacture of specific products. More than 200 processes are described. The partial table of contents below gives chapter headings and **examples of some** subtitles. Numbers in parens give numbers of processes in chapter.

ISBN 0-8155-0776-3

372 pages

WATER AND SOIL REPELLENTS
FOR FABRICS 1979

by Charles S. Sodano

Chemical Technology Review No. 134

Readers will find in this book many ways to impart water and soil repellency to all types of fabrics. Well over 200 processes for producing and prolonging such properties are described.

Because durable press fabrics have a propensity towards staining by oil- or color-bearing substances, special attention is given to overcoming problems caused by durable press curing treatment. Achieving satisfactory soil release, so often aggravated, too, by the soil repellents, can be complex and frustrating. When an aqueous medium cannot wet a fabric to release its soil, new approaches are needed. One class of compounds with excellent soil release properties is the acrylic polymers, among the many compounds treated here at length.

Compositions acting as water repellents include silicones, fluorochemicals, quaternary ammonium compounds, and organometallics. Their application sometimes imparts soil repellency and release at the same time.

Chapter headings and **examples of some** subtitles follow, with the number of processes per topic in parentheses.

1. **FLUORINATED ACRYLIC POLYMERS (17)**
   Polymers Containing Nitrogen, Sulfur
   Acrylic Thiol Esters
   Hybrid Tetracopolymers
   Electrolytically Fluorinated Polymers

2. **FLUORINATED POLYMERS (25)**
   Perfluoroalkyl Thioether Alcohols
   Urea-Linked Condensates
   Alkyl Vinyl Ether Copolymer
   Esters of Polymeric Phosphonitrilic Acid
   Synergism with Polyethers

3. **FLUORINATED POLYMER PRODUCTS (16)**
   Dry Cleaning with Insoluble Polymers
   Maleamic-Siloxane Copolymer Extender
   Aerosol Starch Preparation
   Coating with Glass-Reinforced Spheres

4. **FLUORINATED COMPOUNDS WITH SILICONE OR SULFUR (12)**
   Perfluoroalkyl Silanes
   Vinyl Sulfone Adducts

5. **FLUORINATED AMINES (18)**
   Pyridinium Salts
   Quaternary Ammonium Sulfates

Adduct with Poly(alkoxymethyl)melamine
Fluoroamine Tetrakis(hydroxymethyl)-
   phosphonium Chloride

6. **FLUORINATED AMIDES (14)**
   Preparation via Hydrocarbon Acyl Halide
   N-Heterocyclic Acid Esters
   Hexahydrotriazine Amides

7. **OTHER FLUORINATED COMPOUNDS (15)**
   Phosphine Oxides
   Polyfluoroisoalkoxyalkyl Isocyanates
   Sulfonamide Urethane Adduct
   Grafting with Perfluoroalkyl Vinyl Ether
   Chromium Acid Complexes

8. **ACRYLIC POLYMERS (31)**
   Copolymers with Polyvalent Metals
   Synergistic Acrylic Mixtures
   Waterproof Breathable Textile
   Radiation Process
   Heat Treatment with Steam
   Laundry Mixture with Alkyl Trimellitate
   Treatment with Acrylonitrile
   Aldoximes

9. **AMINOPLASTS (17)**
   Free Radical Polymerization in Situ
   Polymer Hydrosol Mixture
   Treatment of Elastic Fabrics

10. **SILICON COMPOUNDS (6)**
    Epoxysilicone Emulsions

11. **POLYESTERS (9)**
    Ethylene Terephthalate Copolyesters
    Renewable Finish Process at Acid pH

12. **MISCELLANEOUS TREATMENTS (56)**
    Metallic Borate Esters
    Fatty Acid Zirconia Absorbate
    Betaine-Styrene Copolymers
    Cyclic Imide Waterproofing Compositions
    Maleic Anhydride Interpolymers
    Cellulose Derivatives
    Polyoxyalkylene Thiol Resins
    Surface Modification with Sulfur Trioxide
    Condensation Products of Lactams
    Propylene Glycols with Crosslinking Agents
    Citrate Esters
    Alkali Treatment of Textiles
    Chlorinated Paraffin Mixture
    Polyaryl Polyisocyanate
    α-Pyrone Waterproofing

ISBN 0-8155-0761-5

395 pages

WATER-SOLUBLE POLYMERS

Recent Developments 1979

by Yale L. Meltzer

Chemical Technology Review No. 126

There is a wide interest today in the field of water-soluble polymers. In virtually every area of home and office are products dependent for their existence on these polymers—wallpaper, insecticides, lotions, carpet cleaners, ceramics, cereals and cigarettes—to name a few. Industry relies upon them for electrocoating, rubber processing, waste treatment and other innumerable applications. This small sampling of the markets for water-soluble polymers indicates how extensive the research must be to produce these versatile substances.

This review details approximately 250 recent processes developed from such research. It covers the processing and applications of the less expensive natural starches, as well as the more versatile semisynthetics and synthetics. Scientists will find the work described in this book of tremendous value in tailoring traditional starch products and synthetics to meet a special end use.

A partial, condensed table of contents is given below, with chapter headings and some subtitles. The number of processes discussed in each chapter is given in parentheses.

ISBN 0-8155-0742-9

496 pages